Contributions to Economics

More information about this series at http://www.springer.com/series/1262

Sarah Debor

Multiplying Mighty Davids?

The Influence of Energy Cooperatives
on Germany's Energy Transition

Sarah Debor
Wuppertal Institute for Climate
Environment and Energy
Wuppertal, Germany

Dissertation University of Wuppertal, Germany

ISSN 1431-1933   ISSN 2197-7178   (electronic)
Contributions to Economics
ISBN 978-3-319-77627-9   ISBN 978-3-319-77628-6   (eBook)
https://doi.org/10.1007/978-3-319-77628-6

Library of Congress Control Number: 2018937103

© Springer International Publishing AG, part of Springer Nature 2018
This work is subject to copyright. All rights are reserved by the Publisher, whether the whole or part of the material is concerned, specifically the rights of translation, reprinting, reuse of illustrations, recitation, broadcasting, reproduction on microfilms or in any other physical way, and transmission or information storage and retrieval, electronic adaptation, computer software, or by similar or dissimilar methodology now known or hereafter developed.
The use of general descriptive names, registered names, trademarks, service marks, etc. in this publication does not imply, even in the absence of a specific statement, that such names are exempt from the relevant protective laws and regulations and therefore free for general use.
The publisher, the authors and the editors are safe to assume that the advice and information in this book are believed to be true and accurate at the date of publication. Neither the publisher nor the authors or the editors give a warranty, express or implied, with respect to the material contained herein or for any errors or omissions that may have been made. The publisher remains neutral with regard to jurisdictional claims in published maps and institutional affiliations.

Printed on acid-free paper

This Springer imprint is published by the registered company Springer International Publishing AG part of Springer Nature.
The registered company address is: Gewerbestrasse 11, 6330 Cham, Switzerland

Foreword

Energy cooperatives have been playing a central role within the ongoing German energy transition, known as *die Energiewende*, towards low-carbon and environmentally sustainable energy usage. As a specific form of organisation, they have contributed decisively towards promoting and supporting this energy transition "from below". Nevertheless, hardly any systematic analysis of the possibilities for and limits of the influence of energy cooperatives on the energy transition has been undertaken. This is where the present work of Sarah Debor comes in, as she seeks to systematically describe and evaluate the influence of energy cooperatives on the changing German energy system. With its conceptual and empirical treatment of the topic, this work is aimed at developing a differentiated understanding of one of the most important and still-ongoing socio-economic transformation projects in Germany as well as highlighting the role of a specific type of organisation within this process of change. The reader is thus offered an exploration of a highly relevant topic, set within the intersection between (transition) management and organisational theory.

The author first builds a bridge from the global climate and energy policy debate to the special role of Germany and the energy cooperative as a specific organisational actor within the country's present energy transition. Taking a comprehensive look at the existing international literature, she illuminates current scientific discussion and existing empirical findings on energy cooperatives. On this basis, Debor pinpoints two important research gaps in the field: (a) very limited empirical research concerning the effects of energy cooperatives on the current energy transition and (b) underdeveloped embedding of the phenomenon of energy cooperatives within conceptual research into complex transformation processes.

The book contains a description of the historical, political, institutional and technical dimensions of the transformational space in which energy cooperatives operate. This frame of reference then becomes the basis for assessing the actual impacts of energy cooperatives on the transition, with theoretical examination of the phenomenon of energy cooperative in Germany forming the work's conceptual heart. Debor combines the approaches of transitions to sustainable development and transition management with work on structuration theory and develops, on this basis, an independent frame of reference which she calls interactions between agency and structure in transitions (IBAST).

Focusing on the role of niches and niche players in diffusion and transformation processes lays a foundation for gaining theoretical understanding of the impacts of energy cooperatives, wherein a special role is played by a differentiated spatial reach concept. This approach is combined with structural approaches to business based on Giddens' structuration theory. The IBAST framework synthesised by Debor not only creates a basis for understanding energy cooperatives in terms of their structural impacts but also for distinguishing among different structural dimensions. Meanwhile, it also marks an important, independent conceptual contribution towards a transformation- and structural-policy enlightened management theory and decisively develops further the structural-policy work of the St. Gallen School.

A comprehensive survey carried out by the author among German energy cooperatives during the study period provides an instructive overview of the overall dimensions and scope of energy cooperatives in Germany. In particular, Debor focuses on the resources mobilised by energy cooperatives (mobilised capital, number and structure of members), as they are central to the structural analysis built upon them. With this quantitative survey of the German energy cooperative landscape, the author makes a valuable empirical contribution to contemporary research on the energy cooperative and its current effects in Germany.

With three contemporary case studies, Debor provides a comprehensive review of the history, structure and concrete investment projects of German energy cooperatives today. Particularly noteworthy here are her conceptualisation of energy cooperatives as an umbrella for local actor constellations and how they mobilise resources for the energy transition as well as systematisation of the different forms of possible participation in energy cooperatives. The structural policy contribution of energy cooperatives in local contexts is fundamental and extends far beyond the material resources they mobilise. Especially of central importance here are the new value concepts, guiding principles and legitimisation paths that are mobilised by energy cooperatives towards the *Energiewende* project.

In sum, these are highly relevant results for both further energy-cooperative research and the practical design of future energy-cooperative strategies and policies—precisely because recent changes in the German Renewable Energy Act are significantly impacting the incentives and scope of energy cooperatives as well as endangering their potential for social innovation. This work is, therefore, dedicated to examining a quite timely and important issue at the intersection of management, innovation and social transformation. It is an important contribution towards making German experience with energy cooperatives available to international scientific discussion.

Wuppertal Institute for Climate   Uwe Schneidewind
Environment and Energy
Wuppertal, Germany

Abstract

German energy cooperatives have experienced a renaissance over the last decade. Due to their strong growth, concentration on renewable energy and very special business logic, members of this young and innovative enterprise group have received widespread public attention in discussions concerning Germany's ongoing energy-system transition. Energy cooperatives are strongly considered to be a key driving force in this transition, being pioneers in the establishment of a sustainability-oriented energy sector. However, very little is known about the real influence of energy cooperatives within Germany's so-called *Energiewende*. A systematic, holistic and theory-grounded exploration of the contribution and limitations of this organisational group with regard to developing energy system change has been missing. Thus, the question addressed in this book is: How can we reflect—in a systematic and holistic sense—upon the influence of energy cooperatives on Germany's energy transition? A theoretical framework is developed here for assessing the impacts of innovative rising-actor groups within socio-technical system change, based upon interrelations between interaction and structure at three spatial levels: the local, trans-local and wider system. This framework then guides a comprehensive empirical assessment of German energy cooperatives. The mixed-method research design employed here includes a far-reaching quantitative data assessment and a detailed qualitative case study analysis of three cases. It is demonstrated that energy cooperatives stand for a decentralised, regionally embedded and democratically legitimised renewable energy production set-up, representing a completely new combination of technology, values and guiding principles in Germany's energy sector. The structuring potential of energy cooperatives lies in their creation *and* multiplication of very particular collaborative interaction models across time and space, potentially leading to diffusion of a new concrete pattern of rules and resources throughout Germany's energy system. In many places, energy cooperatives function as organisational umbrellas, under which local citizens as well as key local decision takers from civil society, politics and business can commit themselves to new symbiotic relationships, with the concrete intention of developing renewable energy in self-defined regions. Energy cooperatives are, therefore, more than citizenship energy. More importantly, they can unify key actors from diverse fields and professions in their focus regions in elaborating renewable energy projects. Energy cooperatives literally function as a proof of concept, demonstrating how decentralised renewable energy can be embedded and work within society across time and space, despite being radically different compared to conventional energy system structures. It is their shared and well-defined collective identity as well as their shared resource base which has turned them into important structuring actors within Germany's energy system, despite only representing a minority of very small to medium-sized organisations. They have created high regard for their way of thinking and acting beyond the boundaries of their own communities, to the extent that the government has explicitly praised and valued their actions in the German Renewable Energy Act 2016. The ability to create such a level of reciprocity between community insiders and outsiders can become a basis for institutionalisation. Energy cooperatives also demonstrate well that structural change can emerge through small social groups in society.

Further, this work shows that implementation of similar interaction models is not only the *greatest strength* of the community but also its *greatest weakness*. Besides intended actions, also unintended consequences and unacknowledged conditions of their own activities are manifested across time and space. The overwhelming majority of energy cooperatives have a risk-averse business approach, due to their sense of responsibility for citizens' capital, prevailing economic principles that remain unquestioned and a lack of internal professionalisation of management structures. Thus, energy cooperatives primarily focus on implementation of small to medium-sized renewable-energy production units, preferably photovoltaic and biomass plants, which receive a secured return on investment through feed-in tariffs under the German Renewable Energy Act. Even though most energy cooperatives aim at realising diverse business activities, a minority of them has become engaged in other areas. As a consequence, energy cooperatives have, thus far, been limited in transferring their potential for structuring change—combining technological energy infrastructures with new societal rules and guidelines—into other renewable energy-related activity areas, such as the mobility sector, being important for achieving a fully renewable energy system. The community of cooperatives has also lost its strong growth dynamic over the last years, because most of them have not managed to develop business models that are robust enough to cope with external changes, such as the strong changes prompted by the German Renewable Energy Act. Foremost, energy cooperatives are having problems facing the new cost-reduction-driven political direction within Germany's energy transition, revealing that politics can have a strong influence on actor-driven socio-technical change processes. A conclusion here is that politicians need to understand that the German Renewable Energy Act has not only been the cornerstone for the development and diffusion of energy technologies but has also provided the fundamental conditions for societal innovation.

Contents

1 **Introduction** 1
  1.1 Research Questions 4
  1.2 Research Approach and Contribution 5
      1.2.1 Research Approach 6
      1.2.2 Contributions of This Work 7
  1.3 Structure of the Work 9

2 **Energy Cooperatives: A 'New' Phenomenon in Germany** 13
  2.1 Defining German Energy Cooperatives 15
  2.2 Current State of Research 15
  2.3 Research Gaps 19

3 **Germany's Energiewende** 21
  3.1 A Transition in the Making 21
  3.2 Energy System Goals 23
  3.3 Decentralised Versus Centralised Renewable Energy Systems .. 27
  3.4 System Challenges 29
      3.4.1 Technical Challenges 30
      3.4.2 Market Challenges 30
      3.4.3 Societal Challenges 32
      3.4.4 Political Challenges 32

4 **Organisations in the Context of Socio-technical Transitions** . 37
  4.1 Understanding Socio-technical Transitions 37
      4.1.1 Socio-technical Transitions Towards Sustainability:
            A New Distinct Research Field 37
      4.1.2 Defining Socio-technical Transitions 38
      4.1.3 Prevailing Concepts for Studying Transitions 41
      4.1.4 The Multi-Level Perspective on Sustainability
            Transitions 43
      4.1.5 Summary 58

4.2	Understanding the Relationship Between Enterprise Interaction and Structural Change		60
	4.2.1	Theories of Structural Change in Organisational Studies	60
	4.2.2	Structuration Theory	64
	4.2.3	Structuring Forces of Enterprises	71
	4.2.4	Critical Reflection	77
4.3	Developing a Framework for Analysing the Influence of Organisations on Socio-technical Transitions		80
	4.3.1	The Need for a New Analytical Framework	81
	4.3.2	Main Conceptual Building Blocks of the IBAST Framework	83
	4.3.3	The IBAST Framework	91
	4.3.4	Summary and Reflection	99

5 Research Design and Methods ... 103
- 5.1 Mixed-Method Research ... 105
- 5.2 Quantitative Assessment of Secondary Data ... 107
  - 5.2.1 Data Collection ... 107
  - 5.2.2 Data Analysis ... 113
- 5.3 Qualitative Case Study Analysis ... 114
  - 5.3.1 Selection of Case Studies ... 116
  - 5.3.2 Data Collection ... 117
  - 5.3.3 Data Analysis ... 122
- 5.4 Evaluation of Results ... 125
- 5.5 Critical Assessment of the Research Design ... 126

6 German Energy Cooperatives: A Rising Cosmopolitan Enterprise Community ... 127
- 6.1 The Enterprise Network ... 127
  - 6.1.1 Growth Dynamic ... 127
  - 6.1.2 Location of Organisations ... 131
  - 6.1.3 Business Goals ... 132
- 6.2 Mobilisation of Substantial Resources ... 135
  - 6.2.1 Development of Cooperative Members ... 136
  - 6.2.2 Development of Financial Capital ... 138
  - 6.2.3 Installed Renewable Energy Production Capacity ... 143
  - 6.2.4 Resource Interrelation ... 145
- 6.3 Associated Actors ... 146
  - 6.3.1 Dominant Collaborative Partners ... 146
  - 6.3.2 Intermediate Actors ... 151
  - 6.3.3 Community Outsiders ... 153
- 6.4 Summary ... 155

7 Analysis of Dominant Collaborative Interaction Models ... 159
- 7.1 Collaboration Between Energy Cooperatives and Energy Related Companies ... 160

	7.1.1	Introduction of the Focus Region	161
	7.1.2	The Energy Cooperative BürgerEnergiegenossenschaft Wolfhagen eG	166
	7.1.3	Collaborative Concept	171
	7.1.4	Mobilisation of Allocative Resources	173
	7.1.5	Creation of Authoritative Resources	182
	7.1.6	Creation of New Regulations	191
	7.1.7	Drawing Upon New Interpretative Schemes	199
	7.1.8	Summary	205
7.2	Collaboration Between Energy Cooperatives and Banks: Energie + Umwelt eG	216
	7.2.1	Introduction of the Focus Region	216
	7.2.2	The Energy Cooperative Energie + Umwelt eG	221
	7.2.3	Collaborative Concept	223
	7.2.4	Mobilisation of Allocative Resources	226
	7.2.5	Creation of Authoritative Resources	236
	7.2.6	Creation of New Regulations	242
	7.2.7	Drawing Upon New Interpretative Schemes	249
	7.2.8	Summary	254
7.3	Collaboration Between Energy Cooperatives and Communities: Neue Energien West eG and Bürgerenergiegenossenschaft West eG	263
	7.3.1	Introduction of the Focus Region	263
	7.3.2	The Energy Cooperatives Neue Energien West eG and Bürgerenergiegenossenschaft West eG	269
	7.3.3	Collaborative Concept	272
	7.3.4	Mobilisation of Allocative Resources	274
	7.3.5	Mobilisation of Authoritative Resources	288
	7.3.6	Creation of New Regulations	297
	7.3.7	Drawing Upon New Interpretative Schemes	305
	7.3.8	Summary	311

8 Discussion of Results ... 323
   8.1 Influence of Energy Cooperatives from a Local Perspective ... 323
      8.1.1 Turning Local Actors into Knowledgeable and Capable Renewable Energy Agents ... 323
      8.1.2 Creating New Legitimation Pathways for Regional Renewable Energy ... 326
      8.1.3 Increasing the Significance of Renewable Energy Within Regions ... 331
   8.2 Influence of Energy Cooperatives from a Trans-local Perspective ... 333
      8.2.1 Changing Prevailing Actor-Sets Across Regions Through Active Networking ... 334

		8.2.2	Providing Decentralised Renewable Energy Production with a Technical and Societal Structure Across Time and Space	337
		8.2.3	Distributing Proven Technologies and 'Secure' Project Types	339
	8.3	Influence of Energy Cooperatives from a Systems Perspective		344
		8.3.1	State of System Integration	344
		8.3.2	Expansion of Trans-local Practices	346
		8.3.3	Stability of Trans-local Practices Over Time	347
9	**Conclusion and Outlook**			353
	9.1	Summary and Suggestions for the Future		353
	9.2	Reflection on Theoretical Framework Employed		358
	9.3	Perspectives for Future Research		360
Annex				363
References				399

List of Figures

Fig. 1.1	Structure of the work	11
Fig. 2.1	Energy value chain	15
Fig. 3.1	Share of renewable energy in relationship to overall energy demand: 1990–2015	24
Fig. 3.2	German primary energy demand 1990–2015, according to energy source	24
Fig. 3.3	Energy demand in Germany differentiated according to heating and cooling, power and transportation	25
Fig. 3.4	Existing and envisioned shares of renewable energy for end energy demand and gross power demand by 2050	27
Fig. 4.1	The multi-level perspective on sustainability transitions	44
Fig. 4.2	Emerging socio-technical path carried by sequences of local projects	53
Fig. 4.3	Dimensions of the duality of structure	69
Fig. 4.4	Origins and purpose of the IBAST framework	83
Fig. 4.5	Analytical dimensions of space for visualising socio-technical transition processes	90
Fig. 4.6	The IBAST framework for systematically conceptualising interrelationships between agency and structure within societal transformation processes	92
Fig. 5.1	General overview of the research design applied in this work	104
Fig. 5.2	Mixed-method research approach followed for assessing the influence of energy cooperatives on Germany's energy transition	106
Fig. 6.1	Annual registration of German energy cooperatives, 2000–2015	128
Fig. 6.2	Distribution of registered energy cooperatives in each German Bundesland, as of December 2015	131

Fig. 6.3	German energy production cooperatives differentiated according to their applied energy resources through 2015	134
Fig. 6.4	Membership distribution of German renewable energy production cooperatives, 2010–2012	136
Fig. 6.5	Membership development (2010–2012) of German renewable energy production cooperatives that were registered in or before 2010	137
Fig. 6.6	Total capital of German renewable energy production cooperatives, 2010–2012	139
Fig. 6.7	Equity ratios of German renewable energy production cooperatives, 2010–2012	140
Fig. 6.8	Profit and loss development of German renewable energy production cooperatives that were registered in 2010, timeframe 2010–2012	141
Fig. 6.9	Return on equity of German renewable energy production cooperatives that were registered in 2010, timeframe 2010–2012	142
Fig. 6.10	Installed solar production capacity among German observed solar producing energy cooperatives, as of May 2015	144
Fig. 6.11	Annual registration of German energy cooperatives collaborating with banks, communities or energy-related companies	151
Fig. 7.1	Applied procedure for analysing the case studies	160
Fig. 7.2	Shareholding concept between BEG Wolfhagen eG and Stadtwerke Wolfhagen	169
Fig. 7.3	Collaborative interaction model between Stadtwerke Wolfhagen and BEG Wolfhagen eG	173
Fig. 7.4	Supervisory board of Stadtwerke Wolfhagen	194
Fig. 7.5	Collaborative interaction model between the Energie + Umwelt eG energy cooperative and cooperative banks	225
Fig. 7.6	Location of the 18 municipalities represented by NEW eG	265
Fig. 7.7	Collaborative interaction model between NEW eG and Bürger eG	274
Fig. 8.1	Visualisation of the energy cooperative as a roof for local actor constellations and interaction in renewable energy project development	325
Fig. 8.2	Typology of participation levels that can be activated through energy cooperatives	327
Fig. 8.3	Network set-up of energy cooperatives across time and space	336
Fig. 8.4	Levels of trans-local engagement of German renewable energy cooperatives in action fields considered central for Germany's energy system transition	343
Fig. 8.5	Annual energy cooperative registrations between 2000 and 2014 in relation to amendments to the GenG, EEG and KAGB	349
Fig. 9.1	New structural properties within Germany's energy system created by renewable energy cooperatives	355

List of Tables

Table 3.1	Energy goals of the German government 2011–2015	26
Table 3.2	Techno-economic parameters constituting decentralised and centralised power systems	28
Table 3.3	Technical, societal, political and market challenges facing the German energy transition	35
Table 4.1	Regime framings applied in the 10 latest research articles that use the regime notion and that are published in the journal *Environmental Innovations and Societal Transitions*, as of April 2016	48
Table 4.2	Ideas from the multi-level and strategic niche management perspectives and the aspects they are missing that are important for explaining the influence of actors on system change	59
Table 4.3	Core notions of the IBAST framework and its three analytical levels	100
Table 5.1	Financial statements from 2010, 2011 and 2012 that were published by registered renewable energy production cooperatives by 13 April 2014	109
Table 5.2	Share of German renewable energy production cooperatives for which founding partners and/or steering board members could be identified	112
Table 5.3	Overview of case-study visit purposes	118
Table 5.4	Overview of interviews carried out for this study and the coded names used for analysis	119
Table 5.5	Interview guidelines	120
Table 5.6	Number of selected articles found for each case study	122
Table 5.7	Number of articles that were considered worthy of closer analysis in each case study	125
Table 6.1	Business goals of German energy cooperatives through 2015	133

Table 6.2	Socially and politically related business goals of renewable energy production cooperatives	135
Table 6.3	Solar power capacity of renewable energy cooperatives and total installed solar power capacity in Germany, 2015	144
Table 6.4	Institutional and organisational collaborative partners of German renewable energy production cooperatives; more than one category per cooperative is possible	147
Table 6.5	Overview of the six identified collaborative partner groups for German renewable energy production cooperatives, ordered according to their dominance; more than one category per cooperative is possible	149
Table 6.6	Dominant collaborative partners among the largest German renewable energy production cooperatives	150
Table 7.1	Geographical character of the municipality of Wolfhagen as of 2010	162
Table 7.2	The municipality of Wolhagen's population and its distribution as of 2014	162
Table 7.3	Power demand of Wolfhagen in 2010	163
Table 7.4	Heat demand of Wolfhagen in 2009	164
Table 7.5	Key characteristics of Stadtwerke Wolfhagen	172
Table 7.6	Membership development of BEG Wolfhagen eG 2011–2014	174
Table 7.7	Location of cooperative members as of September 2014	175
Table 7.8	Financial development of BEG Wolfhagen eG, 2011–2014	177
Table 7.9	Core renewable energy power projects in Wolfhagen	181
Table 7.10	Planned and installed projects of BEG Wolfhagen in relation to the municipality of Wolfhagen's energy transition goals	182
Table 7.11	Share of members who attended the general assemblies of BEG Wolfhagen eG, 2012–2014	195
Table 7.12	Allocative and authoritative resources mobilised through collaboration between BEG Wolfhagen eG and Stadtwerke Wolfhagen	206
Table 7.13	Rules jointly created through cooperation between BEG Wolfhagen eG and Stadtwerke Wolfhagen	207
Table 7.14	Articles about BEG Wolfhagen eG published in the *Hessische/Niedersächsische Allgemeine—District Kassel und Wolfhagen* in 2012–2014 that mention new rules created between the cooperative and Stadtwerke Wolfhagen	208
Table 7.15	Influence of collaboration between BEG Wolfhagen eG and Stadtwerke Wolfhagen on the local energy structure	214
Table 7.16	Geographical character of the study region of Neckar-Odenwald and Main-Tauber	217
Table 7.17	Population of Neckar-Odenwald and Main-Tauber and its distribution in 2012	217
Table 7.18	Power demand of the Main-Tauber and Neckar-Odenwald districts in 2011	218

List of Tables

Table 7.19	Heat demand of the Main-Tauber and Neckar-Odenwald districts in 2011	218
Table 7.20	Potential renewable power supply in the Main-Tauber and Neckar-Odenwald districts	219
Table 7.21	Key data from the collaborative banks owning shares in Energie + Umwelt eG as of 2013	224
Table 7.22	Key characteristics of Volksbank Main-Tauber, Volksbank Mosbach and Volksbank Franken	225
Table 7.23	Membership development of Energie + Umwelt eG 2011–2014	227
Table 7.24	Financial development of Energie + Umwelt eG, 2011–2014	229
Table 7.25	Projects realised by Energie + Umwelt eG as of December 2014	233
Table 7.26	Planned and installed projects of Energie + Umwelt eG in relation to the energy transition goals of the Neckar-Odenwald and Main-Tauber districts, as of December 2014	235
Table 7.27	Share of members that have visited the general assemblies of Energie + Umwelt eG, 2012–2014	244
Table 7.28	Allocative and authoritative resources mobilised through collaboration between Energie + Umwelt eG and cooperative banks in the study region	255
Table 7.29	New rules created through collaboration between Energie + Umwelt eG and the cooperative banks in the study region	256
Table 7.30	Share of articles about the energy cooperative published in *Fränksiche Nachrichten* in 2010–2015 according to the respective new societal rules mentioned or discussed	256
Table 7.31	Impacts of collaboration between Energie + Umwelt eG and cooperative banks in the study region on local energy structure	262
Table 7.32	Land use characteristics of the administrative region of Operpfalz, as of December 2012	264
Table 7.33	Population and size of all communities being represented NEW eG, as of 2013	265
Table 7.34	Total power demand for the western part of the district of Neutstadt an der Waldnaab in 2009	266
Table 7.35	Total heat demand of the western part of the district of Neutstadt an der Waldnaab in 2010	267
Table 7.36	Renewable energy potential of the western part of Neustadt an der Waldnaab	268
Table 7.37	Distribution of the political affiliations of encumbent mayors in the municipalities being represented by NEW eG as of 2015	272
Table 7.38	Key characteristics of the 18 cities and municipalities that are members of NEW eG	273
Table 7.39	Membership development of NEW eG, 2010–2014	275

Table 7.40	Membership development of Bürger eG, 2010–2014	275
Table 7.41	Economic development of NEW eG, 2010–2014	279
Table 7.42	Economic development of Bürger eG, 2010–2014	280
Table 7.43	Distribution of member shares attracted by NEW eG, as of December 2014	281
Table 7.44	NEW eG project locations according to installed capacity, as of January 2015	284
Table 7.45	Planned and installed projects of NEW eB and Bürger eG in relation to the municipal energy transition goals	287
Table 7.46	Share of Bürger eG members who have attended its general assemblies, 2010–2014	301
Table 7.47	Allocative and authoritative resources mobilised through collaboration between Bürger eG and NEW eG	312
Table 7.48	New societal rules that have been created through collaboration between NEW eG and Bürger eG	313
Table 7.49	Share of articles written about the energy cooperatives in *der neue Tag* (2009–2013) that mention or discuss new societal rules	314
Table 7.50	Impacts and limitations of collaboration between NEW eG and Bürger eG regarding regional energy structure	320

Chapter 1
Introduction

Development of renewable energy systems stands as one of the great challenges of this century. Contemporary energy systems act as a central driver of anthropogenic climate change. In fact, two thirds—about 35%—of all greenhouse gases (CO_2, CH_4, N_2O, H-FKW/HFC, FKW/PFC, SF_6) are emitted by the global energy-supply sector (Intergovernmental Panel on Climate Change 2014, p. 516; International Energy Agency 2015b, p. 20). The main reason for this is intensive consumption of greenhouse gas-generating fossil energy resources, primarily oil, coal and natural gas.

The Intergovernmental Panel on Climate Change (IPPC)—a large body of scientific researchers from around the world—emphasises that a continuous increase of greenhouse gases will accelerate global warming, leading to fundamental long-lasting changes for all components of the climate system and with severe and irreversible impacts for people and ecosystems (Intergovernmental Panel on Climate Change 2015a, pp. 8ff.). Consequently, reducing the danger of climate change requires a substantial reduction of greenhouse gas emissions (Intergovernmental Panel on Climate Change 2015b, p. 8) and, therefore, changes within the global energy system. Not only the risk of climate change but also the danger of resource scarcity is challenging today's societies. Fossil energy resources are not endlessly available. A shift from fossil to renewable energy resources is unavoidable (German Advisory Council on Global Change 2003). In line with this situation, many industrial countries have been developing long-term renewable energy goals. The European Commission, for example, has elaborated an energy roadmap according to which the European Union must supply about 55% of its final net energy demand through renewable resources by 2050 (European Commission 2011a, p. 7). On a broader scale, at the Paris Climate Conference in 2015, the world community re-emphasised the global aim of increasing climate protection and to keeping the global average-temperature change to well below 2 °C compared to pre-industrial levels. Furthermore, governments worldwide have acknowledged that global emissions must reach their peak as soon as possible and, thereafter, need to be strongly reduced by 2050 (United Nations 2015).

© Springer International Publishing AG, part of Springer Nature 2018
S. Debor, *Multiplying Mighty Davids?*, Contributions to Economics,
https://doi.org/10.1007/978-3-319-77628-6_1

Much research has have shown that the technological potential for achieving a renewable resource-driven and climate-friendly energy system already exists (see for example, European Commission 2011b; German Advisory Council on Global Change 2003, 2011; Intergovernmental Panel on Climate Change 2014, pp. 516ff.; International Energy Agency 2016). From a technical point of view, renewable energy-based systems are already achievable today. However, 78.3% of global total energy consumption was still covered by fossil fuels in 2014 (Renewable Energy Policy Network for the 21st Century 2016, p. 28); worldwide greenhouse gas emissions are still rising and the use of fossil energy resources is actually constantly increasing. Global CO_2 emissions have risen by 50% over the last two and a half decades. Between 2000 and 2014, the annual average increase rate of emissions rose to 2.3%, particularly driven by CO_2 emissions from global power generation (International Energy Agency 2015a, pp. 26 f.).

The German Advisory Council on Global Change has pointed out that a system transformation, such as envisioned in the energy sector, will certainly not take place automatically but, rather, requires a shaping of the "*un-projectable*" (German Advisory Council on Global Change 2011, p. 1). Necessary changes need to be made to go beyond technological adaptions. This means tackling business, social and regulatory reforms, which require new market dynamics, new production processes, as well as change of consumption patterns and lifestyles (German Advisory Council on Global Change 2003, pp. 1ff.). The German Advisory Council even compares the dimensions of the envisioned re-organisation of today's society towards a renewable energy-based and climate-friendly economy with the massive changes that came along with the industrial revolution (German Advisory Council on Global Change 2003, p. 5). The Advisory Council refers to Karl Polanyi (1944) who described the shift from an agricultural economy to an industrial economy as 'The Great Transformation'. Polanyi outlined how modern industrial societies first evolved after embedding uncontrolled market processes into new far-reaching societal structures, including constitutional legality, social services and democracy (see also, German Advisory Council on Global Change 2003, p. 2). Thus, technology and technological innovation cannot be seen separately from social dynamics. Geels (2002) speaks, therefore, of *socio-technical* transitions. Such deep and fundamental change is often blocked by many interdependencies between established technologies and societal structures, explaining the great difficulties presently observable in meeting the commitment towards developing renewable energy systems (Geels 2002, 2004).

For effecting such change, the German Advisory Council on Global Change (2011, pp. 255ff.) has pointed towards the fundamental role of innovative actors in transition processes: so-called pioneers of change. Transitions research claims that forward-looking individual actors or new innovative actor groups can have higher impacts on changing socio-technical (sub)-systems than previously expected by technology-driven researchers or those focusing on systemic path dependencies (Grin et al. 2010c). Departing from this premise, it is necessary to identify innovative actors and new actor constellations that can help to catalyse energy-transition processes which, as we have seen, are not only technological but also societal in character, requiring actors that have the capacity to try out new energy technologies and who, at the same time, are able to challenge prevailing social, economic and political set-ups that are deeply connected to the dominant fossil-energy infrastructure.

Seeking such innovative actors, here we turn our attention towards German energy cooperatives, which have experienced a renaissance over the last decade. In Germany, 1055 such organisations were officially registered by the end of 2015, with the overwhelming majority—92%—having been founded from 2006 onwards. In line with the general legal standards of the cooperative business approach, energy cooperatives are characterised by open membership structures and democratic decision-taking set-ups. They differ from conventional energy companies in that they follow specific cooperative principles, such as self-help, self-responsibility and equal decision rights. Also of note is that the majority of today's registered energy cooperatives—about 88%—exclusively focus their business activities on developing renewable energy sources. Thus, today's registered German energy cooperatives can be described as a newly emergent innovative enterprise group.

Due to their strong growth, their concentration on renewable energy and their different business logic, German energy cooperatives have received much attention in current debates regarding Germany's energy system dynamics. This young enterprise group is associated with a future-oriented business model, which has been claimed to be able to trigger not only technically but also socially oriented change within Germany's energy system, mainly because they have implemented renewable energy technologies throughout Germany while employing a completely different organisational self-understanding than conventional firms. German energy cooperatives have been characterised as being essential carriers of Germany's *Energiewende* (Bührle 2010; Klemisch 2014a, c, p. 35), key promoters of a decentralised renewable energy system (Klemisch 2016), organisational innovators who have greatly contributed towards climate protection (Schröder and Walk 2014) or as pioneers in the development of a sustainably oriented energy sector (for example, Flieger 2011b, pp. 306 f., 2014; Maron 2009).

However, little is known about the real influence of energy cooperatives within Germany's energy system. They are perceived as being organisational pioneers, but their concrete potential and scope for shaping Germany's ongoing energy transition process has not yet been fully understood. Against the background of a general need for more scientific knowledge about how and to what extent innovative actors and actor groups may be able to accelerate wider socio-technical energy transitions, this work will critically explore the role of energy cooperatives in supporting Germany on its way towards developing a future renewable energy system.

Germany belongs to the countries that have been taking a lead in developing renewable energy and has received international recognition for its desired path of developing a renewable energy system by not only leaving behind fossil resources but also national nuclear power (Morris and Pehnt 2015, p. 4). This ongoing energy transition process—which had already led to a renewable power share of more than 32% in the country by 2015—is internationally known as Germany's Energiewende. It is seen as being an important 'real life laboratory' with the potential to uncover how far-reaching systemic change can be accelerated and how challenges can be solved along the way (Morris and Pehnt 2015). Consequently, it is of fundamental importance to gather detailed empirical knowledge concerning the current energy-system dynamics occurring in Germany. Analysing the impact of energy cooperatives and their interconnection with current and anticipated future changes can provide important insights in this regard.

1.1 Research Questions

Based on the foregoing contextual analysis, the main goal of this study is to evaluate the existing and potential impacts of energy cooperatives on developing renewable energy in Germany by applying a systematic and holistic perspective on the phenomenon. Consequently, the core question addressed in this book is:

How can we reflect—in a systematic and holistic manner—upon the influence of energy cooperatives on Germany's energy transition?

Research interest departs from a wider system understanding. The particular focus here lies in examining the role of rising innovative actors, in particular enterprises, in directing not only technical but also societal innovations that may support ground-breaking changes towards a renewable and climate-friendly energy system. The scope of this work thus goes beyond classical enterprise research issues, such as service and product development or internal firm and external market progressions. The research presented here draws greatly upon transitions science, which seeks to explain and uncover how technology is embedded within social contexts and how novelties arise and become strong enough to cause a break from technological concepts that are strongly interrelated with the prevailing societal routines of a system (Grin et al. 2010c; Markard et al. 2012). Of primary interest here is reflecting upon the role of energy cooperatives in combining new technological concepts with new social heuristics, as the creation of a new socio-technical structure is central for a transition such as the Energiewende requires (Geels and Schot 2010).

Core reference point of research interest is Germany's energy system. Germany's energy system already undergoes a process of change. Analysing a transition in the making, such as the one taking place in Germany's energy system today, is fruitful for future research, as many aspects tend to be underexplored in such ongoing dynamically driven change processes. At the same time, current transitions need to be considered to an extent as something of a black box, making it challenging to conduct research. It is not clear how, to what extent and ultimately in what direction these processes will further develop. However, Germany's Energiewende is strongly driven by a desired system outcome: a system based upon renewable energy. In order to assess the relevance of energy cooperatives for achieving Germany's anticipated renewable energy system, it is therefore important to relate their activities not only to current system dynamics but also to the country's long-term energy system vision as well as to challenges standing in the way of these goals. Hence, it is asked:

- What is the current status of Germany's ongoing energy transition?
- What are Germany's long-term energy system goals?
- What observable challenges are related to achieving the envisioned energy system change?

The core research content being pursued here is the interaction within and between energy cooperatives. The theoretical framework developed in this study is aimed at guiding holistic exploration of interrelationships between innovative actor group interactions and ongoing transitions. The central question from which elaboration of the framework departs is, therefore:

- How can the interactions between rising innovative actors and actor groups be systematically related to the direction and scope of ongoing socio-technical transitions?

This question is not only relevant for the exploration of energy cooperatives with regard to energy-system change but, further, may be able to guide reflection upon other actor groups having the aim to make a difference in other systemic contexts, for example in the food sector, the mobility sector or biodiversity. The theories selected to answer such a question need to (a) explain the process of socio-technical transitions beyond pure market logics and incremental innovations and (b) provide a general understanding of the relationship between interaction and system structures.

1.2 Research Approach and Contribution

This thesis is situated within the emerging research field of sustainability transitions science (Markard et al. 2012). Transitions are far-reaching, radical change processes during which the set-up of a socio-technical system is fundamentally altered. Transitions are complex, uncertain developments taking place over several decades, characterised by co-evolutionary processes between technological innovations and changes to societal practices in which technology is deeply embedded. Transitions are driven by multiple actor dynamics within and between many different social groups, across all of the societal arenas belonging to a system, such as the market, lifestyles, science, politics and law (Geels and Schot 2010, p. 11). Socio-technical transitions research is generally interested in better understanding how unsustainable system structures may be transformed[1] into more sustainable ones (Markard et al. 2012, p. 955; van den Bergh et al. 2011, pp. 3 f.). Drawing on the notion of sustainable development, the main task here is to meet the needs of the present without risking the livelihoods of future generations (Brundtland and Khalid 1987, p. 41). The struggle to adapt to and inhibit climate change via the development of renewable energy sources and technologies has become a major strand of research

---

[1]The terms 'transition' and 'transformation' are used simultaneously in this thesis.

within sustainability transitions science. Scholars have been questioning the deep gap between scientific knowledge about the negative impact of ongoing anthropogenic climate change on future society (Intergovernmental Panel on Climate Change 2015a), on the one hand, and the very limited progress thus far in actually reducing global greenhouse gases and in developing renewable energy systems, on the other hand (Sustainability Transitions Research Network 2015, p. 6). The central idea of sustainability transitions research is that socio-technical systems change occurs through the interplay of technical and social innovations and that this process is carried out by the interaction of actors. Transitions research seeks to address the challenge of setting the complex, and to a great extent uncontrollable, character of system change in relation to social-actor dynamics. Socio-technical transition is therefore considered as a fruitful concept for addressing the influence of energy cooperatives within Germany's energy system, as it can help to sketch the complex interrelations that exist between energy technologies and society and to discuss the current and potential roles of actors in addressing fundamental system change.

1.2.1 Research Approach

This research is focused primarily on organisations. Taking sustainability transitions science as its point of departure, it applies an interdisciplinary perspective on the firm. The thesis follows the argument of sustainability transitions research, according to which disciplines, such as economic studies, science and technology studies, innovation studies or approaches from sociology actually do have a long history of studying change processes, but none of them alone can grasp the complexity of today's societal challenges (Geels 2005, pp. 6ff.; Smith et al. 2010). The thesis greatly draws upon the multi-level perspective and, particularly, strategic niche management as core analytical concepts for studying sustainability transitions (Geels 2002, 2004). The multi-level perspective and strategic niche management particularly aim at explaining the co-evolution of new technological infrastructures and new social heuristics by pointing towards the role of actors in their development (Geels and Schot 2007, p. 402; 2010, pp. 35ff.).

Following a socio-technical systems understanding, the research approach of this thesis applies a wider interpretive heuristic compared to most organisational studies that concentrate on the impact of innovative enterprises, the majority of which focus on the market and market issues (for example, Toften and Hammervoll 2010, 2013). Wüstenhagen (1998) and Schneidewind (1998) belonged to the early contributors to German organisational studies, which emphasised the central role of companies in striving towards socio-ecologic development within whole economic sectors. Wüstenhagen (1998) sketched out a path of *"Multiplying Davids"* (Wüstenhagen 1998, p. 1), where innovative companies start to operate with high credibility in the ecologic food market with an eventual impact on the complete food sector. The title of the book refers to Wüstenhagen's idea by viewing system-challenging innovative enterprise groups as 'Multiplying *Mighty* Davids'. Meanwhile, Schneidewind

(1998) outlined the potential of firms to change whole mass markets by integrating societal and ecological content through cooperative interaction and uncovered the great potential of the structuration theory of the British sociologist Anthony Giddens (1984) for explaining and describing firm activities from an enhanced societal perspective. Even though having its origin in sociology, structuration theory appears to be quite applicable for examining the influence of organisations on the wider society. Structuration theory deconstructs and refocuses analysis of the fundamental relationship between agency and societal structure. Accordingly, the organisation is generally understood as a collective societal actor in this study, which follows Schneidewind (1998) and greatly draws upon structuration theory. The interdisciplinary socio-technical perspective on the firm applied in this work majorly helps to widen the view on innovative enterprises and relate their activities to social system dynamics. A central concern is, therefore, discovering in what ways and to what extent rising innovative company groups are able to influence ongoing transition processes in their technological and societal senses.

The research approach is exploratory in kind, with the aim of constructing a holistic understanding of the influence of energy cooperatives on Germany's energy transition, which to a great extent remains unaddressed. The work seeks to produce (1) system knowledge—meaning knowledge about 'what is' (Dubielzig 2004, p. 26)—by uncovering the complex interrelations between the activities of energy cooperatives and their actual contributions towards the development of new renewable energy technologies as well as to societal rules that may help to make them properly function in society. The work also seeks to produce (2) transition knowledge—meaning knowledge about how to achieve change from a current stage to an anticipated future stage (Dubielzig 2004, p. 26)—by discussing in what ways energy cooperatives have the potential to help Germany in accomplishing its anticipated shift towards a renewable energy system.

Exploring the influence of innovative actors on system dynamics through a systematic and holistic attempt requires a comprehensive and far reaching empirical research design. Hence, the work applies a mixed method research design, involving quantitative data assessment and qualitative case study analysis.

1.2.2 Contributions of This Work

The overall contribution of the work is twofold.

1. Empirically, it seeks to significantly contribute towards understanding the role of energy cooperatives in Germany's energy transition.

    - The concrete potentials and limitations of energy cooperatives with regard to supporting Germany's energy system change have not yet been reflected upon in depth. The majority of studies about energy cooperatives have been focused on discussing certain aspects of the phenomenon, without considering the phenomenon itself, as a whole. However, research about the general cooperative sector has already discovered that cumulative foundations of cooperatives have often been related to societal or economic insecurities or crises, settings characterised by high levels of transformative pressure (for example, Schwendter 1986). The important role of cooperatives for social and economic change processes has also been underlined by the United Nations during its International Year of Cooperatives in 2012 (United Nations 2013, pp. 4ff.).

- Most studies on energy cooperatives have not generated their observations against the background of a theory-based understanding of interaction, structural change and systems. Hence, very little conceptually rigorous and theory-based knowledge exists concerning energy cooperatives in Germany. The importance of developing better theoretical frameworks for analysing energy cooperatives has recently been emphasised by Yildiz et al. (2015).[2] In conducting a more consciously theory-based reflection upon the phenomena outlined above, this study seeks to arrive at a more comprehensive understanding of the actual and potential role of Germany energy cooperatives for the success of Germany's energy transition.
- This holistic empirical and theory-based exploration is also geared towards improving the knowledge base for the energy policy discourse in Germany, seeking to uncover the potential of German policy for influencing the role of energy cooperatives in the country's energy transition. In the end, suggestions for future policy consideration will be provided.
- This study helps to better understand the role of cooperatives and the process of energy transitions in other countries. The ongoing energy transition in Germany may help us to uncover how a far-reaching system change could potentially be accelerated. In this vein, it is a concrete aim of this work to make its insights available to the international research community—this being the reason why the work is written in English. The overwhelming majority of studies and articles about German energy cooperatives have to date been written in German.

2. Theoretically, this work contributes towards substantially supplementing actor research, in general, and enterprise research, in particular, in the field of sustainability transitions science within which the work is positioned.

Transitions are explained as occurring through multiple interrelated interactions, taking place within and between different societal groups. However, thus far transitions research has failed to systematically deconstruct actor influence within (on-going) transitions, and the rather blurred position of actors in transitions has been a long-discussed research gap in the science field (Smith et al. 2005). Core analytical frameworks for transitions research, such as the multi-level perspective and strategic niche management, have failed to provide a precise and consistent explanation of agency and societal structure as well as how these are interrelated,

---

[2]The author of this thesis is also co-author of this journal article.

beyond the sole fact that they are connected. In this vein, the main conceptual notions of transitions research—the niche and the regime—have been applied in a very confusing and inconsistent way in existing transition studies. It has not been clear what they mean, what they entail and in what ways agency should be embedded within them (see Sect. 4.2 for more details). As a result, transition studies often appear to create a superficial distance between the big abstract system, on the one hand, and small innovative haptic spaces, on the other. Transitions studies seem to either focus on overall and abstract structuring forces, instead of providing answers on substantive mechanisms of interaction (for example, Geels and Schot 2010, p. 19), or they focus on small innovations created by a few individual actors (for example, Seyfang and Smith 2007).

The work aims at contributing towards closing this long-existing conceptual gap within actor and interaction research in transitions science by elaborating a framework that can systematically deconstruct the interrelations between agency and structural change in socio-technical transitions along three different spatial dimensions: the local, the trans-local and the wider system level. Furthermore, this book provides a concrete, coherent and theory-driven understanding of agency and structure within the context of sustainability transitions, achieved by reemphasising Anthony Giddens' structuration theory (Giddens 1984) as an important conceptual building block for transitions research. By more strongly relying on the notions of structuration theory, the book further contributes towards specifying the concrete meanings of niches and regimes and provides guidance with regard to how they should be used in empirical studies on transition processes.

Enterprises are positioned at the centre of theory-elaboration, drawing upon Schneidewind (1998), who systematically delineated the structuring forces of the firm, based upon Giddens (1984). By integrating Schneidewind's ideas into transitions research, this work substantiates the enterprise perspective within the field of socio-technical system change, as, until now, enterprises have remained something of a black box there.

1.3 Structure of the Work

This chapter has provided an introduction to the main research interest of this book as well as presenting its research questions, contributions and structure.

Chapter 2 introduces energy cooperatives as a new phenomenon in Germany. After providing a brief overview of energy cooperatives and their special mode of action, a summary is given about the current state of research regarding this special organisational group and key research gaps are identified.

Chapter 3 outlines Germany's energy transition. It starts with providing a brief introduction of its historical development, continues by describing the dominant energy system's goals and also discusses decentralised and centralised system approaches. At the end of the chapter, aligned system challenges are outlined.

Chapter 4 elaborates the theoretical perspective adopted for this work and discusses organisations in the context of socio-technical transitions in great detail. It begins by defining, describing and reflecting upon the theory of sustainability transitions and continues by discussing the relationship between enterprise interaction and structural change. The second principle theory used here—structuration theory—is then introduced. In the last part of the chapter, a framework for visualising the **I**nterrelations **B**etween **A**gency and **S**tructure in **T**ransitions (IBAST framework), through which the influences of organisations on socio-technical transitions can be examined, is elaborated.

Chapter 5 describes the work's research design and methods applied. A wide-ranging empirical assessment of energy cooperatives was conducted, based upon a mixed-method approach, including quantitative and qualitative aspects. Due to the complexity of the matter under investigation, all analytical steps of the study are described in detail throughout the chapter. In the end, the chosen research design is critically evaluated.

Chapter 6 present the results of a quantitative assessment of German energy cooperatives, which are introduced as a rising cosmopolitan community through identification of associated actors, descriptions of their enterprise networks and how they mobilise substantial resources.

Chapter 7 presents in great detail a comprehensive analysis of three case studies regarding the dominant collaborative interaction models used by German energy cooperatives in order to conduct business.

In Chap. 8, the results of the empirical assessments are merged and jointly discussed. Through the analytical lens of the IBAST framework, the influence of energy cooperatives is considered from local, trans-local and wider system perspectives.

Chapter 9 offers an overall summary of results, including suggestions regarding the future of energy cooperatives in Germany, a final reflection upon the elaborated theoretical framework and an outline of possibilities for future research.

Figure 1.1 displays the structure of the work.

1.3 Structure of the Work

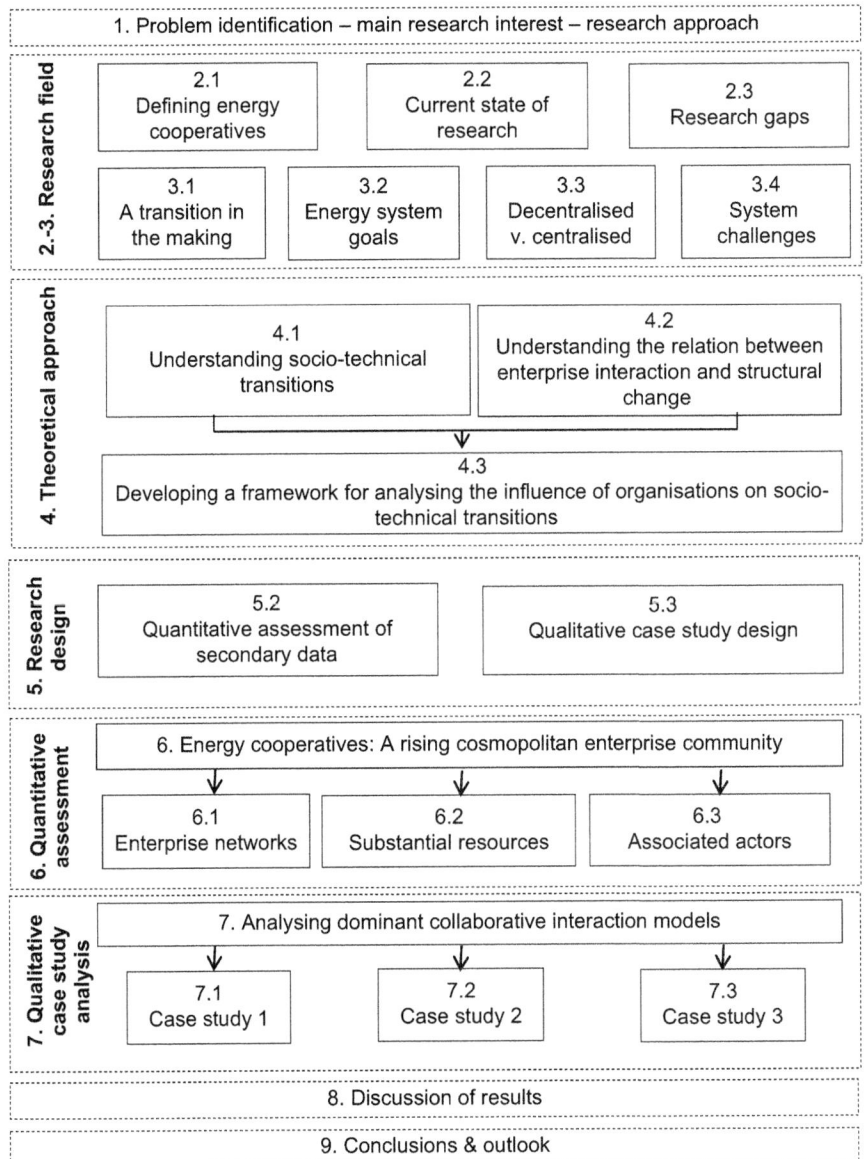

Fig. 1.1 Structure of the work

Chapter 2
Energy Cooperatives: A 'New' Phenomenon in Germany

The cooperative represents a legal business form that exists in many different countries and can be found in many different sectors. As of 2014, about one billion individuals were active in cooperatives all over the world (The International Co-operative Alliance 2014), and more than 5600 cooperatives were operating in Germany as of 2015, bringing together about 19.4 million members (German Cooperative and Raiffeisen Confederation—reg. Assoc. 2015). These figures demonstrate that cooperatives present a relevant business group in Germany and worldwide.

According to the international cooperative alliance:

"A cooperative is an autonomous association of persons united voluntarily to meet their common economic, social, and cultural needs and aspirations through a jointly-owned and democratically-controlled enterprise" (The International Co-operative Alliance 2014).

The modern cooperative form as it is still being practised today[1] evolved in the middle of the nineteenth century, at a time during which simple people, such as small farmers, craftsmen or small traders, had to cope with many business challenges, for example, high interest rates for borrowed capital, low business security and high product prices. Hermann Schulze-Delitzsch and Fridrich Wilhelm Raiffeisen, the two core founders of the German cooperative idea, noticed the hardship of simple people and had the vision of supporting them in organising self-help. People should have the opportunity to solve economic and social problems out of their own strength and by joining forces while, at the same time, staying independent. Based upon this idea the first farmers cooperatives, credit cooperatives and purchase cooperatives were founded between 1850 and 1860. They supported farmers in acquiring animals or land and they helped craftsmen to borrow capital on reasonable terms or to buy necessary resources at acceptable prices (Grosskopf et al. 2012, pp. 12ff). German cooperatives have been regulated by the German cooperative law

---

[1]The guild is seen as an early type of cooperative form (Grosskopf et al. 2012, p. 12).

(GenG: *Gesetz betreffend die Erwerbs- und Wirtschaftsgenossenschaften*) since it was introduced in 1867 (Grosskopf et al. 2012, p. 18).

Cooperatives follow four main principles that are combined with concrete business activities:

1. **Creating member value**: The main purpose of cooperatives is the social, cultural and economic support of their members (§1GenG), with an organisational approach. They are driven by their interests and needs of their members. The business goals of cooperatives are thus balanced between achieving economic benefits, on the one hand, and taking social responsibility, on the other (Grosskopf et al. 2012, p. 19). Cooperatives are normally based upon an open membership structure—open to private persons as well as to associations and firms (German Cooperative and Raiffeisen Confederation—reg. Assoc. 2013b). Yet cooperatives are also allowed to align membership to specific member characteristics, which are then defined in their particular statutes.
2. **Self-help**: The organisational approach of cooperatives is built upon the idea of achieving goals through the joining of forces, with basic guidance being provided through the principle of subsidiarity, so as to empower individual players to economically realise projects they would not be able to achieve by themselves (basic guidance through the principle subsidiarity) (German Cooperative and Raiffeisen Confederation—reg. Assoc. 2013b). According to GenG §§5–8, German cooperatives need to define their own individual statutes, providing fundamental guidance for all cooperative activities and offering a high degree of flexibility for the setting up cooperative businesses. Though a large part of the content is standardised and required according to GenG §§5–18. Due to their member-focused character, cooperatives usually conduct their business activities at the community level (German Cooperative and Raiffeisen Confederation—reg. Assoc. 2013b).
3. **Democratic administration**: Cooperatives are based upon democratic administration. Basic decisions are taken in general assemblies, where all members have equal voting rights. Each member has one vote independent of their financial involvement (The International Co-operative Alliance 2014). This fundamental principle of 'one person, one vote' protects co-operatives from the dominance of single shareholders. Furthermore, it makes hostile takeovers almost impossible (German Cooperative and Raiffeisen Confederation—reg. Assoc. 2013b). The supervisory and executive board form the steering committee. The supervisory board—elected by the general assembly—appoints and controls the executive board, which organises operational activities and represents the cooperative in public.
4. **Self-responsibility**: In general, cooperative members are also responsible for their organisation and its business activities. In the early days, cooperative member could be made personally liable for all organisational losses (Grosskopf et al. 2012, p. 21) but, today, German cooperative members have limited liability, which usually only involves the capital that they have actually invested in the cooperative (Zerche et al. 1998, pp. 13f.).

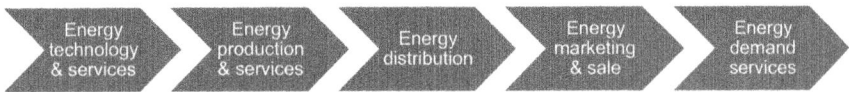

Fig. 2.1 Energy value chain

2.1 Defining German Energy Cooperatives

An official definition for energy cooperatives does not exist. Departing from Holstenkamp and Müller (2013), however, energy cooperatives are understood in this study as legally registered cooperatives that manage activities along the energy value chain, such as those displayed in Fig. 2.1. They may provide technological services, produce energy, be in charge of energy distribution, operate their respective infrastructure, market and sell energy or offer energy demand and supply services.

Energy cooperatives have a relatively long history in Germany, with electricity cooperatives already playing a fundamental role in electrifying the rural regions of Germany in the early 1900s. About 6000 of such electricity cooperatives existed by in the mid-1930s. Most of them vanished between 1930 and 1970, due to growing competition from larger power stations or often related to financial difficulties originating from their small size. Consequently, by 2015, only 44 of these original organisations were still operating (Holstenkamp 2015, p. 2).

Thus, today's German energy cooperatives can be described as a 'new' phenomenon of the last two decades. In fact, 92% of all the 1055 companies registered by December 2015 were founded from 2006 onwards. Today's German energy cooperatives belong to a new generation because the overwhelming majority of them focus on renewable energy, as outlined in detail during the empirical assessment in Chaps. 6 and 7. Their open membership approach, their democratic decision-taking structure, as well as their focus on values beyond economic benefit are core differences to other business forms. In this sense, energy co-operatives follow a different logic of operation than conventional enterprises in the energy sector.

2.2 Current State of Research

Many articles have been published about energy cooperatives, painting a picture of a strong general interest among the German public in this new rising enterprise group (for example, Flieger 2011c; Flieger et al. 2015; Kaehlert 2011; Klemisch 2014b; Lautermann 2016; Münkner 2011; Radtke 2014; Rutschmann 2009). Yet, despite the strong public interest in energy cooperatives, only a few scientific studies exist that explicitly explore and provide concrete empirical evidence about them. Most of these studies analyse particular aspects of the field, with dominant topics including the following:

- Financial focus: Degenhardt and Holstenkamp (2011), for example, observe the financing models of energy cooperatives that invest in renewable energy production (see also, Holstenkamp and Ulbrich 2010). Meanwhile, Sagebiel et al. (2014) reflect upon whether consumers are willing to pay more for electricity from cooperatives (Sagebiel et al. 2014).
- Member focus: Several studies observe members of energy cooperatives, outlining their motivations and reasons for joining them (for example, Rauschmayer and Masson 2015; Schulze 2015).
- Internal enterprise focus: Several studies observe the internal dynamics of energy cooperatives, such as their conflict management methods or investment behaviour of members (for example, Brummer et al. 2016; Höfer and Rommel 2015). Lautermann (2016) provides concrete future orientation for operating energy cooperatives.

Energy cooperatives are particularly associated with fostering citizen participation in renewable energy projects (for example, Flieger 2011a; Kayser 2014; Radtke 2016). Flieger even sees the approach of energy cooperatives as *"a first step towards a more complex form of active citizen participation with both an eco-political and economic orientation"* (Flieger 2011a, p. 58). In this vein, energy cooperatives are regarded as a cornerstone of the German 'citizenship energy' or 'community energy' movement (see for example, Hauser et al. 2015; trend:research GmbH and Leuphana Universität Lüneburg 2013). Both—citizenship energy and community energy—are buzzwords, which generally stand for the (financial) participation of citizens in renewable energy projects (Hauser et al. 2015, pp. 3f.) or for the (financial) participation of communities or, actors that live therein, in renewable energy projects[2] (Hieschler 2011). Although scientific studies on citizenship or community energy often use energy cooperatives as their main empirical examples (for example, Bauwens et al. 2016; Becker et al. 2012; Becker and Kunze 2014; Hieschler 2011; Kalkbrenner and Roosen 2016; Seyfang et al. 2014; van der Schoor and Scholtens 2015; van der Schoor et al. 2016), it is striking that most of them do not explicitly reflect upon the interrelationships between energy cooperatives and community- or citizen-focused energy projects. It seems to be regarded as self-evident that a strong conformity between energy cooperatives and citizens or community energy exists. Consequently, the special set-up of energy cooperatives as well as their particular contributions and limitations to these unique movements remain almost completely unexamined in these studies.

In all of the above-mentioned cases, the studies do generally provide important insights into the characteristics and logic of energy cooperatives, but they do not explain the broader influences of energy cooperatives on Germany's energy system. Very few studies explicitly address energy cooperatives and approach them in a systematic and holistic sense. Nonetheless, those that do provide valuable entry points for better understanding the contributions and limitations of energy

---

[2]Many works interpret citizens and community energy as the same phenomenon (for example, van der Schoor and Scholtens 2015).

cooperatives in relation to Germany's ongoing energy transition are outlined and discussed in more detail in the following.

Two surveys offer an empirical overview of the overall field of German energy cooperatives. The German Cooperative and Raiffeisen Confederation (DGRV: *Deutscher Genossenschafts- und Raiffeisenverband*) (2014) annually conducts a survey among energy cooperatives that are members of the association, with the results including growth dynamic rates, membership development and overall investment activities. The latest report, published in 2014, is based upon a questionnaire which was answered by 216 energy cooperatives in 2014 (German Cooperative and Raiffeisen Confederation—reg. Assoc. 2014, p. 4). The second report was put together by Holstenkamp and Müller (2015), whose primarily focus lies in assessing the number of registered energy cooperatives in Germany over time. In 2013, they also presented a classification of business goals existing among energy cooperatives (Holstenkamp and Müller 2013), drawing upon secondary data from official cooperative registers and internet websites. Their data set is, therefore, much larger than that of the DGRV, which is limited to contributions from its own members. However, both studies offer valuable information regarding the status quo of the whole organisational field. Amongst other insights, they outline how the number of new registrations developed over time and describe in which particular fields of the sector energy cooperatives are active. Their contributions are limited to providing and summarising the data. A coherent interpretation of their results with regard to what they could tell us about the influence of energy cooperatives on Germany's energy system is missing. In addition, both reports remain at the field level. The large quantitative data bases that they use, does not allow them to further dive into the dynamics, potentials and limitations of immediate interaction between them and other social actors.

Volz (2012) sees energy cooperatives as a new field of economic activity, discussing their status quo and potential future development. Through a survey, he not only collects data concerning objective characteristics, such as growth dynamics and business goals. He also gathers information about general attitudes and undertakes a first network analysis of founding partners. The collected data allows him to reflect upon the overall mind-sets of energy cooperatives and to identify supporting actors. However, Volz's main interest is to characterise the field of energy cooperatives as such and, even though he makes some references to Germany's changing energy system, they remain on the surface. In addition, Volz (2012) does not add a critical review of the ways in which energy cooperatives act and think, as this would require a more detailed analysis of their backgrounds and interaction patterns.

Kaphengst and Velten (2014) approach energy cooperatives by conducting a detailed single case study analysis of one region in Bavaria in which several interrelated energy cooperatives were founded. The study aims to investigate how energy cooperatives may support sustainable implementation of renewable energy and whether they can trigger behavioural changes in this respect. They understand energy cooperatives as a transition experiment and are particularly interested in their immediate activities. Based upon findings from the case study, Kaphengst and Velten (2014, pp. 53ff.) underline the high value of energy cooperatives for the economic and social development of rural areas. Accordingly, energy cooperatives provide potential for social inclusion, participation, capacity building, as well as for fostering local economies and community activities—characteristics that can also be transferred to other transition processes (Kaphengst and Velten 2014, p. 64). They thus conclude that energy cooperatives are:

> "A promising model [...] for using [the] resources and capabilities of local actors and residents, which together could pave the way for a successful energy project" (Kaphengst and Velten 2014, p. 57).

The study also outlines several pre-conditions that led to the initiation of energy cooperatives in Northern Bavaria (Kaphengst and Velten 2014, p. 53). However, as pointed out by Kaphengst and Velten (2014, p. 56) themselves, critical reflection of the role of energy cooperatives within Germany's changing energy system cannot be accomplished purely based upon insights from one best-practice case study. Thus, more research with a wider focus is necessary in order to approach this question. The study does, however, demonstrate the high value of detailed case study analysis for understanding and interpreting the activities of energy cooperatives.

The Research Institute for Cooperation and Cooperatives at the Vienna University of Economics and Business (2012) describes energy cooperatives as a form of social innovation process and has been primarily interested in local ownership structures in the field of renewable energy technologies. Researchers there have conducted several case studies and have so far belonged to the few ones, which provided an international perspective (with a European focus) on energy cooperatives. They analysed cooperatives in Germany and Austria and comparing results with evidence from the literature on Denmark and Great Britain. Since their focus did not lie specifically on Germany, it was not their intention to discuss the overall impact of energy cooperatives on Germany's changing energy system. Nevertheless, the study provides valuable information about the field of European energy cooperatives.

The research interests of Maron and Maron (2012) appear to be most similar to those guiding this work, as their main aim has been to observe the impact of energy cooperatives and they depart from Germany's changing energy system as their starting point. However, their primary focus has been on geographical aspects of this topic, with their main contribution to date being a detailed overview of the growth and geographical distribution of energy cooperatives in Germany up to 2012 (Maron and Maron 2012, pp. 93ff.). Amongst other things they show that most German energy cooperatives have been founded in rural areas with low population density.[3] Outlining how energy cooperatives adapt to the local needs of the regions in which they operate, they conclude that energy cooperatives are the best applicable organisational form for decentralising Germany's energy system (Maron and Maron 2012, p. 19). In line with this, they argue that energy cooperatives express an overall trend within society towards increased demand for citizen-focused self-organisation

---

[3] As demonstrated in Sect. 6.1.2, I have interpreted these results in a slightly different way, coming to the conclusion that energy cooperatives are also quite present in big cities.

and more direct control of infrastructure—such as energy—as a general societal interest (Maron and Maron 2012, p. 12). However, the study does not elaborate a concrete understanding about the energy system or about energy system change. Nor do they critically discuss what the decentralisation of energy systems actually means, what it requires and what kinds of contributions energy cooperatives may offer in this regard. Furthermore, their assessment appears to be lacking a theoretical foundation that could have helped to better structure their results so as to evaluate them against the background of system change.

Nevertheless, the far-reaching analysis of Maron and Maron (2012) does provide a promising starting point for the present work, especially because their study demonstrates that multiple data sources are necessary in order to analyse the potential of energy cooperatives. First, they conducted an assessment of secondary data from cooperative registers. They were the first in the field to apply this methodological approach, followed by Holstenkamp and Müller (2015). Then, similarly to Volz (2012), they combined their assessment with a survey (Maron and Maron 2012, pp. 55ff.). Even though the survey was only answered by about 15% of the approached energy cooperatives, it offered valuable information about the field. Furthermore, they analysed public articles and conducted several random interviews. This mix of data offered unknown insights to the field of German energy cooperatives that had not been provided by other studies prior to 2012. However, as with the other studies presented above, Maron and Maron (2012) did not dive into local patterns of energy cooperative interaction. Yet, gathering information about how energy cooperatives act and how their members personally interpret their actions is fundamental in order to better understand the role of energy cooperatives in Germany's energy system.

Lautermann (2016) provides an interesting and comprehensive set of guidelines for energy cooperatives which provide valuable insights into current challenges and how energy cooperatives should develop in the future. He begins by characterising the cooperative as an organisational form and draws an interesting picture of the German energy cooperative as such. However, exploration of the influence of energy cooperatives does not stand at the centre of the report, as its main aim, rather, is to offer guidance for the organisations themselves. Furthermore, given this practical orientation, the guidelines are not evaluated from a theoretical perspective.

2.3 Research Gaps

The foregoing analysis of existing studies on energy cooperatives has uncovered two major research gaps, empirical and conceptual in kind:

Research Gap 1 Existing scientific studies about energy cooperatives do not appear to have yet reflected upon the influence of these organisations with the necessary empirical depth. Most studies focus on particular aspects in order to discuss certain characteristics of energy cooperatives, or they discuss energy cooperatives as part of a larger field of study, such as citizen-produced energy. The few studies that have an explicit and more expansive research interest do not relate their results to Germany's energy system in an adequate manner. Furthermore, existing studies tend to either stay at the field level, without looking at concrete modes of interaction, or they purely focus on individual activities by analysing one or two best practice examples.

An explicit, systematic and holistic analysis of the impact of energy cooperatives on Germany's energy system is at present missing. In order to understand the influences of energy cooperatives on Germany's energy system—to critically observe their contributions and their limitations—it is crucial to conduct much more in-depth empirical research on them.

Research Gap 2 The majority of studies on energy cooperatives have not embedded their observations into explicitly theoretically based understandings of system change and actor dynamics. However, a theory-based approach is key for adequately reflecting upon energy cooperatives and their role in Germany's energy system in a scientific manner. The strong need for elaborating better theoretical frameworks for analysing energy cooperatives has already been pointed out by Yildiz et al. (2015).[4]

Due to this lack of empirical depth and paucity of theoretical concern, little knowledge exists about the concrete influences that energy cooperatives have been having on Germany's energy system. Thus, the present work aims to close these two research gaps, with the primary research interest being to deconstruct energy cooperatives' structural change potential by drawing upon a holistic, analytically sound and theoretically grounded research approach. Point of departure are the studies outlined above, in particular from Holstenkamp and Müller (2015), Kaphengst and Velten (2014), Volz (2012), as well as Maron and Maron (2012), and the central points of previous findings are discussed throughout the book.

---

[4]It should be noted that I am a co-author of this journal article.

Chapter 3
Germany's Energiewende

The impacts of Germany's energy cooperatives today can only be observed against the background of a detailed understanding of the country's energy system, current dynamics, desired energy system goals and aligned challenges. Hence, this section provides an overview of Germany's ongoing energy transition. Section 3.1 outlines the historical context, Sect. 3.2 describes Germany's energy system goals, and Sect. 3.3 discusses the difference between decentralised and centralised system approaches. Meanwhile, Sect. 3.4 identifies system challenges.

3.1 A Transition in the Making

The current national government of Germany names the achievement of a renewable energy system as one of its most central political goals (Governing coalition of Germany 2015, p. 1). Thus, Germany's *Energiewende* (Germany's energy transition) has become a synonym for this ambitious and far-reaching aim to change Germany's fossil fuel-based energy system into that is primarily based upon renewable energy (Federal Ministry for Economic Affairs and Energy 2015a, b; German Advisory Council on Global Change 2011, p. 3; Schneidewind 2015, pp. 8f.). The Energiewende stands for more than just a simple change of energy sources. Rather, according to the German Ethics Commission for a Safe Energy Supply, it is a:

> "Composite work for the future [...] that can only succeed through collective effort spanning all levels of politics, business and society" (Ethics Commission for a Safe Energy Supply 2011, p. 5).

Meanwhile, Friends of the Earth Germany (BUND) and the Association of German Scientists (VDW) describe this massive change of the country's energy system as a:

"Socio-ecological transformation process, a national and social testing ground for the ability to reform a highly developed industrial country in the age of globalisation" (Bartosch et al. 2014, p. 9).

As outlined in the following, the Energiewende is a transition in the making, for which no historical model so far exists. Up to now, there has been no country that has sought to almost fully supply its energy demand for power, heat and transportation with energy from renewable resources (International Energy Agency 2015c). Germany provides fundamental insights into the potential for and challenges being faced on the way towards a society that primarily draws upon renewable energy.

The phrase Energiewende was already introduced in the 1980s, in the course of discussing alternatives to nuclear power production in Germany. In 1981, the Institute for Applied Ecology (Öko-Institut e. V.) published an energy system study in which a national energy scenario without nuclear power and fossil fuels is outlined (Krause et al. 1981). This study, which was titled *Energie-Wende, Wachstum und Wohlstand ohne Erdöl und Uran* (Energy transition, growth and prosperity without petroleum or uranium), became the scientific cornerstone for one of the four energy system scenarios that were discussed during the German Parliament's inquiry commission on future nuclear power policy from 1979 to 1983 (Bartosch et al. 2014, p. 16; Deutscher Bundestag 2015).

The roots of the Energiewende, however, actually go back to the 1970s (Morris and Pehnt 2015, pp. 55ff.), when two oil crises demonstrated the consequences of Germany's strong dependence on fossil fuel imports. At the same time, the government's plans for increasing national production of nuclear power were severely questioned through social protests against the construction of new nuclear power plants in places like Wyhl, Brokdorf, Kalkar and Gorleben (Bartosch et al. 2014, p. 13; Morris and Pehnt 2015, pp. 58ff.). Nuclear power was strongly criticised for its high risks and potential damage it can bring to the environment and humanity. Meanwhile, in 1972, the Club of Rome published its report about the limits of growth (Meadows et al. 1972), triggering a far-reaching debate about resource scarcity and its socio-economic impacts on the world (Club of Rome 2015). These events provided the basis for a growing societal movement that began to fight for climate protection, the abolition of nuclear power and the sustainable use of natural resources (Bartosch et al. 2014, pp. 13f.). The nuclear power accident in Chernobyl, Ukraine in 1986 further strengthened this broad movement and intensified political discussions about the future of energy production in Germany (Bartosch et al. 2014, p. 15). A fundamental goal of the movement was to stop nuclear power. 14 year later, in 2000, the German government coalition between the SPD (Social Democratic Party of Germany—Sozialdemokratische Partei Deutschlands) and the Green Party Bündnis 90/die Grünen finally decided to fully bail out on nuclear power production in Germany by 2022, as specified in the respective law that was passed (Bartosch et al. 2014, p. 15). At the same time, the coalition started to strongly support the production of energy from renewable resources by passing the German Renewable Energy Act (EEG) (Gesetz für den Vorrang Erneuerbarer Energien (Erneuerbare-Energien-Gesetz-EEG) sowie zur Änderung des Energiewirtschaftsgesetzes und des

Mineralölsteuergesetzes 2000). Thus, for the first time, the national government demonstrated a will to oversee a far-reaching change of Germany's energy system towards renewable energy.

However, the subsequent government coalition, formed by the Christian Democratic Union of Germany (CDU: Christlich Demokratische Union Deutschlands) and Free Democratic Party (FDP: Freie Demokraten), revised the political bail-out in 2010 and extended the operational lifetime of German nuclear power plants to 2036 (Bartosch et al. 2014, p. 20). In their opinion, Germany should benefit from low-cost energy from existing nuclear plants as long as possible (Bartosch et al. 2014, p. 19). At the same time, the coalition supported the vision of a future renewable energy system and passed ambitious renewable energy goals to be achieved by 2050 (Federal Ministry for Economic Affairs and Technology, and Federal Ministry for the Environment, Nature Conservation, Building and Nuclear Safety 2010). However, the nuclear power accident in Fukushima Daiichi Japan in 2011 led to a sudden change of the government's position, as the accident demonstrated once more the great risks of nuclear power for humanity (Hennicke and Welfens 2012, pp. 91ff.). In June 2011, soon after the accident, the government decided to shut down eight of the oldest German nuclear power plants at once and to go back to the initial plan of phasing out nuclear power by 2022 (Bartosch et al. 2014, p. 22). Their decision was guided by the Ethics Commission for a Safe Energy Supply, composed of 17 members of leading German associations and institutions who had been appointed by the German *Bundeskanzlerin* (Federal Chancellor), Angela Merkel, after the accident in order to discuss a way out of nuclear power and its consequences for society (Ethics Commission for a Safe Energy Supply 2011).[1] This change in position of the conservative CDU and FDP parties signaled the achievement of a major milestone for the Energiewende, because it created a broad political consensus across all leading German political parties against nuclear power and for the development of a renewable energy system. Such overall agreement had not existed before, though political opinions still vary strongly with regard to how this goal can and shall be achieved.

3.2 Energy System Goals

Since 1990, social, technical and political developments in Germany have been accompanied by a continual increase of renewable energy production in the three main sectors of heating and cooling, power and transportation, as can be seen in Fig. 3.1.

---

[1] The appointment of the Ethics Commission for a Safe Energy Supply, as well as its member composition, was also criticised because, among other reasons, it was argued that the full abandonment of nuclear power had already become a society-wide consensus. Furthermore, the commission did not include representatives of the anti-nuclear power movement or leading nature conservation associations (see for example, Füchs, 2011).

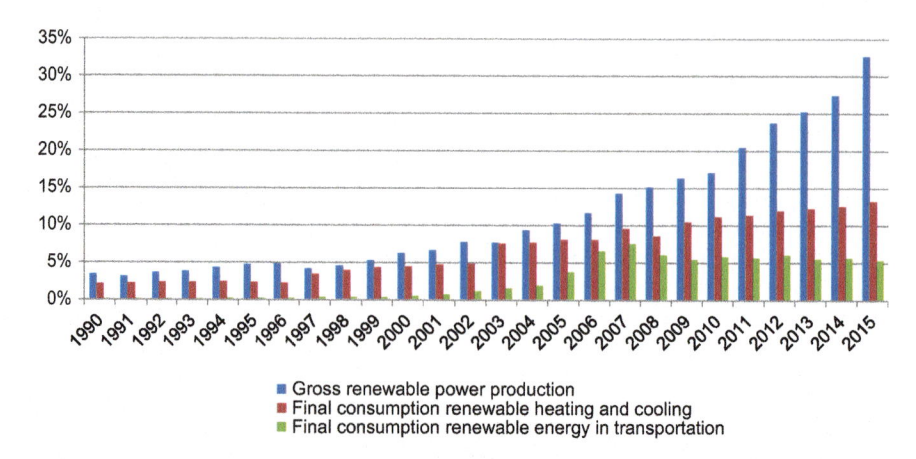

Fig. 3.1 Share of renewable energy in relationship to overall energy demand: 1990–2015 (Federal Ministry for Economic Affairs and Energy 2016c)

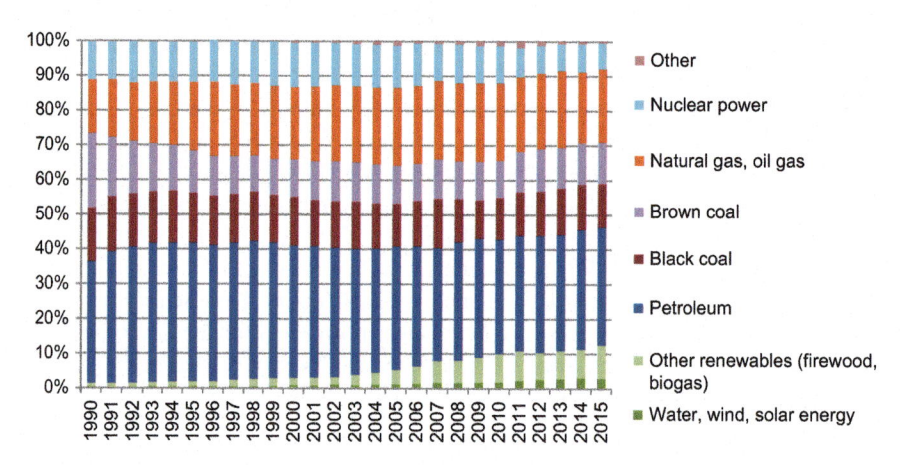

Fig. 3.2 German primary energy demand 1990–2015, according to energy source (Federal Ministry for Economic Affairs and Energy 2016a)

The strongest growth of the renewable energy share can be observed in the power sector, especially after receiving support through the German Renewable Energy Act, which came into law in 2000 (see also Sect. 6.1). By 2015, renewable power already made up 32.6% (195,882 GW h) of national gross power production. In 2014, renewable energies sources surpassed brown coal, which had been the most important energy resource for many decades in Germany (Federal Ministry for Economic Affairs and Energy 2015a, p. 4). However, the share of renewable energy on primary energy consumption is still low and only reached 12.54% (1669 petajoule) in 2015, as shown in Fig. 3.2.

Non-renewable energy sources, including petroleum, coal, natural gas and nuclear power, still dominate the German energy mix. Especially important are

3.2 Energy System Goals

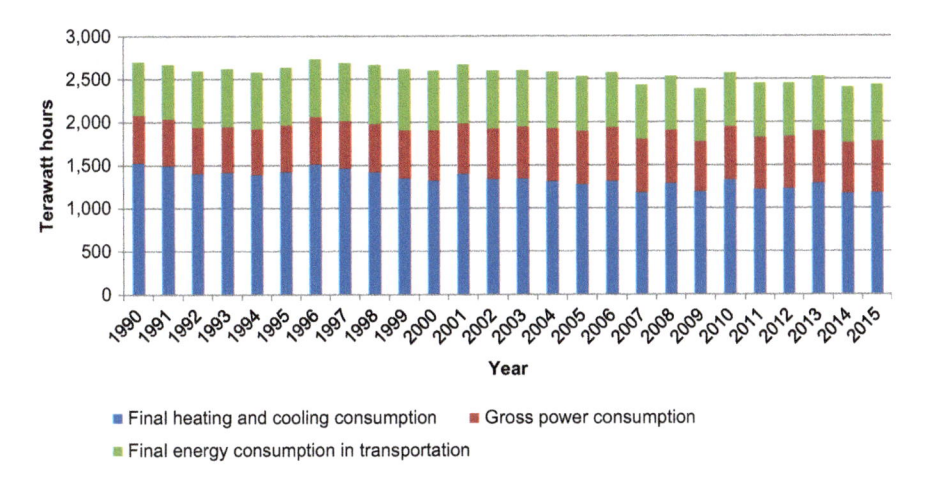

Fig. 3.3 Energy demand in Germany differentiated according to heating and cooling, power and transportation (Federal Ministry for Economic Affairs and Energy 2016c)

increases of renewable energy in the heating and cooling and transport sectors. Most energy—about 52% of total end energy—is actually demanded for heating (mostly space, process and warm-water heating). However, between 2011 and 2015, the share of renewable heating and cooling energy within primary energy demand remained practically stable at around 11–13% (see Fig. 3.1). Renewable sources in the transport energy mix reached a share of 5.3% (34,263 GW h) in 2015 and, thus, actually decreased compared to figures from 2006 (Federal Ministry for Economic Affairs and Energy 2016c).

The regular increase of renewable resources in the energy supply mix since 1990 shows that Germany's energy system has already been in the process of change for well over 15 years. However, Fig. 3.2 reveals that the transition towards a renewable energy system is actually only at its very beginning, as fossil fuels, with a combined 89% share for supplying primary energy demand, are still the leading sources in the national energy mix.

In addition, energy demand has remained almost stable over the last two decades in Germany, as can be seen in Fig. 3.3.

In June 2011, the German government passed a package of several laws[2] presenting a political energy road map (The Federal Government 2011) that is still the basis for the energy policy of the current national Government, with some updates from 2015 (Federal Ministry for Economic Affairs and Energy 2015b, d). Departing from the overall goal to shut down all German nuclear power plants by 2022 and to reduce greenhouse gases by −80 to 95% based upon 1990 values

---

[2]The energy concept of 2011 is based upon the energy concept from 2010 and the resolutions of the *Bundesregierung* (German Cabinet) from 2011 (Federal Ministry for Economic Affairs and Technology and Federal Ministry for the Environment, Nature Conservation, Building and Nuclear Safety 2010; The Federal Government 2011).

Table 3.1 Energy goals of the German government 2011–2015[a]

Energy goals 2011/2015	2020	2025	2030	2035	2040	2050
Greenhouse gas reductions, based upon 1990 values (%)	−40		−55		−70	−80 to 95
Share of renewable energy in primary or (gross) end energy demand (%)	18		30		45	60
Share of renewable energy in gross power demand (%)	35	40–45	50	55–60	65	80
Reduction of primary energy demand, based upon 2008 values (%)	−20					−50
Reduction of gross power demand, based upon 2008 values (%)	−10					−25
Reduction of energy demand in the transport sector, based upon 2008 values (%)	−10					−25
Reduction of heating demand in the building sector, based upon 2008 values (%)	−20					−80

[a]Based upon Federal Ministry for Economic Affairs and Energy (2014b, 2015b), Federal Ministry for Economic Affairs and Technology, and Federal Ministry for the Environment, Nature Conservation, Building and Nuclear Safety (2010), The Federal Government (2011)

(Federal Ministry for Economic Affairs and Technology, and Federal Ministry for the Environment, Nature Conservation, Building and Nuclear Safety 2010, p. 5), this energy concept involves a mix of various energy system goals that are outlined in Table 3.1.

With these plans, the German government aims to reach a renewable energy share of 60% of gross energy demand and a renewable energy share of 80% of total power demand by 2050 and, furthermore, anticipates ambitious reductions of energy demand by then. Many leading German institutions[3] have elaborated energy system scenarios demonstrating how the government's goals can be reached by 2050 or even outperformed. Greenpeace, World Wildlife Fund (WWF) Germany, Umweltbundesamt and German Advisory Council on the Environment even go so far as to argue that Germany will be able to reach a 100% renewable energy share for Germany's gross power demand by 2050. WWF Germany further claims that it is possible to achieve up to a 70% renewable energy share for end energy demand in the country by 2050. (Annex 1 provides a summary of four prevailing energy system

---

[3](See for example, Federal Ministry for the Environment, Nature Conservation, Building and Nuclear Safety 2012; Umweltbundesamt 2010; WWF Deutschland et al. 2009; Greenpeace 2009; The German Advisory Council on the Environment 2011; Peter 2013; Umweltbundesamt 2014; Fraunhofer Institute for Solar Energy Systems 2012; Nitsch 2014).

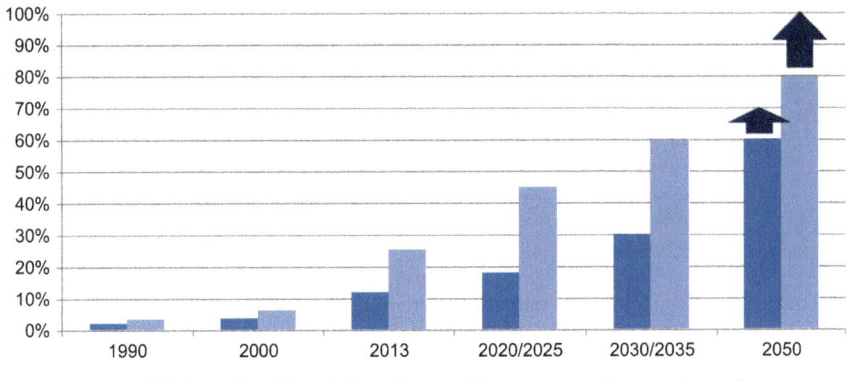

Fig. 3.4 Existing and envisioned shares of renewable energy for end energy demand and gross power demand by 2050 (based upon Federal Ministry for the Environment, Nature Conservation, Building and Nuclear Safety 2012, Greenpeace and Friends of the Earth Germany 2013; The German Advisory Council on the Environment 2011; Umweltbundesamt 2010; WWF Deutschland et al. 2009)

studies and compares their anticipated energy system changes with the goals of the German government.)

However, compared to the current renewable energy share, renewable energy production needs to be strongly increased in order to achieve the anticipated system goals. Figure 3.4 visualises the large gap that exists between current renewable resource levels and future levels envisioned by the government and leading institutions.

3.3 Decentralised Versus Centralised Renewable Energy Systems

Despite the long way that still needs to be gone, as we have seen, the German government and leading institutions widely agree that a renewable energy system can be reached in the country by 2050. However, it is to a great extent unclear how a renewable energy system will eventually look. A central aspect of the current debate is deciding how a future energy system based upon renewable resources needs to be structured in order to function properly. In general, a distinction is made between predominantly centralised and predominantly decentralised system approaches. In essence, these two system modalities refer to different ways of connecting and coordinating energy supply and energy demand. However, discussions about centralised and decentralised system setups often lead to confusion, because they address many different aspects and perspectives on the topic as well as the differing

Table 3.2 Techno-economic parameters constituting decentralised and centralised power systems (Bauknecht et al. 2015, p. 9; based upon Bauknecht and Funcke 2013)

	Production plants	Geographical distribution	Flexibility	Coordination
Decentralised	Energy production plants are connected to the distribution grid	Energy production plants are distributed according to energy demand	Flexibility options, such as for energy storage, are connected to the distribution grid	The system is coordinated in a decentralised way (e.g. through decentralised markets)
Centralised	Energy production plants are connected to the transmission grid	Energy production plants are concentrated in one location, independent of energy demand	Flexibility options, such as for energy storage, are connected to the transmission grid	The system is coordinated in a centralised way (e.g. through spot market prices)

interests of various actor groups. Bauknecht et al. (2015) use four techno-economic parameters in order to specify the two poles of the power sector, as outlined in Table 3.2.

The first and second parameters represent the size and geographical distribution of production plants. In a centralised energy system, renewable energy is supplied by large central energy plants, such as coal, gas or nuclear power plants, which are connected to national transmission grids. Energy is produced independently of energy demand locations and is transported via a well-developed energy transportation infrastructure to energy demand hot spots. In a decentralised energy system, renewable energy is produced in widely distributed small energy production units, which are connected to many national distribution grids, more or less in relation to locations of high energy demand. The third parameter specifies different flexibility options in order to coordinate fluctuating renewable energy supply with energy demand. Storage concepts and IT technology play an important role here. Centralised flexibility options balance energy demand and supply at the level of the transmission grid through, for example, large pump storage plants and central control centres. Decentralised flexibility options lie at the level of the distribution grid. One approach here is intelligent coordination of flexible energy supply, energy load management and the operation of small storage plants based upon web-based IT technology. This concept is summarised under the buzzwords 'smart grid' or 'virtual power plant' (see for example, Servatius et al. 2012). The fourth parameter has to do with the steering of the energy market (Bauknecht et al. 2015, p. 9). A mainly centralised approach, such as existsing today, is primarily controlled by a wholesale power market via the stock exchange, whereas a mainly decentralised approach would operate through decentralised over-the-counter market approaches and self-consumption concepts.

Centralised and decentralised approaches are often discussed as being two contradicting models (for example, Breyer et al. 2013, p. 11). This is particularly revealed in debates around 'energy autarky', 'energy autonomy' or 'energy independency' (McKenna et al. 2015). A set definition for these terms does not exist, but actors supporting the ideas behind them generally strive to achieve a situation in which a specified unit, for example, a house, a village or a certain geographical region, completely self-supplies itself with 100% renewable energy.

> "Energy services used for sustaining local consumption, local production and the export of goods and services are derived from local renewable resources" (Müller et al. 2011, p. 5800).

However, many studies have strongly questioned whether a system exclusively formed by many self-coordinated and autonomous small energy production regions is technically and economically reasonable (for example, Peter 2013; see also, McKenna et al. 2015). Energy autarky may work in individual cases, but it may not work for the whole energy system of Germany and, further, the European Union. Consequently, currently prevailing energy system studies often envision a mix of both configurations. All agree that the increase of renewable energy will be necessarily accompanied by higher decentralisation of the prevailing mode of energy production, which is still primarily based upon large coal and nuclear power plants. For the foreseeable future, the most dominant types of renewable power will be produced by decentralised onshore wind, photovoltaic and biomass plants (see Annex 2). But energy production and demand is levelled across regions, and even across European countries, which in some ways requires centralised flexibility options and centralised coordination.

The central question may, therefore, not be whether Germany's renewable energy will be primarily decentralised or centralised but, rather, which combination of the two approaches will be necessary in order to achieve a system that is completely based upon renewable resources.

3.4 System Challenges

At this point in time, it is completely unclear which development path the ongoing energy system change will follow in the end. The direction in which Germany's energy system will further develop strongly depends upon technological developments, political guidance and society to foster and adapt to the emerging needs of a renewable energy system. The Ethics Commission for a Safe Energy Supply has described the currently ongoing energy system change as a fundamental decision of society with far-reaching consequences for the economy, politics and the lifestyles of citizens. Questions regarding the composition of the energy supply and the roles of technical infrastructure, research and development and society need to be carefully considered during the next few decades (Ethics Commission for a Safe Energy Supply 2011, p. 17). It is therefore important to better understand the multiple challenges involved in achieving a renewable energy system and in finding an appropriate system configuration that can successfully reach the goals set for 2050.

3.4.1 Technical Challenges

Germany's envisioned change of energy supply and demand entails the need to fundamentally change its technical energy infrastructure. Existing renewable energy production and energy-efficiency technologies have to be further developed and more strongly diffused throughout the country. At the same time, new innovative technologies need to be developed. The production of wind and solar energy is strongly fluctuating and not (yet) 100% predictable, so new energy storage options or smart technology solutions for better demand and supply management need to be integrated in order to balance times of excess wind and solar energy versus times in which produced power does not fully meet demand (Federal Ministry for Economic Affairs and Technology 2009).

A renewable energy system requires a new energy transmission infrastructure. According to the national grid-development plan for power, between 3700 and 4300 km of new transmission grid infrastructure needs to be installed, and about 5200–5800 km of existing grid has to be reinforced. This is a massive undertaking. In fact, the investment costs for (re)developing the transmission power grid over the next 10 years are expected to be around 27–34 billion euros (50Hertz Transmission GmbH et al. 2016, p. 160).

Bartosch et al. (2014, p. 39) point out that the German Energiewende can only be achieved if not only the power sector but also the heating and cooling and transport sectors are fundamentally transformed. In this vein, Greenpeace (2013) calls for a share of almost 80% renewable energy in the heating and cooling sector by 2050. New solutions for the heating and cooling and transport sectors are also required. Leading energy system scenarios propose that electrical vehicles and plug-in hybrid vehicles (powered by an electric motor and a secondary motor, such as a combustion engine) must reach a share of between 50 and 100% in the personal automotive sector by 2050 (see Annex 1). Long-term solutions for the mobility sector may also include hydrogen mobility, and new heating solutions are likely to involve power-to-heat technologies (for example, Greenpeace 2009, pp. 24ff.; Nitsch 2014).

3.4.2 Market Challenges

The increase of renewable energy has challenged the energy market. This can be especially observed in the power sector. Conventional power capacities are mainly coordinated through a centralised market instrument: the wholesale electricity market. Stock power prices are formed through the merit order,[4] where available power

---

[4]The merit order is a concept used in the German energy wholesale market for ranking available power capacities according to marginal costs. The price of electricity is determined on the spot market through the most expensive power plant that is still needed to satisfy current electricity demand (merit order) (Sensfuß and Ragwitz 2007, p. 2). The merit order seeks to ensure that overall

3.4 System Challenges

capacities are ranked according to their marginal costs, with only the most convenient power plants going online in order to meet national demand. Furthermore, power production is normally differentiated into basic-load plants, providing a continual amount of power (long-term controllable production plants, such as coal or nuclear powered), and peak-demand production units, which produce additional power in times of high demand (short-term controllable power production units, such as pumped-storage hydroelectricity or gas turbine plants).

Under current conditions, renewable power cannot be fully integrated into this conventional centrally oriented market structure. Since the dominant renewable power resources today, mainly wind and solar, are strongly fluctuating and not predictable, they can neither fully replace basic-load plants nor can they fully meet peak-demand requirements. In addition, they fundamentally challenge the merit order concept. On the one hand, renewable energy production plants have different cost structures and operate at extremely low marginal costs.[5] On the other hand, they are currently receiving financial support under the German Renewable Energy Act (see below). The recent increase of renewable energy supply, with its very low marginal costs, has already changed the merit order and led to a strong reduction of stock power prices, to a level at which refinancing conventional power plants has become extremely challenging (Kemfert and Traber 2013, p. 17).

Yet, the energy market needs to be redesigned if the share of renewable power is to further increase. In July 2015, the national government published concrete suggestions for a 'power market 2.0' (Federal Ministry for Economic Affairs and Energy 2015c). However, ideas about necessary changes vary strongly. Large prevailing energy providers, such as E.ON SE and RWE AG, suggest the introduction of an additional centralised capacity market where coal and gas turbine plants would get additional payments for being able to provide power during hours of low power production from renewable energy units (Schultz 2014).[6] Meanwhile, renewable energy providers support a market model with a more strongly decentralised orientation that allows operators to market their renewable energy directly to the customer, without making their way through the stock market (Clean Energy Sourcing et al. 2015). These contradictory positions reveal the great challenge of finding a market design which will be able to mediate between the interests of various market actors.

---

costs of power generation are minimized, because only the most convenient power production plants are used in order to meet national power demand.

[5]Once installed, renewable power plants can be operated at very low costs, as they draw upon energy resources, in particular wind and solar radiation, that can be used at no cost.

[6]Leading associations and initiatives, such as Greenpeace, Friends of the Earth Germany (BUND) and German Institute for Economic Research, are against such a capacity market, because they assume that it may subsidise fossil-fuel power plants that have become inconvenient in the context of increasing renewable energy (German Institute for Economic Research 2013; Greenpeace and Friends of the Earth Germany 2013).

3.4.3 Societal Challenges

Germany's energy system goals can only be met if increased renewable energy production is attended by strong efforts towards lowering overall energy demand. Thus the government aims to reduce primary energy demand by 50% by 2050 (see Table 3.1), and efficiency potentials are to be mobilised in all sectors. The highest reduction potential is foreseen in the building sector through reducing the demand for heat (see Table 3.1), with the national government aiming to decrease heat demand by 80% in the building sector by 2050 (Federal Ministry for Economic Affairs and Energy 2015b, p. 12). However, as can be seen in Fig. 3.3, historical energy demand has been quite stable over the last 25 years, revealing that efforts towards reducing energy consumption will need to be strongly intensified in all areas if the 2050 goals are to be met. One of the great societal challenges in this domain, then, is how to trigger far-reaching reductions of energy use in all areas, which may only be possible if people make changes in their lifestyles (Ethics Commission for a Safe Energy Supply 2011, p. 19).

Yet such far-reaching system changes cannot be dictated via a top-down process. As the Ethics Commission for a Safe Energy Supply has indicated, active participation of citizens in the transition process and the creation of acceptance amongst the population need to be considered central components of achieving societal consensus for the Energiewende (Ethics Commission for a Safe Energy Supply 2011, pp. 18f.). Bartosch et al. (2014) add:

> "Again and again, it has been forgotten to involve citizens in the democratisation and decentralisation of the energy system and to actively encourage their participation. This is not merely a decorative accessory but, rather, a fundamental pillar of the Energiewende and its social acceptance" (Bartosch et al. 2014, pp. 9f.).

The decentralised character of renewable energy production also creates new potential relationships between energy consumers and producers, as renewable energy can be directly consumed by actors that are located near respective production units. However, such concepts can only be realised if market and political actors provide the necessary instruments and if such new direct consumer–producer relationships are also desired by society.

3.4.4 Political Challenges

The main responsibility of those in the political sphere is to govern the ongoing energy system transition and to translate energy system goals into feasible legal guidelines. Yet the latest far-reaching amendments of the German Renewable Energy Act (EEG) show how difficult it is to find the right balance between multiple interests within the shift from one system to the other. The EEG is one of the most

3.4 System Challenges 33

important legal guidelines[7] for the transition and regulates the production of renewable power (§1 EEG). The fundamental principle of the EEG used to be increasing power from renewable resources through guaranteed feed-in tariffs and a guaranteed connection to the power grid. Amendments of the EEG in 2014 and 2016,[8] however, changed the overall focus towards limiting the costs of energy system change[9] by integrating renewable energy into the energy market and having full government control over increases of power capacity (Federal Ministry for Economic Affairs and Energy 2014a; The Federal Government 2016). These amendments included the reduction of feed-in tariffs for almost all renewable energy technologies, including solar, biomass and on-shore wind power, as well as the introduction of a new direct-marketing model.[10] The most central elements of EEG 2014 and EEG 2016 are the introduction of a tendering model and definition of limits for annual increases of power capacity. The aim of the tendering model is to shift from guaranteed tariffs defined by the government to technology-specific tariffs formed by the market—through bids and calls for permission to install certain renewable energy capacities throughout Germany (see more details about tenders in Sect. 6.1). In this way, the government wants to achieve installation of a certain amount of capacity at the lowest possible price. For each renewable energy technology, EEG 2014 and EEG 2016 have defined so-called corridors of capacity expansion. For example, according to EEG 2016, the *net* expansion of onshore wind power shall not exceed 2800 MW per year up through

---

[7]Other important regulations include the Renewable Energy Heat Act (EEWärmeG 2015), which guides the use of renewable heating and cooling in buildings; the cogeneration law (KWKG 2016), which regulates implementation and use of highly efficient combined heating and power plants; the energy reduction decree (Energieeinsparverordnung EnEV 2014), which supports energy efficiency; and the bio-fuel quota law (Biokraftstoffquotengesetz BioKraftQuG), which coordinates the national introduction of bio fuels.

[8]Amendments to the EEG were passed on 8 July 2016. Even though the amendments did not take effect until January 2017, they are called EEG 2016 throughout the thesis, following the official draft being passed by the German Bundestag (Deutscher Bundestag 2016).

[9]The costs for promoting renewable power are paid by energy consumers in the form of an EEG surcharge (*EEG-Umlage*), which makes up a large part of individual electricity bills. The EEG surcharge was increased from 0.19 cents/kW h in 2000 to 6.24 cents/kW h in 2014 (Mayer and Burger 2014, p. 2). Criticism arose, however, claiming that renewable energy development would lead to an unreasonable rise of individual electricity costs. It was not, however, only the continual increase of renewable power plants that led to raising the EEG surcharge, as other factors, such as the aligned decrease of energy stock exchange prices and the exemption of companies with high power costs from paying the EEG surcharge (§ 64 EEG 2014), strongly influenced it (Mayer and Burger 2014).

[10]According to EEG 2014, operators of renewable power plants with a capacity above 500 kilowatts until 2016, and above 100 kW from 2016 onwards, have to market their power directly on the stock exchange. The final tariff that they receive for their power is formed of two components: the stock exchange price and a market premium, which is equal to the difference between the stock exchange price and the guaranteed feed-in tariff. This way, operators of bigger power plants still receive a guaranteed return on investment for their power over 20 years but are, at the same time, responsible for marketing their power. The aim of this regulation was to foster the integration of renewable energy into the conventional energy market (§§ 40–54 EEG 2014). EGG 2016 combines the direct marketing concept with a tendering model (Deutscher Bundestag 2016).

2019 and 2900 MW per year from 2020 onwards. Meanwhile, the *gross* expansion of solar power shall not exceed 2500 MW per year (Deutscher Bundestag 2016, p. 17). At the same time, the installation of greater capacity automatically leads to reduced guaranteed compensation for new plants in the following year (for an overview of legal changes in EEG 2014 and EEG 2016, see Dagasan et al. 2014; Deutscher Bundestag 2016; Federal Ministry for Economic Affairs and Energy 2014a). These changes to the laws governing the German energy transition have been controversial, with renewable energy representatives, such as the National Association for Renewable Energy, the German Wind Energy Association and the German Solar Association, as well as various eco-power providers, being especially opposed to the amendments. They feel that these amendments put them at a disadvantage, threatening the further development of renewable power in Germany (Bundesverband Erneuerbare Energie e.V. 2014; Bundesverband Erneuerbare Energien e.V. 2016; German Solar Association 2014). They argue that decreased feed-in tariffs, the new tender model and the direct-marketing concept will lower financial support for renewable energy provision, on the one hand, while unreasonably increasing project risks on the other. Such an approach particularly penalises small actors, such as citizen-driven energy projects, because they are not be able to cope with the new complexity of, for example, the tender model as well as large project developers can (Greenpeace Energy eG 2014).

The current government is composed of a grand coalition between the CDU/CSU and the SPD; it has been called the *Grosse Koalition*, or GroKo for short. Nitsch (2014) has reflected upon the extent to which the GroKo energy policy really fosters the country's energy system goals for 2050 by extrapolating from the current development level of the energy system based on current regulations. He comes to the conclusion that neither a 60% share of renewable energy for primary energy demand nor the reduction goal of −50% in overall demand will be met by the GroKo scenario. It has, therefore, been argued that political action would need to be designed in such a way that the transition dynamic strongly increases, if the national government actually wants to take the energy system goals seriously (see also, Bartosch et al. 2014, pp. 31f.).

Despite these criticisms, the government appears to be planning to keep its general direction, with planned amendments focusing once again on capacity control. The central element of the EEG 2016 is, therefore, amplification of the tendering model to solar parks and rooftop solar energy, onshore and offshore wind energy.[11] Only plants with a capacity below 1 MW are to be excluded. In this way, about 80% of produced power from new plants will be controlled and managed through tenders, with a dynamic capacity limit being annually defined (Federal Ministry for Economic Affairs and Energy 2016b, p. 2). In addition, the EEG 2016 plans further reduction of feed-in tariffs for certain projects types, such as onshore wind.

---

[11]Whether or not the tendering model is amplified to biomass capacities is in the process of analysis (Federal Ministry for Economic Affairs and Energy 2016b, p. 1).

3.4 System Challenges

Table 3.3 Technical, societal, political and market challenges facing the German energy transition

	Main challenges
Technical	• Diffusion and elaboration of existing renewable energy technologies in the power, heating and cooling and transport sectors • Development of new innovative technologies, such as energy storage options or smart solutions, for better demand and supply management
Market	• Creating a new energy market design • Stronger coordination of energy demand and renewable supply
Societal	• Energy efficiency in all areas, particularly in the heating and cooling sector • Acceptance of energy system change
Political	• Governing the energy transition at different political levels • Balancing the shift from one system to another

Table 3.3 summarises some of the most pressing challenges that need to be solved in order to achieve a renewable energy system in Germany.

Chapter 4
Organisations in the Context of Socio-technical Transitions

The central aim of this section is to develop a theoretical framework for analysing the impact of organisations within a far reaching system change. The work shall be embedded in the context of sustainability transitions research and will contribute to actor research in general and enterprise research in special in this emerging research field.

In Sect. 4.1 the theory of socio-technical transitions is outlined and reflected. Section 4.2 discusses the relationship between enterprise interaction and structural change based upon structuration theory. Section 4.3 elaborates a new framework for analysing the influence of organisations within socio-technical transitions.

4.1 Understanding Socio-technical Transitions

4.1.1 Socio-technical Transitions Towards Sustainability: A New Distinct Research Field

A growing body of research literature has been considering the transition processes of socio-technical systems (Markard et al. 2012). Core aim is to comprehend how unsustainable system structures may be transformed into more sustainable ones (Grin et al. 2010a, p. 2; Rip and Kemp 1998; Verbong and Loorbach 2012b, p. 7).

In the early 2000s, sustainability transitions research evolved as a distinct scientific direction out of a shared motivation among scholars with different theoretical backgrounds to better understand why it is so hard to solve the grand challenges of the current century, such as climate change and resource scarcity. Against this background, questions of how to stimulate and govern wider transition processes towards a more sustainable future have received increasing attention (Markard et al. 2012, p. 955; van den Bergh et al. 2011, pp. 3f.). Within this context, sustainable development refers to:

"Development that meets the needs of the present without compromising the ability of future generations to meet their own needs. It contains [...] the idea of limitations imposed by the state of technology and social organization on the environment's ability to meet present and future needs" (Brundtland and Khalid 1987, p. 41).

Sustainable transitions research can still be seen as an emerging field that is undergoing a process of maturation. As of August 2015, the official Sustainability Transitions Research Network (http://www.transitionsnetwork.org) reached about 1000 members, with associated scholars distributed all over the world; the majority, however, are situated in the United Kingdom, the Netherlands and Germany (Sustainability Transitions Research Network 2015, p. 5). Up to September 2015, about 1400 peer-reviewed articles had been written, receiving about 25,000 citations by that time (Sustainability Transitions Research Network 2015, pp. 4ff.). The community launched its own journal Environmental Innovation and Societal Transitions in 2011 (van den Bergh et al. 2011, p. 2).

Important for this work is that the topic of transforming energy systems has garnered the highest amount of empirical interest in the field of sustainability transitions research (Sustainability Transitions Research Network 2015, p. 6). This can be partly explained by the circumstance that research scholars from the Netherlands had been strongly involved in elaborating transitions research as a discipline when working together with the Dutch Ministry of Environmental Affairs in the field of renewable energy (Grin et al. 2010b, p. xvii; Verbong and Loorbach 2012b, p. 14). At the same time, many technical, political and societal attempts can be observed around the world that have endeavoured to support the development of renewable energy. A growing number of researchers have become engaged in analysing and reflecting upon these ongoing activities and dynamics, since they are seen as seeds of a wider energy system transition process (Sustainability Transitions Research Network 2015, pp. 4ff.).

Crucial to note here is that, although the process of system transformation or system transition is being explored and discussed by many other research fields, it is not the aim of this work to provide a full overview of the manifold research strands concerned with societal transformation; a fruitful attempt to map the various directions within this domain of research can be found in Kristof (2010), for example.

4.1.2 Defining Socio-technical Transitions

Studies of socio-technical transition towards sustainability generally focus on the development of key socio-economic domains, including energy, mobility, transportation, water and sanitation, agriculture and waste. In transitions research, a system is delineated along a specific societal function in terms of meeting a human need, such as providing food, energy or water (Geels 2002, p. 1257; 2004, p. 898; Rip and Kemp 1998, p. 330). These systems are highly influenced by technology and are, thus, framed as socio-technical systems (Markard et al. 2012, p. 956; Sustainability Transitions Research Network 2015, pp. 4ff.). Socio-technical systems enclose all

4.1 Understanding Socio-technical Transitions

elements necessary for the system to work and to fulfil its system function. Accordingly, they comprise of actors embedded in social groups (citizens, consumers, firms and politicians), formal and informal rules (guidelines, standards and regulations), as well as material artefacts, including technical infrastructure (Geels 2002, p. 1258; 2004, pp. 898ff.; Kemp et al. 2012, pp. 3ff.; Markard et al. 2012, p. 956). All system elements are tightly interconnected, such that aligned technologies should actually be seen as *"technologies-in-context"* (Rammert 1997, p. 176; see also, Geels and Schot 2010, pp. 13) or as *"embedded technologies"* (Schneidewind 2011). By turning attention towards the societal structure in which technologies are embedded, socio-technical system analysis highlights the ways in which the success and failure of technologies are eventually determined by how they work within society. This system interpretation goes beyond the focus and scope usually applied in organisation studies, in which structures used to be mainly delineated along the technology itself or along industry sectors (for example, Porter 1990).

Following Verbong and Loorbach (2012b, p. 9), the energy system can thus be defined as a complex whole that includes all actors, artefacts, rules and social heuristics, which together produce the societal function of providing energy. They note that:

> "Our society is fully adapted to the fossil-based energy system, which in turn is deeply entrenched in all social domains and practices" (Verbong and Loorbach 2012b, p. 5).

Energy technologies are strongly rooted in the present socio-technical context and have emerged with modern society. At the same time, the socio-technical energy system is aligned to other systems, such as the mobility or agricultural systems (Schneidewind et al. 2013; Schneidewind and Scheck 2012; Verbong and Loorbach 2012b, pp. 5ff.).

In this domain, transitions are defined as major *"shifts from one socio-technical system to another"* (Geels and Schot 2010, p. 11 based upon Kemp 1994). Accordingly, a system in transition undergoes a radical change. The particular system function, such as supplying energy, may stay the same but the system structure is fundamentally altered (Verbong and Loorbach 2012b, p. 7). The notion behind it is that sustainability problems arising from energy, mobility or agriculture are deeply rooted in today's system configuration. Established technologies are strongly connected to policies, user practices, business models and values (Geels 2004, p. 900; Kemp 1994, p. 1024; Markard et al. 2012, p. 955; Markard and Truffer 2008, pp. 598ff.). Consequently, solving fundamental problems arising from such broad and stable socio-technical set-ups, such as climate change, may thus require more than incremental, narrowly focused changes, which have been shown to rather stabilise the establishment and allow the overall system to move on with business as usual. Instead, the system itself may have to undergo a far-reaching transition, tackling problems from technical, political, economic and societal dimensions (Kemp et al. 2012, pp. 3ff.; Markard et al. 2012, p. 956; Verbong and Loorbach 2012b, pp. 1ff.).

According to Geels (2010, p. 11), transitions have four main characteristics. (1) They are co-evolutionary processes, technical and societal in kind, with a high number of driving factors. (2) Transitions involve multi-actor dynamics in and between many different social groups, including business, market users, scientists, policymakers and citizens. (3) Transitions are understood as radical shifts during which the set-up of a socio-technical system fundamentally changes. (4) Transitions do not happen in a linear way with a clear causality between associated events and progressions. They are complex, uncertain developments that take place across a long time horizon of several decades.

Some important conceptual frameworks for understanding socio-technical transitions have been elaborated through studying historical transitions. Examples here include the shift of personal mobility from horse and carriage to the individual car (Geels et al. 2012), the change from sailing to steam ships (Geels 2002), the transition from cesspools to sewer systems (Geels 2006), the change from extensive to intensive agriculture, and the rise of electricity (van der Vleuten and Högselius 2012). Compared to these historical cases, the nature of the transition processes surrounding today's mobility, energy and agricultural systems—the centre of current sustainability transitions research—is fundamentally different. First of all, they represent developments that are still ongoing and, thus, are called *"transitions in the making"* (Farla et al. 2012, p. 991). As they have an open-end character, it is virtually impossible to predict or direct a total system transition in such way that one can say when it will take place, in what form and at what speed. Transitions in the making are determined by many different pathways that co-evolve and compete against each other as they evolve (for example, Geels and Schot 2007). Furthermore, unlike many historical cases, today's transition processes are mainly aligned to the conscious aim of addressing sustainability problems (Smith et al. 2010, p. 437; van den Bergh et al. 2011, p. 3) and are, therefore, based upon a desired outcome, an overall long-term vision (Farla et al. 2012, p. 992).

One could, then, say that the general idea of achieving a desired and norm-driven system change towards sustainability, such as that associated with ongoing transitions today, stands somewhat in contrast to the understanding of transitions as long-term and complex system dynamics which cannot be controlled because they involve so many multiple processes. Transitions research points out that, even though transitions cannot be guided as a whole, actors do have the power to influence transition dynamics, especially in terms of their direction (Verbong and Loorbach 2012b, p. 10). Central here is the idea that change happens through the interaction of actors (Geels 2004, p. 902; 2005, pp. 16ff.). Therefore, an important difference between studying historical transitions ex-post and reflecting upon sustainability transitions ex-ante is taking an explicit interest in how and to what extent it may be possible to support and guide transition processes. Looking at the present challenges to achieving sustainability, it is decisive to accumulate more scientific knowledge in this regard. System configurations which cause these problems are considered to be particularly persistent because they are extremely widespread and strongly interlinked with modern society (German Advisory Council on Global Change 2011).

The present work departs from this premise and systematically reflects upon the actual and potential influence of an innovative enterprise group—German energy cooperatives—that is currently involved with such a complex and long-term socio-technical transition. Analysing Germany's energy system as a socio-technical system helps to sketch the big picture, which seems necessary in order to understand the scope of the multiple system changes that are likely to be involved in realising Germany's energy vision. The current change of Germany's energy system outlined in Chap. 3 represents a socio-technical transition process in the making. Furthermore, employing the notion of transitions addresses the challenge of setting the complex, and to a great extent uncontrollable, character of system change in relation to actor dynamics. Sustainability transitions research has been chosen to guide the core research interest of this work via a relevant set of concepts and a fruitful analytical framework.

4.1.3 Prevailing Concepts for Studying Transitions

Transitions research draws on a variety of scientific fields, including complexity theory, science and technology studies, innovation studies, and governance studies as well as on approaches from history, political science and sociology (Geels 2005, pp. 28ff.; Geels and Schot 2010, p. 12; Grin et al. 2010b; Markard et al. 2012). While these disciplines themselves have in fact a long history of studying change processes, proponents of transitions research argue that none of them alone can grasp the complexity of today's grand challenges. Smith et al. (2010) and Geels (2005, pp. 6ff.), for example, point out that a broader analytical perspective is needed in order to comprehend all the challenges involved in the kinds of dynamics that may lead to a system transformation in favour of sustainable development, as material, economic, social, cultural and political processes have to be jointly observed in context. Building upon this variety of different research domains, transitions research has developed its own conceptual frameworks in order to observe how radical transitions unfold and what kinds of factors drive or constrain them. The most central of these frameworks are introduced in the following.

A core analytical approach of socio-technical system analysis is the **multi-level perspective**, which is an overarching framework through which the dynamics behind a transition process can be described and explained. Explicitly paying attention to societal structures in socio-technological development, it conceptualises change along three levels of structuration: innovative niches, the prevailing all-encompassing regime and the superior landscape level. Novel configurations are created within protected niches. Their successful adoption and diffusion depends upon the conditions and characteristics of the selection environment, meaning here the incumbent regime, as well as upon landscape events (Geels 2002, 2004, 2005, pp. 85ff.; Geels and Kemp 2012, pp. 50f.; Geels and Schot 2010, pp. 18ff.).

Strategic niche management is closely aligned to the multi-level perspective framework. Its main interest is to analyse how socio-technical novelties created in niches become stronger in order to survive beyond the experimentation stage and diffuse so as to eventually change the regime (Geels and Schot 2010; Kemp et al. 1998; Raven 2012; Schot and Geels 2008).

Transition management is a conceptual approach which focuses on the question of how to actively steer transitions (Loorbach 2007; Loorbach and Rotmans 2010; Rotmans et al. 2001) and is primarily associated with the transition management cycle (Loorbach and Rotmans 2010, p. 2). This involves several main action steps, such as problem structuring, developing transition agendas, mobilising actors and monitoring. The aim of studying the cycle is to support key actors in structuring anticipated transition processes (Loorbach and Rotmans 2010, p. 2).

Transition governance represents another perspective that is experiencing increasing interest (Docherty and Shaw 2012; Grin 2010; Smith et al. 2005; Verbong and Loorbach 2012a), although a prominent conceptual framework has not been elaborated within it thus far. Instead, scholars within this subfield primarily draw on the concepts of the multi-level perspective, strategic niche management and transition management.

The approach of **technological innovation systems** has been set in relation to the multi-level perspective (Markard and Truffer 2008; Suurs and Hekkert 2012), although its proponents apply a different conceptual understanding. The technological innovation system approach uses a special set of system functions (system motors) for describing the functional pattern of a system (Bergek 2002; Bergek et al. 2008; Hekkert et al. 2007). Earlier works tended to have had a narrow focus on technological systems and sectors (for example, Carlsson and Jacobsson 1997; Edquist 2004; Ehrnberg and Jacobsson 1997). The latest studies contribute towards a better understanding of the systemic contexts of technological change in society (Binz et al. 2014; Jacobsson and Bergek 2004, 2011; Tigabu et al. 2015; Wieczorek et al. 2015a).

Another concept currently receiving increased attention is the **sectoral innovation system** approach. Initially introduced by Breschi and Malerba (1997), and Malerba (2002), it was developed further by Dolata (2009) and Geels (2014). This approach relates a system to the structure and boundaries of a sector and, thus, to the environment of firms within it. Here, the focus is more specifically on industries though the concept provides a multidimensional perspective on sectors. The interest here is to better understand how a technology, deeply aligned to the nature of a sector, can be subject to changes, by taking into account the wider environment of industries.

Long wave theory (Freemann and Perez 1988; Kondratieff and Stolper 1935; Perez 1983; Rosenberg and Frischtalk 1984) and the **S-curve** (Rogers 1962) represent earlier concepts which helped to form transitions research. Long wave theory explains shifts of techno-economic paradigms through change of key factors which have stabilised a system over long time periods. Even though both conceptual frameworks have been fundamentally important for transitions research, it has been argued that they do not provide enough insights into how a transition process actually unfolds (Geels and Schot 2010, p. 17).

All of the frameworks presented above apply a holistic, complex system perspective and focus on the interconnection between multiple system elements. However, they differ in terms of their analytical approaches, with the question of which framework(s) could be best applied for a particular study mainly depending upon research interest. In order to address the research questions of this work, it is primarily drawn upon the multi-level perspective and strategic niche management. As explained in the following, the multi-level perspective helps to analyse the structure of a system, while strategic niche management enables analysis of the development of innovation projects. Both concepts have been commonly applied in order to study energy transitions, so there is a rich store of scientific reflection on energy transitions that can be drawn upon.

4.1.4 The Multi-Level Perspective on Sustainability Transitions

The multi-level perspective understands socio-technical systems as configurations of multiple, interrelated structures and differentiates between three levels of their structuration (1) technological niches, (2) socio-technical regimes and (3) the external landscape (Geels 2002, pp. 1259ff. based upon Rip and Kemp 1998, pp. 337ff.; Schot 1998, pp. 175ff.). Socio-technical regimes are broad and well-established all-encompassing arrangements, providing overall guidance for groups of actors. These arrangements are deeply rooted in society, in the market, in policy, and it is, thus, difficult to deviate from them because they are constantly being reproduced. Since these ongoing processes take place across various dimensions, however, there is room for change. Regime structures can thus be described as 'dynamically stable' (Geels and Schot 2010, p. 21). Innovations can evolve but may rather remain incremental in kind. Niches are theorised as small and unstructured new 'spaces' in which innovations are created outside the mainstream context. Meanwhile, the landscape level forms the autonomous, exogenous environment that can hardly be influenced (Geels 2002, pp. 1259ff.; 2005, pp. 75ff.; Geels and Kemp 2012, pp. 57f.; Geels and Schot 2010, pp. 23ff.; Grin 2008, pp. 52ff.). The multi-level perspective helps to identify different patterns and pathways, as well as typical characteristics and underlying mechanisms, making up a transition process (for example, Geels 2002, 2005, pp. 103ff.; Geels and Kemp 2012, pp. 59ff.; Geels and Schot 2007; Grin 2008, 2010; Rotmans and Loorbach 2010; Schot and Geels 2008; van der Brugge 2009). Figure 4.1 displays the multi-level perspective of sustainability transitions research.

Socio-technical transitions unfold through the co-evolutionary interplay of multiple developments within and between the three structural system levels: niche, regime and landscape (Geels and Kemp 2012, pp. 50f.; Geels and Schot 2010, pp. 18ff.). Simplified, slowly emerging or sudden landscape events, such as climate change or resource scarcity, can put pressure on an existing regime. Maladjustments or tensions between the existing 'way of life' and emerging social, economic or ecological challenges trigger critical reflection upon prevailing technologies, market set-ups or existing societal principles. Such regime destabilisation opens up

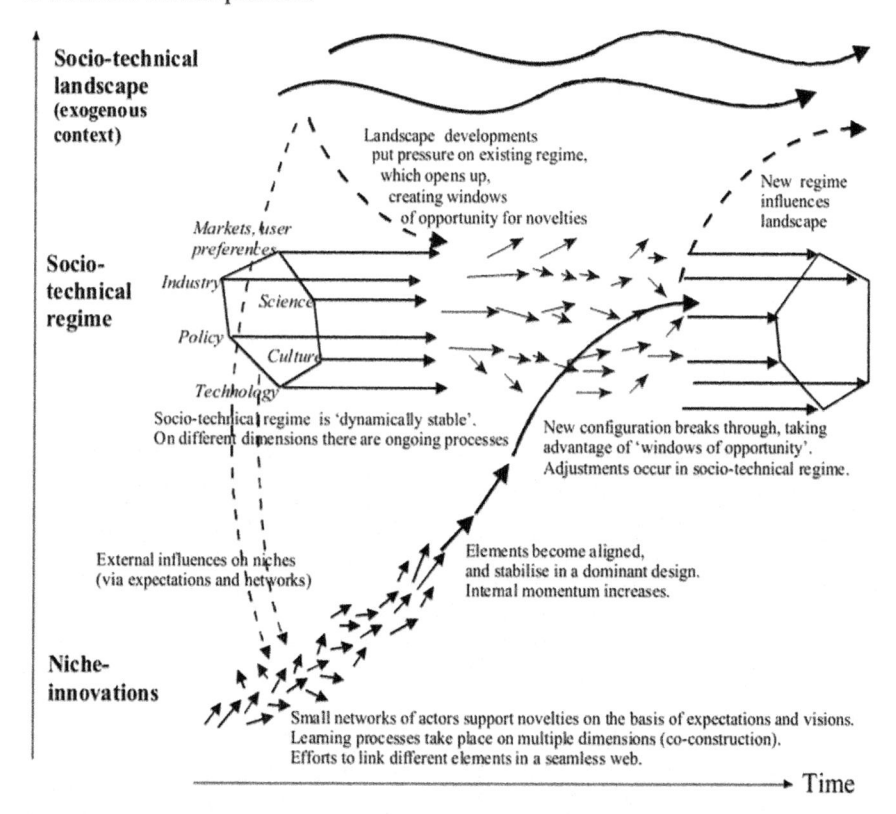

Fig. 4.1 The multi-level perspective on sustainability transitions (Geels and Schot 2010, p. 25 based upon Geels 2002, p. 1263)

windows of opportunity for new socio-technological concepts to emerge. Technological niche innovations may then evolve, driven by the expectation of finding better solutions for rising socio-technical problems. Innovative niche activities are, thus, directly inspired by ongoing dynamics at the regime level. However, niche innovations may only be able to break through if they diffuse and become accepted by a vast majority of society. This happens once niche elements begin to form a stable design. The increasing dominance of such new configurations over time leads to adjustments and changes in the existing regime. Niche novelties can enrich present socio-technical regimes or even replace them by eventually becoming part of a new regime themselves (Geels 2002, pp. 1259ff.; 2004, pp. 910ff.; 2005, pp. 75ff.; Geels and Kemp 2012, pp. 59ff.; Geels and Schot 2007, pp. 405ff.; Grin 2008, pp. 52ff.).

The multi-level perspective heavily builds upon science and technology studies, evolutionary economics and social theory (Geels and Schot 2010, pp. 30ff.). Science and technology studies has become an important research domains that has been seeking to demonstrate the complexity of technological change since the 1980s and 1990s. By characterising technology development as *"heterogeneous engineering"* (Law 1987, p. 111) or as building a *"seamless web"* (Hughes 1986, p. 281), scholars began to turn their attention towards the interactions between society and technology, arguing that technology and its environment need to be framed as larger systems than previously (see also, Bijker et al. 1987; Bijker and Law 1992). The prevailing idea at this time was that technology develops autonomously from society and, once introduced, becomes a main driver of societal change (Geels and Schot 2010, p. 31). In contrast to this rather linear view on technological development, in the early 1980s evolutionary economics (Dosi 1982; Dosi et al. 1988; Nelson and Winter 1982) offered alternative conceptual directions that became important principles of the multi-level perspective. The central premise of evolutionary economics is that populations of different components—including agents, institutions and technology—and their environment develop together through innovation and selection (adoption, diffusion) (Nelson and Winter 1982). In line with this idea, the multi-level perspective highlights the co-evolution of new technical infrastructures and new social heuristics at different levels of structuration (Geels and Schot 2007, p. 402; 2010, pp. 35ff.). Variation and selection of socio-technical concepts are central explanatory foci. Variation takes place at the niche level where novelties arise, and the adoption and diffusion of these novelties are guided by regime and landscape dynamics. Depending upon the adaptive capacity of regimes, the capabilities of niches and landscape pressures, various transition pathways may emerge or be inhibited (Geels and Schot 2007). Unlike evolutionary economics, however, the multi-level perspective does not stop at the boundaries of users and markets. It takes societal rules and policies as well as infrastructures and cultural discourses as important elements under consideration.

The multi-level perspective is further inspired by structuration theory of Anthony Giddens (Giddens 1984, see Geels 2002; Geels and Schot 2010, pp. 29ff.). The multiple developments in and between the niche, regime and landscape levels come about through the interaction of actors (Geels 2004, p. 902; 2005, pp. 16ff.). Giddens (1984) provides his own ontology of agency and structure, two fundamental building blocks for approaching the research interest here: analysing the influence of energy cooperatives on energy systems. The relationship between structuration theory and organisation theory is outlined and reflected upon in great detail in Sect. 4.2. For the argumentation in this chapter, the core notion is briefly outlined at this point.

The central idea of structuration theory is the duality of structure (Giddens 1984, pp. 25ff.). Social structures are the primary medium for as well as the outcome of social interaction. All actors are embedded in the existing structures that make up their society. Social structures form the context of daily life and provide the framework for social interaction (medium). At the same time, social structures only exist through interaction, because they are constantly being reproduced or transformed by interacting actors (outcome). Giddens (1984) emphasises the constraining as well as the enabling character of social structure. Structures would not exist without interaction yet, at the same time, they also determine social interaction. The multi-level perspective generally draws upon Giddens' (1984) understanding of agency and structure when explaining how transitions take place. However, as argued throughout this work, Giddens' ideas need to be re-emphasised as being essential for sustainability transitions research.

4.1.4.1 The Regime

Transitions are often described as non-linear regime shifts (Geels 2011, p. 29; Raven 2007, p. 2390; Verbong and Loorbach 2012b, p. 9). Socio-technical regimes are understood as very broad, all-encompassing, established arrangements that consist of multiple elements. They refer to the prevailing system structure that coordinates the activities of social groups. They represent unquestioned social patterns of daily life, urging actors to act in certain ways and to adapt particular perspectives on problems, conflicts, beliefs and agenda settings. It is important to note that regime structures seek to stabilize an existing system independent of its actual quality or functionality (Geels 2005, pp. 85f.; 2011, p. 27; Geels and Kemp 2012, pp. 54ff.; van Bree et al. 2010, pp. 532f.; Verbong and Geels 2007, p. 1026).

The regime concept goes back to Nelson and Winter (1982) and Dosi et al. (1988), who argue that every industry has a dominant technological framework providing general guidance for all present and future activities related to that industry. In this way, common technological designs are formed and stabilised because they are shared among engineers and other industry relevant actors. Rip (1992), Kemp (1994), Schot (1998), Rip and Kemp (1998) amplified the regime heuristic by strengthening its social dimension. Rip and Kemp (1998), for example, define regimes as a:

> "Rule-set or grammar embedded in a complex of engineering practices, production process technologies, product characteristics, skills and procedures, ways of handling relevant artefacts and persons, ways of defining problems—all of them embedded in institutions and infrastructures" (Rip and Kemp 1998, p. 338).

Technical components are aligned to societal dimensions. Together they form a configuration that works (Rip and Kemp 1998, p. 330). The car cannot be viewed as a simple artefact. Rather, it is connected to a wide infrastructure of roads, traffic regulations and lifestyle aspects. Hence, technology cannot only be seen in the context of fulfilment of certain technical functions, as the social dimension of technologies plays a decisive role for their functioning. Geels (2002, 2004) further elaborated upon the sociological understanding of regimes, describing them as all-encompassing societal structures that permeate all areas of life, including market practices, policy, user preferences, technology, aspects of culture, social guidelines and policy. The result of this interconnection between technologies and society is a socio-technical trajectory: a prevailing pattern that dominates a socio-technical system and its development (Raven 2005, p. 28).

Common notions related to regimes in socio-technical transitions studies are path-dependency and lock-in effects, which may allow incremental innovations to rise along narrow trajectories but prevent broad radical system change. A dominant explanation here is that strong linkages between societal rules, social groups and technical infrastructure, which have emerged over long time periods, make regime changes—and thus the transformation of a socio-technological system—extremely complex, challenging and time intensive. A shared interest of transitions scholars is to explain regime persistence by identifying conditions that trigger regime amendments or even a regime shift. The co-existence of stability and change is thus an important aspect of regime logic in transitions research (for example, Geels 2004; Geels and Schot 2007; Hoogma et al. 2002; Markard and Truffer 2008; Monstadt and Wolff 2015; van de Poel 2003; Vergragt 2012).

Geels (2011) points out that:

> "[A] regime is an interpretative analytical concept that invites the analyst to investigate what lies underneath the activities of actors who reproduce system elements" (Geels 2011, p. 31).

The boundaries of socio-technical systems—and thus also the boundaries of regimes—are rather more empirically than theoretically defined (Geels 2004, p. 901; 2011, pp. 31f). Hence, the framing and delineation of regimes goes in line with identification of the research scope of a given study. Regime descriptions vary strongly in terms of their empirical focus and extension. This is also the case for energy regimes, which have been framed in many different ways. They have been described around a certain energy resource, such as coal (Turnheim and Geels 2012) or natural gas (Arapostathis et al. 2014; Raven and Verbong 2007). They have been framed around a certain kind of energy, such as electricity (Hofman and Elzen 2010; Raven and Verbong 2009; Raven 2006; Verbong and Geels 2007, 2012). Energy regimes have also been characterised with a primary focus on a certain market sector, such as the utility sector (Konrad et al. 2008). They have been described by concentrating on how regime elements are primarily coordinated, for example, whether or not they are mainly centrally ruled (Hughes 1983, 1987). Energy regimes have also been conceptualised by focussing on subdomains, such as energy policy or governance patterns (Bolton and Foxon 2015; Kern and Smith 2008; Thue 1995; van der Loo and Loorbach 2012) or are framed as established arrangements for energy supply (Raven 2004). Such diversity in empirically framing energy regimes can also be found amongst the latest energy research articles, which were published in the journal Environmental Innovations and Societal Transitions as of April 2016 and which apply a variety of regime framings to socio-technical regimes, as demonstrated in Table 4.1.[1]

---

[1] The basic article review was executed on the 08 April 2016 as follows: The review was focused on articles published in the journal *Environmental Innovations and Societal Transitions*, the official journal of the sustainability transitions research network. As outlined in Sect. 4.1.1, it was launched by the transitions research community in 2011. For the article search, the journal's own search platform was used, which is publically accessible under the following address: http://www.sciencedirect.com/science/journal/22104224. The word 'regime' was searched for in 'search all fields', and 84 articles were found for the years 2014 through 2016. They were then ordered according to date of publication. For the review, the first 10 articles were selected, which used

Table 4.1 Regime framings applied in the 10 latest research articles that use the regime notion and that are published in the journal *Environmental Innovations and Societal Transitions*, as of April 2016

Kind of applied regime framing	Number of articles
Energy policy regime	4
General energy regime	3
Electricity regime	5
Heating regime	1
Oil and gas regime	1

Reviewing energy regime conceptualisations in the literature not only demonstrates a rich diversity in characterising regimes along given lines of empirical research interest but also that the overwhelming majority of studies only briefly introduce energy regimes as part of incumbent socio-technical energy constellations. Most energy transitions scholars agree that the technological energy infrastructure, organisations and other actors, as well as leading ideas and principles, mutually reinforce each other, giving energy systems a strong stability and a common resistance to change (for example, van den Bergh and Bruinsma 2008; Verbong and Loorbach 2012a). However, beyond this general and rather brief characterisation of regimes as prevailing system patterns, energy regime descriptions remain at the surface and tend to lack analytical depth.

In fact, the regime concept has been criticised for not being tangible enough in order to be well applied empirically (Fuenfschilling and Truffer 2014; Konrad et al. 2008, p. 1193), and a need for developing a more defined regime understanding has been explicitly spelt out for studying ongoing energy transitions (Loorbach and Verbong 2012, p. 321). Various works attempt to present general categories along which the analytical and empirical scale of regimes should be defined. For example, Holtz et al. (2008) propose using five aggregation points: (1) Regime purposes in the form of its societal functions, (2) degree of coherence of all elements, (3) stability of their interrelationships, (4) autonomy between elements and (5) non-guidance by single actors. Fuenfschilling and Truffer (2014, p. 776) propose to use as empirical entry points (1) the content and coherence of structures as well as (2) their degree of institutionalisation. Even though these categories remain abstract and may not work as a blueprint for all studies, they reveal that it is not so much the elements themselves but rather the relationships between them that should primarily be spelt out empirically. This goes as well for energy transition studies, where scholars have used the regime notion to emphasise the stability of prevailing structural system patterns and not in order to systematically list all elements that make up a particular regime.

The regime is a central analytical concept for understanding socio-technical transition processes, and thus the way in which a regime is described has an impact

---

socio-technical transition theory in the context of regimes and focused on energy. An overview of the 10 selected articles can be found in annex 3.

4.1 Understanding Socio-technical Transitions

on framing research—in this case about the influence of energy cooperatives within Germany's energy system. Instead of avoiding empirical delineation of such a regime altogether, in the following it is indicated and reflected upon particular analytical areas for which a better regime definition is necessary.

The first aspect that needs to be elaborated is a clear configuration of regime elements. As outlined by Markard and Truffer (2008, p. 605) (see also, Augenstein 2014, pp. 66f. and Best et al. 2012), regime descriptions differ with regard to what a regime actually comprises. One group of scholars understands regimes purely as established societal rules (for example, Geels 2004, 2005, pp. 13ff.; Hoogma et al. 2002). Interconnected by a common logic, they represent the deep structure of a socio-technical system that coordinates all social activities (Geels 2002, p. 1260; 2004, p. 910; 2011, pp. 31f.). They are supported by incumbent social groups and embedded in dominant artefacts (Verbong and Loorbach 2012b, p. 9). Meanwhile, a second group of scholars conceptualises regimes as whole sets of rules and material elements, including technological infrastructures (for example, Markard et al. 2012; Smith et al. 2005, p. 1493). Others also define actors as being part of a regime (for example, Holtz et al. 2008, p. 629; Konrad et al. 2008, p. 1191). All of these different strands regarding the configuration of regimes can also be found in studies on energy transitions (examples for a broad definition Foxon 2013; Hofman and Elzen 2010; Shackley and Green 2007; van der Loo and Loorbach 2012; van der Vleuten and Högselius 2012; Verbong and Geels 2012; Vergragt 2012, p. 117); (examples for a narrow framing Raven and Verbong 2007, 2009; Turnheim and Geels 2012). Such mixed and often unclear understandings of regimes are also displayed within the 10 latest energy transition studies published in the journal *Environmental Innovations and Societal Transitions* as of April 2016 (see annex 3 for more details).

Central research interest of this work is to analyse the influence of innovative enterprises on transitions. It is thus important to have a clear understanding of how systems, actors, structures and structural change are best related to each other. Otherwise it will not be possible to describe in detail interrelationships and dynamics between actors, their interactions and transitions. The multi-level perspective has been criticised for primarily focussing on the "*structuring forces*" (Geels and Schot 2010, p. 28) underlying a socio-technical system. It rather tend to describe very broad macro-specific patterns in transitions without leaving enough room for analysis of the roles of actors and their practices (for example, Shove 2012; Smith 2012). As Geels and Schot (2010, p. 19) have pointed out, the multi-level perspective is indeed understood as an overarching framework or middle-range theory that tends to only provide answers on general transition mechanisms rather than on substantive mechanisms of interaction. Thus, the multi-level perspective has largely been elaborated based upon technologically oriented system studies that stay at the level of (industry) sectors, as shown above in Sect. 4.1.4. This may explain why so many transition studies remain unclear regarding the positioning of actors in regime descriptions, as pointed out above in this section.

However, the multi-level perspective is also inspired by Giddens' (1984) theory of structuration (Geels 2004, pp. 906ff; Geels and Schot 2010, pp. 42ff.). As has been pointed out by Augenstein (2014, p. 68), the strength of structuration theory is its clear analytical separation between social structure and social practices. Actors draw upon these rules and resources when they interact. Agency can thus, not be part of social structure from a conceptual point of view. According to Augenstein (2014, pp. 69ff.), reconsideration of Giddens' structuration theory as a theoretical foundation for the multi-level perspective can help to overcome criticism of the multi-level perspective being overly structuralistic, enabling a differentiated understanding of the role of actors in stabilising or changing regime structures. The present work departs from this idea, with an attempt to explain in more detail how Giddens' idea of structure and interaction can help to set systems, actors, structures and structural change in a very clear relationship to each other, thus providing a basis for analysing the influence of innovative enterprises within socio-technical transitions. This is done in detail in Sect. 4.2.

The second aspect that needs to be differentiated in more detail when describing the relationship between regime elements is the dimension of space. Energy regimes are commonly delineated as geographical entities. In particular, they have been framed as national units (Bruno Turnheim and Geels 2012; Foxon 2013; Hofman and Elzen 2010, pp. 658ff.; Konrad et al. 2008, pp. 1193ff.; Raven 2005, 2007, p. 2398; Shackley and Green 2007; Späth and Rohracher 2010, 2012; Verbong and Geels 2007) as well as European or international entities (Hofman and Elzen 2010, pp. 661ff.; Nilsson 2012; van der Vleuten and Högselius 2012). These studies visualise multiple geographical levels involved in energy transitions. What is striking, however, is that up until recently sustainability transitions research has largely ignored the spatial scope of sustainability transitions (Hansen and Coenen 2015, p. 93). Several scholars have pointed out the need for a spatial conceptualisation of transitions in general (for example, Binz et al. 2014; Coenen et al. 2012; Raven et al. 2012; Späth and Rohracher 2010; Truffer and Coenen 2012 and for including a spatial dimension in energy transitions in particular Loorbach and Verbong 2012, p. 322; van der Vleuten and Högselius 2012, p. 99).

For the present work, geographical concerns play a decisive role because whether Germany will eventually achieve a renewable energy system and whether it will be dominated by a centralised or decentralised system set-up depend upon multiple societal, economic, political and technical dynamics at different geographical levels. In order to understand the influence of energy cooperatives on these developments, it is important to track their activities within the different spatial scales of the German system.

Fortunately, interest in the spatial aspects of transitions research has increased lately, and several conceptual contributions have been made (Hansen and Coenen 2015, p. 93). However, most studies have focused on the geography of niches, and less attention has been given to regimes (Hansen and Coenen 2015, p. 104). Furthermore, such energy transition studies have tended to only analyse one spatial level, such as that of a nation state, largely ignoring that, within this scope, prevailing regime structures can be further broken down into smaller spatial entities with similar institutional structures, such as municipalities. Consequently, empirical studies on transitions have been criticised for describing regimes as being too "*'monolithic' and 'homogenous', not adequately considering persistent institutional*

4.1 Understanding Socio-technical Transitions

tensions and contradictions" (Fuenfschilling and Truffer 2014, *p. 772 see also, Hansen and Coenen* 2015, *p. 104)*. As a consequence, institutional regime variations or conflicts, as well as their effects on niche–regime connections have been neither empirically nor theoretically adequately studied (Hansen and Coenen 2015, p. 95). Evaluation of the influence of actors on system change requires a more differentiated perspective on the spatial scale of regimes. Therefore, this work aims here at conceptualising a regime perspective which takes into account their spatial/geographical dimensions.

4.1.4.2 Explaining the Emergence and Stabilisation of Socio-technical Niche Novelties

Niches represent the level at which novelties arise (Geels and Schot 2010, p. 22; Hegger et al. 2007; Kemp et al. 1998; Raven 2012, p. 130; Schot et al. 1994). Niches have been described as small and protected arenas in which proactive actors create and foster innovative ideas with the aim of finding solutions for rising problems at the regime level (Geels and Schot 2010, p. 22; Hegger et al. 2007; Kemp et al. 1998; Raven 2012, p. 130; Schot et al. 1994). Niches provide a space where deviations from the existing regime are possible (Smith and Raven 2012). They emerge outside the mainstream context. In this sense, small is not related to a concrete measurable size but rather to a concept that is 'not common sense' and thus not as widespread as existing practices within the broader regime (Geels and Schot 2010, p. 22; Raven 2012, p. 130). The innovation process taking place within niches has two dimensions that strongly underline the co-evolution of technology and society:

(a) *The technological dimension of innovation*: Actors foster the development of new technologies.
(b) *The social dimension of innovation*: In a more or less consciously directed process, new principles and ways of thinking necessary for the use and diffusion of new technological concepts are tried out, reflected upon and elaborated (Geels and Schot 2010, p. 22; Hegger et al. 2007, p. 732; Raven 2012, p. 130).

During the innovation process, actors continually challenge prevailing socio-technical patterns, trying out different directions of thinking and acting, independent from what is normally expected (Geels 2004, pp. 912ff.). But the recurring process of self-reflection makes niche novelties unstable and vulnerable to selection pressures arising from the surrounding mainstream regime. New ideas must, therefore, gain power and support in order to survive and become more influential (Smith and Raven 2012). Consequently, niche development is essential for the diffusion of innovative socio-technological arrangements.

Studies on strategic niche management have observed how niche novelties are created and what characteristics can especially lead to an up-scaling of socio-technical innovations. Three internal niche processes are considered to be of primary importance for effective niche-building, supporting innovations until they can leave the creation stage with a better chance to survive (Geels and Schot 2010; Hoogma et al. 2002; Kemp et al. 1998; Raven 2012; Schot et al. 1994). They entail (1) establishment of broad and deep social networks, (2) creation of specific and qualified visions and (3) second-order learning. A first postulate of strategic niche management is that the establishment of networks—comprised of multiple stakeholders (broad) and committed to the exchange of substantial resources (deep)—can help niche ideas to develop. New socio-technological arrangements may only diffuse into wider regime structures if they are eventually accepted by a diverse and powerful group of people (Geels and Schot 2010, p. 82; Raven 2012, p. 139). The growth of niches is particularly stimulated by visions (Borup et al. 2006; Raven et al. 2008) that help to set a concrete agenda for which resources are to be mobilised and help to legitimise on-going innovations (Raven 2012, p. 135). Visions can improve their usefulness when they are substantiated in realistic concepts (Raven 2012, p. 135). Learning plays a key role in the niche development process, differentiated between first- and second-order learning (Byrne 2009, pp. 23ff; Schot and Geels 2008). Whereas first-order learning is mostly focused on technical and economic improvements, second-order learning is concentrated on user preferences, cultural and symbolic meaning, policy issues, as well as societal and environmental consequences of innovations (Raven 2012, p. 142). Existing norms, laws, societal routines and heuristics have to be adjusted in order to diffuse new technologies. In this sense, second-order learning helps to align technical concepts with social characteristics.

While social network building, vision creation and second-order learning have become important building blocks for explaining the rise of niche novelties, they do not explain very well how the process of niche development actually takes place. Niches influence the wider society to the extent that the three processes are said to become *"robust"* (Seyfang et al. 2014, p. 21). But what does that mean?

Deuten (2003), Geels and Raven (2006), as well as Geels and Deuten (2006) differentiate between local practises and an emerging global niche level in order to better analyse the relationships between first-mover activities at the local level and the growth of niche innovations into more and more stable designs. The local niche level is the site of concrete local practices in the form of innovation projects, whereas the global niche level describes a new socio-technical field, populated by various local projects that are applying the same rules (Geels and Deuten 2006, pp. 266ff.; Geels and Raven 2006, p. 378). Geels and Raven (2006, pp. 378ff.) add a network perspective by differentiating between local networks, consisting of those actors who actually realise projects at the concrete local level, and a global network, framed as a new *"cosmopolitan community"* (Deuten 2003, p. 2). The community emerges because local networks start to share the same rules and can, thus, identify themselves as a single network existing in many locations beyond the boundaries of their individual projects. Eventually the rules of the cosmopolitan community start to guide local practices, as they become standard in new projects. In consequence, new projects are implemented in similar ways, which in turn leads to the stabilisation and diffusion of socio-technical novelties in more and more regions. Raven (2012, p. 131) defines the global niche level as an *"emerging institutional field"*. This process of niche development is visualised in Fig. 4.2.

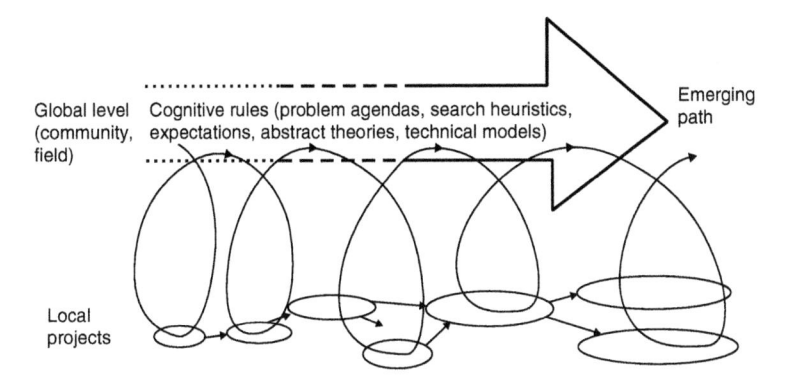

4.1 Understanding Socio-technical Transitions

Fig. 4.2 Emerging socio-technical path carried by sequences of local projects (Raven and Geels 2010, p. 91)

Geels and Deuten (2006, pp. 266ff.) point out that the aggregation of local knowledge, visions and experiences into standardised rules does not simply happen automatically but, rather, requires dedicated aggregative activities. Here, aggregation means the transformation of local knowledge and visions into generic rules, the production of a collective good. Generic rules are sufficiently abstracted and de-contextualised so that they can travel between locations (Geels and Deuten 2006, pp. 266f.). Aggregation is particularly supported by the work of intermediate actors, such as standardisation organisations, industry associations or research institutes, that have distance to local projects and a primary interest in creating a collective reservoir (Geels and Deuten 2006, pp. 267f.).

In sum, wider diffusion and stabilisation of innovative designs take place when experiences undergone by local networks in their local projects are actively aggregated into generic rules which can then be applied by other local networks in other locations. A new socio-technical field then becomes robust, as it is now being advocated by a broader cosmopolitan community.

While strategic niche development research sheds light on important dynamics in niche formation, it also raises several questions. Those that are most decisive for this work and for analysing the impacts of innovative enterprise groups within system change are spelled out in the following.

Interplay Between Niches and Regimes

Whereas the multi-level perspective became the dominant framework for explaining a transition process from a structural systems perspective, strategic niche management turned its focus towards concrete innovations at the local level to explore how novelties are created and how they can grow to the extent that they become robust enough to survive. However, due to this niche-centred research interest, associated studies tended to stay at the level of niches and, consequently, do not explain how niche structures may actually influence the regime structure of a system (for example, Raven et al. 2008; Raven 2005; exceptions do exist, such as Hegger et al. 2007). Furthermore, the prevailing explanatory frameworks used in strategic niche management have led to misleading interpretations with regard to how niches and regimes are related to each other. The dominant description of niches as *"incubation rooms"* (for example, Geels and Schot 2010, p. 22) or as *"protected spaces"* (for example, Fudge et al. 2016, p. 6; Hegger et al. 2007, p. 730; Kemp et al. 1998, p. 185; Raven 2012, p. 130; Schot et al. 1994, p. 1061; Smith and Raven 2012, p. 1025) in which innovative actors interact suggests that niches and regimes are different in kind: On the one side, the large rigid regime, a *structure* which covers the whole system; on the other side, the small innovative niche, a *space or room* where concrete interaction takes place. The local–global perspective on niche development claims that these local spaces of interaction first need to be turned into a more abstractly global niche structure before niches can begin to *face* the regime (Geels and Deuten 2006; Geels and Raven 2006; Raven and Geels 2010, p. 90; Raven 2012, p. 132). In other words, niches are seen as developing almost independently and 'protected' from the regime at the local level until they are strong enough to 'stand up' against or 'overthrow' the dominant regime: the wider system structure that is in a very abstract sense exclusively located only at the wider system level (for example, Smith and Raven 2012). What that means empirically, however, does not seem clear. There is, nonetheless, a clear and strong difference between the sides of the equation: The large, abstract and homogenous regime on the one side, and the small haptic niche on the other. More importantly, this rather mechanical view fails to give credit to the process of co-evolution. Yet the multi-level perspective does speak of co-evolutionary developments within and between the levels of structuration (Geels and Kemp 2012, pp. 50f.; Geels and Schot 2010, pp. 18ff.). Co-evolution occurs *within* a niche between new technologies (technical dimension) and social heuristics (societal dimension), and it also happens *between* niche and regime in the form of variation and selection (Geels and Kemp 2012, pp. 50f.; Geels and Schot 2007, p. 402; 2010, pp. 18ff.). However, as argued here, the co-evolution between niches and regimes is not well explained in current transition studies. Departing from this point, the work aims to better conceptualise their simultaneous development by building a more consistent understanding of niches and regimes and how they shape each other.

Focus on Actors and Agency

Strategic niche management has always acknowledged the important role of actors and agency in niche development. As we have seen the local–global perspective points out that niche development involves two interrelated processes with concrete actor dynamics: (1) Multiple innovation projects are realised at the local level and are fostered by local actors being directly involved in them. (2) Simultaneously, a global niche level emerges, because a cosmopolitan community starts to draw upon the same rules aggregated from the local level (Geels and Raven 2006, p. 378).

However, the influence of actors on wider transition processes is still rather more implicitly than explicitly discussed.

One of the main reasons for this could be the circumstance that most studies—also on energy—are framed around innovation projects or innovative technologies and not around innovative actors (for example, Fontes et al. 2016; Geels and Raven 2006; Hegger et al. 2007; Hoogma et al. 2002; Lopolito et al. 2011; Raven and Geels 2010; Raven 2005; van Eijck and Romijn 2008; Verbong et al. 2008). Even though both actor and project perspectives are closely interrelated—innovative projects are realised by innovative actors, after all—it makes a difference whether projects or actors are chosen as the main entry point for conducting research. Adopting a concrete actor perspective allows the scholar to analyse the effects of innovation on those actor groups that developed and supported them. In this vein, more recent studies have applied the concept of strategic niche management to entrepreneurs, members of civil society and communities (Faller 2016; Forrest and Wiek 2015; Seyfang et al. 2014; Seyfang and Haxeltine 2012; Smith 2012; Witkamp et al. 2009, 2011). From an empirical point of view, these studies emphasise the need for a stronger actor perspective in work on system transitions.

The need for niche creation, diffusion and stabilisation implies that there are different levels of agency that are decisive for niche development. However, another problem here is that the vast majority of studies on niche development—in particular, on energy niche development—exclusively focus on detailed case study analysis (for example, Faller 2016; Gottschick 2015; Lopolito et al. 2011; Späth and Rohracher 2010; van Eijck and Romijn 2008; Verbong et al. 2008). Consequently, although they have contributed towards generating important knowledge about how innovations are created, they have tended not to reflect upon their observations against the dynamics of diffusion and stabilisation carried out by a cosmopolitan community. Case studies that observe local innovative dynamics in one or a few cases, but without embedding them in a wider assessment of their roles within trans-local dynamics of agency, entail the risk to be overly optimistic regarding the change potential of niches, because the importance, difficulties and effects of wider diffusion and stabilisation of local activities at the global niche (network) level are marginalised. In this sense, Farla et al. have proposed that:

> "Further research may thus try to uncover how much leeway actors really have in pursuing sustainability transitions within an existing system. Because joining forces between actors can be important to that end, it is also promising to address the issue of collective action [...]" (Farla et al. 2012, p. 996).

Studies on trans-local networks exist. However, they mainly apply an internal network perspective, seeking to show how such networks function and evolve (for example, Caniëls and Romijn 2008; Hermans et al. 2013; Law and Callon 1992; Lopolito et al. 2011), or they analyse what is needed to foster the creation of such networks (for example, Hendriks 2008; Khan 2013). Questions regarding how these networks may influence wider transition processes and how they are backed by local interaction do not form their central research interest.

In addition, strategic niche management has remained particularly vague on the concrete roles and potential of enterprises to create, diffuse and stabilise new socio-technological niche structures. Thus far, niche research has mostly described enterprises either as creators of pure technologies in pilot projects or as path-dependent actors reluctant to adjust to innovations (Geels and Schot 2007, p. 407; Penna and Geels 2012; Rothaermel and Boeker 2008; Rothaermel and Hill 2005).

I aim here to take an explicit and explorative look at the influences of a rising enterprise group on wider transition processes, taking into account the need to understand their concrete practices at the local level and their role in diffusing innovative concepts at a global level. Such a perspective also implies examining actors at different levels of agency.

Interplay Between the Creation of Rules and the Mobilisation of Resources

Socio-technical systems are made up of multiple elements which, next to actors, include material artefacts, technologies, regulations, guiding principles, agenda settings and user preferences (Geels 2002, p. 1258; 2004, pp. 898ff.; Kemp et al. 2012, pp. 3ff.; Markard et al. 2012, p. 956). The multi-level perspective and strategic niche management have made the co-evolution of these elements into a central conceptual premise of socio-technical transitions; following Giddens (1984), these elements can be differentiated and grouped into societal rules and resources (Giddens' 1984 understanding of structure is outlined in detail in Sect. 4.2.2).

Earlier, strategic niche scholars tended to empirically focus on technological innovation projects and the question of how to nurture them for further growth beyond the experimentation stage (Hoogma et al. 2002, pp. 12ff.). But the dominant focus on technology as the entry point in technological innovation research was criticized by several scholars (for example, Hegger et al. 2007; Verheul and Vergragt 1995; Witkamp et al. 2011) because it puts too much attention on technological experimentation and underestimates the need for co-evolution of technology and society.

Lately, a growing body of literature has demonstrated empirically how niche development can be guided through visions (Lovell et al. 2009; Späth and Rohracher 2010, 2012), joint values (Witkamp et al. 2011), expectations (Bakker et al. 2012; Budde et al. 2012; Geels and Raven 2006; Raven et al. 2008), as well as societal learning effects (Brown and Vergragt 2008; Gottschick 2015) and abstract knowledge flows (Geels and Deuten 2006). While these studies underline the importance of developing new societal rules that are shared within a growing niche community, derived from those applied in local practices, they do not sufficiently visualise the interplay between the creation of new rules and the mobilisation of resources. Although many of these studies do observe the mobilisation of resources, such as technology and money, the interplay between creation of new guiding principles and mobilisation of such resources is still rather implicitly and inadequately discussed (for example, Seyfang et al. 2014; Späth and Rohracher 2010; Trutnevyte et al. 2012). An explicit focus on resource formation was, however, undertaken by Musiolik et al. (2012), though they show no recognition of the simultaneous creation of guiding principles. Overall, it can be said that the creation and diffusion process of combined sets of new rules and resources is still under-theorised.

The local–global perspective on niche development puts a strong emphasis on the need to theorise a new generic rule level for stabilising niche innovations. In fact, Geels and Raven (2006, p. 378) argue that global rules form resources at the local level which means that technological development would follow new social rules. Giddens' (1984) interpretation of social structure and social agency reveals, however, that a linear view on the interplay between the development of rules and the mobilisation of resources in the form of 'local resources follow global rules' is too short-sighted, because it does not give enough credit to the role of power in niche development. Resources are structural elements that empower actors to act (Giddens 1984, pp. 14ff.). Consequently, we can now see that mobilisation of resources is an important basis for factually creating new rules within concrete practices. At the same time, mobilisation of resources is strongly influenced by rules, which give actors the legitimation to act (Giddens 1984, pp. 17f.). So rules, such as new guiding principles, cannot be developed independently from resources nor can resources, such as technology or money, be mobilised without social heuristics.

The strong interdependency between rules and resources as elements of social structure and their joint role in limiting or enabling social agency makes it is important to shed more light on the interplay between rule and resource formation in niche development. With a focus on actors it is particularly decisive to visualise relations of domination, underlining the need to analyse a given resource level in combination with the appropriate rule level. Therefore, in this work I adopt Giddens' (1984) understanding of structure and agency (see Sect. 4.2) and turn my attention towards the interplay between creating new societal rules and mobilising material resources through social practices in the course of niche development.

The Spatial Dimension in Niche Development

Transitions research has been criticised for not sufficiently paying attention to geography (Binz et al. 2014; Coenen et al. 2012; Hansen and Coenen 2015, p. 93). This is not only the case for regimes, as outlined earlier but also applies to research on niche development as well. For example, Shove (2012, p. 72) points out that, in order to understand how energy intensive practices, including the use of air-conditioning, take hold in different social situations, one has to pay much more attention than previously to issues of social and spatial scale. Dewald and Truffer (2012) argue that an a-spatial perspective on the formation of photovoltaic in Germany may lead to the conclusion that national regulatory incentives have been the only factor leading to a diffusion of this technology. Lately, however, several works have emerged that have observed a spatial dynamic in niche development (for example, Fontes et al. 2016) and, from an empirical point of view, underline the need for a stronger conceptualisation of the dimension of space in niche development. Hence, I aim here to theorise the spatial dimension of the rise of a global community and a global niche level.

4.1.5 Summary

In this section, I have outlined how sustainability transitions research has sought to explain the complex process of socio-technical system change and have proposed that the multi-level perspective can help in identifying the characteristics as well as underlying mechanisms that make up a transition process. Central here is bringing to light interrelationships between the three different levels of structuration: niche, regime and landscape. Whereas the regime represents prevailing system structures, niches are innovative spaces created beyond the mainstream context. A transition is understood as a regime shift mainly triggered by the rise and development of niches. Actor dynamics play a decisive role in socio-technical transitions, as niches are created through innovative actors who want to make a difference. A transition is the result of multiple, parallel existing interaction patterns between many different actors (for example, Geels and Schot 2010, p. 11). However, the ways in which the notions of regime and niche have been interpreted and applied in transitions research has led to several short-comings, making explicit analysis of actor influence within transitions very challenging.

A review of studies on the multi-level perspective and strategic niche management has revealed the following salient points: Socio-technical transitions are explained through actor interaction but without providing concrete systematisation of agency in relation to the process of structuration taking place during transitions. Studies strongly differ in their applied regime descriptions. Regimes basically involve everything that is 'established'. Whether this includes rules, artefacts and/or even actors strongly varies. Such unclear use of the regime concept has led to a situation where the multi-level perspective was criticized for being overly structurally oriented, not paying enough attention to actors (for example, Shove 2012; Smith 2012). Energy regimes have generally been understood as homogenous patterns that exclusively exist in an abstract sense at the wider system level, usually that of a nation state. As a consequence, geographical and institutional regime variations, as well as their effects on niche–regime connections have not been adequately discussed. The regime concept has been criticised for being too monolithic and too abstract (for example, Fuenfschilling and Truffer 2014, p. 722). Niches are generally described as local interactive 'spaces' which first have to generate generic rules supported by a cosmopolitan community before being able to face the much stronger all-encompassing regime structure. Such an understanding not only involves confusing definitions of spaces and structures but also suggests a distance between niches and regimes, thereby not considering the co-evolutionary processes that, according to the multi-level perspective, take place between the different structural levels of a system. However, these dynamics need to be taken into account when analysing the interactions between actors and their potential influences.

Furthermore, strategic niche management has been primarily addressing certain types of technologies or innovation projects, which have been analysed in detailed case studies. Actors or actor groups have usually not been the point of departure for such research. Thus, explicit and explorative focus on the influence of certain actors

4.1 Understanding Socio-technical Transitions

Table 4.2 Ideas from the multi-level and strategic niche management perspectives and the aspects they are missing that are important for explaining the influence of actors on system change

Prevailing viewpoints of leading studies using the multi-level and strategic niche management perspectives		Aspects that are important for assessing the influences of actors within system change
Main focus on three different levels of structuration—niche, regime and landscape—...	misses	... a systematic description of different interaction levels in relation to different levels of system structure.
The understanding of niches as innovative 'spaces' and the understanding of regimes as prevailing system patterns incorporating everything that is 'established'...	misses	... a consistent conceptualisation of new niches and established regimes, as well as of their relationship.
The notion of innovative niches to develop to a certain point before being integrated into the regime or before being powerful enough to replace the regime...	misses	... a niche and regime perspective that gives concrete consideration to co-evolutionary niche–regime dynamics.
A primary focus on innovation projects or on innovative technology fields...	misses	... an explicit and explorative perspective on actors.
The non-spatial understanding of regimes and niches ...	misses	... a niche and regime perspective which acknowledges internal heterogeneity and structural development in space.

or actor groups is not well conceptualised. This is particularly the case for enterprises, which are mostly reduced to either technology creators or incumbents that fight against change. The need for niche creation, diffusion and stabilisation implies that there are different levels of agency, as well as different dimensions of space, within niche development. When analysing actor influence these levels and dimensions need to be clearly separated in an analytical sense. Otherwise there is a risk of mixing variables of different development levels, leading to short-sighted conclusions and the potential identification of false causalities. However, analytical dimensions of agency are somewhat implicitly acknowledged in the multi-level perspective and in strategic niche management, such as when explaining the rise of networks and visions or discussing the generalisation of rules. However, these dimensions need to be more explicitly brought to the fore.

Summarising ideas from the multi-level and strategic niche management perspectives, Table 4.2 provides an overview of features that are important for taking a systematic view on the influence of actors within transition processes but which are missing from existing studies.

In this work, I argue that the conceptual short-comings identified in the existing transition studies literature can be overcome by re-emphasising structuration theory as a theoretical building block of transitions research. As mentioned in Sect. 4.1.4, transitions research in general and the multi-level perspective in particular are actually inspired by Giddens' (1984) structuration theory (Geels 2004, pp. 906ff; Geels and Schot 2010, pp. 42ff.). However, the integration of structuration theory has been somehow relegated to the overall idea that structure is both a reference point and result of interaction at the same time. Consequently, the far-reaching explanatory power behind the concept of the duality of structure has to a great extent been neglected. Section 4.2 describes the notions of structuration theory in detail, with the aim of embedding Giddens' (1984) comprehensive understanding of agency and structure into a framework which aims to visualise the influence of actor interaction within transition processes. In addition to the work of Giddens, Schneidewind is also drawn upon (1998). Since his work about enterprises as structuring actors has outlined how and to what extent structuration theory helps to systematically delineate structuring forces of the firm. Schneidewind's (1998) work is crucial for conceptualising the transformative potential of enterprises.

4.2 Understanding the Relationship Between Enterprise Interaction and Structural Change

Structuration theory is a leading contemporary social theory, elaborated by the British sociologist Anthony Giddens in the late 1970s and early 1980s (Reckwitz 2007, p. 314), which proposes an explanatory framework for understanding the relationships between the agency of social actors and social structure. Giddens outlined his ideas in a series of books and essays (Giddens 1976, 1979, 1981, 1982) and then merged them in his central work about the constitution of society (Giddens 1984).

In order to position Giddens' structuration theory within organisational research, an overview of organisation theories related to structural change is given at the beginning of this section. Then, structuration theory is described in great detail in Sect. 4.2.2 with the aim of displaying its deep and comprehensive understanding of agency and structure. Against the background of the research interest—analysing the influence of rising innovative enterprise groups within transitions—Schneidewind's (1998) concept of the structuring enterprise, based on structuration theory, is then introduced in Sect. 4.2.3. At the section's end, structuration theory is critically evaluated (4.2.4).

4.2.1 Theories of Structural Change in Organisational Studies

Significant for the present work is that structuration theory has been commonly applied in economics and organisational science. Since the early 1980s, much works have shown that Giddens' (1984) ideas can be a fruitful theoretical basis for approaching research questions related to the enterprise (for one of the earliest works, see Ranson et al. 1980). In this vein, Whittington (2010), based upon Parker (2000), notes that:

"To some extent the basic idea of structuration has become a conventional wisdom of organization studies [...]" (Whittington 2010, p. 116).

Organisation research strands that have employed structuration theory include the following:

- general management studies, including information and production management, as well as accounting research (for example, Busco 2009; Jones 2011; Ortmann 1995, for an overview see Whittington 1992);
- research on enterprise strategies (for example, Becker 1996; Capallo 2008; Ortmann and Sydow 2001; Pozzebon 2004, for an overview on prevailing strategy work applying Giddens, see Whittington 2010);
- organisation network research (prevailing studies are Sydow 2005; Sydow and Duschek 2010; Sydow and Windeler 1998); and
- organisation research examining a systemic context (for example, Schneidewind 1998).

Giddens' (1984) structuration theory continues to be a source of inspiration in organisation research, as indicated by recent publications such as (Englund and Gerdin 2014) and (Hauptmann 2015).

Before outlining the theory of structuration in more detail, however, I would like to show that enterprise interaction and structures have been subject to many rich discussions in organisation research over the last few decades. It is not the aim here to examine each theory in great detail but, rather, to underline the important role of research on enterprises and structural dynamics.

Evolutionary economics seeks to explain organisational change through the co-evolution of selection and adaption as an overall structuring dynamic (Dosi 1982; Nelson and Winter 1982). Structural change-oriented organisation research has been greatly influenced by the broad research agenda of the economist Joseph Alois Schumpeter, who is seen as an early pathbreaker for evolutionary economics and entrepreneurship research (Swedberg 2007). Schumpeter drew for his time a new picture of overarching societal models, such as the economy, capitalism and democracy (for example, Brinckerhoff 2000; Schumpeter 1942). He also coined the widely recognised term *"creative destruction"* (Schumpeter 1934), which indicates the *"process of industrial mutation [...] that incessantly revolutionizes the economic structure from within, incessantly destroying the old one, incessantly creating a new one" (Schumpeter 1942, p. 83)*. Schumpeter's work was path breaking because of his focus on the underlying structural dynamics of systems. Accordingly, he understands human behaviour as a process that *"[...] includes not only actions and motives and propensities but also the social institutions that are relevant to economic behaviour such as government, property, contract, and so on [...]" (Schumpeter 1954, p. 21)*. Being an important theoretical conceptual contributor to the multi-level perspective, evolutionary economics has already been outlined in Sect. 4.1.4. At this point, it should be added that organisational interaction in evolutionary economics is mainly described as a uni-directional process in which firms are primarily seen as adaptive entities that have to cope with the external selection pressure of the market in which they compete for limited assets (see also, Geels 2014, p. 263).

New institutional theory directs the attention to the forms of external legitimacy organisations need to act (DiMaggio and Powell 1983; Powell and DiMaggio 1991; Scott 1995). Organisations adjust their structures to institutional requirements in order to gain or sustain societal legitimacy for their actions. According to institutional theory, organisations are thus not primarily influenced by their competitive environment but more so by expectations and constraints that arise from the institutional environment (DiMaggio and Powell 1983). Scott (1995, pp. 35ff.) describes institutions[2] as cognitive, regulative and normative rules that are industry wide, nationally or internationally accepted. Scott's definition has also been used by leading transitions scholars in order to define regimes (for example, Geels 2004, p. 905).

In this sense the set-up of an organisation is not seen as a rationally established basis for the coordination of market activities but, rather, the materialisation of institutional expectations and constraints (Walgenbach 2006, pp. 353f.). New institutional theory has very much contributed to directing the understanding of organisational interaction towards a new perspective—one that is not solely dominated by competitive and market theory concepts of organisations as rational actors that continually strive for technical and economic efficiency and a unique competitive position. Early work on new institutional theory primarily focused on how the institutional environment influences the firm, where organisations were seen as passive elements that continually needed to adapt to institutional expectations. Here, the actions of organisations seemed to be over-socialised and almost detached from interest-driven decision making processes (Powell 1991, p. 194). One of the problems was that the actor itself was not seen as the centre of attention but institutional structures instead (Powell and DiMaggio 1991). Later work turned towards examining institutional change, arguing that the general potential for change is an elementary character of institutions (for example, Powell 1991, pp. 195ff.; Scott 1994). However, an overall framework that would constructively harmonise the influence of individual organisational behaviour on its institutional environment was never really developed. Several studies, however, have developed more substantial arguments by combining new institutional theory with other theories (for a selection, see Walgenbach 2006, p. 394).

The **resource-based view** is a major concept in organisational literature seeking to explain the competitive advantage of enterprises (Barney 1991, 2001; Peteraf 1993; Wernerfelt 1984). It concentrates on the firm's internal capabilities in the form of tangible (material assets) and intangible (immaterial assets) resources. Crucial firm-specific resources are valuable, rare, hard to imitate and non-substitutable (Barney 1991; Das and Teng 1998; Hall 1992; Wernerfelt 1984). The resource-based view explains firm interaction as an adaptive process in which organisations

---

[2]There is a wide range of interpretations about what institutions are exactly. Some even describe established organisations as institutions. In general, however, they are understood as established procedures constituted within society (Jepperson 1991, p. 143; Scott 1995, pp. 35ff.).

must have the capability to absorb and apply resources in order to create and sustain their position in an externally given market environment, though the theory has also been lately used in order to explain change dynamics within the firm's environment. Musiolik et al. (2012), for example, have analysed how firm resources can affect a technological innovation system. Since the focus here lies on resources, such studies tend to leave the rules level under-explored.

Organisation research on **entrepreneurship** conceptualises a proactive understanding of firm interaction. Schumpeter (2000) had already claimed that entrepreneurial activities belong to the core mechanisms creating economic development and that they can creatively disturb economic systems. Organisation research on entrepreneurship has been strongly committed to better understanding the firm-related process of innovation as well as aligned market dynamics (for example, Audretsch et al. 2008, 2011; Bönte et al. 2016; Braukmann et al. 2008, 2012; Braukmann and Schneider 2010; Witt 2016). Entrepreneurship has not only been addressed from a purely market perspective, however, as new research fields were established in which entrepreneurial activities have been analysed in relation to societal and environmental change. Sustainable entrepreneurship focuses on entrepreneurship and sustainable development (for example, Dean and McMullen 2007; Schaltegger and Wagner 2011), whereas ecopreneurship refers to research on environmental protection through entrepreneurial activities (for example, Schaltegger 2002; Schaper 2002). The field of social entrepreneurship observes entrepreneurship which aims to create social values (for example, Brinckerhoff 2000; Carolis and Saparito 2006; Steven et al. 2014; Volkmann and Tokarski 2010). Scholars have also evaluated entrepreneurship in relation to institutional change (for example, Battilana et al. 2009). As outlined in Sect. 9.3, it would be very fruitful to draw upon such grounded understandings and knowledge of entrepreneurship in order to analyse the personalities behind energy cooperatives. An important demand for future research identified in this work is the necessity to better understand the pioneers who found, drive and guide energy cooperatives. Crucial guidance in this regard can be provided by the special research field of entrepreneurship and education (Braukmann 2012; Braukmann and Bartsch 2014; Braukmann et al. 2008).

Research on **market niches** is closely aligned to entrepreneurship. Studies that analyse market niches are largely interested in the influence of innovative enterprises on markets or in understanding the activities of enterprises in certain market niches[3] (for example, Toften and Hammervoll 2010, 2013). Research on market niches often deals with change and innovation processes, though the central point of departure for such research is a market perspective and not wider system one. Wüstenhagen (1998) provides a concrete bridge between entrepreneurship and market niches, discussing in detail the challenges of entrepreneurial up-scaling and putting the

---

[3]Market niches are not to be confused with the socio-technical niches described in the previous section. Even though they may have similarities, they are used for different explanatory purposes. The greatest and most obvious difference is that market niches refer to the market and socio-technical niches refer to a socio-technical system.

development of ecological mass markets out of market niches at the centre of his investigation. Even though his focus lies on the market, he partly includes societal aspects in his discussions, making his article title 'Greening Goliaths versus Multiplying Davids' an interesting point of departure for the present work.

This brief overview of theories and research strands shows that organisation research offers many different approaches for examining organisations and structural dynamics. However, since the market occupies the centre of most research here, concrete and well-elaborated explanatory concepts for wider system perspectives have been rare. Those theories that go beyond the market, such as new institutional theory, tend to focus on adaptation processes of firms with regard to their external environment. Meanwhile, structuration theory takes a completely different approach, providing a differentiated view on the relationship between agency and structure as such by taking into account both their constraining as well as their enabling character. Thus, compared to the other theories just described, structuration theory, to which we now turn, fits very well with the aims of this work.

4.2.2 Structuration Theory

The central claim of structuration theory of Giddens (1984) is that structures and agency are mutually interrelated with each other. Social structures are the primary medium for, as well as the outcome of, social interaction. They also form the contexts of daily life and provide frameworks for social interaction (medium). At the same time, structures do not exist by themselves. They are present because they are continually produced or reproduced through interaction (outcome).

One of Giddens' main intentions was to offer an integrative and more differentiated view on the relationship between agency and social structure in order to overcome a traditional dualism that existed between objective or functionally focused and subjectively oriented theoretical approaches (Giddens 1984, pp. 1ff.; Reckwitz 2007, pp. 314f.). Functional approaches, such as developed by for example, Parson (1951), tend to position society as a whole over the social subject, departing from the assumption that actors are mainly controlled and constrained by pre-existing societal norms (imperialism of the object). In contrast, subjectively oriented theoretical models, such as Schütz (1974), predominantly claim a prioritisation of the social individual by drawing upon the idea that individual actors are generally capable of determining their own behaviour (imperialism of the subject) (Giddens 1984, p. 2). Giddens saw the need to unite these two fundamental viewpoints (Giddens 1984, pp. 1ff.; Reckwitz 2007, pp. 314f.).

According to Giddens (1984, p. 29), agency and structure cannot be seen as two autonomous concepts, where one may be superior over the other one, as agency and structure have the same ontology and can only be analytically differentiated. The process of continuous structuration in society is immediately embedded in the process of social agency. Equally, structural conditions that make actions possible are reproduced through actions and within actions. Thus, actions and the conditions

4.2 Understanding the Relationship Between Enterprise Interaction and... 65

for action are recursively connected with each other (Giddens 1984, p. 2). They are two sides of the same coin (Capallo 2008, p. 112). Giddens defines this integrative essence of structure and agency as the *"duality of structure"* (Giddens 1984, pp. 25ff.). The three central building blocks of structuration theory—agency, structure and the duality of structure—are now explained in further detail.

4.2.2.1 Meaning of Agency

Agency is generally defined as a snapshot of a *"durée, a continuous flow of conduct"* (Giddens 1984, p. 3) attached to the surrounding world, as well as to the coherence of an acting self. In other words, agency *"is a constant intervention in the natural and social world of happening" (Kießling 1988, p. 4)*. Actors are described as socially operating agents with intentional and reflexive abilities, though Giddens (1984, p. 3) does not interpret reflexivity in the sense of pure self-consciousness. Reflexivity rather implies that all actors—being interacting humans—inherently have a general capability to influence the ongoing flow of social life. Important here for Giddens (1984, p. 5) is that the continuous guidance of activities in daily life is not merely based upon a cognitive and discursive process of reflection but, rather, follows routines. According to his stratification model of the agent, in which Giddens (1984, p. 5) outlines the process of action, agency is mainly formed by the rationalisation of action, the reflexive monitoring of action and the motivation of action.

The reflexive monitoring of action is most central for the formation of action and points towards its continuous guidance (Giddens 1984, p. 5). Giddens distinguishes between a discursive and a practical consciousness in order to explain why the reflexive monitoring of action mainly follows routines and is, thus, usually not based upon cognitive mechanisms. The discursive consciousness makes agency explicit, because it enables actors to discursively explain the intentions and reasons for their actions (Giddens 1984, p. 6). Discursive mechanisms enable explicit reflection on action. The act of agency as such, as well as the produced or reproduced structure, is therefore only present in a discursive moment of attention (Giddens 1984, p. 4). However, agency is not primarily guided through discursive consciousness. It is, rather, practical consciousness that normally directs us through daily life. Practical consciousness hosts the implicit knowledge of actors, which is usually eluded from the process of discursive reflection (Giddens 1984, p. 7). Practical consciousness supports the routine guidance of action and can be compared to the grammar of a language. Here, actors are not primarily aware *that* they follow grammatical rules but more or less *how* (well) they follow them (Reckwitz 2007, pp. 319f.). Such routines facilitate daily life because those actions that we routinely carry out are not fundamentally questioned on a continuous basis. Practical consciousness is placed in between discursive consciousness, where action as such is made explicit, and unconscious cognition, which cannot be easily made visible. Important is that the boundary between the discursive consciousness and the practical consciousness is permeable (Giddens 1984, p. 7), providing the basis for change, as explained in detail further below.

The *rationalisation of action* describes intentionality as a process, meaning that competent actors can maintain a theoretical understanding of the general reasons why they act during the moment of action (Giddens 1984, pp. 3f.). It is, like the reflexive monitoring of action, predominantly a routine characteristic of agency that is mostly conducted in a 'taken-for-granted fashion' (Giddens 1984, p. 4). The routine monitoring and rationalisation of action not only leads to a reproduction of the same actions but also to a reproduction of the social structures in which they take place (Giddens 1984, p. 5).

The *motivation of action* is related to the needs that actors have and reflects the general potential for action. Although motivations are rather visualised in broader overall action plans, based upon which a variety of actions can be realised, they only tend to have a direct impact on a concrete sequence of action in unusual situations, where actors break with routines (Giddens 1984, pp. 6f.).

Since neither routine actions nor their contexts are generally discursively reflected upon, a major aspect of action is its unacknowledged conditions (Giddens 1984, pp. 41ff.). According to Giddens (1982), the routine guidance of actions and the accompanying reproduction of structures are deeply integrated in our society. At the same time, Giddens points out the enabling character of action. 'Knowledgeability' and 'capability' are key characteristics of human behaviour (Giddens 1982, p. 29; 1984, pp. 281ff.). To be capable means to be able to make a difference within a social context and to be able to influence the direction of certain events (Giddens 1982, p. 30). According to Giddens, any agency involves a certain level of power, such that actors have the general opportunity to act non-deterministically. Giddens calls this the *"dialectic of control in social systems"* (Giddens 1984, p. 16), which basically means that no actor is completely powerless over their own action. Giddens (1984, p. 15) understands power as a 'transformative capacity'. Knowledgeability refers not only to routines but also to explicit awareness of social conventions. The knowledgeability of actors incorporates reflexive abilities based upon discursive and intentional processes, as well as reflection on routines (Giddens 1982, p. 30). With this characteristic of knowledgeability, Giddens emphasises that even though human actors primarily tend to follow routines, they carry the ability to make practical routines explicit, provided that they become the subject of a discursive reflection process. Hence, the ability to act creatively is inherent in any human being (Giddens 1984, p. 7). Knowledgeable and capable agents are able to turn interaction into a discursively conscious process with active and influential consequences on structures. With this, Giddens (1984) highlights the constraining and enabling character of agency and structure—a central aspect for the argument laid out in this work.

Coming to the quick conclusion that human actors can easily control their actions as soon as they reflect upon them through a discursive process may lead to an inadequate understanding of Giddens' views on agency and structure, especially as the conscious realisation of action is often accompanied by unintended consequences (Giddens 1984, pp. 5f.). Giddens (1984, p. 8) provides the example of the English language. The intention to speak English correctly may be a conscious process. However, at the same time, the practice of speaking 'correct English' leads to the unintended reproduction of the existing linguistic structure.

Due to the *unacknowledged conditions of action* and the *unintended consequences of action* the production or reproduction of structure may appear to the actor as an external and autonomous process (Schneidewind 1998, p. 139). However, the production and reproduction of structure is identical to the process of action.

4.2.2.2 Meaning of Structure

Structures are understood as sets of rules and resources that are recursively implicated in social systems (Giddens 1984, p. 25; see also, Schneidewind 1998, p. 137). Giddens (1984, p. 377) differentiates between structures as individual rules and resources and structures as sets of rules and resources. They provide the meaningful and material conditions for competent action, out of which they are at the same time created (Giddens 1984, pp. 16ff.). Because actors draw upon rules and resources when they interact, the character of social structure becomes real, though structure only exists as memory traces or when being instantiated in action (Giddens 1984, p. 377). 'Structuration' turns our focus towards the conditions governing the reproduction of the social practices that make up a social system (Giddens 1984, p. 25).

Rules are understood as *"methodological procedures of social interaction"* (Giddens 1984, p. 18) and are primarily placed within the implicit knowledge of practical consciousness in the form of mental traces (Giddens 1984, pp. 17ff.). This way they have direct effects on actors' routine guidance of their own and others' activities without undergoing additional reflection. Rules make action possible in the first place, because they provide practical knowledge of how to adequately act within a society (Giddens 1984, p. 23). At the same time, rules limit interaction because they control, in a practical sense, what actions are generally imaginable (Reckwitz 2007, p. 317). Rules cannot be grasped apart from resources (Giddens 1984, p. 18) which are, in turn, used in order to apply rules (Giddens 1984, pp. 14ff.). Resources can be allocative (domination over objects, for example, technology) or authoritative (domination over humans, for example, know-how) in kind. They represent material and immaterial elements actors may dispose of, providing them with capabilities to act and constituting their levels of power (Giddens 1984, p. 33).

An understanding of time and space is central to structuration theory and the meaning of structure. Giddens differentiates between 'structure' and 'system' (Giddens 1984, pp. 16f.). He describes structures in its most elementary form as 'simple' sets of rules and resources (Giddens 1984, p. 17). Meanwhile, systems are defined as *"the patterning of social relations across time-space, understood as reproduced practices"* (Giddens 1984, p. 377). Giddens further points out that systems are highly diverse in their 'degree of systemness', as they *"rarely have the sort of internal unity which may be found in physical and biological systems"* (Giddens1984, p. 377). With his understanding of structure and system, Giddens points towards two central aspects: (1) the bounding across time and space gives structure and agency a systemic form and (2) expansion in time and space is accompanied by a high variability of structures and interactions in which structures are embedded. As noted by Geels and Schot (2010, p. 45), Giddens pays more attention to structures than to the make-up of systems. Within transitions research, which is process-oriented and focused on system change, structuration theory is mainly devoted to describing the elementary constitution of structure and agency—how action has to be specified and how it has to be related to structure, change and constraint (Bryant and Jary 1991, p. 10).

4.2.2.3 Duality of Structure

Giddens (1984) does not imagine agency and structure to be two separate dimensions of the social. Instead, they are recursively connected. There is no other structure than the one that is produced or reproduced in interaction. Structures are, thus, not provided from the outside and do not constitute an external source of determination for spontaneous action (Giddens 1984, p. 16). Recursivity means in this regard that *"in and through their activities agents reproduce the conditions that make these activities possible"* (Giddens 1984, p. 2). Recursivity is *"the iterative application of one operation/transformation on its own result" (Schneidewind 1998, p. 145)*.[4] With the duality of structure, Giddens (1984, pp. 25ff.) emphasises that structuration is not a static property of social systems but an ongoing process (Jones 2011, p. 116).

Having established this concept of duality, Giddens (1984, p. 29) then differentiates structure into three analytical sub-dimensions: *signification, domination* and *legitimation*. The dimension of signification defines what is important in society. Signification originates from the Latin word *significare*, which means to name. Legitimation is the dimension of authorization in society. Achieving legal legitimation means *to gain the official, legal right to do something*. Achieving normative legitimation means *to gain normative permission to do something*. Giddens (1984) associates societal legitimation with the constitution of society. Central to this understanding is that societal legitimation is created by a group (for example, the majority of voters in a democracy). Legitimation must therefore be differentiated from acceptance. To accept is a voluntary act of individuals, where a person actively agrees with something or someone. In this sense, acceptance can rather be seen as an

---

[4]Schneidewind (1998, p. 145) defines recursivity based upon (Ortmann 1995, p. 81; Zimmer and Ortmann 1996, p. 90) and indirectly upon (Foester 1992).

4.2 Understanding the Relationship Between Enterprise Interaction and... 69

Structure	Signification ⟷	Domination ⟷	Legitimation
	⇅	⇅	⇅
(Modality)	Interpretative scheme	Facility	Norm
	⇅	⇅	⇅
Interaction	Communication ⟷	Power ⟷	Sanction

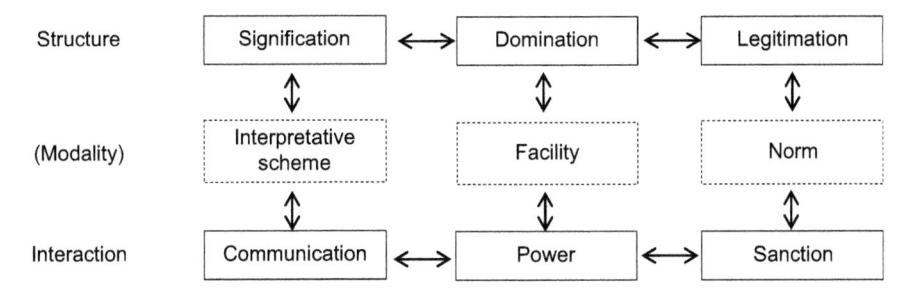

Fig. 4.3 Dimensions of the duality of structure (Giddens 1984, p. 29)

authoritative resource individuals may or may not offer, whereas legitimation is a structural dimension.[5] Domination means to have authority, the power.

These three sub-dimensions of structure can only be analytically separated. In practice, structures always contain, as sets of rules and resources, all three dimensions. Along the line of these sub-dimensions, action is itself divided into three sub-dimensions: *communication, exertion of power and realisation of sanctions*. Based upon this differentiation, Giddens (1984, p. 29) visualises the duality of structure as displayed in Fig. 4.3.

The dimensions of interaction and structure are combined to form (1) signification–communication, (2) domination–power, as well as (3) legitimation–sanction (Giddens 1984, p. 29). As represented in Fig. 4.3, the modalities of structure function as mediators between the general structures that exist beyond the moment, on the one hand, and quite concrete interactions taking place in particular situations, on the other (Schneidewind 1998, p. 143). Actors draw upon interpretative schemes when they communicate and, in this way, reproduce the context of signification in society. Actors also express the mobilisation or prevention of sanctions against others by drawing upon norms. At the same time, the prevailing structures of legitimation are reproduced (or, more rarely, not). Power is executed by allocating resources within particular situations, based upon which actors constitute the dominant force in a society (Schneidewind 1998, p. 140).

The systematisation of the duality of structure in the form of the above-presented dimensions rather remains abstract. In order to fully operationalise the concept, I draw upon the interpretations of Schneidewind (1998, pp. 139ff.) and Ortmann (1995, p. 60). Differently to Giddens, they position the modalities of structure in the centre of their conceptual reflections, as they see the modalities providing the necessary operationalisation of structure as well as giving agency a necessary degree of abstraction. The modalities concretise rules and resources. At the same time, they help to abstract agency from the very specific moment in which it acts, so that specific aspects of a particular act of agency as such can be faded out (Schneidewind 1998, pp. 143f., based upon Duschek 1996 and Zimmer and Ortmann 1996, p. 93).

---

[5]This differentiation between legitimation and acceptance is central to my assessment of energy cooperatives and, therefore, is introduced at this point.

Accordingly, interpretative schemes and norms form the concretised rule level that makes up social structure, specifying the structural dimensions of signification and legitimation within particular contexts. Interpretative schemes provide an understanding of what is important (signification), whereas norms provide an understanding of what is acceptable (legitimation). Allocative and authoritative resources, encapsulated under the term *facility*,[6] form the concretised resource level of social structure. Allocative and authoritative resources are tied to the structural dimension of *domination* and express the exercise of societal authority in concrete situations (Schneidewind 1998, p. 143, based upon Duschek 1996 and Zimmer and Ortmann 1996, p. 93). In this sense, agency is understood in this work as:

the creation or re-creation of interpretative schemes and norms (rule dimension of social structure) as well as the mobilisation or re-mobilisation of allocative and authoritative resources (resource dimension of social structure).

The modalities of structure play a crucial role throughout the book and are, therefore, introduced in more detail here.

Interpretative schemes represent informal rules that become particularly visible via the ways in which actors communicate. Actors draw upon interpretative schemes in order to constitute a meaning for something (structural dimension of signification) and are needed in order to understand each other and interpret what others mean. With regard to enterprises, informal rules play a decisive role in the constitution of business goals, business relationships and in the formulation of strategies. Interpretative schemes not only guide the ways in which customers interpret the quality and significance of products and services but also how they see enterprises in relation to society. At the same time, interpretative schemes provide a basis for enterprises to define themselves in relation to society and politics. Interpretative schemes can also reveal what actors informally expect from themselves and from others (Schneidewind 1998, p. 142).

Norms can be understood as formal rules that define the ways in which agency is evaluated, that is whether actions are seen as acceptable or unacceptable. They are used by actors in order to constitute a degree of conformity with someone or something (structural dimension of legitimation). Norms can represent official regulations and laws but also, for example, rules of courtesy and respect. Enterprises create or recreate norms in many ways. On the one hand, the market as well as the agency of enterprises is strongly regulated by a diverse set of laws. On the other hand, enterprises draw upon norms such as courtesy and respect when they offer their products and services to customers (Schneidewind 1998, p. 142).

Allocative resources involve power over physical artefacts, such as technology, natural resources, human resources, products or money (Giddens 1984, p. 258). Actors draw upon allocative and authoritative resources in order to execute power (structural dimension of domination). Allocative resources play a crucial role for enterprises, as they constitute the options of economic action within the market and

---

[6]The conceptualisation of Giddens' (1984, p. 29) modality of facility as allocative and authoritative resources is part of Schneidewind's interpretation (1998, pp. 139ff.) and Ortmann (1995, p. 60).

how much input an enterprise may be able to give in order to pursue its goals. Allocative resources represent all the material resources offered by enterprises as well as a great part of the technical and economic infrastructure they use (Schneidewind 1998, p. 142).

Authoritative resources involve power over humans through for example, cultural aspects or know how (Giddens 1984, p. 258). With regard to enterprises, they are present at any time in the form of company credibility, trust or courage. They can be seen as immaterial resources needed for the exercise of agency and the attainment of power (Schneidewind 1998, p. 142).

According to Giddens (1984, p. 29), the interaction–modality–structure dimensions of the duality of structure can be analytically separated from each other. However, as shown by the arrows in Fig. 4.3, they should not be interpreted independently from each other (Schneidewind 1998, p. 142), as it is the interplay between them that provides insights into the dynamics between agency and structure. For example, the overlap of interpretative schemes and norms expresses the space in which responsibilities and decision-making competencies are constituted (Giddens 1984, p. 30). Facility, in the form of authoritative and allocative resources, reveals whether agents are capable of realising these responsibilities and competencies or not. Negotiations, for example, are intensive interaction processes that involve all three of these interactive dimensions.

The routine reproduction, as well as the creative change, of norms, interpretative schemes, authoritative and allocative resources is inherent in Giddens concept of the knowledgeable and capable agent. Rules influence and are influenced by the knowledgeability of actors, whereas resources influence and are influenced by their capability to act. Patterns of rules and resources can be formed because they are embedded in the interactions of knowledgeable and capable actors. According to this line of argument, the concept of the duality of structure provides a structural interaction reference framework that incorporates entry points for voluntarism as well as determinism. Social structures constrain and enable interaction at the same time such that, even though actors are directed by structures, they are not completely determined by them. In order to analyse process of change, conditions need to be identified through which actors become *discursively* knowledgeable and capable agents.

4.2.3 Structuring Forces of Enterprises

In his book *The enterprise as a politically structuring actor*, Schneidewind (1998) explains in great detail how and to what extent structuration theory can help to describe the interrelationships between the activities of enterprises and societal structures. Based upon Giddens (1984), he offers a theoretical approach for systematically delineating the structuring forces of firms. The scope and perspective of Schneidewind (1998) fits well with the research interest of this book, because his work provides a crucial basis for understanding the influence of innovative firms in relation to socio-technical transitions.

Schneidewind (1998) understands structuring forces as a firm's capacity to change the societal structures it draws upon (Schneidewind 1998). Instead of focussing on a particular playing field, he seeks to take a holistic perspective, pointing out that the structuring activities of enterprises cannot be limited to the market but also need to be embedded in the overall context of society. A theory of the enterprise as a structuring actor must therefore be generalisable with regard to policy, the market and social arenas (Schneidewind 1998, pp. 207ff.). In order to meet this claim, Schneidewind (1998, pp. 140f.) closely follows Giddens' (1984, p. 29) differentiation of structures in the three analytical dimensions discussed above: signification, legitimation and domination.

With his holistic approach, Schneidewind (1998) made an important contribution to research on organisations and structuration theory. Even though there is a large number of works analysing organisations through the lens of Giddens' (1984) structuration theory (see Sect. 4.2.1), only a few apply a systems perspective. Most studies primarily focus on market interactions (for example, Becker 1996; Jones 2011; Ortmann 1995; Pozzebon 2004; Sydow and Duschek 2010; Whittington 2010), while other works analyse interaction process within the enterprise itself (for example, Ortmann et al. 2000). The lack of studies applying Giddens' concerns with regard to enterprises in terms of the plurality of social systems was already noticed by Whittington (1992, p. 694). Subsequently, Schneidewind (1998) helped to close this research gap.

4.2.3.1 Enterprises as Collective Actors

According to structuration theory, any actor can affect social structures through interaction. With this view, Giddens (1984) regards the actor as a social individual. In order to apply structuration theory to enterprises, Schneidewind (1998, p. 136) conceptualizes them as 'collective actors'. The essence of agency and structure, such as it is understood by Giddens (1984), becomes relevant when noting that firms and organisations are managed and operated by a collection of social individuals. In this sense, the agency of enterprises is enabled and constrained by structures in ways similar to the situation for human individuals (Schneidewind 1998, p. 136). Firms may follow routines and reproduce structures or they may change them under certain circumstances.

Describing enterprises as collective actors draws attention towards the interrelationships between the firm as such and its environment. This clear analytical focus facilitates reflection on the structuring forces of enterprise, though, at the same time, it reduces the firm to a 'black box', leaving aside the internal firm perspective (Schneidewind 1998, p. 445). It also ignores how a firm's actions are the sum of multiple, sometimes conflicting, interactions of employees, managers and directors working within it. As noted by Schneidewind (1998, pp. 445f.), the notion of the firm as a structuring actor can be further developed by acknowledging the internal heterogeneity of firms, such as by taking into account the individual character of aligned workers. In this work, I partly follow this suggestion by considering internal and external dimensions of firm cooperation during the empirical analysis of Chap. 7.

4.2.3.2 The Concept of the Structuring Enterprise

Schneidewind (1998, pp. 204ff.) takes as his point of departure three central notions of structuration theory. (1) The duality of structure, which refers to the same ontology of structure and agency. Schneidewind (1998, p. 140) argues that enterprises play an important role in reproducing or changing market, societal and political structures, because they belong to a group of actors who draw upon respective rules and resources when they interact. (2) Recursivity: Action and the conditions for action are recursively related. In consequence, the structuring forces of enterprises have a highly recursive and multi-dimensional character (Schneidewind 1998, pp. 205f.). (3) Reflexivity: Even though structures are normally reproduced through habits, any enterprise incorporates certain amounts of knowledgeability and capability that can make a difference towards change and acting non-deterministically (Giddens 1982, pp. 29ff.; Schneidewind 1998, pp. 206f.).

Duality, recursivity and reflexivity represent the three general mechanisms of structuring activities (Schneidewind 1998, p. 140), based upon which the structuring forces of enterprises can be systematically delineated (Schneidewind 1998, p. 141). Schneidewind differentiates between one direct and two indirect options firms have in order to exert influence on structures.

(1) At the centre of his interpretative scheme stands the circumstance that firms can *exert direct influence on the modalities of structures*[7] in a concrete social context through their own interaction with other firms. Schneidewind (1998, pp. 200f.) outlines how rules and resources can be directly influenced in two ways. Either norms, interpretative schemes, allocative or authoritative resources can be changed as such or in terms of their relative meaning in relation to each other. For example, if enterprises market their products as environmentally friendly items, they may challenge existing guiding principles underlying how customers choose products (change of rule level). However, in times of increased commercialisation, norms and guiding principles may lose part of their relative importance in favour of monetary resources (change of weighting between the existing rule and resource levels) (Schneidewind 1998, p. 201).

---

[7]Schneidewind focusses on Giddens' (1984) modalities of structures. As explained in Sect. 4.2.2.3, norms, interpretative schemes, as well as allocative and authoritative resources mediate between Giddens very abstract understanding of structure and his very specific understanding of agency (see also, Schneidewind 1998, p. 194). This interpretation has been adapted for use in this thesis.

(2) *Indirect effects on structuring forces through change of actor sets*: The entry of new actors or a scarcity of prevailing actors in certain firm constellations can have an immediate effect on practiced norms, interpretative schemes or on the management of resources. For example, a strong reduction of the number of firms in a particular sector can turn a market into an oligopoly with effects on product prices, cost structures, product quality and available services. It can also change costumer perceptions. Similarly, the creation of firm lobbying groups can influence policy agendas and external perceptions regarding certain products (Schneidewind 1998, pp. 201f.).

(3) *Indirect effects on structuring forces through change of mechanism-reproducing structures*: Enterprises can become structuring actors in the sense of changing the structures if they take notice of the mechanisms that are mostly responsible for reproducing social structure (and agency) (Schneidewind 1998, pp. 202f.). A key step towards actually changing structures and towards breaking with routines consists of turning the reflexive monitoring of action into a discursive and intentional process. To become aware of those norms, interpretative schemes and resources that one draws upon during interaction is crucial for being able to change them. Routines generally become visible through conscious discourse between actors (Schneidewind 1998, p. 203).

Even though the three options that enterprises have for exerting influence on structures have been analytically differentiated, they are strongly interrelated in practice and can be said to have their own recursive character. To become aware of the reproduction of structures can enable actors to reflect upon and change rules and resources in a reflexive and discursive process. Equally, a change of actor sets may lead to discursive reflection and the empowerment of actors, with subsequent effects on norms, interpretative schemes and resources (Schneidewind 1998, pp. 205f.).

4.2.3.3 Reflexive and Discursive Character of Collaborative Interaction

Schneidewind (1998, pp. 286ff.) identifies cooperation[8] as a concrete space in which the three options for activating structuring forces can be successfully operationalised. Central here is that the general mechanisms of structuring activities (duality, recursivity and reflexivity) are particularly apparent in collaborative interaction. Gray (1991) defines cooperation as:

> "A process through which parties who see different aspects of a problem can constructively explore their differences and search for solutions that go beyond their own limited vision of what is possible" (Gray and Wood 1991, p. 4).

---

[8]The literature refers to cooperation, as well as to alliances and networks. Cooperation and alliances may already be created between two parties, whereas networks usually involve more than two parties (Zentes et al. 2005b). Collaborative interaction is the act as such that primarily takes place in cooperation, alliances and networks.

According to this definition, cooperation presents a dynamic process related to change. It is often oriented towards a particular goal, a joint vision or the resolution of major problems. Essential for cooperation is constructive interaction between at least two parties in a way that may enable them to go beyond their own individual viewpoints and action spaces (Gray and Wood 1991).

Organisations in cooperation continually need to negotiate, since a hierarchical order is not automatically foreseen between collaborating partners. Cooperation can thus be characterised as the conscious and communication-oriented coordination of organisational activities (Schneidewind 1998, p. 324). Whereas the hierarchical coordination of business, which particularly exists in the classic market model, generally supports routine control of action, cooperation provides a model that can enhance discursive interaction through which conscious guidance of agency can go beyond the pure reproduction of routines (Schneidewind 1998, p. 325). Engaging in reflexive discourse can help to make mechanisms of reproduction visible and enable collaborators to review or question existing interaction patterns. Through cooperation, knowledge and capabilities of involved agents can become explicit. Collaborative interaction can, thus, be an important starting point for turning interaction into an intentional process. Cooperation can help to mobilise allocative and authoritative resources and provide an important basis for drawing upon new norms and interpretative schemes. Actors in cooperation can directly affect the modalities of structure (Schneidewind 1998, pp. 325ff.), and cooperation can bring together new actor sets. Schneidewind (1998, pp. 328ff.) explicitly points out that firm cooperation may not only involve other firms but also politicians, citizens, and societal or environmental organisations (see also, Schneidewind and Petersen 1998). The need for discursive exchange is increased when actor constellations undergo change. At the same time, bringing together the knowledge and capabilities of many different actors can strongly enhance potential for breaking with routines. For example, firms that collaborate may end up having more money (allocative) and technical know-how (authoritative) at their disposal, potentially giving them the power to change technical set-ups (Schneidewind 1998, pp. 331ff.). Collaborative interaction between firms and politicians may help to create new product standards for certain firm activities, affecting the structural dimension that legitimises interaction (Schneidewind 1998, pp. 363ff.). Collaboration between organisations and citizens may result in new consumer guiding principles, changing the ways in which certain products or services are signified (Schneidewind 1998, pp. 378ff.).

In sum, collaboration is an interactive, discursive and reflexive medium through which involved actors can become able to directly and indirectly change structures of legitimation, signification and domination. The relevance of collaborative interaction for transformative processes has also been acknowledged in transitions research. As outlined in Sect. 4.1.4, one of the three central factors considered to be responsible for the development of niche innovations is the establishment of broad and deep social networks. Schneidewind (1998) complements strategic niche management by providing a theorised and coherent explanation for the underlying mechanisms thought to be responsible for the high structural-change potential of collaborative interaction.

However, cooperative interaction also involves challenges. Empirical research has revealed that collaborative interaction can be highly complex, long-lasting and, thus, not always as effective and efficient as desired (for example, Rosenfeld 1996). Sydow has identified eight lines of conflict that collaborations may have to deal with, including autonomy versus dependency or levels of control versus levels of trust (Sydow 2010b, p. 404). Despite these difficulties, however, the overall innovative potential of cooperation is widely acknowledged in organisation research (for example, Hauschildt and Salomo 2011, pp. 174ff.).

The present work is based on Schneidewind's (1998) ideas and primarily focuses on collaborative interaction. Cooperative interaction models are the core entry point for developing a new framework for analysing actor influence within socio-technical transitions in Sect. 4.3 as well as for conducting the empirical analysis of energy cooperatives in Chap. 7. My interest here is to explore the structuring effects of collaboration. The management and internal handling of cooperation has already been analysed by other authors from many different perspectives (examples for German organisation research on cooperation include Sydow 2010a; Sydow and Duschek 2010; Zentes et al. 2005a). Though, internal factors of success and failure are taken into account while reflecting upon the structuring forces of cooperative interaction.

4.2.3.4 Scope of Structuring Forces

Schneidewind (1998) draws a positive picture of the enterprise as a structuring actor. Instead of focussing on how structures may generally constrain and limit firm activities, he systematically describes how enterprises may exert direct and indirect influence on structures.

The three general mechanisms of structuring activities (duality, recursivity, and reflexivity) are central for such conceptualisation (Schneidewind 1998, pp. 204ff.), as they make clear that structuring forces do not follow a simple causal process, where enterprises face an externally given environment that can be changed with selected interventions. Structure is not a concrete design object (Schneidewind 1998, p. 140). Structure can neither be seen as a purely deterministic force or set of conditions nor solely as a voluntaristic result of interaction. It is always both. When drawing upon the ideas of structuration theory, it is thus highly important to always take into account the limiting as well as the enabling character of structure (and agency). Schneidewind (1998, pp. 204ff.) fully incorporates the mediating idea of structuration theory by pointing out that structuring forces of enterprises are always aligned to unknown conditions for action and unintended consequences of action—making action highly unpredictable.

1. Unpredictability through unknown conditions for action: Since structure is at the same time both result and medium, change of structures is always accompanied to a certain degree by structural reproduction. In the course of changing certain norms, interpretative schemes, allocative or authoritative resources, enterprises

also have to draw upon existing rules or resources; otherwise they would not be able to act. For example, a firm may introduce an ecologically friendly product with the aim of strengthening the guiding principle of environmental protection. The firm does so by using its existing monetary resources and the existing external perceptions about the firm. The mix of structural change and reproduction can neither be exactly predicted nor controlled (Schneidewind 1998, pp. 204f.).

2. Unpredictability through unintended consequences of action: Recursivity characterises the strong interrelationship between agency and structure, between norms, interpretative schemes and resources, as well as between the three options for activating structuring forces. Recursivity means that a change of one aspect always effects a change of other aspects (see Sect. 4.2.2.3 for the definition of recursivity). Rules are applied through the use of resources. Challenging norms, interpretative schemes, allocative or authoritative resources is always also accompanied by a change of other respective modalities. In which ways a change of actor sets may influence the modalities of structures or in which ways rules are affected by changing resources is never fully clear and, consequently, cannot be precisely planned or controlled. The recursivity of structures and agency thus entails the creation of unintended consequences of action, making intentional change highly unpredictable (Schneidewind 1998, p. 201).

The unpredictable and uncontrollable character of unacknowledged conditions of action, as well as of unintended consequences of action, frame the structuring forces of enterprises and limit their scope. It is the highly dual, recursive and reflexive character of agency and structure which make their manifestation so challenging (Schneidewind 1998, pp. 204f.).

4.2.4 Critical Reflection

As a grand social theory, structuration theory seeks to provide a 'social ontology'—a basic vocabulary of the social and how it works—which can be applied universally to any social system (Reckwitz 2007, p. 323). The idea of the duality of structure—as medium for and outcome of interaction—is so powerful because it provides a clear idea of the linkage between the interaction of actors and social structure. Giddens (1984) does not see them as independent elements that stand in conflict with each other but as a parts of a mutually recursive duality. Structuration theory combines the individual and institutional level of social practices and turns our attention towards the recursive nature of social life (Jones 2011, p. 119). As noted by Whittington (2010): *"the structurationist framework can handle both creativity and circularity, agency and structure"* (Whittington 2010, p. 118).

The functioning of structuration theory as a mediator between micro and macro perspectives, between agency and structure, as well as between deterministic and voluntaristic sociological approaches has been commonly noticed as a central strength of the concept—a unique value added for many theoretical and empirical discussions (for example, Jones 2011; Reckwitz 2007; Schneidewind 1998, p. 144; Whittington 2010). At the same time, it is precisely this aspect of the theory that has been subject to criticism. Some scholars contend that structuration theory is particularly rooted in visualising the constraints of structures and social systems by mainly focussing on the reproduction of social practices and by emphasising the routine guidance of action (for example, Bauman 1989, pp. 42ff.). Other scholars criticise Giddens' theory for conceptualising agency and structure in an overly optimistic way, as he traces the existence of social systems all the way down to the agency of the social individual (for example, Callinicos 1985). Yet, both strands of criticism are short-sighted, since they neglect the mediating character of structuration theory. However, together the two viewpoints underscore that one must be careful not to apply Giddens' ideas in a lopsided manner, meaning here that, when employing the theory, it is decisive to always take into account both the enabling and the constraining character of structure (and agency).

A second criticism addresses the suitability of structuration theory for pursuing empirical inquiries. It has been said that, in its attempt to be a meta theory, it remains overly abstract, lacking the conceptual precision and relevant degree of specification that is needed for empirical work (for example, Gregson 1989; Jones 2011, p. 114; Sewell 1992, pp. 5ff.). In other words, it is thought that structuration theory would be of no use to empirical research. Giddens, however, points out that empirical inquiries cannot be conducted without abstract theoretical concepts and, from his perspective, structuration theory works out *"the substantive connotations of the core notions action and structure"* (Giddens 1989, p. 296). The high level of abstraction gives structuration theory its universal character and leaves the necessary room for interpreting the diversity of social life in any and every context of action (Giddens 1989, p. 295). At bottom, Giddens hold that the theory *"should be seen more as a sensitizing device than as providing detailed guidelines for research procedure"* (Giddens 1989, p. 294). In this vein, many scholars have pointed out that structuration theory can unfold its fruitful value for empirical enquires in particular when being combined with other theories (for example, Capallo 2008, pp. 118f.; Pozzebon 2004, pp. 266ff.). The present book takes such an approach by re-emphasising that structuration theory can be a fundamental building block for empirical research within sustainability transitions research through a merging and synthesis of their concepts (see Sect. 4.3 for details).

As we have already seen above, based upon Giddens' (1984) ideas, Schneidewind (1998) elaborates a theory of structuring forces of the enterprise. Structuring forces are understood as the process of interlinked (re)creation of norms and interpretative schemes and the (re)mobilisation of allocative and authoritative resources that shape enterprises and are recursively shaped by them. The changes achieved through such forces can be achieved directly, by immediately exerting influence on the modalities of structure, or indirectly, by changing actor sets or by making the mechanisms responsible for the reproduction of structures visible. Adopting the three general mechanisms of structuring forces—recursivity, duality and reflexivity—Schneidewind emphasises the following:

- The concept of structuring forces describes the ways in which enterprises directly or indirectly draw upon a recursively connected mix of structural modalities.
- Structuring forces of enterprises must be analysed in relation to the three structural dimensions of legitimation, signification and domination, adding that it is short-sighted to focus on only one of them.
- Interaction needs to be discursively reflected upon order to become explicit and to create the conditions for breaking with routines, such as what can be achieved through cooperative interaction. It is the analytical differentiation between the discursive and practical consciousness of agents that makes structuration theory into such a strong explanatory concept.
- Structuring forces of enterprises should not merely be seen a set of causally interlinked management assignments (Schneidewind 1998, p. 444), as the recursive nature of the interrelationships between agency and structure often leads to unacknowledged conditions and unintended consequences of action, making the structuring activities of enterprises highly unpredictable and uncontrollable.

Schneidewind illustrates his reconceptualisation of structuration theory through a collection of many vivid examples about enterprise collaborations being engaged in that have attempted to make the mass market more ecological with selected products (Schneidewind 1998, pp. 266ff.). As he points out, however, further empirical research is necessary in order to analyse the ways in which enterprises can change rules and resources and what kinds of effects this may have (1998, pp. 448f.). The assumption here is that empirical review of actual instances of cooperative interaction can generate important conclusions about how collaborators deal with interpretative schemes, norms and translated resources. The analysis of 'playing fields' of cooperative interaction, based upon concrete contexts, requires concrete methodological approaches and the selection of specific situations (Schneidewind 1998, p. 331). The present work does this by selecting energy cooperatives as enterprise actors with concrete collaborative interaction models and by asking what kinds of influences renewable energy cooperatives have been having on the German energy system.

Schneidewind (1998, pp. 196f.) acknowledges that structuration theory primarily provides a general logic of the social. Accordingly, he mainly discusses the interactions of firms, and detailed reflection on structuring forces in relation to system change is not pursued. Based upon Giddens (1984, pp. 16f.), Schneidewind (1998, pp. 137f.) points out that, being diffused in time and space, the influence of single actors on structures is limited, so that one must be careful not to draw an overly optimistic picture of firm influence. However, he does not outline what implications his approach could have for achieving a deeper and more differentiated understanding of structuring enterprises forces within a transition process. That Schneidewind focuses on structures as such and not on changing systems is in line with Giddens (1984), who pays more attention to social structure than to social systems. The particular constitution of agency and structures in the process of socio-technical transitions thus remains rather blurred. The present work aims to fill this gap by elaborating a conceptual framework that explicitly outlines the interrelationship of agency and structure within socio-technical transitions.

The question remains whether other social theories would have been as adequate or even more suitable for conceptualising the interrelationships between enterprises and social structures. It is not the aim to add at this point a detailed comparison of structuration theory and other social theories against the background of my research interest. Schneidewind (1998, p. 148) and others, such as Kieser and Ebers (2006), have already pointed out that other sociologists—including Jürgen Habermas, Niklas Luhmann and Max Weber—also offer promising concepts for understanding, explaining and analysing enterprise activities. For my work, the structuration theory of Anthony Giddens is of central interest because it forms a crucial building block of sustainability transitions research and offers a mediating perspective on the enabling and constraining character of agency and structure. In addition, my research interest lies in the influence of a particular enterprise group on societal structures, and Schneidewind's (1998) conceptualisation of structuring forces of enterprises clearly shows the potential of structuration theory for comprehending enterprise activities in relation to structural change.

4.3 Developing a Framework for Analysing the Influence of Organisations on Socio-technical Transitions

The central aim of this section is to develop a comprehensive framework for systematically describing and assessing the influences of innovative enterprise groups in relation to socio-technical transitions. This framework will then be used to guide empirical research on the impacts of energy cooperatives on Germany's energy system. Framework building is an important part of theory elaboration and crucial for producing new empirical results. According to Porter:

> "Frameworks identify the relevant variables and the questions which the user must answer to develop conclusions tailored to particular [phenomena]" (Porter 1991, p. 98).

Geels (2014, p. 262) adds that conceptual frameworks provide a language for describing and discussing variables and their relationships. Framework building not only strives towards capturing the full richness of an empirical phenomenon through focus on a limited number of key dimensions but also allows scholars to systematically approach their research topics from a theoretical perspective. Frameworks therefore "help the analyst to better think through the problem" (Porter 1991, p. 98).

In Sect. 4.3.1 I outline why a new framework is needed for analysing actor influence in transitions, even though actor dynamics already play an important role in transitions research. Then Sect. 4.3.2 introduces the main conceptual building blocks for my IBAST framework—a framework for visualising the Interrelations Between Agency and Structure in Transitions—which is elaborated in Sect. 4.3.3 in terms of three spatial levels. A summary is then provided in Sect. 4.3.4.

4.3.1 The Need for a New Analytical Framework

As I already explained, assessing the influence of innovative actors in general and innovative enterprises in particular within socio-technical transitions requires a clear theoretical understanding of the interrelationships between agency and social structure. This requires explicit conceptualisation of the scope of agency in relation to emerging system change and an explicit focus on actors.

In Sect. 4.1 I outlined how transitions research not only offers a well-elaborated picture of socio-technical transitions but also provides valuable entry points for setting the complex and to a great extent uncontrollable character of system change within the context of actor dynamics. A transition is understood as a radical system change, characterised through non-linear, co-evolutionary socio-technical developments that are the result of multiple interactions in and between many different social groups (Geels and Schot 2010, p. 11). One of the core interests of transitions research is to gain scientific knowledge about the underlying mechanisms of transitions. Actor dynamics are seen to be the main driver for transitions, because system change happens through interaction (Geels 2004, p. 902; 2005, pp. 16ff.). While the multi-level perspective and strategic niche management, two core concepts of sustainability transition research, provide appropriate conceptual building blocks for relating actors and their interactions to wider transition processes, they lack depth in important areas, as socio-technical transitions are explained through actor interaction without providing a concrete systematisation of agency in relation to different and heterogeneous levels of structural change taking place in socio-technical systems (see Sect. 4.1.4).

The multi-level perspective fails to align a concrete conceptualisation of interaction with the structural levels of niche, regime and landscape. As a result, the framework has been criticised for being overly structurally oriented not giving enough attention to actors (Shove 2012; Smith 2012). Meanwhile, strategic niche management does integrate actor interaction as a natural part of niche development in highlighting how networking, visioning, learning, as well as the creation of generic rules being fostered by a rising cosmopolitan community, are central niche development factors (Geels and Raven 2006; Kemp et al. 1998; Schot et al. 1994). Yet, niche research is mainly interested in examining innovation projects themselves and, explicit and explorative focus on innovative actors and their influences is still underdeveloped (see Sect. 4.1.4).

Furthermore, the description of niches and regimes, as well as of their interrelationships, has been inconsistent within prevailing transition studies. A regime is usually understood as the overall established structure of a system, but there is variation among authors in terms of how they define structures. A review of notions of energy regimes within the current literature (see Sect. 4.1.4.1) reveals that some limit regimes to the underlying rule level of a system, whereas others take rules, artefacts and actors as regime structures. Although regimes remain abstract, niches are seen as protected haptic spaces (for example, Kemp et al. 1998, p. 185; Schot et al. 1994, p. 1061). The description of niches as spaces and of regimes as structures with unclear elements makes it challenging to elaborate a precise understanding of co-evolutionary niche–regime dynamics and to gauge the scope of agency in relation to emerging niche–regime change.

As outlined in Sect. 4.1.4, the multi-level perspective and strategic niche management both draw upon structuration theory. Giddens (1984) offers a fundamental ontology of agency and structure as such. However, the integration of structuration theory has been somehow marginalised in sustainability transitions research to the overall idea that structure is simultaneously both a seemingly stable reference point and changeable result of interaction. The far-reaching explanatory power behind the concept of the duality of structure, introduced in Sect. 4.2, has been to a great extent neglected. However, as I have outlined above, structuration theory entails valuable notions for

- elaborating a sound and consistent definition of niches and regimes as well as of their interrelationships;
- systemising interaction in relation to different levels of structure, based upon taking into consideration the dimension of space; and
- conceptualising the co-evolutionary dynamics between niches and regimes through a focus on actors and their interactions.

It thus seem sensible to reclaim structuration theory as a theoretical building block for transitions research (see also, Augenstein 2014, pp. 69ff.) by merging the comprehensive explanations of agency and structure provided by structuration theory with the latter's descriptions of socio-technical transition.

Throughout the following sections (4.3.1 to 4.3.4), I develop a new conceptual framework—Interrelations Between Agency and Structure in transitions (IBAST)—which will then be applied to guide empirical research on structuring forces related to actors within system change. The IBAST framework is conceptualised in a general way here, seeking to aid assessment of the influences of innovative actors in general, whereas my particular empirical research interest is the interaction of innovative enterprise groups. Thus, as explained above, I also drawn upon Schneidewind (1998), who theorised the enterprise as a structuring actor based upon Giddens' structuration theory. Schneidewind (1998, p. 136) shows, that even though Giddens (1984) primarily refers to the social individual, structuration theory is quite applicable for theorising the enterprise as a structuring actor with effects on the market as well as political and social environments. This interpretation emphasises the high diversity of interaction models that can exist between private persons, firms, politicians or scientists.

Figure 4.4 provides an overview of the theoretical origins of the IBAST framework.

4.3 Developing a Framework for Analysing the Influence of Organisations... 83

```
┌─────────────────────────────┐  ┌─────────────────────────────────────┐
│  Multi-level perspective    │  │      Structuration theory           │
│  Strategic niche management │  │  Theory of the firm as a structuring actor │
└─────────────────────────────┘  └─────────────────────────────────────┘
                    │                      │
                    └──────────┬───────────┘
                               │
┌──────────────────────────────────────────────────────────────────────┐
│                          IBAST framework                             │
│  Systematic visualisation of the Interrelations Between Agency and Structure in Transitions │
├──────────────────────────────────────────────────────────────────────┤
│  Guidance for empirical research on the influence of innovative actor groups, such as │
│         enterprise groups, within socio-technical transitions         │
└──────────────────────────────────────────────────────────────────────┘
```

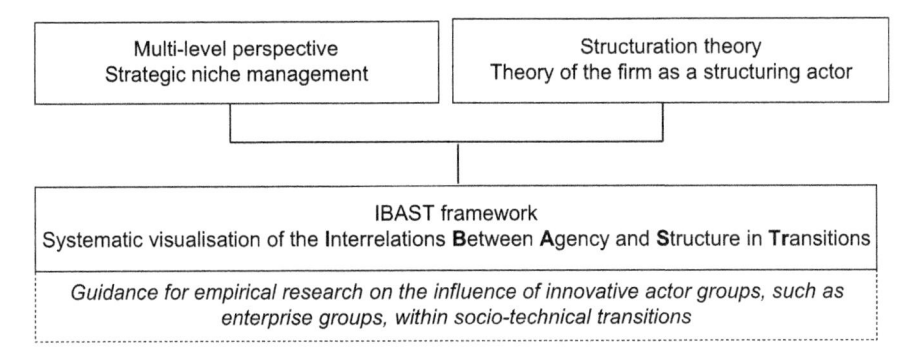

Fig. 4.4 Origins and purpose of the IBAST framework

4.3.2 Main Conceptual Building Blocks of the IBAST Framework

Before outlining the IBAST framework itself, its main conceptual building blocks—(1) agency and structure as well as (2) the dimension of space—are discussed in relation to socio-technical transitions.

4.3.2.1 Agency and Structure

In order to achieve a sound theoretical understanding of the constitution of niches and regimes as well as of their interrelationships it is necessary to outline their fundamental ontology. This will be done in the following by systematically integrating the ideas of structuration theory into transitions research, as I seek to show that agency plays the most central role for developing a precise theoretical meaning of niches and regimes. In this and the following sections, I articulate and then explain several guiding statements regarding the study of regimes and niches that form the basis for the IBAST framework.

1. Niches, as well as regimes, are best theorised as societal structures consisting of rules and resources.

Niches and regimes are understood as structural levels of a system (Geels 2002, pp. 1259ff., based upon Rip and Kemp 1998, pp. 337ff.; Schot 1998, pp. 175ff.). As we have seen, according to Giddens (1984, p. 29), societal structures are always comprised of the three dimensions of *signification*, characterising what is important; *legitimation*, characterising what is acceptable; and *domination*, implying the existence of various levels social authority. In theory they can be separated, but in practice they are always united. It is therefore not possible to comprehensively explain the constitution of social structure if one of the three dimensions is neglected. Rules guide actors in how to act within a society because they define what is acceptable and important (Giddens 1984, p. 23). In concrete situations, they consist of norms and interpretative schemes. Official norms can be statutes, proverbs, liturgies, contracts or laws. Interpretative schemes are principles of action, habits, conventions or ways of thinking. Resources constitute the power of actors and define their capabilities for action and are strongly linked to the dimension of domination (Giddens 1984, p. 33). Allocative resources are material in kind, such as technology or money, whereas authoritative resources are immaterial in kind, such as credibility or know-how. Resources are used in order to apply rules (Giddens 1984, pp. 14ff.) and rules are used in order to mobilise resources. In order to grasp the fundamental meaning of niches and regimes as social structures, they must be described as patterns of rules *and* resources.

Crucial to this understanding is that agency and structure are recursively connected (Giddens 1984, p. 2). Actors draw upon niche and regime structures, but they are not part of them in an analytical sense. Niches and regimes provide meaningful and material conditions for competent actors to act. At the same time, niches and regimes are shaped by actor interaction, being embedded in a continuous flow of conduct (based on Giddens in Kießling 1988, p. 4). Therefore, the present work generally defines niches and regimes in their most elementary form as socio-technical structures that are comprised of rules and resources, shaped via the three structural dimensions of legitimation, signification and domination. Niche and regime structures are recursively connected to agency.

Having briefly clarified the fundamental constitution of niches and regimes, it is now necessary to explain, from an abstract point of view, how they differ from each other.

2. In their most elementary form, niches and regimes must be differentiated according to levels of discursivity in aligned agency.

According to the multi-level perspective, niche structures are new and innovative. They are not at this point part of the mainstream common-sense and emerge outside of normal regime contexts. Meanwhile, regime structures are those underlying patterns (of rules and resources) that characterise an existing system (Geels 2002, pp. 1259ff.; 2005, pp. 85f.; 2011, p. 27; Geels and Kemp 2012, pp. 54ff.; van Bree et al. 2010, pp. 532f.; Verbong and Geels 2007, p. 1026).

Giddens (1984, pp. 5ff.) differentiates between the continual reproduction of structures based upon unquestioned routines and the production of new structures based upon processes of discursive reflection. Routine interaction follows the implicit knowledge of practical consciousness and is, therefore, eluded from a discursive reflection process. Because they are neither recognised nor questioned, the routine reproduction of the same interactions and the same structures facilitates daily life, similarly to the grammar of a language (Giddens 1984, pp. 4ff.; Reckwitz 2007, pp. 319f.). Regimes form such social structures and are continuously reproduced through unquestioned routine interaction patterns. Regime structures represent the underlying patterns of rules and resources that express, in their continual and unquestioned reproduction, a socio-technical system. The routine guidance of action and reproduction of structures is a crucial part of every functioning society.

Even though human actors primarily reproduce structures and patterns of interaction, they all carry the knowledgeability and capabilities necessary to break with their practical routines (Giddens 1984, pp. 281ff.). This happens when agency and structure become part of a discursive reflection process that takes place within discursive consciousness. Discursivity makes interaction as such explicit. Intentional interaction based upon discursive reflection leads to the questioning of prevailing structures and the potential creation of new patterns of rules and resources (Giddens 1984, pp. 5ff.). One of the most fundamental preconditions for the evolvement of niche structures is, therefore, that agents need to discursively reflect upon their normal interaction patterns. Such a standpoint suggests that, in order to observe the development of niche structures, the analysis of creativity and structural change should begin with the social actor and not with the innovation project, as is predominantly being done in the strategic niche management literature (see Sect. 4.1.4).

3. Innovative niche and established regime structures co-exist in concrete situations of interaction.

Taking our point of departure from Giddens (1984), it is important to understand that, independent of whether innovative or well-established in kind, structures are contextualised in the very moment of interaction. There is no other structure than the one that is created or recreated through the interaction of actors. Regime and niche structures only exist because they are continually produced or reproduced through immediate interaction, taking place in families, between consumers, enterprises, politicians, scientists or any other actors. Established regimes and innovative niches represent structures as well as concrete interaction models that co-exist through and within interaction.

Change and reproduction of structures takes place simultaneously within concrete situations of the exercise of agency (Schneidewind 1998, pp. 204f., based upon Giddens 1984). Due to structure always being simultaneously both result and reference point of interaction, the creation of new structures is always accompanied by the reproduction of other, prevailing structures (Giddens 1984). Actors partly draw upon established regime structures in order to develop innovative sets of rules and resources. Niche and regime structures co-evolve during the very moment of interaction. Giddens describes this process as follows: "*All social life, of course—even in the most radical phases of social change, like revolutions—involves continuities*" (Giddens 1989, p. 277). Drawing attention to the simultaneous co-evolution of niches and regimes within and through concrete situations of interaction can enable the kind of symmetrical analysis of stability and change that has been called for by Loorbach and Verbong (2012, p. 321). But how can co-existence and co-evolution of niches and regimes be conceptualised?

Although it has been completely neglected in transitions research so far, Giddens' idea of the unacknowledged conditions and unintended consequences of action (Giddens 1984, pp. 41ff.) can provide guidance for understanding key aspects of action. As certain routines are made explicit and become subject of change, the majority of other routine interaction patterns remain invisible, functioning as the unacknowledged or unquestioned basis for the development of new sets of rules and resources. Actors may only be aware of the creation of new rules and the mobilisation of new resources. In this sense, regime structures always partly shape unacknowledged conditions of innovative action, and innovations are built into the very nature of routines. The full innovation process is, thus, never completely discursively controlled and monitored.

In this sense, it is misleading to speak of niche or regime actors, as has been widely done in transition studies (for example, Hansen and Steen 2015, p. 3; Hofman and Elzen 2010, p. 659; Raven 2007, p. 2397). Usually, regime actors are described as big, powerful players, who mainly benefit from the existing regime and, thus, do not have an interest in any change. In contrast, niche actors are described as small and innovative agents that are active at the local level. However, there is no such actor who purely draws upon newly created rules and newly mobilised resources. When entrepreneurs mobilise venture capital for their ideas, they always also need to partly follow the prevailing structures of the capital market. This shows once more how important it is to analytically separate actors and their interactions from social structures. Otherwise, there is the risk of drawing misleading conclusions.

In addition, the intentional realisation of action based upon discursive reflection and the aligned intentional creation of innovative niche structures is always accompanied by unintended consequences of action (Giddens 1984, pp. 5f.). As explained in the following, it is this unpredictable mix of intentional interaction, unacknowledged conditions of action and unintended consequences of action which characterises co-evolution between niches and regimes and is to a great degree responsible for the highly non-linear development of transitions.

4.3.2.2 The Dimension of Space

In the previous section I have argued that, in their most elementary form, niches and regimes should be differentiated according to their levels of discursivity in agency. While this understanding recognises that niche and regime structures co-exist and co-evolve in concrete situations of interaction, it does not yet explain how they are related to time and space.

One of the challenges entailed by my research focus here is how to analytically link the micro-specific interaction of innovative energy cooperatives to the macro-specific character of the long-term, non-linear energy transition process currently taking place in Germany. On the one hand, change is associated with social action. On the other hand, transitions take place beyond the control of individual or small groups of actors (Geels and Schot 2010, p. 11). A fundamental step in conceptualising the IBAST framework is, therefore, to choose a sound systematisation of the interrelationships between structural change and actor interaction along which the influence of actors can be delineated. In the following, then, I seek to show that actor influence within a transition process is strongly aligned to the dimension of space.

Giddens (1984) criticises the meanings generally attributed to the concepts of micro and macro. In his view, micro-analysis and macro-analysis create misleading associations, because they tend to suggest one has to choose between them, separating the social into intentional micro-activities and constraining macro-patterns (Giddens 1984, pp. 139f.). To remedy this, Giddens draws upon Collins, who called for the 'micro-translation' of 'structural phenomena' (Giddens 1984, p. 140, based

4.3 Developing a Framework for Analysing the Influence of Organisations...

upon Collins 1981). Giddens, however, replaces the terms micro- and macro- with 'social integration' and 'system integration' (Giddens 1984, p. 139). Social integration indicates the *"reciprocity of practices between actors of co-presence, understood as continuities in and discontinuities of encounters"* (Giddens 1984, p. 376). System integration describes *"reciprocity between actors or collectivities across extended time-space, outside conditions of co-presence"* (Giddens 1984, p. 377). He then merges social and system integration by placing time and space at the very heart of agency and structure (Giddens 1984, pp. 110ff.). The dimensions of space and time are constitutive of the concept of structuration (Giddens 1984, p. 3). It is not possible to clarify why and how events happen without taking into account where and when they happen. The situated contexts of interaction, consisting of rules and resources, entail temporal as well as spatial dimensions (Giddens 1984, pp. 110ff.). Accordingly, Saunders (1989) points out that *"the problem of order can be conceptualized as the problem of how social systems are bounded or integrated over time and across space"* (Saunders 1989, p. 218).

While the dimension of time is generally acknowledged in the multi-level perspective, forming the x axis (see Fig. 4.1) of most studies, the dimension of space has not been explicitly elaborated in a theoretical sense. In fact, Grin et al. (2010a, p. 4) propose that the levels of niche, regime and landscape refer to different scales of structuration and not to spatial or geographical dimensions. Feeling the need to emphasise the 'non-spatial character' of the niche, regime and landscape levels may be in line with Urry, who argues that:

> "It is impossible and incorrect to develop a general science of the spatial [...]. This is because space per se has no general effects. The significance of spatial relations depends upon the particular character of the social objects in question" (Urry 1981, p. 458).

If one takes 'space' in a purely geographical sense, Grin et al. (2010a, p. 4) may be right. However, Giddens (1989, p. 276) who considers Urry's (1981, p. 458) argument to be the starting-point of his space–time reflections, describes space as a spatial, social and physical construct. Time and space do not influence interaction and structure from the outside. Spatial as well as temporal parameters are mobilised as aspects of rules and resources through and within interaction (Giddens 1984, pp. 116ff.). It is time and space that contextualise agency and structure.

According to Giddens' viewpoint, it is important to understand how niches and regimes are formed over time and how they are integrated in space through modes of interaction. At the same time, one must also be able to comprehend how niches and regimes form time and how they shape space within and throughout their co-evolutionary development. Based upon these assumptions, not only time but also space—being the intersection of the social, the spatial and the physical—becomes a contextual parameter of transitions and, thus, an important analytical dimension in theory building. Empirical studies within transitions research have implicitly demonstrated that space plays a decisive role, as they have generally described regimes through *spatial boundaries* and niches as innovative ideas first created in *local* projects (see Sect. 4.1.4).

Giddens' contribution to elaborating an understanding of time and space is widely acknowledged in social theory (Urry 1991, p. 160), although he has also been criticised for presenting ideas and concepts of time and space that are lacking in precision and rigor throughout his work (for example, Gregory 1989; Saunders 1989, pp. 223ff.; Urry 1991, p. 166). It is, therefore, not the aim of this work to apply Giddens' descriptions *en bloc* but to point out those notions that can help to make the dimension of space a central notion for analysing transition processes.

Taking my cue from Giddens (1984, pp. 110ff.), I hold that the co-evolutionary production and reproduction of niches and regimes is contextualised in time and space, along with the diversification of agency and structure. Again, it is important to note that the dimension of space is not to be seen as purely geographical but rather as a social, spatial and physical construct embedded in the rules and resources that are produced and reproduced in and through interaction.

1. The co-evolutionary production and reproduction of niches and regimes is contextualised in space (and time).

According to Giddens, the process of structuration is a patterning of presence and absence that links the components of social and system integration with each other (Giddens 1984, p. 16). Social integration is a process with a high level of co-presence or, at least, situations of high presence availability (concrete interaction through emails, phone calls, or letters), often involving face-to-face interactions in daily life. Meanwhile, system integration refers to the reproduction of system structures across time and space and is, therefore, characterised by a high degree of absence. Social relations are maintained between actors despite great time-space distanciation. Actors behave in similar ways at different times and in different places and can, therefore, relate their interactions to each other even in their absence (Giddens 1984, pp. 139ff.; Saunders 1989, pp. 218f.).

Giddens links social integration occurring in co-presence with system integration taking place under conditions of great time-space distance by introducing the 'locale' and the region (see also, Saunders 1989, p. 219): *"Locales refer to the use of space to provide the settings of interaction, the settings of interaction in turn being essential to specifying its contextuality"* (Giddens 1984, p. 118). Although these categories describe the basic spatial dimensions of face-to-face interaction, they vary in terms of their extension. Locales, for example, can include a room, a street corner, a town or a city (Giddens 1984, p. 118). They are *"the meeting place of social structure and human agency"* (Thrift 1983, p. 38); (see also, Saunders 1989, p. 230). Yet Giddens sharply differentiates the meaning of locale from the pure meaning of places by characterising them, in line with space, as the intersection of the social, spatial and physical (Giddens 1989, p. 280). Meanwhile, beyond the locale, regions are zones which are extended in time and space. Again, Giddens emphasises that regions as he understands them include more than physically bounded areas and are, rather, made of particular sets of rules and resources being reproduced across time and space (Giddens 1984, p. 118). Giddens (1984) argues that:

> "The connections between social and system integration can be traced by examining the modes of regionalization which channel, and are channelled by, the time-space paths that members of a community or society follow in their day-to-day activities. Such paths are

4.3 Developing a Framework for Analysing the Influence of Organisations...

strongly influenced by, and also reproduce, basic institutional parameters of the social systems in which they are implicated" (Giddens 1984, pp. 142f.).

Social integration cannot be separated from system integration because social systems are reproduced when interpersonal routines are preserved (Saunders 1989, p. 220). However, Giddens would be misunderstood if one concluded that system integration could be simply described through the aggregation of processes of social integration (Giddens 1984, p. 140). The act of regionalisation, the reproduction of individual interaction patterns across time and space, requires more than mere duplication. A region is formed through the creation of an 'ontological security' between actors at great space time-distanciation (Giddens 1989, p. 278). Ontological security means that actors feel secure in doing the 'right thing' in response to others, even when they are not present (Giddens 1989, p. 278). Social integration in locales are strongly characterised by particular individual interaction patterns that are taken for granted in everyday life. These patterns are regionalised when actors create a form of collective identification in absence from each other. Ontological security is one of the main drivers for the reproduction of the same patterns across time and space.

Lately, the interest in geographical aspects in transitions research has increased, and several conceptual contributions have been made (for an overview see Binz et al. 2014; Hansen and Coenen 2015). Hansen et al. (2015), for example, claim that:

"The geography of sustainability transitions captures the distribution of different transition processes across space" (Hansen and Coenen 2015, p. 95 based upon Bridge et al. 2013).

Accordingly transitions are not pervasive. Transitions are geographical processes that happen in concrete geographical locations (Hansen and Coenen 2015, p. 95). Späth and Rohracher (2010, p. 453) make an important contribution in showing that the two important niche development factors of visioning and learning can happen at different spatial scales. However, as mentioned earlier, space is still largely neglected in framework building within transitions research. At first sight, the local-global concept of niche development from Deuten (2003), Geels and Raven (2006), and Geels and Deuten (2006) seems to direct theory towards integrating spatial dimensions into transition analysis, as niche innovations develop once a global niche level stabilises local practices by providing standardised guidelines for the set-up of new projects. Local projects then start to follow the same rules being generalised from individual locations. It is not just the accumulation of similarly operating projects but the creation of generic rules and the establishment of a strong network—a cosmopolitan community—which comprise niche development. However, Geels and Raven (2006) emphasise that the concepts 'local' and 'global' should be understood as socio-cognitive concepts and not as geographical dimensions, degrading space again to an implicit connotation. Other work on transitions that acknowledges spatial specifications often only focuses upon two space-related levels—the local and/or the system—mostly national in kind (for example, Späth and Rohracher 2010, p. 453), creating a huge gap between local projects and the patterns that make up a system. In consequence, such work tends to empirically approach only one of these levels at a time.

Fig. 4.5 Analytical dimensions of space for visualising socio-technical transition processes

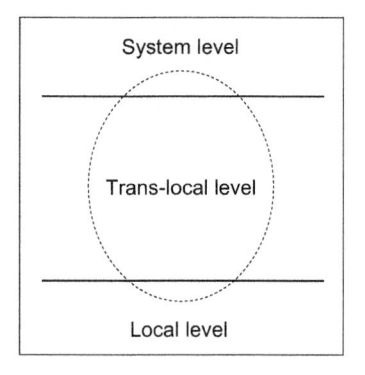

| System level |
| Trans-local level |
| Local level |

I have been arguing that the co-evolutionary production and reproduction of niches and regimes is contextualised in space (and in time) and that the interrelationships between agency and structure within a transition process can be coherently systemised based upon the dimension of space. For my purposes, three levels of space need to be analytically differentiated, although they are strongly connected in practice: the local, trans-local and wider system levels. Figure 4.5 visualises how the three levels of space are related.

The local is the level at which the creation of niche structures and the continual reproduction of regime structures are contextualised within social agency. Main interest with regard to this dimension is to understand how niches are created and how this interrelates with the simultaneous reproduction of regime structures in concrete situations of interaction.

The trans-local level is where analysis of agency patterns being formed via increasing space-time distanciation between actors takes place. The trans-local level stretches from the local to the system level, building a bridge between social and system integration. Main interest with regard to this dimension is to grasp the interactions of a particular actor group that makes up a cosmopolitan community and analyse how it reproduces established as well as innovative structures across time and space.

The wider system level shifts analytical attention towards the maximised expansion of rules and resources in time and space. Analysis of the wider system level is fully concentrated on understanding the institutionalisation of interaction patterns, resulting in the accumulated sum of regime principles in a maximised expansion over time and space.

2. Agency and structure are diversified in space.

In the following it is argued that niches and regime structures involve a high internal diversity which can be best grasped through their co-evolutionary development across space (and time). Giddens points out that: *"Social reproduction is uneven [...]"* (Giddens 1989, p. 277). The more different actor groups are involved in reproducing certain rule and resource patterns over time and space, the more these patterns are subject to local variation. Hence regimes and niches do not have an internal unity but comprise of a high diversity. Späth (2012, p. 474), one of the few scholar acknowledging the local within regimes, point out that when looking at regimes in 'high resolution', strong regime inconsistencies and variations can be identified (Späth and Rohracher 2012, pp. 475:466).

According to Giddens (1984, p. 376), regime structures can be described as collections of diverse, uneven and sometimes also conflicting structural principles—sets of rules and resources—which differ in their level of institutionalisation. With regard to the interests of the present work, such structures are involved in the overall institutional alignment of a socio-technical system. For example, 'the central production of fossil energy' can be described as a structural principle of the prevailing German energy regime, which is comprised of a set of rules (for example, regulations concerning large-scale centralised power production) and resources (for example, fossil power production technology). Central production of fossil energy has been reproduced in time and space to the extent that it basically stands for the current German energy system. Co-evolving niche structures can be described as rising structural properties that are becoming part of the system while also pointing beyond it, towards a new regime.

Why is conceptualisation of the spatial diversity of niches and regimes important when analysing actor influence within transitions? It is of great importance, because it allows the researcher to better locate actor influence within transition processes and can help to better approach questions such as: What rules and resources patterns are being challenged and how are they contextualised in time and space? What does it mean for desired transition outcomes?

4.3.3 The IBAST Framework

In this section, I propose the IBAST framework (see Fig. 4.6) as a means for systematically visualising interrelationships between agency and structure within transition processes. The framework is intended to guide analysis of the influence of innovative actors, in general, and innovative enterprises, in particular, within socio-technical transitions.

Interrelationships between agency and structure can be explored from a local, trans-local or wider system perspective. These three analytical levels of the IBAST framework are explained in detail in the following.

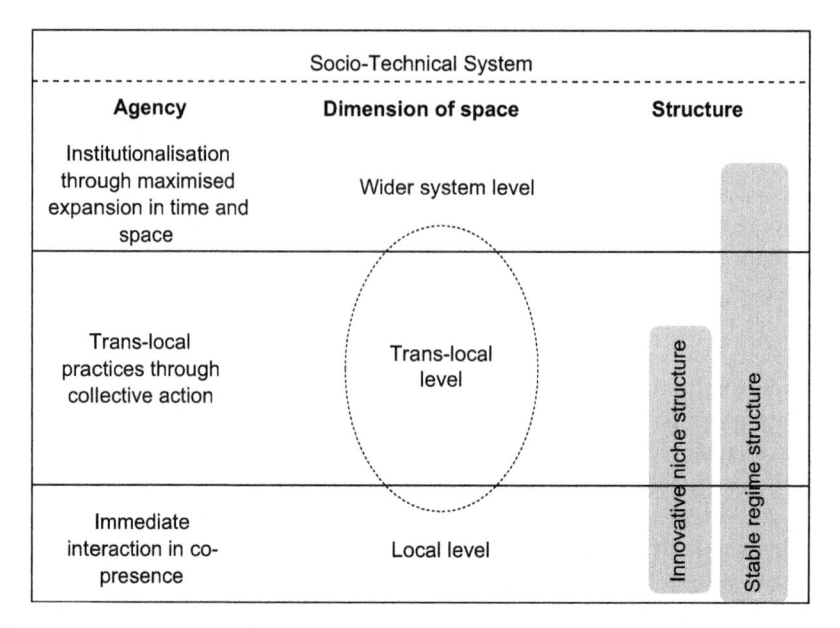

Fig. 4.6 The IBAST framework for systematically conceptualising interrelationships between agency and structure within societal transformation processes

4.3.3.1 Interrelationships Between Agency and Structure at the Local Level

Interrelationships between agency and structure at the local level refer to social processes with face-to-face interactions or with interactions of high co-presence (Giddens 1984, p. 118). Actors stand in direct contact with each other, making immediate interaction possible. Thus, the local level is a space of immediate encounter. It can be, for example, a room, a house, a village or a city. Innovative niche structures as well as prevailing regime structures are both contextualised at the local level, as actors produce or reproduce norms, interpretative schemes, as well as allocative and authoritative resources (Giddens 1984, p. 29). Exploring the influence of enterprises on structural change from a local perspective generally means analysing their concrete modes of immediate interaction.

In daily life, enterprises usually monitor and realise their interaction in a 'taken-for-granted-fashion' (Giddens 1984, p. 4). Day-to-day interaction relies on the implicit knowledge of practical consciousness rather than discursive reflection processes and, consequently, supports routines. Such routine guidance of action not only leads to continual reproduction of the same interaction patterns but also to the reproduction of societal, political and market structures in which action takes place (Giddens 1984, p. 5). In routine interaction, enterprises draw upon those norms and interpretative schemes that they already know and activate those resources that they already dispose of. One of the central ideas guiding this work is that prevailing

4.3 Developing a Framework for Analysing the Influence of Organisations...

regime structures need to be continually renewed in local, immediate and routine interaction, even though they may represent the underlying rule and resource patterns of a whole socio-technical system. Without continual reproduction at the local level, regime structures would simply not exist. Yet, only Späth and Rohracher (2010, 2012), Geels and Schott (2010, pp. 44ff.), and Essletzbichler (2012) belong to the few scholars who generally acknowledge that regime structures have a local character. A consequence of the spatial dispersion of localities is that regime structures vary strongly within their local instantiations, because actors within a system reproduce similar but not identical rule and resource patterns. Actors in one hospital, for example, may follow slightly different norms and guiding principles than actors in another hospital, due to very context-specific conditions. Citizens of one village may draw upon slightly different rules and resources in their daily lives than citizens of another village, again due to very context-specific conditions. Local regime diversity occurs due in part to the many different combinations of social processes at the local level. Local regime diversity is an important source for change, because it also increases the diversity of local interaction models (see also, Späth and Rohracher 2012, p. 467).

Knowledgeable and capable agents can interact in innovative ways, potentially with the consequence of transforming the very structures that originally gave them the capacity to act (Sewell 1992, p. 4). The development of innovative niche structures requires a change of routines. Innovative enterprises exert influence on prevailing regime structures in immediate interaction when they draw upon new rules and resources in their daily business practices. Precondition is that they become aware of those mechanisms leading to the continual reproduction of established interaction patterns over and over again (Schneidewind 1998, pp. 202f.). Practical routines can be made explicit through cognitive and discursive reflection on one's own interactions (Giddens 1984, p. 6). In addition, actors not only need to have the intention to act differently and must also be empowered to do so (Giddens 1982, p. 29; 1984, p. 7). Power arises from controlling resources as well as through obtaining formal authority and legitimacy (Powell and DiMaggio 1991). Power is always locally expressed.

When exploring the influence of actors, for example of enterprises, within transitions at the local level, it is therefore essential to gain an understanding of the conscious intentions of enterprises during interaction, to analyse in what ways they draw upon new norms and interpretative schemes while pursuing these intentions (visualising their knowledge), and to learn how and to what extent they mobilise new allocative and authoritative resources in order to implement their new business projects. New goals and strategies involve new guiding principles, and the creation of new projects may lead to the mobilisation of new allocative and authoritative resources (visualising their capabilities).

Immediate collaborative interaction models are an important entry point for analysing structuring enterprise forces at the local level (Schneidewind 1998, pp. 286ff.), as collaborative interaction is likely to be driven by intense and reflexive communication, in which prevailing interaction patterns may be discursively questioned, helping or even forcing involved actors to change their day-to-day modes of interaction. Critically constructive reflection enables collaborative parties to go beyond their own viewpoints and individual action spaces. Collaboration is, therefore, a central medium in which the creation of niche structures can be observed. In cooperation, firms may also be able to establish substantial resources that can empower them to follow newly created rules. The need for empowerment particularly applies to young firms, which may have normatively driven ideas but are not yet the position to pursue them. Levels of power in cooperation play a fundamental role in mediation between the differing rules and resources that are brought into a cooperative effort.

Understanding interrelationships between agency and structure at the local level means to analyse the unacknowledged conditions and unintended consequences of innovative interaction. They make the co-existence of local niche creation and local regime reproduction explicit, which have been widely neglected in transitions research (see Sect. 4.1.4). Examining the unacknowledged conditions and unintended consequences of action can provide valuable insights regarding the potentials and limitations of structuring forces of enterprises, because they turn immediate discursively reflected agency into a non-linear and unpredictable process. The reproduction of routines and the creation of innovative ideas occur simultaneously during moments of the exercise of agency. Since structure is always both medium and result (Giddens 1984, pp. 25ff.), it is simply not possible to develop new niche structures completely independent from regimes or to keep regime structures separated from niche structures. The mix of structural change and reproduction in concrete situations of interaction can neither be predicted nor controlled. This recursive connection within and between agency and structure makes the process of production and reproduction—the simultaneous formation of niche and regime structures in interaction—highly volatile.

In sum, innovative action is based upon the discursive and reflexive capacities of agency, the conscious intentions of the agent to act differently, and the agent's capacity to act differently. Primary niche development factors, such as visioning and second-order learning, can be also be characterised accordingly. Immediate (collaborative) models of innovative enterprise interaction, as well as aligned unacknowledged conditions and unintended consequences of action unfold within concrete spaces. Niche variety arises through different combinations of innovative interaction in space with different social, spatial and physical natural intersections. A fundamental requirement for exploring the influences of enterprises is, therefore, not only the analysis of immediate interaction but also characterisation of the local spaces involved.

4.3.3.2 Interrelationships Between Agency and Structure at the Trans-local Level

Interrelationships between agency and structure at the trans-local level have to do with how innovative niche structures and prevailing regime structures co-evolve across time and space. The reproduction of similar structures requires the reproduction of similar interaction patterns, which applies not only to prevailing regimes but also to innovative niches. From the perspective of actor influence within a transition process, a key interest is to understand how new rules and resources—first mobilised in local immediate interaction—can subsequently be reproduced beyond individual boundaries and how the trans-local recreation of niche structures goes in line with, or departs from, the simultaneous recreation of regime structures.

The reproduction of similar local interaction patterns in more than one place can be characterised as collective action across time and space, leading to the diffusion of certain rules and resources beyond individual locations—an important condition for system integration (Giddens 1984, pp. 139ff.). Collective action is not the simple multiplication of interaction. It does not happen automatically. Collective action needs to be formed by superordinate interaction processes, including networking and acts of trans-local aggregation. Socio-technological structures will only tend to become diffused if they are supported by a growing number of actors that actively unite across time and space. Networking has been identified as one of the most important niche development factors (Geels and Schot 2010, p. 82; Raven 2012, p. 139, see also Sect. 4.1.4.2). In this vein, Binz and Truffer (2011) argue that the geographical scale of structures follows along with social networks of actors involved in them, thus turning our attention towards social groups that are connected across different levels of spatial scale (see also, Coenen et al. 2012; Raven et al. 2012; Truffer and Coenen 2012). The spatial scale of transitions is, therefore, influenced by the extent to and ways in which actors develop relationships across space (Coenen et al. 2012, p. 977). Coenen et al. (2012, p. 976) argue that a closer look needs to be taken at the interrelationships between global networks and local nodes of transition processes, which can be done by setting interaction at the local level in relation to interaction at the trans-local level.

A cosmopolitan community emerges when actors start to share the same rules (Geels and Raven 2006, pp. 378ff.). Shared rules help actors in different places to identify themselves as one network, existing beyond the boundaries of individual locations. Shared rules can bring about the 'ontological security' which, according to Giddens (1984, p. 125), is desired in order to feel confident in 'doing the right thing' in response to others, even when they are physically absent (Giddens 1984, p. 125; 1989, p. 278). Creating ontological security is central for the emergence of system processes (system integration). Individual interaction patterns are therefore multiplied when actors create a collective identification in the absence of others. Geels and Deuten (2006, pp. 266ff.) point out that the creation of shared rules requires active aggregation, as rules need to be sufficiently abstracted and de-contextualised from the local level so that they can travel between locations (Geels and Deuten 2006, pp. 266ff.). However, the primary focus on the rule level with regard to the process of aggregation may be short-sighted, because it ignores the importance of power. In each new location, actors must also be empowered to act; otherwise they will not be capable of actually applying shared rules in real-life contexts. Domination of some actors over others is constituted through disposition of resources (Giddens 1984, p. 33).

Multiple actors apply shared rules by using similar resources. The understanding of collective action should, therefore, be extended as follows. Collective action means the creation of a shared identities and shared resource bases. Both are formed through ongoing selection processes taking place within a community with regard to which patterns of rules and resources are to be multiplied. Not all rules and resources mobilised and created at the local level in immediate interaction are reproduced across time and space. Only those which appear to work in different locations are adopted by more actors. Whether rule and resource patterns 'function' in multiple locations depends upon the knowledge and capabilities of local actors as well as upon locally concrete formation of regime structures (unacknowledged conditions of innovative action). Both of these vary across time and space, making the adoption of certain rules and resources in some locations easier and the adoption of other rules and resources harder. Those rules and resources that eventually become aggregated enough to form a multiplied pattern can then become a new structural property: a structured feature of a social system stretching across time and space (based upon Giddens 1984, p. 377).

Understanding interrelationships between agency and structure at the trans-local level requires analysing the manifestation of unacknowledged conditions and unintended consequences of innovative action in multiple places. The reproduction of interaction patterns across time and space detaches unacknowledged conditions and unintended consequences of action from their individually located manifestations. Certain regime structures provide the unreflected basis for the trans-local reproduction of new structural properties. Reproducing similar innovative sets of rules and resources across time and space also leads to the diffusion of similar intended and unintended consequences of action. It is the trans-local diffusion of structural properties through a whole community, based upon networking, a shared identity and a shared resource base, that characterises interrelationships between agency and structure at the trans-local level. The accompanied multiplication and manifestation of unacknowledged conditions and unintended consequences of action across time and space make the co-evolution of niche and regime structures, as well as of collective (innovative and routine) action into a highly uncontrollable and unpredictable process.

Exploring the influence of enterprises within transitions at the trans-local level means analysing the collective action of a newly rising cosmopolitan community. The subject of analysis at the trans-local level is, therefore, not the single enterprise within a community but the community as a whole. A key entry point for research here can be the identification and characterisation of a rising innovative enterprise group. Observing networking activities and remapping act of trans-local aggregation can reveal how such a community has emerged, with the main interest of such research being to identify dominant structural properties that make up the community. These rule and resource patterns are reproduced in local interaction by the majority of actors that belong to a particular trans-local group. Comparing the local activities of actors with those dominant rule and resource patterns that are eventually represented by the whole group can reveal to what extent the local influences of individual actors can be multiplied across time and space and to what extent they cannot be diffused because interaction is too focused on the particularities of local space. Uncovering unacknowledged conditions and unintended consequences of action aligned with the trans-local reproduction of dominant structural properties can give us important insights with regard to the potentials and limitations of the influence of whole groups, as well as to conflicts between their visions and their realisation.

4.3.3.3 Interrelationships Between Agency and Structure at the Wider System Level

Interrelationships between agency and structure at the wider system level include the reproduction of practices and structures over a widely extended span of time and space. Giddens (1984, p. 17) defines those practices that are most widely extended, and therefore embedded in the reproduction of societal totalities, as institutions.[9] Maximised expansion in time means that such forms of interaction are deeply integrated in social life, to the extent that aligned rule and resource patterns are continually reproduced over generations (Giddens 1984, p. 16). At this stage, interaction patterns have reached a level of embeddedness in the routines of daily life that makes it very hard to deviate from them. Maximised expansion in space means that similar interaction models can be found in many different intersections between the social, spatial and physical. Multiple actors that belong to many different societal groups in different places reproduce similar rule and resource patterns, including consumer groups, firms, politicians, scientists and citizens, (based upon Giddens 1984, pp. 110ff.). The process of institutionalisation, therefore, means the transfer and diffusion across great temporal and spatial distances of new practices to the many different societal groups that make up a system.

Regime structures are the medium for and the result of institutionalised interaction, representing the sum of the many different rule and resource patterns that organise the totality of a socio-technical system (based upon Geels and Schot 2010, pp. 49f.; Giddens 1984, p. 377). Regime structures influence and are influenced by multiple interactions in multiple social, natural and physical settings and stand for a whole system, because they create reciprocity between all the different actor groups that belong to a system through a state of maximised expansion of time and space outside conditions of co-presence. Many groups draw upon the same norms, the same interpretative schemes and therefore mobilise similar resources, leading to ontological security and a shared resource base across very great temporal and spatial distances. Once innovative structural properties have expanded so much through time and space that they have become part of the practical routines of the many different social groups that make up a socio-technical system, they can then become part of the overall existing regime structure, with the intensity of regime change depending upon the radicalism of co-evolved niche structures.

In order to explore the influence of innovative organisations on transition processes from a wider system perspective, it is crucial to reflect upon the development stage in which innovative niche structures actually begin to conglomerate into the prevailing regime. However, when analysing transition processes in the making, it is neither possible to make a final reflection with regard to the co-evolution of niche

---

[9]Jepperson (1991, pp. 143f.) points out that various understandings of institutionalisation and institutions exist. The definition provided by Giddens (1984, p. 17) complements the core notion of institutions used in general sociology, according to which an institution is an organized, established procedure (Jepperson 1991, p. 143).

and regime structures, nor can the overall contribution of certain actor groups be evaluated in a final sense, as it is highly uncertain in which direction system change is heading and what scales new interaction models and new rule and resource patterns may reach. The role of innovative enterprise groups in this process cannot be predicted with an acceptable degree of accuracy, though the understanding of institutionalisation drawn from structuration theory and transition theory outlined above can provide valuable entry points in order to assess the potentials of actors to support the institutionalisation of particular niche structures. The entry point for exploring actor influence within transitions at the wider system level taken here is twofold.

1. First, actor influence from a wider system perspective is constituted by a community's potential to maximise the diffusion of its interaction models. One indicator here is the growth dynamic of a particular enterprise community, while another important indicator is the community's capability to transfer its interaction patterns to other social actor groups, as only then will structures be diffused beyond the boundaries of the community. The transfer of similar interaction models to other groups does not happen automatically but, rather, requires that connections be created between community insiders and outsiders. A new innovative community needs to be able to create reciprocity between the insiders and outsiders of its enterprise network. The work of intermediate actors can provide valuable insights in this regard. Intermediate actors are boundary operators whose primary interest lies in creating a collective good (Geels and Deuten 2006, p. 267; Moss 2009, p. 1482). Their work fields are characterised by their 'in-betweeness' (Moss 2009, p. 1481). Intermediate actors can be organisations or individuals, such as firm associations, temporal societal committees, professional societies, or standardisation organisations (Geels and Deuten 2006, p. 267). According to the definition of Geels and Deuten, intermediate actors can also include journalists. The typical work[10] of intermediate actors includes standardisation, writing of handbooks, formulation of best practices, as well as writing and distributing general information, for example through published articles (Geels and Deuten 2006, p. 267). New innovative enterprise communities are much more likely to receive broad attention—indicating that their ways of thinking have left the internal boundaries of the community—if their activities become adopted, used and passed on by intermediate actors.

2. Second, from a wider system perspective, the degree of actor group influence is constituted by the level of long-term stability reached for the new practices that it seeks to foster. Widespread durability of influence over a long expanse of time is only possible if interaction patterns become stable. Despite being diffused in space, new models of interaction can remain unsteady and may vanish over time.

---

[10]As with any other actors, intermediate actors interact at the local level in places such as editorial offices of journals and newspapers, in seminars, conferences, strategic workshops or in political meetings, where interaction is exclusively targeted to enabling collective action (Geels and Deuten 2006, pp. 267f.).

4.3 Developing a Framework for Analysing the Influence of Organisations... 99

One valuable option for better understanding the process of institutionalisation at the wider system level within an ongoing transition process seems to be to observe emerging interrelationships between new structural properties and existing national regulations. National regulations represent regime principles which play a fundamental role in organising procedures at the wider system level. Important to note is that both increase as well as reduction of regulative support can provide valuable insights with regard to degrees of structural stability. Consequently, both continuing spread of new socio-technical practices despite reduction of legislative support, such as subsidies as well as the introduction of new legislation in order to provide legal guidance for new practices can be taken as signs of rising stability.

The more actors are involved in the reproduction of structure and the higher the time-space distanciation between them, the greater is the self-dynamic of collective interaction. Giddens (1995, p. 173) uses the example of the juggernaut of modern society in order to visualise such dynamics[11] (see also, Schneidewind 1998, pp. 400ff.). A juggernaut in Giddens' sense is a wild and very powerful vehicle that can only be collectively controlled to a certain extent by one or more groups of actors, as it continually defies control and risks smashing itself up (Giddens 1995, p. 173). A single actor may be able to discern the direction in which the juggernaut is travelling without being able to purposefully influence it (Schneidewind 1998, p. 401). The juggernaut can neither be stopped nor can it be forced to return to a previous point. However, it can be stabilised in a positive way (Schneidewind 1998, p. 413). Modern societal systems are similar to Giddens' juggernaut. High levels of time-space distanciation, leading to a great variety of interaction patterns, and the close, often unacknowledged as well as unintended links between these patterns make full control of modern society impossible. Consequently, innovative enterprise or other actor groups cannot completely strategically manage the expansion and stability of their new interaction patterns. However, they can play a decisive role in such expansion and creating stability for them, as they are the ones who support these interaction patterns.

4.3.4 Summary and Reflection

The aim of this section has been to conceptualise a more systematic and explicit perspective on the interrelationships between agency and structure within socio-technical system change in order to better understand and guide analysis of actor influence on transition processes. A new analytical framework—the IBAST—was elaborated by integrating the comprehensive and far-reaching notions of structuration theory into sustainability transitions research. Giddens' (1984) concept of the

---

[11] According to Giddens (1995, pp. 72ff.), modern societies are characterised by a very high level of space-time distanciation, leading to a great variety of interaction patterns.

Table 4.3 Core notions of the IBAST framework and its three analytical levels

Agency/Structure Dimension of space	Agency	Structure
Wider system level Space covering an entire system	**Institutionalisation** Transfer of trans-local practices to many societal groups that make up a system, across great time-space distanciation.	Maximised expansion and embeddedness of societal **structural principles** (regime structure).
Trans-local level Extendable space characterised by distanciation between actors	**Trans-local practices** Collective action across time and space based upon networking, a shared identity and a shared resource base.	Diffusion of **structural properties** (co-evolution of niche and regime structures) beyond localised time and space.
Local level Space of co-presence	**Immediate interaction** Both producing new and reproducing established norms, interpretative schemes, allocative and authoritative resources.	**Local contextualisation** of innovative niche and prevailing regime structures at the moment of interaction (co-existence of niche and regime structures).

duality of structure has been a crucial, but not yet well-elaborated, building block of transitions research. The IBAST framework seeks to close this conceptual gap by articulating a concrete understanding of the interrelationships between interaction and structure in transitions, based upon three dimensions of space—the local, trans-local and wider system levels—which are seen as intersections of the social, spatial and physical. Table 4.3 summarises the core notions of the three analytical levels of the IBAST framework.

The following aspects of how innovative actors can influence transitions are crucial to the IBAST framework:

1. Innovative niches as well as prevailing regimes are best theorised as societal structures consisting of rules and resources, expressing the structural dimensions of legitimation, signification and domination. Agency, niches and regimes are recursively connected to each other, and new innovative niche and prevailing regime structures co-exist in and through concrete agency. Change and reproduction of structure occurs simultaneously in immediate situations of interaction, taking place at the local level within families and between consumers, enterprises, politicians, scientists or any other actors.
2. Exploring actor influence within transitions from a localised perspective means trying to understand how actors exert influence on those structures that they immediately draw upon in their daily lives by (re)mobilising or (re)creating norms, interpretative schemes, and allocative and authoritative resources in concrete interactions. Innovative action takes place based upon discursive reflection, the agent's conscious intentions to act differently and the agent's degree of capability to act differently.
3. Actor influence on societal structures reaches a trans-local dimension when they contribute to the reproduction of similar interaction models beyond the boundaries of individual agency. From a trans-local perspective, actor influence is

4.3 Developing a Framework for Analysing the Influence of Organisations... 101

characterised by collective action across time and space. Collective action happens through networking as well as through the creation of shared identities and shared resource bases within rising communities. This takes place through acts of aggregation. A key object of interest here is understanding how and why new rule and resource patterns, at first mobilised through local immediate interaction, are later selected, aggregated and reproduced by whole actor groups beyond the local level. The trans-local level builds an analytical bridge between social integration and system integration, between changing structures in individual and local agency and the changing structures of a whole system.
4. System change happens when innovative niche structures are reproduced in and through similar models of interaction by multiple social actor groups and in multiple social, spatial and physical settings over time, to the extent that aligned new rule and resource patterns become part of the practical routines that characterise a whole socio-technical system. The success of transition processes are, thus, aligned to with the transferability of new but similar (collaborative) local interaction models to multiple localities and multiple groups over time.
5. A crucial point here is that niche and regime structures continually co-evolve across time and space and co-evolution of niches and regimes is best understood through examining the unacknowledged conditions and unintended consequences of action, as intentional action is always accompanied by unintended consequences and unknowledged conditions. They make the mechanisms of the co-existence of local niche creation and local regime reproduction more explicit. They are responsible for the highly non-linear development of transitions.

I argue that the IBAST framework can capture the richness of actor interaction within the process of system change and aid in the identification of the relevant variables that need to be analysed in order to assess the influences of social actors, in general, and innovative enterprises, in particular, within transition processes. Important to note is that differentiating three dimensions of space (local, trans-local and wider system levels) helps to analytically reveal interrelationships between agency and structure with the intention of understanding actor impacts within transitions. In practice, however, the spatial dimensions within a system, as well as correlated dynamics of interaction and structural change, are highly recursively connected. None of the three levels can thus be seen as fully separate and autonomous areas of the social. Consequently, what takes place at the wider system level is the result of interaction at the local level. At the same time, local interaction is guided by processes at the trans-local and wider system levels. An ontological hierarchy between the levels does not exist, and the dynamics being analytically differentiated through the IBAST framework often occur simultaneously within transitions. As a consequence, niche and regime structures need to be seen as consisting of multiple rule and resource patterns with different institutional intensity grades leading to structural diversity within a socio-technical system.

In line with Schneidewind (1998), I have sought to draw a positive picture of the enterprise as a structuring actor within socio-technical transitions. Instead of focussing on how structures may generally constrain and limit firm activities, however, the present work attempts to systematically describe how innovative enterprises may exert influence on structures within the processes of system change. However, actor influence should not be understood as a one-dimensional and uni-linear process, as the co-existence of unacknowledged conditions and unintended consequence of action underlines that it is not possible to completely and confidently guide and control a fully-fledged transition process. Yet, actors do have the power to influence the direction of transition dynamics, as structural change happens through interaction. However, theories seeking to assume a mediating perspective on the micro- and macro-levels of analysis regarding agency and structure need to cope with the risk of being applied in lop-sided ways (as shown in Sect. 4.2.2, structuration theory has been criticised in this regard). This risk is even higher when investigation concentrates on development processes that are still in the making, such as an ongoing transition. When using the IBAST framework, particularly within contexts of normatively driven desired transition outcomes, it is therefore important to take into account the challenge of taking a levelled perspective between the uncontrollable character of transitions, on the one hand, and the structuring forces of actors, on the other. A crucial aspect of the research being presented here, then, is critical reflection upon the identified potentials and limits of actors against the background of the enabling and constraining character of the structures (and agency) within which they operate.

Chapter 5
Research Design and Methods

In this chapter, I outline the research design and empirical methods applied for the work being presented here. The research design refers to how the research topic was empirically approached, whereas the methods are the procedures used for collecting and analysing my empirical data (Creswell and Plano Clark 2011, p. 53). The core research question that needs to be addressed is the following: How can we assess—in a systematic and holistic sense—the influence of energy cooperatives on Germany's energy transition?

My research design is based upon the theoretical model elaborated in Sect. 4.3, which seeks to demonstrate how the interaction of rising innovative actors and actor groups can be systematically related to the direction and scope of ongoing sociotechnical transitions. With this model, the impact of innovative enterprise groups on socio-technical system transitions is analysed by examining the dynamic interrelationships between agency and structure at three different system levels: the local, trans-local and wider system perspectives. With this in mind, the core research question has then been divided into several sub-questions.

From a local perspective, my main interest is to understand the impact of energy cooperatives on their immediate local environment. In order to achieve a detailed understanding about interaction that can be said to be typical for the whole group, my analysis focuses on collaborative interaction models that are used by the majority of energy cooperatives in order to observe whether and how dominant collaborative interaction models create new patterns of rules and resources and what effects they may have on their immediate environment. Consequently, my main sub-questions are:

(a) How do energy cooperatives activate new rule and resource patterns in collaboration with others in their immediate environment?
(b) What aligned potentials and boundaries can be observed?
(c) What are important preconditions for immediate interaction at the local level?

From a trans-local perspective, the analytical interest lies in understanding the status quo and the dynamics of the whole organisational group of energy cooperatives, asking whether and how they have emerged as a new cosmopolitan community (Geels and Deuten 2006, p. 269). The aim here is to understand the consequences of their collective action that takes place across time and space. Therefore, the relevant sub-questions are:

(d) What has the growth dynamic of energy cooperatives been during the last years?
(e) What are main business goals of the enterprise group?
(f) What kinds of substantial resources has the group of energy cooperatives collected as a whole so far?
(g) Who have been their main collaborative partners and associated actors? How are they seen by community outsiders?

From a wider system perspective, the research interest here lies in achieving a better understanding of levels of institutionalisation, with the sub-question here being:

(h) In what ways and to what extent are individual and collective action related to dynamics that take place outside the enterprise group?

In the end, an aggregated conclusion from analysing energy cooperatives at three different system levels is drawn in order to answer the main research question. Figure 5.1 provides an overview of the research design guiding the work done for this work.

Having an understanding of the preferred future energy system is an important conceptual guideline for the whole book. Then, focussing on the desired outcomes of Germany's system transition and its challenges seems quite relevant for shedding light on the impacts of energy cooperatives on country's energy system, because it allows the setting of prevailing energy visions in relation to the current activities of such organisations. Regarding these points, I have already provided an overview of Germany's current energy system and answered the following questions in Chap. 2:

- What is the current status of Germany's ongoing energy transition?
- What are Germany's long-term energy system goals?
- What challenges exist related to achieving the envisioned energy system change?

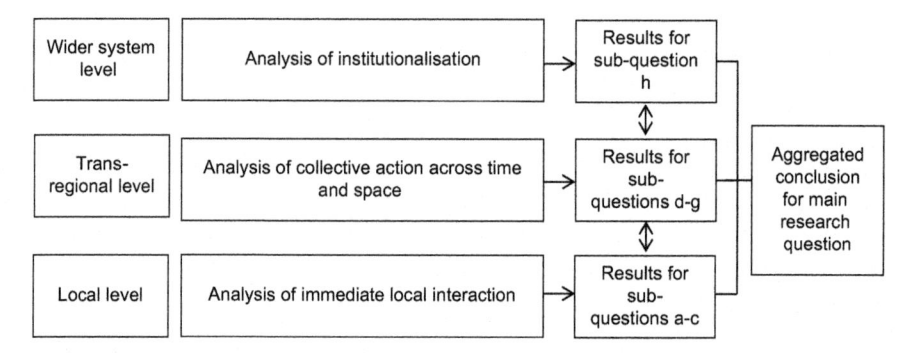

Fig. 5.1 General overview of the research design applied in this work

5.1 Mixed-Method Research

Due to the chosen approach of different assessment levels, it is not feasible to apply just one research method. Thus, a mixed-method approach has been taken, involving the assessment of quantitative and qualitative data. According to Creswell and Plano Clark (2011), a mixed-method research approach:

> "Focuses on collecting, analysing and mixing both quantitative and qualitative data in a single study or series of study. Its central premise is that the use of quantitative and qualitative approaches, in combination, provides a better understanding of research problems than either approach alone" (Creswell and Plano Clark 2011, p. 2).

The aim of a mixed-method approach is to overcome the boundaries of single quantitative or qualitative methods and to observe a research topic from many different angles so that scientific knowledge generation may be increased (Flick 2011). In line with my research design, the mixed-method approach applied in this book involves two steps during which quantitative and qualitative data were collected and analysed one after another. In a third step, the results were assessed together (Creswell and Plano Clark 2011, pp. 81ff). In this way, the advantages of quantitative and qualitative data were combined such that, for example, results from a large-sized sample like the identification of trends could be set in relation to in-depth and explorative insights from smaller samples derived from empirical study.

My assessment began with a quantitative assessment of secondary data in order to achieve a detailed overview of the field of energy cooperatives in Germany. Large sets of secondary data were accumulated and examined (Creswell and Plano Clark 2011, p. 8; Kühl et al. 2009, pp. 14ff.), including the number of registered energy cooperatives up to 2015, their business goals, their dominant collaborative partners, installed renewable power production capacity, as well as the development of their members and their capital over a timeframe of 3 years. Data analysis involved the categorisation and explanation of variables, as well as the identification of possible causalities between them (Creswell and Plano Clark 2011, p. 8; Kühl et al. 2009, pp. 14ff.). In a second step, a detailed qualitative case study analysis was conducted. The aim here was to assess in-depth the activities of energy cooperatives and their partners, aligned effects on their immediate environment, as well as important pre-conditions for their business development. During this phase, my empirical research needed to be able to unfold the multifaceted—and to a great extent underobserved—interrelationships between energy cooperatives, their partners, and the creation of new rule and resource patterns. Qualitative case study analysis is considered to be a well applicable method for 'diving' into a system, because it enables researchers to closely explore social action and social structure in their natural settings (Baxter and Jack 2008, p. 556; Orum et al. 1991, p. 7). Qualitative case studies are a generic method that captures the *"complexity of social behaviour"* (Gerring 2007, p. 4) by observing individual cases in great depth. The explorative and holistic approach of case studies makes them particularly suitable for analysing contemporary system transition processes. Three case studies were conducted,

Step 1 — Quantitative assessment of secondary data

Data collection
- Identification of all operating energy cooperatives in Germany until 2015
- Identification of their business goals
- Identification of actor network composition
- Assessment of installed power production capacity
- Assessment of their membership and capital development over a timeframe of three years

Data analysis
- Applied descriptive statistics, categorised variables

Step 2 — Qualitative case study analysis

Conduction of three detailed case studies – one for each identified dominant collaborative interaction model being favoured by energy cooperatives

Data collection
- Execution of 15 semi-structured interviews
- Review of regional newspaper articles
- Additional field research

Data analysis
- Qualitative content analysis

Step 3 — Concluding assessment
- Interpretation of combined quantitative and qualitative results
- Discussion of relationships between quantitative and qualitative results
- Complementation and validation of findings
- Reflection of theory construction

Fig. 5.2 Mixed-method research approach followed for assessing the influence of energy cooperatives on Germany's energy transition

selected based upon the results of the prior quantitative assessment. In a fourth step, results of both assessment steps were merged into a concluding analysis where the findings were analysed together, complemented, validated and contrasted (Creswell and Plano Clark 2011, pp. 62f.). The combined interpretation step also helped me to recursively reflect upon the theoretical framework I was applying. A high level of interaction between quantitative and qualitative data was therefore supported throughout the complete research design. Figure 5.2 provides an overview of the chosen mixed-method approach that I adopted.

5.2 Quantitative Assessment of Secondary Data

Quantitative assessment of German energy cooperatives was focused on the collection and analysis of secondary data. In the following, each data type is outlined and each step of data collection explained. Then I outline how the data was analysed.

5.2.1 Data Collection

5.2.1.1 Identification of Operating Energy Cooperatives

At first a thorough assessment was conducted of all energy cooperatives that were officially operating in Germany by the end of 2015. Organisations were identified based upon their being entered in the German register for cooperatives, which is a valid data source because it lists all cooperatives that are officially registered at one of the German registry courts. Registration is compulsory for each German cooperative after its foundation (§10; §11 GenG). The registry, which is administrated by the federal government and the German Bundesländer (federal states), is publically accessible via the internet at https://www.handelsregister.de/. By 31 December 2015, 1055 registered energy cooperatives had been identified.

The assessment was conducted as follows: Registration lists were extracted for each registry court through online register access. The lists included all registered cooperatives, because it is not possible to automatically select cooperatives according to their business purpose. In these lists, energy cooperatives were identified based upon three information sources:

1. Their business name, as most cooperatives use their business name in order to indicate their business approach. Names of most energy cooperatives already involve the word *Energiegenossenschaft* (energy cooperative). At the same time, other non-energy cooperatives are often named *Raiffeisen-* or *Volksbank*, referring to a cooperative bank, or *Wohnungsgenossenschaft*, referring to a residential building cooperative.
2. The business approach published in the registry, which lists the business approach of each cooperative, such as stated in its statutes.
3. Secondary information that is available online. If the registry did not provide enough information, secondary information was drawn upon. Core information sources consisted of cooperatives' business websites, press releases or published articles.

With this method, the business approach of all cooperatives was identifiable, and there was no cooperative for which the operating sector remained unclear. During this process, core data was extracted for each energy cooperative from the registry into my own data list that included the business name, registration number, registry court, firm headquarters and registration date.

Full assessment of the German registry for cooperatives was conducted two times in order to ensure that all energy cooperatives were identified. The first full review was done between October and November 2013 and the second in December 2013. Since registry announcements are usually published a couple of days after the day of a cooperative's official registration, a third review focused on cooperatives registered at the end of 2013 and was conducted in February 2014. A fourth review with a focus on energy cooperatives registered in 2014 was conducted in May 2015 and a last review with a focus on energy cooperatives registered in 2015 was conducted in July 2016. Register entries can be ordered chronologically according to their registry numbers so that newly registered energy cooperatives are clearly identifiable, without having to re-examine all previous entries.

Earlier assessments applied a similar empirical approach for identifying energy cooperatives, such as Maron and Maron (2012) and Holstenkamp and Müller (2013, pp. 5ff.). However, as I have noted in Sect. 2.2, these studies had different research focuses.

5.2.1.2 Identification of Business Goals

In a next step, the main business goals for all registered energy cooperatives were identified. Core business goals are business activities or projects that have been either factually realised or actively planned. Business goals were mainly determined based upon information derived from cooperatives' official internet websites, as 68% of all registered energy cooperatives have a website wherein they describe their business purpose, as well as their planned and implemented projects. In addition, most energy cooperatives provide a link to their statutes, which have to contain a description of their business approach (§6 GenG). If no website was available, or if the website did not contain enough information in order to identify the cooperative's core business goals, I drew upon third-party information available online, such as press-releases, newspaper articles or information from municipalities and districts. In some cases, cooperatives were contacted by phone. In this way, the business goals of 98.8% of all registered energy cooperatives were identified. Only for 12 organisations did their business approach remain fully unknown to me. Business goals were reviewed twice, seeking to ensure that the information was correctly interpreted.

Subsequent assessment steps concentrated on energy cooperatives whose main business goal is the production of renewable energy, which at 73% represent quite a large majority of registered energy cooperatives. I have, thus, considered it sufficient to focus on this group in order to better understand the influence of energy cooperatives on Germany's energy system, especially as it is easier and more revealing to compare collected data for energy cooperatives with similar goals.

5.2.1.3 Assessment of Cooperative Members and Financial Capital Development

Membership and financial capital development were analysed for renewable energy production cooperatives over a timeframe of 3 years, from 2010 through 2012. For this, data was collected from annual financial statements. Financial statements are a valid data source in order to assess the development of capital and members because German cooperatives are required to publish them annually, at the latest 12 months after the end of a business year (§325; §339 HGB). All available financial statements covering the business years 2010, 2011 and 2012 were analysed. Published financial statements are accessible via the national electronic registry (*bundesanzeiger*; www.bundesanzeiger.de), which is administrated by the German Federal Ministry of Justice and Consumer Protection. The assessment included all balance sheet figures, including equity and borrowed capital, profit and loss, short- and long-term invested capital, as well as number of members and number of members' shares. Relevant data from each financial statement was then extracted into my own separate data sheet.

As of 13 April 2014, 1021 financial statements for 2010, 2011 and 2012 had been published by all renewable energy production cooperatives that were registered in or before the respective business year. My assessment covered all 1021 of these financial statements. Table 5.1 displays the number of published financial statements for each observation year in relation to the number of registered renewable energy production cooperatives as well as showing how many financial statements included additional membership information.

As can be seen, the majority of organisations followed the legal requirement of publishing their annual reports, and between 79% and 92% of all renewable energy production cooperatives that were registered in or before the respective business year published financial statements for 2010, 2011 and 2012. Meanwhile, between 69% and 80% of all renewable energy production cooperatives that were registered in or

Table 5.1 Financial statements from 2010, 2011 and 2012 that were published by registered renewable energy production cooperatives by 13 April 2014

Number of published financial statements as well as number and share of renewable energy production cooperatives	Business year			
	2010	2011	2012	Total
Total number of renewable energy production cooperatives that were registered in or before the respective business year	249	407	541	
Total number of published financial statements from registered renewable energy production cooperatives	229	366	426	1021
Respective share of renewable energy production cooperatives that published financial statements	92%	90%	79%	
Total number of published financial statements that included additional membership data	199	325	374	898
Respective share of renewable energy production cooperatives that provided additional membership data	80%	80%	69%	

The table was first published by the author in (Debor 2014, p. 7)

before the respective business year provided additional membership information in their financial statements. In general, cooperatives must specify membership data in the appendixes of their annual reports (§ 338 HGB).

Financial statements were analysed in two assessment rounds in order to inhibit errors. The first assessment was done between October 2013 and February 2014. Since most business years last from January to December, the latest possible date for publishing an annual report for 2012 was 31 December 2013. However, some energy cooperatives tend to publish their financial statements after the deadline. Therefore, the second assessment round was conducted in April 2014, in order to maximise the collection of available reports for 2012.

5.2.1.4 Assessment of Installed Renewable Energy Production Capacity

Assessment of the renewable energy production capacity for all renewable energy production cooperatives was also carried out. This was not difficult, as most energy cooperatives list their realised projects on their websites, and project achievements are often summarised in articles or press releases.

However, only data on photovoltaic power was finally used for this book in order to formulate conclusions about energy capacities installed by energy cooperatives. Collected data on solar power is representative, as solar power-producing energy cooperatives tend to provide satisfactory information about their projects as well as detailed data for each installed technology plant. Consequently, capacity could be identified for 357 organisations, representing 66% of all registered solar power-producing energy cooperatives.

Many energy cooperatives that produce biomass energy do not list their project capacity online. Since most of them also operate their own small district heating grids, it is assumed that their activity radius is focused on those actors that are connected to the grid, which is usually only a couple of kilometres long. Due to their very small action area, they may predominately use local and face-to-face communication channels instead of online based information platforms. Nonetheless, some data was collected from these cooperatives and added to the analysis if seemed to complement the results on solar capacity.

Energy cooperatives that produce wind power are often partial shareholders within large wind parks, and only a few have provided easily accessible information about the concrete size of their shareholding position. Hence, it was often not possible to identify the respective capacity share for which a particular energy cooperative would have been responsible, and the capacity of a complete wind park, which is often quite large, is not representative of energy cooperatives. Hence, collected data on wind capacities were not used in the analysis for this book.

In consequence, data on solar power capacity were primarily used for further research, complemented by data on biomass capacities. Since, with a 70% share, solar power-producing energy cooperatives represent the great majority of renewable energy production cooperatives, and since the desired capacity for achieving a 100% renewable energy system is separated according to energy resources, it is sufficient to focus on solar power and, to some extent, on biomass energy capacities in order to assess the share of energy cooperatives in installed renewable energy in Germany's energy system.

5.2.1.5 Identification of Associated Actors

In order to observe the influence of energy cooperatives on system structures, it is important to shed light on their collaborative partners and to gain an overview of additional actors who can be associated with the enterprise community.

A basic actor network analysis was conducted for identifying network composition around energy cooperatives in Germany. Actor network analysis has been applied within many different theories, with many different intensity grades (for example, Jackson 2011; Scott 2013; Wasserman and Faust 1994). However, three aspects seem to be fundamental for defining a social network: (1) Actors of a network interact; (2) their interaction is visualised through structural linkages between them (Wasserman and Faust 1994, p. 4); and (3) their relationships influences one another's behaviours (Knoke and Yang 2008, p. 3). For this book, a basic network analysis was deemed sufficient.

Collaborative partners were identified for all renewable energy production cooperatives registered by December 2014.[1] Collaborative partners are those actors that have a particularly intense and influential relationship with, in this case, energy cooperatives through direct interaction on a continuous and long-term basis. Identifying collaborative partners can shed light on collaborative interaction models that exist within the community and based upon which energy cooperatives may primarily realise their business activities. Collaborative partners of energy cooperatives were identified by concentrating on two groups:

1. Founding partner(s): Founding partners strongly influence energy cooperatives because they have guided and supported their foundation process and, thus, usually play a crucial role in developing the organisational set-up. In many cases, founding partners are the ones who actually had the idea of creating an energy cooperative to begin with.
2. Members of a cooperative's steering board: The executive board and the supervisory board represent the operational heart of an organisation and, thus, hold powerful positions. Next to the general assembly, they are responsible for all business activities and strategic decisions. To a great extent, it is the steering board which shapes the character of an energy cooperative.

---

[1]The actor network analysis was done in mid-2015, so all renewable energy production cooperatives registered through December 2015 could not yet be included. Hence, I decided to focus on all cooperatives registered until the end of 2014. Results from 2014 seem to be sufficient, since only 42 additional organisations were registered in 2015.

Official company websites represent the main data source for identifying founding partners and steering board members. Many energy cooperatives describe their foundation history on their website and clearly name their initiators, founding members or founding supporters, along with a description of the company. Management and supervisory board members are usually listed with their personal names on separate subpages. The leaders of the executive and supervisory boards are, further, listed under a website's legal notice. In cases where company websites did not exist, or if they did not provide enough information, I drew upon press releases, newspaper articles or information from municipalities and districts. Especially press releases or articles about a cooperative's foundation often provided sought-for information about initiators, founding partners and named the members of newly elected management and supervisory boards. In most cases, company websites or third-party sources further stated whether a cooperative's steering board members also represent other companies, associations or political committees. In this way, I was able to identify whether an energy cooperative was managed by private persons and/or by other outside organisations. If the professional background of steering board members was unclear, additional internet research was undertaken, based upon their personal names. If they had a leading position, for example, if they were managing director of a company or mayor of a city, they were categorised according to their professional background. It was assumed that actors with leading positions most likely act professionally. If they did not have a leading position, or if no clear reference was made to another organisation or institution on the cooperative's website or in articles, it was assumed that such individuals acted privately.

Table 5.2 indicates the share of German renewable energy production cooperatives for which founding partners and/or steering board members could be identified.

As can be seen, for 512 renewable energy production cooperatives it could be identified whether or not they have founding partners—representing 68.7% of all renewable energy production cooperatives that were registered by 31 December 2014. For 233 organisations, however, it remained unclear who the founding members were. Members of the steering board could by identified for 517 organisations—representing 69.4% of all registered renewable energy production cooperatives, leaving 228 unidentified. It should be noted that a representative tendency for certain actor groups is sufficient in order to provide insights into the overall network composition and power constellations that make up the cosmopolitan enterprises

Table 5.2 Share of German renewable energy production cooperatives for which founding partners and/or steering board members could be identified

Energy cooperatives	Unit	Figure
All German renewable energy production cooperatives registered through December 2014	Number	745
Cooperatives for which founding partners could be identified	Number	512
	Percent	68.7%
Cooperatives for which steering board members could be identified	Number	517
	Percent	69.4%

community around energy cooperatives. Hence it was not crucial to identify the existence of collaborating partners for all energy cooperatives.

In a second step of the actor network analysis, intermediate actors were identified who could be directly associated with the enterprise community. As outlined in Sect. 4.3.3.3, intermediate actors are boundary operators whose primary interest lies in creating a collective good through, for example, standardisation, formulation of best practices, as well as distributing general information and representing the interests of a particular group against others (Geels and Deuten 2006, p. 267; Moss 2009, p. 1482). The work of intermediate actors provides valuable insights into the degree of unification and strength of the overall group. Intermediate actors were identified via web-based research as well as being part of the general empirical work carried out for the work.

In a third step, actors were identified who were not part of the community but were still beginning to acknowledge and officially value the existence and activities of energy cooperatives. The aim here was to analyse how energy cooperatives have been able to achieve respect from community outsiders for their ways of thinking and operating. This approach seems well-suited for providing indications of the transfer of new rule and resources patterns into other societal groups. I primarily focused on two actor groups that particularly function as multipliers and opinion leaders in society: (1) Governmental representatives and politicians and (2) societal intermediate associations. It was not the aim to provide a complete list of all group representatives who recognise the benefits of energy cooperatives but, rather, to demonstrate that such recognition exists more generally. It was, therefore, not considered necessary to conduct this step in such a detailed manner as when identifying collaborative partners and associated actors. Spot samples of quotes, statements or articles of key actors belonging to the two groups were deemed sufficient in order to identify and evaluate supportive tendencies and increasing respect from outside of the enterprise community.

5.2.2 Data Analysis

The collected secondary data sets were analysed using basic descriptive statistics, and ranges were formed for the categories of business goals, number of mobilised members, amount of allocated capital, installed solar capacity, and collaborative partners. The distribution of all analysed organisations along these categories provided a comprehensive overview of the whole group of German energy cooperatives, in general, and renewable energy-production cooperatives, in particular.

Furthermore, number of members, amount of total capital, amount of member shares and installed solar capacity were extrapolated for the total number of renewable energy production cooperatives registered by 31 December 2015. For each category, the extrapolation was done as follows:

- First, a mean value was calculated for each range. For example, the category members was divided into seven ranges: (3–100 members, 101–200 members, 201–300 members, 301–400 members, 401–500 members, 501–700 members, and above 700 members). For each range, a mean value was calculated which covered all cooperatives which belonged to that range in 2012. For example, the mean value for all cooperatives which had 3–100 members was 51.9 members in 2012.
- Second, the distribution of energy cooperatives in each range was related to the number of energy cooperatives registered through 2015. For example, 50% of all renewable energy production cooperatives registered through 2012 had between 3–100 members. Accordingly, 389 of the 773 renewable energy production cooperatives registered in 2015 most likely had between 3–100 members.
- Third, the mean value of each range was multiplied by the respective number of renewable energy production cooperatives registered in 2015. For example, in 2015, 389 renewable energy production cooperatives had about 51.9 members.
- Fourth, the calculated number of energy cooperatives for each range was added.

5.3 Qualitative Case Study Analysis

In order to assess the qualitative influence of energy cooperatives on Germany's changing energy system, a detailed case study analysis was conducted for each dominant collaborative interaction model being favoured by the majority of renewable energy production cooperatives in order to realise their business activities.

Prominent case study designs have been elaborated by (Baxter and Jack 2008; Gerring 2007; Stake 1995; Yin 2009), amongst others. This book follows the approach of Yin, who defines case studies as a form of:

> "Empirical inquiry that investigates a contemporary phenomenon in depth and within its real-life context, especially when the boundaries between phenomenon and context are not clearly evident" (Yin 2009, p. 18).

Accordingly qualitative case studies focus on the analysis of social processes observed in individual cases. This definition emphasises that case studies acknowledge the strong interrelationships between phenomenon and context. They help to visualise and understand complex dynamics, as well as interrelationships, between actors and their environment. Due to their strong explorative approach, case studies generally take a more holistic approach than quantitative methods and draw upon information from a large variety of data sources, such as group discussions, interviews, field observations or secondary data, often over a longer time period (Orum et al. 1991; Yin 2009, p. 20). As Baxter and Jack propose:

> "[This] ensures that the issue is not explored through one lens, but rather a variety of lenses which allows for multiple facets of the phenomenon to be revealed and understood" (2008, p. 544).

5.3 Qualitative Case Study Analysis

In this way, case studies are capable of providing a richness and depth to the description and analysis of the kinds of events that constitute social life (Orum et al. 1991, p. 6) that often cannot be provided by other research approaches, such as surveys. Missing comprehensive knowledge about energy cooperatives and their partners, the in-depth research interest with respect to exploring their activities as well as the close interrelationships between energy cooperatives and their environment made a case study analysis to the most applicable research method in order to assess the qualitative influence of energy cooperatives within Germany's changing energy system.

As noted by Baxter and Jack (2008, p. 565), prominent case study approaches, such as Yin (2009), draw upon the constructivist paradigm, according to which reality is a social construct and must, therefore, always be seen in relation to the humans who have created its meaning (for example, Berger and Luckmann 1966). The strength that this paradigm brings to doing case studies is that it permits the researcher to reveal social action in ways that come very close to the ways it is understood by the involved actors themselves (for example, Orum et al. 1991, p. 8). Case studies are thus able to bring to light the impacts of underlying beliefs, guiding principles, as well as personal agendas on social action in an effort *"to grasp the total complex world of social action [and social structure] as it unfolds"* (Orum et al. 1991, p. 9). This appears to be particularly important in order to comprehend transition dynamics between actors in socio-technical systems.

Case studies can be designed for single or multiple cross-case applications (Yin 2009, pp. 46ff.). Here, I have chosen to pursue a multiple cross-case study, in which three different cases—one for each dominant collaborative interaction model identified during the quantitative study—are jointly compared, analysed, and assessed. The identification of similarities and replication among several cases is considered to be a convincing means for indicating the existence of general phenomena (Yin 2009, p. 55). Thus, the results of multiple-case studies are often considered to be more robust than examination of single cases (for example, Herriott and Firestone 1983).

According to Yin (2009, pp. 27ff.), the following five steps are especially important in order to conduct qualitative case studies:

(1) Defining the research question: Sub-questions a–c, presented at the beginning of this section, are to be answered through the case study analysis applied in this book.

(2) Defining propositions if needed: Since an explorative case study approach was chosen, defining propositions was not applicable.

(3) Defining the unit of analysis and the means of collecting data: The unit of analysis and the ways in which data was collected are explained in detail in Sects. 5.3.1 and 5.3.2.

(4)–(5) Analysing data and interpreting case study findings: The data analysis approach is explained in Sect. 5.3.3.

5.3.1 Selection of Case Studies

Each case study is *"a complex entity located in its own situation"* (Stake 2005) and, if one chooses to analyse a particular case, one always also needs to study its context. Hence, one of the most crucial aspects in qualitative case study research is defining the unit of analysis and selecting the case. The unit of analysis can, for example, be individuals, small groups, organisations or partnerships (Yin 2009, p. 33). Cases can also be very different in kind, representing for instance typical, diverse, extreme, deviant, influential, most-similar, or most-different examples (Seawright and Gerring 2008, p. 297).

Collaboration between energy cooperatives and their dominant partners is the unit of analysis in all three cases, which were selected based upon the results of the quantitative assessment of business goals, network composition, power capacities and the development of members and capital. The aim in case selection was to choose typical, as well as powerful, examples for the whole niche community. Three dominant collaborative partner groups were identified during the quantitative assessment phase. As outlined in Sect. 6.3, 72% of all analysed renewable energy production cooperatives collaborate with other energy-related companies, banks or communities in order to realise their own business. One case study was chosen for each of these three dominant collaborative interaction models, and the cases are considered to represent concepts that are *typical* for the community, enabling references to be made regarding the whole organisational field when discussing results. Based upon the quantitative assessment phase, three cases were chosen in which partners managed to mobilise a great amount of resources, including members, capital and installed energy production capacity. The chosen cases are thus considered to also be *powerful* examples that can hopefully provide insights into best practise activities as well as the aligned potentials, challenges and important pre-conditions of the entire community. Accordingly, the following cases were chosen:

1. **Dominant collaborative interaction model I**: The first dominant interaction model represents collaboration between energy cooperatives and other energy-related companies, in particular municipal energy providers. As will be explained in further detail in Sect. 7.1, collaboration between *Bürgerenergiegenossenschaft Wolfhagen eG* (BEG Wolfhagen eG; The Citizens' Energy Cooperative of Wolfhagen) and *Stadtwerke Wolfhagen* (the municipal energy provider for the city of Wolfhagen) was chosen for the first case study. In 2012, the energy cooperative was among the largest renewable energy production cooperatives in Germany with respect to members and total capital. The organisation also became, next to the city of Wolfhagen, the largest shareholder of its municipal energy provider—the Stadtwerke Wolfhagen—and thus has a particularly intense collaboration with this other energy company. They worked together to help Wolfhagen reach its goal of establishing a 100% renewable and regional energy structure by 2015.

5.3 Qualitative Case Study Analysis

2. **Dominant collaborative interaction model II**: The second dominant interaction model represents collaboration between renewable energy production cooperatives and banks, in particular cooperative banks. The collaboration between Energie + Umwelt eG (Energy + Environment eG) and 11 cooperative banks was selected for the second case study. In 2012, the energy cooperative belonged to the largest renewable energy production cooperatives in Germany with respect to members and total capital, and it has particularly intense collaborative interaction with cooperative banks. Energie + Umwelt eG was founded by three cooperative banks and is also fully managed by them. In addition, eight other cooperative banks became members of the energy cooperative and work together closely. They have actively supported the districts of Main-Tauber and Neckar-Odenwald in becoming a zero-emissions region.
3. **Dominant collaborative interaction model III**: The third dominant interaction model represents collaboration between renewable energy production cooperatives and communities. The collaboration between *Neue Energien West eG* (NEW eG, New Energies West eG), the aligned *Bürgerenergiegenossenschaft West eG* (Bürger eG, Citizen Energy Cooperative West eG) and the communities they serve were the object of the third case study. The NEW eG unifies 18 municipalities and cities as well as two municipal energy and water providers which have the joint goal of making their region—parts of the districts Neustadt an der Waldnaab, Amberg-Sulzbach and Tirschenreuth—independent from fossil energy resources by 2030. The citizens energy cooperative Bürgerenergiegenossenschaft West eG has brought together citizens who want to invest in the energy projects realised by the inter-municipal energy cooperative NEW eG. In 2012, the Bürgerenergiegenossenschaft West eG belonged among the largest renewable energy production cooperatives in Germany with respect to members and total capital.

5.3.2 Data Collection

Three main data sources were used in order to analyse the cases: (1) Semi-structured interviews with key actors, (2) newspaper articles, and (3) additional field data. In order to collect data, each of the case areas were visited two times. Table 5.3 provides an overview of the main purposes of the case study visits.

During the first visit, I conducted interviews in each case study region. The interviews for the first case were done 1 year before doing the interviews for the second and the third cases, as I wanted to fully test the research and interview setup in the first case before continuing with the others. This way, experience from the first case could be used in order to improve the overall approach. During the second visit, which took place at least 1 year later, results were assessed with interview partners in order to make sure that their stories, opinions and perceptions were accurately captured. The long time frame between the first and second visits further allowed

Table 5.3 Overview of case-study visit purposes

Case study	Visit 1 Execution of interviews	Visit 2 Reflection of results with interview partners and newspaper article research
Case 1: Bürgerenergiegenossenschaft Wolfhagen eG	April 2013	July 2015
Case 2: Energie + Umwelt eG	August 2014	July 2015
Case 3: Neue Energien West eG/ Bürgerenergiegenossenschaft West eG	August 2014	July 2015

me to learn more about organisational developments and to discuss the interview results against the background of these developments. In addition, the second visits were used for conducting regional newspaper-article research, for which the interview results provided the basis.

5.3.2.1 Execution of Interviews

The heart of the qualitative case study analysis was the conducting of semi-structured interviews. Following Baxter and Jack (2008, p. 565) and Orum et al. (1991, p. 9), the main purpose of studying individual cases was to explore the individual perspectives of key actors regarding the development, activities and effects of each energy cooperative and its collaborative interactions. The interviews were of great empirical value, because the interview partners were able to provide opinions, viewpoints and facts based upon their personal experiences. They transported their particular views concerning the reality of the energy cooperative they were associated with or connected to by telling their own stories. These included recounting the complete foundation process of the respective energy cooperative as well as describing current activities and planned activities. Furthermore, the interview partners gave detailed information about rule and resource patterns that they thought had been created through their energy cooperative and its main collaborative partner. Hence, the interviews provided comprehensive insights about the self-awareness of the energy cooperatives and their partners.

Interview partners were selected according to their relation to the energy cooperative and its collaborative partner. In each case study, the following mixture of people was interviewed:

- At least one member of the executive board of the energy cooperative who could describe and evaluate organisational activities and collaborative interaction from the perspective of the energy cooperative.

5.3 Qualitative Case Study Analysis

Table 5.4 Overview of interviews carried out for this study and the coded names used for analysis

Case study	Number of interviews	Codes for interview partners during case study analysis
Case study R1	5	R1N1, R1N2, R1N3, R1N4, R1N5
Case study R2	5	R2N1, R2N2, R2N3, R2N4, R2N5
Case study R3	5	R3N1, R3N2, R3N3, R3N4, R3N5

- At least one high-ranking decision maker of the collaborative partner who could describe and evaluate organisational activities and collaborative interaction from the perspective of the collaborative partner.
- At least one high-ranking regional politician or a high-ranking actor from a regional political agency who could evaluate activities of the energy cooperative against the background of political energy goals and, thus, provide a political perspective regarding the energy cooperative and its activities.
- If possible, one actor who had a critical perspective on the energy cooperative and its collaborative interaction model. Opinions from critical actors were valuable in order to better evaluate the perceptions of those actors who support the energy cooperative.

In all, 15 interviews were conducted: five for each case study. The interviews were carried out bilaterally, consisting of the interviewer and one interview partner. Thirteen interviews were done face-to-face, whereas two were done by phone in cases where it was not possible to organise a personal meeting. Each interview was recorded and lasted between 60 and 120 minutes; all were later transcribed. All interview partners officially approved of the use of the interview content for the analysis presented in this work, and their anonymity was guaranteed. Table 5.4 provides an overview of the number of interviews per case study, as well as of the codes being used for the interviews during the case study analysis in Chap. 7.

Furthermore, seven additional interviews were conducted with relevant actors from three other energy cooperatives with similar collaborative interaction models, in order to make sure that the case selection and the case study set-up were reliable. However, these interviews do not form part of the case study analysis.

All interviews were carried out based upon a set of semi-structured interview guidelines. The set up and content of the questions were based upon the research themes and theoretical framework elaborated in Sect. 4.3. The guidelines included narrative questions, which anticipated the description of activity patterns and experiences, as well as expert questions, which anticipated critical reflections, opinions and evaluations (Kruse 2008, pp. 59ff.). The interviews were executed in a flexible manner—the questions were not, for example, always asked in the same order—to allow for the individual development of each interview process, with great attention being paid to following the thematic prioritisation of the interview partner. In this way, it was possible to cover the themes that I was interested in while, at the same time, giving the interview partner enough space to communicate their own

Table 5.5 Interview guidelines

Target interview partner	Central interview questions	Thematic block
All	(a) Question about the background of the interview partner (name and position)	• Ice-breaker question
All	(b) Please describe the position of the [organisation] in the region	• Existing action mandate and strategic considerations of the organisation • Integration of the organisation in niche or regime structures
All	(c) How was the energy cooperative founded?	• Foundation process of the energy cooperative • Creation of niche actor • Existing regime conflicts
Political actors	(d) Please describe the energy goals of current regional policy.	• Integration of key politicians in niche or regime structures
All	(e) How was cooperation between [organisation] and [organisation] initiated?	• Cooperation initiation process • Existing regime conflicts
All	(f) How is collaboration between [organisation] and [organisation] being carried out?	• Development of rules and mobilisation of resources
All	(g) How do you evaluate the collaboration between [organisation] and [organisation]? (h) What aspects have changed through the energy cooperative and/or the cooperation?	• Consequences of fostered resource and rule patterns
All	(i) How do you evaluate the future of the energy cooperative (j) How do you evaluate the future of the cooperation?	• Challenges to regime change
All	(k) In your opinion, what role does the cooperation between [organisation] and [organisation] play?	• Potential and challenges for ongoing regime change
All	(l) How would you describe a regional energy system?	• Understanding of regional energy system

interpretations, thematic preferences and opinions (Kruse 2008, p. 58). The interviews were characterised by an open-minded and trustful atmosphere. All interview partners were highly interested in the research and, consequently, willing to provide detailed insights for all questions. The questions were asked in German, since all interview partners had a German background.

Central questions for each interview covered one or two specific thematic blocks. They were formulated in a very general manner, in order to give each interview partner maximum flexibility in forming their answers. For each central interview question, additional supporting questions were prepared that focused on more specific details regarding the content of the central question. These latter questions were only asked if they seemed to be useful for gaining further insight or information. Altogether, the questions were intended to help guide each interview through each thematic block without disturbing the individual dynamic of the interview partner. Table 5.5 displays the core questions and thematic blocks. The full interview guidelines, including all sub-questions, can be found in Annex 13.

Prior to conducting the interviews, the guidelines were assessed with the help of other researchers from the Wuppertal Institute for Climate, Environment and Energy. An interview pre-test was also conducted to test the interview guidelines and the questions to be asked. Despite this pre-test, however, it was realised that some central questions needed to be slightly revised after the first real interviews so that they could be better understood by the interview partners. Furthermore, some of the sub-questions were taken out, as they tended to make the interviews too long.

5.3.2.2 Selection of Newspaper Articles

In addition to the interviews, newspaper-article research was conducted for each case study, the main aim of which was to reflect upon and compare the internal perspectives of the interview partners with the external perspectives presented in regional newspapers. Two aspects were important here: (1) A high degree of similarity between the internal and external perspectives was considered to confirm that the individual views of interview partners on their own activities was to a greater or lesser extent shared by outsiders. (2) Newspaper articles were taken as indicating to what extent new societal rules were being diffused beyond the boundaries of the energy cooperative and its partners, as articles from regional newspapers reach a wider public.

For each case study, one regional newspaper was selected. A regional newspaper was seen as being suitable because each of the three energy cooperatives mainly concentrated its activities on a particular (focus) region. For the choice of paper, the interview partners were consulted, and the editorial offices of several newspapers were also contacted. In the end, the following three newspapers were chosen:

Case study I: *Hessische/Niedersächsische Allgemeine—Kreis Kassel und Wolfhagen.*
Case study II: *Fränkische Nachrichten.*
Case study III: *Der neue Tag—Oberpfälzischer Kurier.*

Since none of the three regional newspapers have well-developed online search platforms for their articles, their editorial offices were visited in order to use their internal article-search programs. For each case study, all articles were selected that included the name of the respective energy cooperative, as it was assumed that articles written about energy cooperatives, their partners and their activities would also include the names of the cooperatives. All articles were selected that were published from the beginning of a cooperative's foundation until 5 years afterwards or until 31 December 2014, if the cooperative was younger than 5 years old. Table 5.6 provides an overview of the number of selected articles and the covered timeframes.

As can be seen, 50 articles were collected for case study I, covering a timeframe of 3 years. For case study II, 33 articles were selected, beginning with their

Table 5.6 Number of selected articles found for each case study

Case study/Name of energy cooperative	Number of articles containing the name of the respective energy cooperative	Covered timeframe
Case study I—Bürgerenergiegenossenschaft Wolfhagen eG	50	01.01.2012–31.12.2014 (3 years)
Case study II—Energie + Umwelt eG	33	01.01.2010–31.12.2014 (5 years)
Case study III—Neue Energien West eG	194	01.01.2009–31.12.2013 (5 years)

foundation year of 2010 until the end of 2014, covering a time frame of 5 years. For case study III, 194 articles were identified which contained the name of the energy cooperative, covering a timeframe of 5 years.

5.3.2.3 Additional Field Research

The interviews and newspaper article review were complemented by additional field research throughout the case study analysis. Data sought here included the following:

1. General information about the existing regional energy structure, including energy demand and supply, future energy goals and climate protection plans. Core data sources consisted of documents from the respective political administration of the region and from statistical agencies of the districts or the Bundesländer.
2. Additional information about the energy cooperatives and their collaboration patterns. Primary data sources consisted of press releases from the energy cooperatives and their partners as well as online articles from other newspapers not included in the newspaper review.

5.3.3 Data Analysis

5.3.3.1 Qualitative Content Analysis

The interviews were analysed by applying qualitative content analysis (Mayring 2010). Mayring (2010, pp. 58ff.) differentiates between three methods of data interpretation: (1) *Summarisation analysis*, where empirical raw material is reduced and abstracted until only key statements remain; (2) *explication analysis*, where core statements are identified for which additional material is then gathered in order to interpret their content; and (3) *content structuration analysis*, where certain topics, aspects and content are filtered out of the empirical raw data through theoretically derived categories.

Content structuration analysis was chosen, since the empirical research design of this work follows a theoretical framework developed in Sect. 4.3, so interviews were interpreted by drawing upon categories elaborated for that framework. Following Mayring (2010, p. 89), the analysis was structured as follows. First, the units of analysis were identified, in this case the interviews (step 1). Then categories were defined for the extraction of key aspects from the raw data, deduced from the theoretical framework (steps 2–4). The interviews were then reviewed and relevant text paragraphs were allocated to the categories (steps 5–6). Throughout the review, several reflection loops were undertaken, during which the categories were critically regarded and—if necessary—revised (step 7). Then paragraphs were extracted by concentrating on their key content (paraphrased) (step 8). In the end, the extracted paragraphs were interpreted in accord with their relevant category, and interpretations were enriched with quotes from interviewees (step 9–10) (Mayring 2010, p. 89). The interpretations and relevant quotes can be found in Chap. 7.

In steps 2–4 of the content structuration analysis, seven main categories were developed from my theoretical framework, in order to pursue answers to my research questions:

1. Character of the region
2. The energy cooperative
3. The collaborative concept
4. Exchange of allocative resources
5. Exchange of authoritative resources
6. New norms and regulations
7. New interpretative schemes

Various subcategories were also established for each main category.

Category 1 includes information about the region in which the energy cooperatives and their partners prefer to operate.

Category 2 includes information about the energy cooperatives, their foundation processes, their goals, as well as about their collaborating partners and their motives for collaborating.

Category 3 concerns information about the collaborative concepts developed between the energy cooperatives and their main partners.

Category 4 comprises identified allocative resources that were exchanged in collaborative interaction and also summarises aligned challenges.

Category 5 includes identified authoritative resources that were strengthened or newly created throughout collaboration as well as their challenges and limits.

Category 6 contains all statements with respect to new norms and regulations that were fostered by collaboration as well as associated criticisms, difficulties and unintended consequences.

Category 7 comprises all statements regarding new interpretative schemes that seem to have been introduced through the collaboration as well as those regarding challenges, limits and unintended consequences.

The MAX QDA software was used in order to conduct the structural content analysis.

5.3.3.2 Reviewing Interview Material

Causal interdependencies between the collaborative interaction of energy cooperatives and effects on their immediate environment can only be explained if the individual perspectives of core actors is fully incorporated into the analysis (Baxter and Jack 2008, p. 565; Orum et al. 1991). The main aim of my interview analysis was, thus, to achieve a detailed understanding of how the interview partners perceive and guide their own reality. Making sure that the stories, opinions and experiences of the interview partners were accurately captured is a crucial step for the whole case study analysis, because it helps to ensure the reliability of the case study findings.

After finalising the interview analysis phase, the interview partners from each case study were visited a second time in order to jointly reflect upon the results. During the meetings, the findings were presented in a brief manner and then discussed in detail. The interview partners from all three case studies agreed that their activities and viewpoints were well interpreted and summarised via the interview process and, in their opinion, the interview analysis fully mirrored their own reality, including strengths, opportunities, challenges, boundaries and pre-conditions. In addition, during the review process, information about the latest developments was collected and some data was double-checked.

5.3.3.3 Analysis of Newspaper Articles

The analysis of the selected newspaper articles was focused on comparing them with interview findings. The main aim was to see whether the internal views of interview partners were also shared by outsiders and to what extent newly created rules pointed out by the interview partners were being replicated in the articles. Norms and guiding principles introduced through energy cooperatives are considered to be transported into regional society if they are reflected in regional newspapers. The diffusion of new societal rules into society is a crucial pre-condition for a socio-technical system change (Geels 2002).

Analysis of the newspaper articles was conducted as follows: The content of each newspaper article was reviewed. Then the number of articles replicated those norms and guiding principles that had been pointed out by interview partners during interviews was tallied, and the results were quantified for each case study. Articles that solely contained pure announcements or those which covered a completely different topic (for example, the sudden death of a cooperative member) were excluded from the quantification of results. Table 5.7 provides an overview of the

Table 5.7 Number of articles that were considered worthy of closer analysis in each case study

Case study/Name of energy cooperative	Number of articles containing the name of the respective energy cooperative	Number of articles considered worthwhile for closer analysis of societal rule diffusion
Case study I—Bürgerenergiegenossenschaft Wolfhagen eG	50	50
Case study II—Energie + Umwelt eG	33	26
Case study III—Neue Energien West eG	194	120

number of articles that were considered for further analysis regarding the dissemination of rules.

5.3.3.4 Comparison of Case Studies

In the end, the results of the three case studies were compared with each other. Statements about causal relationships which go beyond an individual case not only indicate causality per se but can point towards repeatable causal relationships in the real world (Mayntz 2009, p. 12). In my opinion, results that were found in all three of my case studies can be generalised, since 72% of all renewable energy production cooperatives use one of the three dominant collaborative interaction models that I observed. This means to me that the general structural change potential of energy cooperatives on their immediate environments is derivable from the case study analysis. Since the cases also represent powerful examples of each collaborative interaction model, important pre-conditions for actor empowerment and for establishing successful collaboration opportunities may also by identified by comparing them.

5.4 Evaluation of Results

In the end, all results were merged in order to answer the main research question: 'What influences do energy cooperatives have on Germany's energy system?' The qualitative and quantitative influences of energy cooperatives are strongly interrelated and can only be analytically separated into different conceptual building blocks. Hence, it seems important to jointly discuss my findings from the quantitative assessment and qualitative case study analysis concerning energy cooperatives. Such examination of a desired energy system transformation can then function as a general baseline in order to critically evaluate other findings about energy cooperatives and their collaborative activities.

5.5 Critical Assessment of the Research Design

In this book, I have chosen to apply a comprehensive research design based upon three empirical assessment levels and a mixed-method approach in order to analyse the influence of energy cooperatives on Germany's energy transition. The selected research methods include (1) quantitative analysis of large secondary data sets, seeking to provide a comprehensive overview regarding the status quo of the cosmopolitan community surrounding German energy cooperatives and (2) qualitative case study analysis, seeking to provide detailed insights on the three dominant collaborative interaction models preferred by the majority of energy cooperatives for realising their business goals.

The chosen research design and the aligned mixed-method approach were well appropriate for advancing my research interest. The combination of quantitative and qualitative empirical data allowed me to analyse energy cooperatives from very different perspectives and, in combination, helped to draw a holistic picture of the phenomenon. As show in the following, the research I have undertaken shed light on the character of the whole targeted enterprise community and help to signal opportunities, challenges, limits, as well as pre-conditions of their activities. The quantitative assessment enabled a well-developed basis upon which to select case studies that can be considered to represent typical and powerful examples of the community. Thus it became possible to generate results applicable to the majority of German energy cooperatives and to identify their core conditions as well as limits for actor empowerment. Merging the different empirical data sets has also created the necessary basis for being able to close the research gaps identified in Sect. 2.3 and has enabled a far-reaching and nuanced discussion about the influence of energy cooperatives and their structuring potential within Germany's energy system.

However, the chosen research design also involved challenges. The mixed-method approach is very time intensive and complex. The great variety of data collected during the quantitative and qualitative assessment needed to be considered throughout the whole research period. The approach also requires knowledge about several empirical methods and how to best combine them. Achieving access to so many data sources was at times difficult and required additional time resources. Even though a holistic approach was chosen, several aspects of the community surrounding German energy cooperatives do not appear to be reflected in the results. For example, this book is focused on renewable energy production cooperatives, since they represent the majority of the community. Hence, energy cooperatives with very special business goals may not have been reflected in an adequate manner here. Thus, their individual character would need to be the subject of another analysis.

Chapter 6
German Energy Cooperatives: A Rising Cosmopolitan Enterprise Community

Throughout this section, German energy cooperatives are presented and evaluated in detail as a rising cosmopolitan community. The main goal is to characterise this enterprise group and achieve an overview of its status quo, based upon quantitative assessment of secondary data on German energy cooperatives. Section 6.1 outlines the enterprise network, while Sect. 6.2 describes substantial resources that have thus far been mobilised by the entire community. Section 6.3 characterises associated actors, and a summary is given in Sect. 6.4.

6.1 The Enterprise Network

In the following, the growth dynamic of energy cooperatives is analysed, their business locations are identified and their business goals are described.[1]

6.1.1 Growth Dynamic

In Germany, 1055 energy cooperatives were officially registered by 31 December 2015.[2] Figure 6.1 presents an overview of annual registrations between 2000 and 2015.

---

[1] In the course of elaborating this thesis, some of the used here data has already been published in Debor (2014).

[2] Since there is no official definition for energy cooperatives, related empirical data, such as total numbers or annual registrations, might slightly deviate from other existing research. The assessment providing the basis for this thesis identifies all cooperative organisations as energy cooperatives if they operate along the energy value chain (displayed in Fig. 2.1).

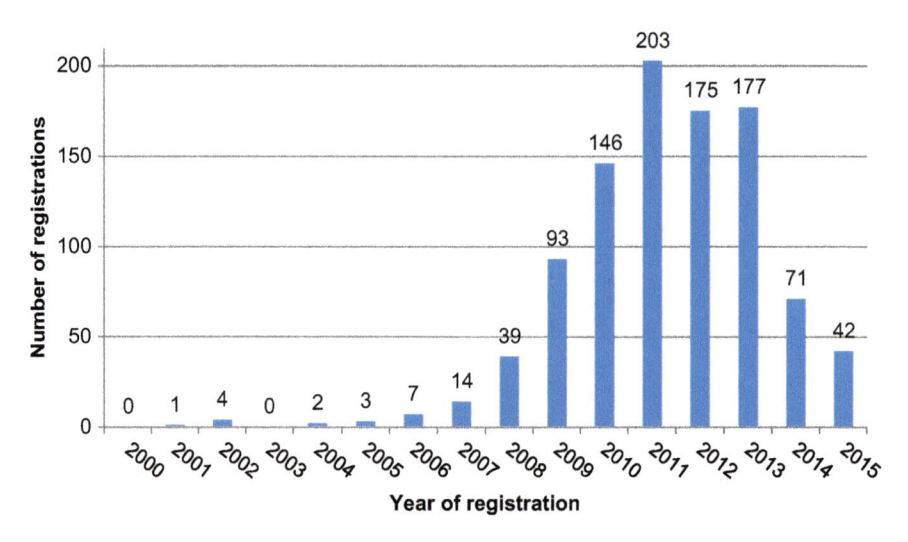

Fig. 6.1 Annual registration of German energy cooperatives, 2000–2015

Summing up the registration data, 92% of all energy cooperatives were registered from 2006 onwards, with the number of registrations continuously increasing and reaching a peak of 203 new energy cooperatives in 2011. This growth dynamic seems to have been predominantly triggered by two events. First, several legal amendments to the German Act for Cooperatives (GenG) were introduced in 2006 that facilitated the general foundation of cooperatives in Germany (Gesetz betreffend die Erwerbs-und Wirtschaftsgenossenschaften (Genossenschaftsgesetz—GenG) 2006; Schaffland and Korte 2006). Second, the foundation dynamic of energy cooperatives correlates with governmental support of renewable energy production. As outlined in the next section, the majority of recently emerging energy cooperatives have focused their business activities on the production of renewable energy. Renewable energy production is presently supported by the German Renewable Energy Act (EEG) and, amongst other benefits, renewable energy plant operators receive a guaranteed feed-in tariff if they produce energy from renewable resources. Nevertheless, the actual growth of energy cooperatives in greater numbers only first started 6 years after the EEG was introduced in 2000 (Gesetz für den Vorrang Erneuerbarer Energien (Erneuerbarer-Energien-Gesetz—EEG) sowie zur Änderung des Energiewirtschaftsgesetzes und des Mineralölsteuergesetzes 2000). However, the strong growth dynamic of energy cooperatives between 2006 and 2011 does directly correlate with a continual increase of the cost-value ratio of photovoltaic power plants. As outlined in Sect. 6.1.3, the majority of energy cooperatives presently install photovoltaic power plants. From 2006 until about 2012, the investment costs of photovoltaic technology experienced a continual decrease. In fact, cost reduction was greater than the annual reduction of the feed-in tariff for photovoltaic power under the EEG (Wirth 2016, pp. 8f.), an interpretation that is in accord with the viewpoint of the Research Institute for Cooperation and Cooperatives from the Vienna University for Economic and Business (2012, p. 13).

After 2011, however, the enterprise network lost this strong growth dynamic, as a decrease of annual registrations can be observed in 2012 and 2013—a decrease that correlates with further amendments to the EEG. Legal modifications that were passed in April 2012[3] involved, notably, a decrease of the feed-in tariff for renewable power production types, including photovoltaic plants. The strongest reduction of annual registrations can be observed in 2014 and 2015, which again correlates with legal amendments to the EEG being passed in 2014 (Erneuerbare-Energien-Gesetz—EEG 2014). The EEG 2014, followed by the EEG 2016, introduced a fundamental change of political orientation, and their main intention has not been the maximisation of renewable energy but, rather, limitation of costs for the overall energy system change, stronger integration of renewable energy into the energy market and greater ability to control increases of power capacity in Germany (Federal Ministry for Economic Affairs and Energy 2014a; The Federal Government 2016). Feed-in tariffs, particularly for photovoltaic plants, were again strongly reduced.

The heart of the EEG 2014 and the EEG 2016 is a new tendering model for renewable energy capacities, which replaces guaranteed feed-in tariffs predefined by the government with technology-specific tariffs that are formed by the market, based upon bids and calls. Participants in such tenders do so in order to receive an allowance to install a certain amount of renewable energy capacities in Germany. According to the EEG 2016, tenders are conducted for on-shore and offshore wind, as well as photovoltaic plants that are larger than 750 kilowatts and for biomass plants that are larger than 150 kilowatts. Within such tenders, the level of compensation for each technology is defined as follows: All tender participants have to offer a price for which they think they are able to install a certain production capacity. All bids are ordered according to the offered price and offered capacity volume, until the capacity limit of the specific tender is reached. All participants within that ordered set are then rewarded and given permission to install the power capacity that they offered to undertake (Deutscher Bundestag 2016, pp. 28ff.). Such a procedure is cost intensive and involves major business risks, such as losing money for pre-development activities when a tender is lost. Many associations and companies have criticised that the tender approach can severely disadvantage small organisations, such as energy cooperatives, as they are not able to cope with the involved risks (Bündnis Bürger Energie e.V. 2015; Peter and Glahr 2015).

The government has tried to provide support for small—particularly citizen-oriented—organisations, and, for the first time, a concrete definition of a citizen-focused organisational approach was introduced in the EEG 2016, which also applies to energy cooperatives (§ 3 paragraph 15 EEG 2016). According to the law, citizen-focused energy organisations are organisations in which at least

---

[3]Gesetz zur Änderung des Rechtsrahmens für Strom aus solarer Strahlungsenergie und zu weiteren Änderungen im Recht der erneuerbaren Energien (2012).

10 natural persons represent members with voting rights and in which at least 51% of the votes are in the hands of natural persons that also live in the community in which projects are installed, as well as organisations in which no member holds more than 10% of the whole organisation. According to the official draft of the EEG 2016 from 8 July 2016 that was passed by the German *Bundestag*, organisations that are covered by the definition of § 3, paragraph 15 of the EEG 2016 follow simplified tendering rules (§ 36a Number 15 EEG 2016). However, risks still remain, and it is highly uncertain whether these exceptions will really and adequately facilitate the participation of small organisations in tenders.

The reduction of the energy cooperative growth dynamic further correlates with the introduction of new regulations for alternative investment funds in 2013, the *Kapitalanlagegesetzbuch* (KAGB),[4] which cover all investment asset organisms that collect capital from a number of investors and represent operative-oriented businesses from outside the financing sector (§ 1 paragraph 1 KAGB). The KAGB act not only involves strict regulation of investment management but also imposes greater requirements and personal liabilities for actors involved in steering positions. Organisations that are covered under the KAGB are controlled by the Federal Financial Supervisory Authority (*Bundesanstalt für Finanzdienstleistungsaufsicht*: BaFin), and the new act is supposed to provide greater protection for investors and greater control of fund activities. Due to the open membership approach of cooperatives, which is geared towards collecting members' financial shares in order to jointly invest in a defined business activity, there was at first general ambiguity whether and to what extent they would be covered under the KAGB. In March 2015, about 2 years later, the BaFin published a clarification according to which officially registered cooperatives are generally not seen as an alternative investment funds, because by definition they are oriented towards member and not shareholder value (BaFin Bundesanstalt für Finanzdienstleistungsaufsicht 2015). This clarification of the BaFin position was the result of strong lobbying activities on the part of superordinate associations for cooperatives and other institutions aligned with them (Zimmermann 2015, p. 58). However, the long period of time during which the legal responsibilities and risks for cooperatives under the KAGB were unclear affected the whole network. This negative influence is palpable in all three analysed case studies outlined in Chap. 7. Up through 2015, the negative trend in annual registrations continued and, with 42 newly registered energy cooperatives that year, reached its lowest level since 2008.

---

[4]Gesetz zur Umsetzung der Richtlinie 2011/61/EU über die Verwalter alternativer Investmentfonds (2013).

6.1.2 Location of Organisations

Figure 6.2 shows how many energy cooperatives were registered in each Bundesland as of December 2015, providing an overview of the distribution of energy cooperatives throughout Germany.

As shown in the figure, energy cooperatives are registered in all 16 Bundesländer. Bayern is the Bundesland that hosts the most energy cooperatives, with 280 registrations, followed by Baden-Württemberg, with 167, and Niedersachsen, with 134. Energy cooperatives can even be found in small Bundesländer, such as Saarland or Bremen. Their geographical distribution throughout the whole country indicates that

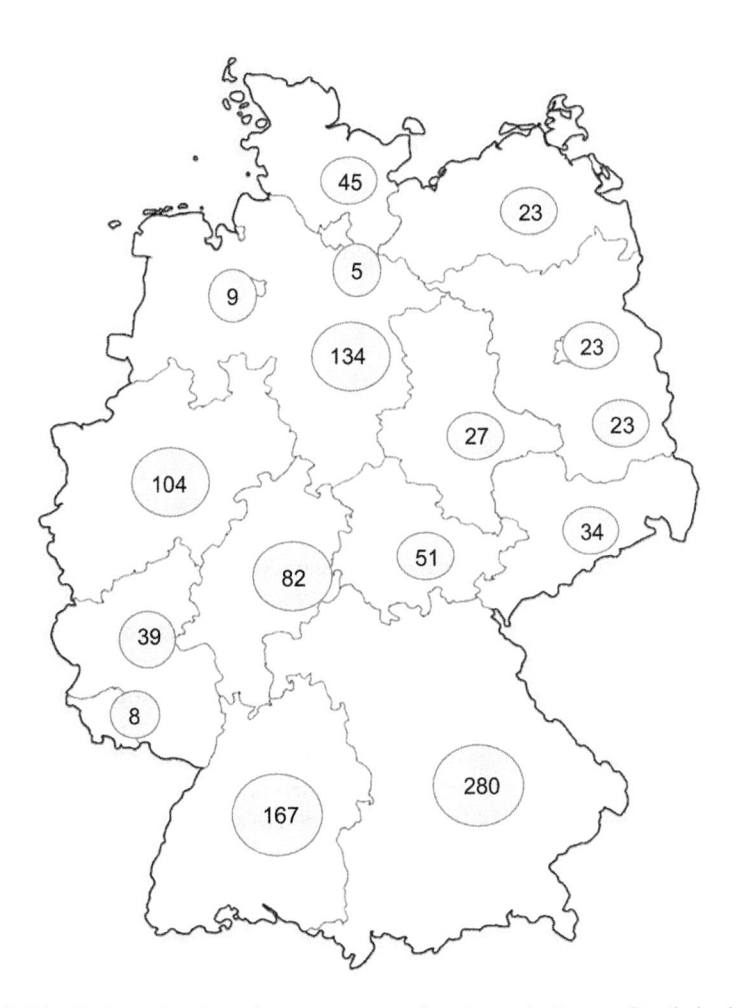

Fig. 6.2 Distribution of registered energy cooperatives in each German Bundesland, as of December 2015

energy cooperatives have left the purely local level and evolved into a trans-local cosmopolitan enterprise community.

Several studies claim that German energy cooperatives are mainly present in towns and small villages. Particularly Maron and Maron (2012, pp. 119ff.) conclude that energy cooperatives primarily operate in rural areas. However, results of the quantitative assessment conducted for this work reveal a slightly different picture. As of 2014, Germany had about 77 large cities with populations of at least 100,000 (Statistisches Bundesamt 2015), and 68% of these cities host energy cooperatives. Focussing on the largest cities, this tendency even increases, as only one of the 14 German cities with at least 500,000 people does not host an energy cooperative. Two aspects may have led to the strong association of energy cooperatives with rural regions. First of all, the majority of energy cooperatives do operate in rural areas, simply because rural areas provide the main geographical context within Germany, with 98% of German cities (about 1986) having populations of less than 100,000. Second, 72% of all energy cooperatives situated within large cities were registered from 2011 onwards. Hence, energy cooperatives first developed in the country and then moved into the larger cities. This development indicates a clear change in geographical distribution compared to the results of Maron and Maron (2012, pp. 119ff.) and other studies from around 2012. All in all, it can be said that energy cooperatives are distributed all over Germany—in rural areas as well as in large cities.

6.1.3 Business Goals

Energy cooperatives represent a heterogeneous group, with many different business approaches. The following business-goal classification draws upon business categories which were elaborated by Holstenkamp and Müller (2013, p. 15), based upon (Flieger 2008; Klemisch and Maron 2010; Theurl 2008), who presented the business goals of all German energy cooperatives that were registered by December 2012. Since the aim of this section is to point out the strong relationship between energy cooperatives and the development of new renewable-energy structures, here Holstenkamp and Müller's prior classification is further developed with regard to three aspects: (1) The number of energy cooperatives with business goals related to renewable energy is clearly specified; (2) the classification is extended using new categories, such as e-mobility; and (3) more than one core business goal has been identified for quite a number of the energy cooperatives.[5] These refinements seem to more accurately represent the reality of a situation where, for example, many energy cooperatives that operate small district-level heating systems also produce biomass

---

[5]One core business goal was identified for 866 energy cooperatives, two core goals for 165 cooperatives, three goals for 11 of them, and four goals were observed for only one cooperative. For 12 of the cooperatives, no business goals were identified.

6.1 The Enterprise Network

Table 6.1 Business goals of German energy cooperatives through 2015 (more than one category per cooperative is possible)

Business goals of energy cooperatives	Total number of energy cooperatives	Number of energy cooperatives focusing on renewable energy
Energy production/investing in energy production	785	773
Implementation and operation of small district-level heating systems	186	186
Marketing and trading of energy	74	23
Marketing and installation of energy technology	52	17
Energy services	44	30
Operation of electricity grid or natural gas net	36	
Acquisition or marketing of biomass	19	19
Business goal could not determined	12	
Production and marketing of biofuel	10	10
E-mobility	10	10
Lobbying and networking	5	4
Operation of fuel station	4	
Shareholder of municipal energy provider	4	4
Research and development	3	3
Total energy cooperatives	1055	933

energy, and energy cooperatives that manage a conventional grid may also market energy as well.

Table 6.1 provides an overview of the core business goals for all of the German energy cooperatives that were registered by December 2015. Next to the total number of energy cooperatives, the number of energy cooperatives that focus on renewable energy is listed for each business goal category.

As can be seen, the overwhelming majority of registered energy cooperatives have a primary focus on renewable energy. With 933 organisations, they represent an 88% share of all 1055 registered German energy cooperatives. Organisations that produce renewable energy or invest in renewable energy production belong to the most dominant group. With 773 entities, they represent 73% of all registered energy cooperatives. Energy cooperatives, which implement and operate small district-level heating systems with biomass heat are, with 186 organisations, the second leading group among energy cooperatives (18% of total). Meanwhile, 74 organisations market and trade energy, with 23 of them having a primary focus on renewable energy; 52 energy cooperatives market and install energy technology, with 17 of them specialising in renewable energy technology, such as the installation of photovoltaic panels. Furthermore, 44 organisations offer energy services, 30 of which claim to primarily support renewable energy by, for example, coordinating energy-efficiency projects. Also of note is that 36 energy cooperatives operate a local

Fig. 6.3 German energy production cooperatives differentiated according to their applied energy resources through 2015 (more than one category per cooperative is possible)

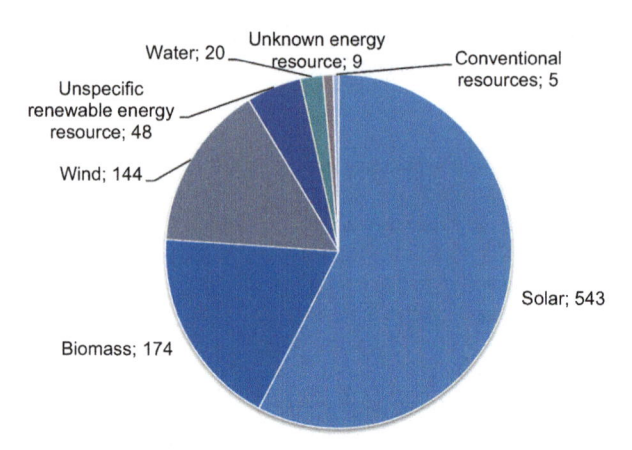

Water; 20
Unknown energy resource; 9
Conventional resources; 5
Unspecific renewable energy resource; 48
Wind; 144
Biomass; 174
Solar; 543

distribution electricity grid or natural gas net, most of them also representing the general energy provider in their region.[6]

The strong increase of energy cooperatives over the last 10 years seems to have been accompanied by a diversification of business settings, with renewable energy cooperatives operating in a variety of additional business areas. Some, for example, market biomass as an energy resource without operating a biomass plant, where as others predominantly acquire biomass. Most of them consist of networks of, for example, biomass plant operators that have unified in order to achieve a better price for biomass. Meanwhile, some energy cooperatives produce and market biofuel for the mobility sector, while others support electro-mobility with a focus on renewable energy in their region. A few energy cooperatives are exclusively engaged in lobbying and networking or in research and development. Some energy cooperatives have even become shareholders of municipal energy providers with the aim of supporting establishment of renewable energy in a region. For 12 energy cooperatives, it was not possible for me to determine their core business goals.

Figure 6.3 differentiates the 785 registered energy production cooperatives according to their applied energy resources.

As shown here, 543 cooperatives operate or invest in photovoltaic plants, representing 69% of all energy production cooperatives, while 174 (22%) produce renewable energy with biomass. Meanwhile, 144 energy cooperatives (18%) operate or invest in wind power plants, and 20 (3%) produce renewable energy with water. Only a very few energy cooperatives (five) produce energy with conventional resources.[7] I was able to identify a further 48 organisations as cooperatives that

---

[6]Organisations that operate a grid below a certain size are excluded from the requirement to separate their energy marketing and grid operation (unbundling) (§7 EnWG).

[7]This category only includes organisations that use conventional resources on a regular basis. Several cooperatives may use conventional resources as a back-up source in peak times. However, since such times are seen as exceptions, these cooperatives have not been counted as organisations that regularly use conventional resources for producing energy.

Table 6.2 Socially and politically related business goals of renewable energy production cooperatives

Social and political business goals of renewable energy production cooperatives	Share of energy cooperatives (n = 531) (%)[a]
Focus on regional energy production	74
Focus on citizen participation	71

[a]Each share here is based on the renewable energy production cooperatives that have an official company website where they can state their own claims regarding such goals

generate renewable energy, without being able to further specify their applied renewable resources. For nine energy production cooperatives, it remains unclear to me whether they primarily use renewable or non-renewable resources.

I have also observed that the majority of renewable energy production cooperatives not only have technology-oriented business goals but also claim to have socio-political ones as well, as listed in Table 6.2.

As displayed, 74% of analysed renewable energy production cooperatives claim to predominantly produce renewable energy in their own region, and 71% have expressed that they strive to actively involve private citizens in the realisation of their main business goals, such as the production or renewable energy.

All in all, it can be said that the majority of all registered energy cooperatives have a focus on renewable energy, in general, and renewable energy production, in particular. In addition to technology-oriented goals, the overwhelming majority of organisations also seek to foster socio-political goals. However, it does not seem to be clear what they actually mean by supporting a regional focus or activating citizen participation. This issue is examined in detail during the case study analysis in Chap. 7.

Further empirical assessment undertaken in the following section focuses on renewable energy production cooperatives, because with 73% they constitute the largest group among all registered energy cooperatives. It is thus sufficient to concentrate on these organisations in order to better understand the status quo of the greater enterprise community. Furthermore, data can be better compared between organisations with similar business goals than between those with differing ones.

6.2 Mobilisation of Substantial Resources

In order to analyse the character and status quo of the German energy cooperative community, it is essential to learn more about their membership (Sect. 6.2.1), capital (Sect. 6.2.2) and technology development (Sect. 6.2.3), as these represent resources which make up their collective power. In Sect. 6.2.4, my observations on each kind of development are set in relation to each other.[8]

---

[8]Some of the data used here has already been published in Debor (2014).

6.2.1 Development of Cooperative Members

Members are a crucial resource for renewable energy production cooperatives, because they can support such organisations in concretely realising their business goals through, for example, providing investment capital. It is thus of interest to observe how many actors generally join cooperatives and to what extent energy cooperatives are able to increase their membership numbers. Figure 6.4 displays the membership distribution of German renewable energy production cooperatives between 2010 and 2012.

As can be seen, the majority of renewable energy production cooperatives are rather small organisations in terms of their membership, and over 80% had between three and 200 members in the years 2010–2012. Nevertheless, within this range there was a shift during this period towards more cooperatives having 101–200 members. Thus, the share of surveyed organisations that had up to 100 members decreased, from 62% in 2010 to 50% in 2012. In turn, the number of analysed cooperatives with 101–200 members increased from 23% in 2010 to 30% in 2012. Although only a minority of surveyed organisations had more than 200 members, their share increased by 4%: from 15% in 2010 to 19% in 2012. Cooperatives with more than 700 members remained at a share of only 2% during the three observation years and, thus, remained an exception. The largest renewable energy production cooperative in terms of members used to be Greenpeace Energy eG, which had about 22,400 members in 2012. Yet, by 2015, it was replaced by PROKON regenerative Energien eG, which achieved about 37,000 members that year. A clear reason for this rapid change was that the PROKON renewable energy production cooperative was

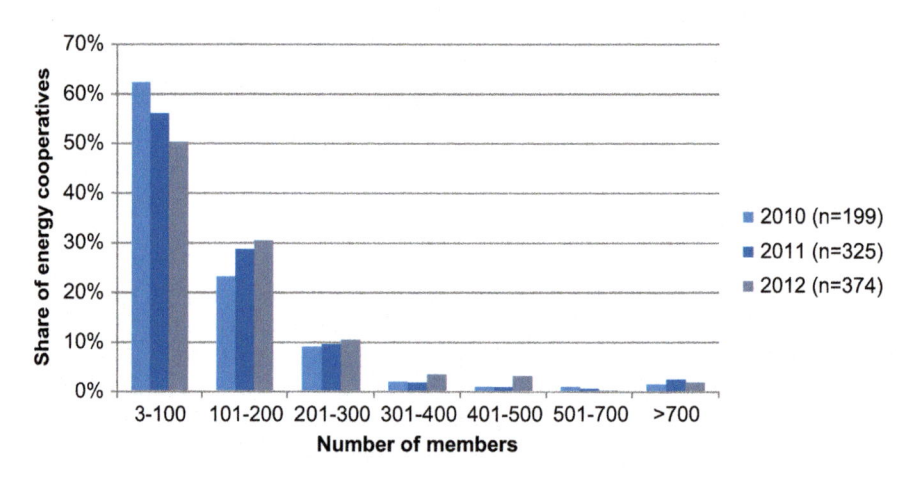

Fig. 6.4 Membership distribution of German renewable energy production cooperatives, 2010–2012

6.2 Mobilisation of Substantial Resources

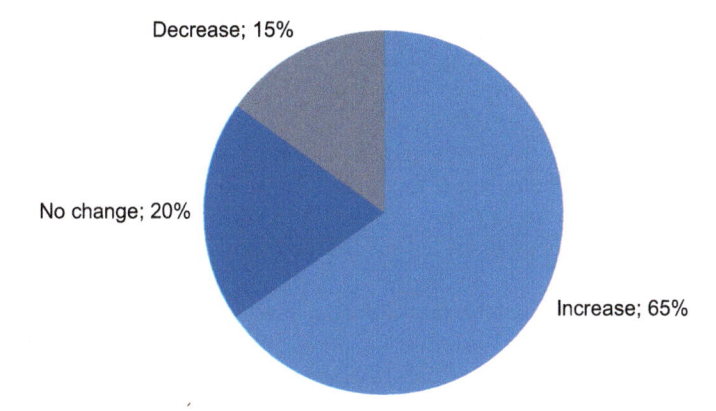

Fig. 6.5 Membership development (2010–2012) of German renewable energy production cooperatives that were registered in or before 2010

founded in 2015 following the insolvency of the former PROKON Regenerative Energien GmbH,[9] which was not a cooperative.

Figure 6.5 displays the distribution of membership increase or decrease among renewable energy production cooperatives between 2010 and 2012, concentrating on cooperatives that were registered in or before 2010, in order to be able to display the membership development of single energy cooperatives over a timeframe of 3 years. Of the 249 cooperatives that were registered in or before 2010, 173 provided membership information for the years 2010 as well as for 2012, representing a share of 69%.

As shown, about 65%—that is, the majority—of analysed renewable energy production cooperatives registered in 2010 increased their number of members by the end of 2012. During the same time period, membership numbers remained unchanged for 20% of the cooperatives, while 15% experienced a decrease in their membership.

Against the background of these results, it can be said that the business activities of renewable energy production cooperatives seem to have been positively recognised by the public, motivating a growing number of actors to become cooperative members. It has also been said that support for the goal of achieving a renewable energy system is a major stimulus for joining an energy cooperative (trend:research GmbH & Leuphana Universität Lüneburg 2014, p. 24). At the same time, renewable energy production cooperatives seem to have been able to

---

[9]PROKON regenerative Energien eG currently represents the largest energy cooperative in Germany, with 37,000 members and about 847 million Euros in total capital in 2015. However, PROKON regenerative Energien eG cannot be compared with the majority of energy cooperatives, since it has a completely different and very special foundation history (PROKON regenerative Energien eG 2016). PROKON used to be an investment company in the form of a GmbH, with a focus on renewable energy. In 2014 it became insolvent. The remaining part was then turned into an energy cooperative in July 2015. Since PROKON regenerative Energien eG was only first founded in 2015, it has not been included as part of the assessment of financial and membership data.

develop and expand their business activities to the extent that they were able to include a growing number of members.

However, existing membership growth rates show that it seems to be challenging for energy cooperatives to achieve membership numbers above 200. Yet, the organisations that have reached more than 200 members demonstrate that energy cooperatives do have the potential to attract a large number of new actors. Hence, stronger membership growth of this cosmopolitan niche community could be possible. Membership growth strategies are analysed in more detail in the case study analysis in Chap. 7.

Extrapolation of the membership numbers given here reveals that, altogether, registered renewable energy production cooperatives most likely gained about 158,000 members by the end of 2015,[10] whereas by comparison the German green political party *Bündnis 90/Die Grünen* mobilised about 59,000 party members during the same period.[11]

6.2.2 Development of Financial Capital

Financial capital is a crucial recourse for this niche community, because it provides the most fundamental basis for renewable energy technology investments. It is thus important to achieve an overview of the capability of energy cooperatives to mobilise financial capital, and examining the development of their total capital, equity ratios, as well as profits and losses may help us to understand to what extent these organisations could able to economically support the development of renewable energy structures.

6.2.2.1 Development of Total Capital

A business organisation's total capital represents the sum of its equity capital and borrowed capital, as published in its annual balance sheets. Figure 6.6 displays the total capital of renewable energy production cooperatives between 2010 and 2012.

As can be seen, the majority of renewable energy production cooperatives are, with respect to their total capital, rather small organisations,[12] as most analysed cooperatives had less than 500,000 Euros total capital in all three observation years (2010–2012). These were followed by renewable energy production cooperatives that had between 500,000 and one million Euros total capital at their disposal. From

---

[10]See Sect. 5.2.2 for further details on extrapolation.

[11]Source: https://de.statista.com/statistik/daten/studie/192243/umfrage/mitgliederentwicklung-der-gruenen/

[12]Small and medium sized firms possess of sales volumes that lay below 50 million Euros (Deutsche Bundesbank 2013, p. 47).

6.2 Mobilisation of Substantial Resources

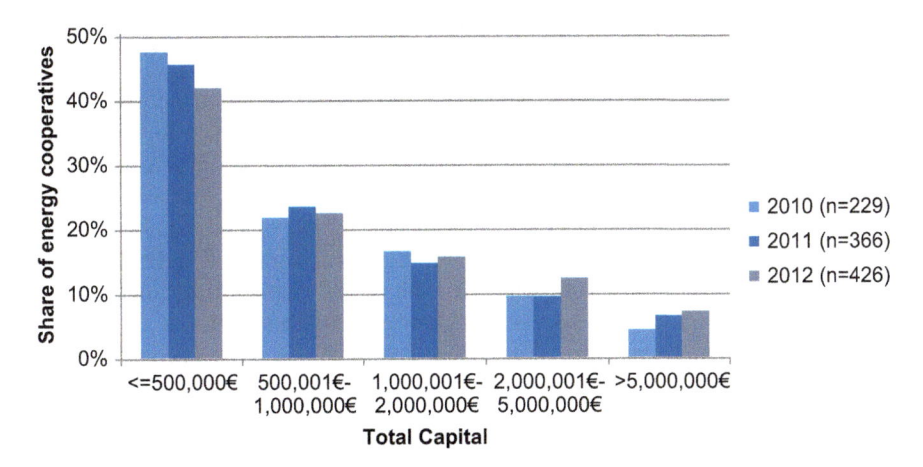

Fig. 6.6 Total capital of German renewable energy production cooperatives, 2010–2012

2010 onwards, there is a shift from low capital rates in the range of one to 500,000 Euros towards higher capital rates that lay above two million Euros, and the share of cooperatives that had a total capital of up to 500,000 Euros decreased from 48% in 2010 to 42% in 2012. Meanwhile, the share of surveyed organisations with a capital of between 500,000 and two million Euros stayed the same at 38% between 2010 and 2012. But the number of analysed organisations with more than two million Euros total capital at their disposal increased from 14% in 2010 to 20% in 2012. The general growth of capital rates indicates that a significant number of renewable energy production cooperatives seemed to have intensified their investment activities by increasing the size or number of their projects.

However, cooperatives that have more than two million Euros capital at their disposal remain a minority, as 80% of surveyed cooperatives still had less than two million Euros total capital in 2012. At the same time, the share of organisations that had more than five million Euros total capital almost doubled from 4% in 2010 to 7% in 2012. These energy cooperatives demonstrate that further growth rates seem to be possible. How such growth rates can be achieved is analysed in detail in the three case studies in Chap. 7.

Extrapolation shows that all renewable energy production cooperatives most likely mobilised about 1.3 billion Euros total capital by the end of 2015.[13] The largest renewable energy production cooperative in terms of total capital used to be Energiegenossenschaft Odenwald eG, which in 2012 had about 27 million total capital. In 2015, however, it was replaced by PROKON regenerative Energien eG, which had about 847 million total capital that year.

---

[13] See Sect. 5.2.2 for further details on extrapolation.

6.2.2.2 Development of Equity Ratio

Equity ratio displays the relation between a business organisation's equity capital and total capital (Heesen and Gruber 2011, p. 281), indicating here to what extent renewable energy production cooperatives are able to finance business activities out of their own internal strength. The circumstance that the analysed cooperatives seem to have slightly similar developments regarding the number of their members and the amount of their total capital may mean that they can primarily finance increasing investment activities through their members. Figure 6.7 displays the equity ratios of renewable energy production cooperatives between 2010 and 2012.

As can be seen, the majority of analysed renewable energy production cooperatives had considerably high equity ratios in all three observation years (2010–2012). Furthermore, the range of their equity ratios remained relatively stable throughout the observation period. For all 3 years, 60% of the surveyed cooperatives had an equity ratio of between 31 and 100%. This is quite high, compared to the average equity ratio of German small and medium-sized firms,[14] which lay at about 23% in 2012, while the figure for large German companies that year was also lower at around 29% (Deutsche Bundesbank 2013, p. 47). These empirical results reveal that financial requirements seem to be met to a great extent by cooperative members, even though investment volumes have increased over the observation period. Financing projects primarily through equity capital complies with fundamental cooperative principles, as through collecting equity from members, cooperatives aim to support actors in the united development of self-defined goals without greatly drawing upon the support of third parties (Zerche et al. 1998, pp. 209ff.).

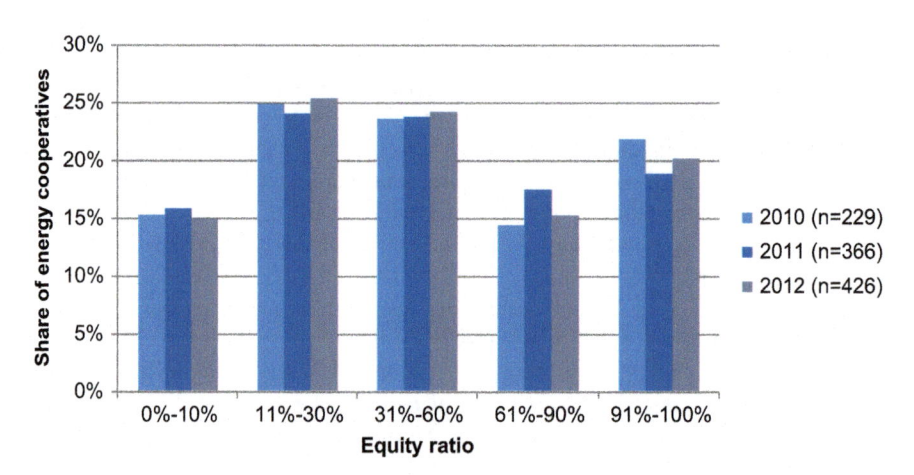

Fig. 6.7 Equity ratios of German renewable energy production cooperatives, 2010–2012

---

[14]Small and medium-sized firms have sales volumes below 50 million Euros (Deutsche Bundesbank 2013, p. 47).

6.2 Mobilisation of Substantial Resources   141

The share of analysed organisations that had very high equity ratios, lying above 90%, was at 22% in 2010 but decreased to 20% in 2012. The limited share and further decrease of very high equity ratios may indicate that, even though renewable energy production cooperatives have a strong equity background, additional capital is needed to realise growing project goals. In total, members of renewable energy production cooperatives provided about 400 million Euros equity in the form of member shares by the end of 2015.[15]

6.2.2.3 Profit and Loss Development

According to general cooperative principles, it is not the maximisation of profit that is central but the support of cooperative members (Zerche et al. 1998, pp. 216ff.). In this respect, cooperatives seem to follow a different business logic than many conventional organisations. Nevertheless, cooperatives do need to generate profit or have to at least avoid losses in order to survive and establish a robust business organisation over the long term.

Figure 6.8 displays the profit and loss development of all analysed renewable energy production cooperatives that were registered in 2010. This particular focus only on organisations that were registered in 2010 enables assessment of their financial development from their founding until their third operating year. Figure 6.9 displays the related returns on equity, which indicates how much income after taxes was achieved per Euro of equity (Heesen and Gruber 2011, p. 285). Out of 125 renewable energy production cooperatives that were registered in 2010,

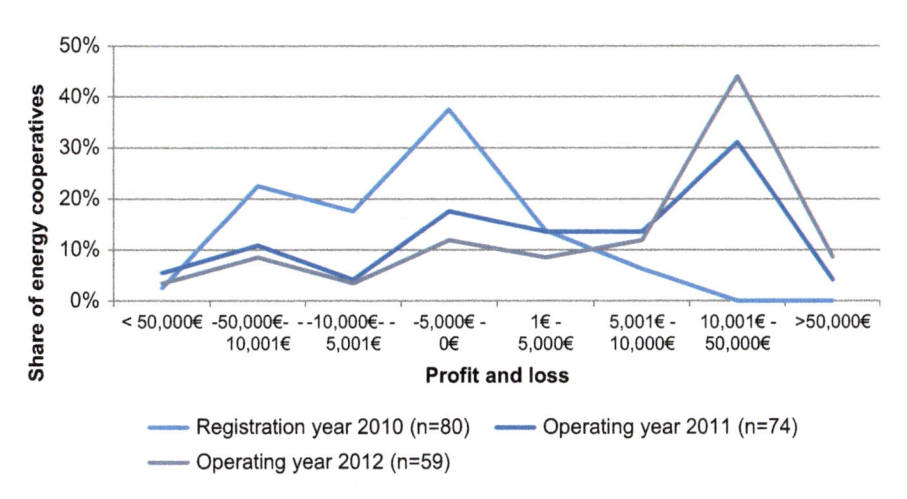

Fig. 6.8 Profit and loss development of German renewable energy production cooperatives that were registered in 2010, timeframe 2010–2012

---

[15] See Sect. 5.2.2 for further details on extrapolation.

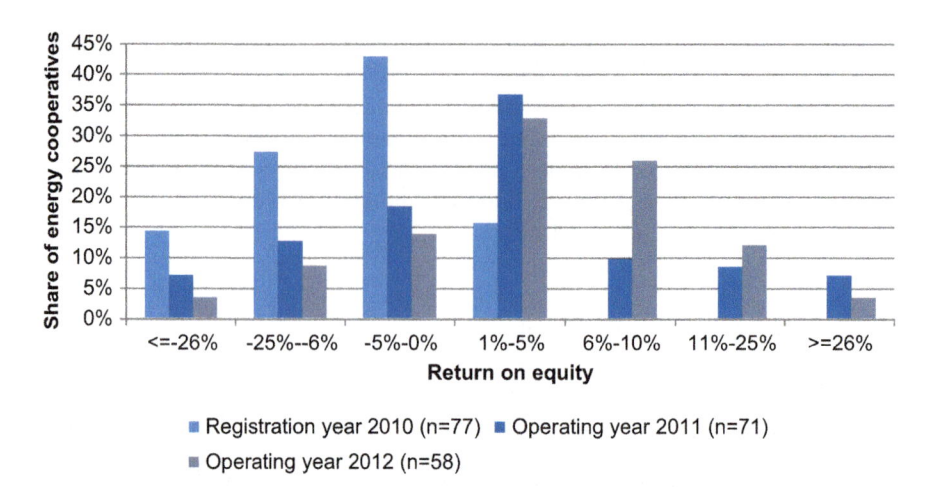

Fig. 6.9 Return on equity of German renewable energy production cooperatives that were registered in 2010, timeframe 2010–2012

80 organisations provided profit and loss information in their balance sheets for the business year 2010 (representing a share of 64%), 74 organisations listed their annual profits or losses for business year 2011 (59%) and 59 cooperatives provided profit and loss data for business year 2012 (47%).[16] The return on equity was not calculated for organisations that have no equity capital. Therefore, the number of organisations for which return on equity is provided in Fig. 6.9 is slightly lower than the number of organisations for which profit and loss data are presented in Fig. 6.8.

As can be seen in Fig. 6.8, about 55% of analysed renewable energy production cooperatives that were registered in 2010 reported between 0 and −10,000 Euros loss in their registration year, and 25% incurred losses below −10,000 Euros. Only a minority of 20% was able to make a profit in the range of up to 10,000 Euros in their first operating year, with no cooperatives making more than 10,000 Euros profit in that year. In line with that development, over 80% of the cooperatives had a negative return on equity in their first business year, as displayed in Fig. 6.9, while a 43% had an equity ratio that ranged between 0 and −5%, and 42% lay below −5%. The circumstance that most renewable energy production cooperatives experienced losses in their first year is not unusual, however, since investments usually need to be made first before profits can be generated.

As displayed in Fig. 6.8, 73% of the surveyed cooperatives that were registered in 2010 were already able to generate revenues 2 years after their registration. A share of 20% made up to 10,000 Euros profit and 44% had profits ranging between 10,001 and 50,000 Euros. Similarly, 59% of all analysed organisations achieved a return on

---

[16]Most energy cooperatives are, per definition of § 267, para. 1 HGB, small corporate entities. Small organisations are not required to publish a separate income statement with detailed information about annual profit and loss figures (§ 339 HGB).

equity that ranged between one and 10% in 2012, while 16% even achieved a return above 10%.

These empirical results reveal that the majority of analysed renewable energy production cooperatives registered in 2010 were able to consolidate business operations within 3 years of their registration. On the one hand, such positive economic development may not be seen as surprising, since renewable energy cooperatives had been receiving a guaranteed feed-in tariff for investments in renewable energy production (Erstes Gesetz zur Änderung des Erneuerbare-Energien-Gesetz 2010). This applies to the 76% of analysed cooperatives that were registered in 2010 generating energy with photovoltaic projects, 10% generating wind power and 20% producing energy from biomass. On the other hand, this positive economic development can also be taken to demonstrate that the new business approach of energy cooperatives seems to be well-suited for professionally operating renewable energy projects. This is especially striking, since energy cooperatives seem to be generally operated on a voluntary basis (see Chap. 7). As energy production projects can be challenging and often involve many project risks, such as the rise of technical problems, these energy cooperatives must have implemented and coordinated their projects well in order to be able to achieve return on investments in the range shown in Fig. 6.9.

6.2.3 Installed Renewable Energy Production Capacity

Installed renewable energy production capacity is a key cooperative resource to examine, because it provides important insights regarding the influence of the community as a whole on the renewable energy technology infrastructure in Germany. As explained in Sect. 5.2, my analysis of installed energy production capacity is focused on solar power producing energy cooperatives, because they generally provide the most reliable data regarding their actually installed capacity. It is sufficient to concentrate on this group in order to understand the technical influence of the cosmopolitan community, because 51% of all registered energy cooperatives produce solar energy. Figure 6.10 presents an overview of the installed solar production capacity among observed solar producing energy cooperatives in 2015.

As we can see, 80% of all analysed solar power producing energy cooperatives had installed up to 1000 kilowatts peak power production capacity by May 2015. Within this range, 24% had installed up to a 100 kilowatts peak, 18% between 101 and 200 and 25% between 201 and 500 kilowatts peak of power capacity. About 15% of all analysed organisations had installed between 1001 and 5000 kilowatts peak solar power, whereas only about 5% had installed more than 5000 kilowatts peak. No energy cooperative could be identified with a capacity greater than 20,000 kilowatts peak solar power at that time. A similar range was observed for biomass heat production, with about 76% of biomass energy production cooperatives having installed up to 1000 kilowatts of heat capacity. Even though the data sources on biomass production were too limited in order to generate robust results,

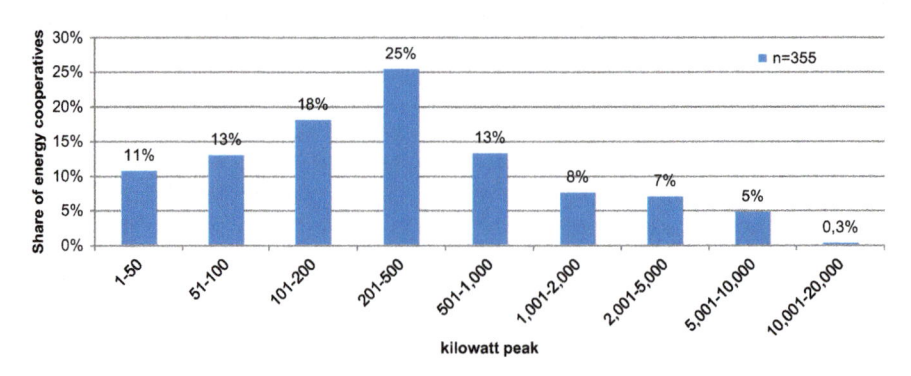

Fig. 6.10 Installed solar production capacity among German observed solar producing energy cooperatives, as of May 2015

the available data does complement the findings on solar power capacity, as the overwhelming majority of renewable energy production cooperatives installed up to 1000 kilowatts peak energy capacity.

Extrapolation of solar power reveals that all solar power production cooperatives that were registered in 2015 installed about 514,580 kilowatts peak solar power capacity as of May 2015.[17] Table 6.3 compares installed solar power capacity with the overall solar power capacity that had been installed in Germany by March 2015.

Looking at these figures, it seems evident that the influence of renewable energy production cooperatives on the distribution of renewable energy technology in Germany is actually quite small, having only a 1.4% share of the total solar power capacity by March 2015. Since the share of renewable energy production cooperatives that is engaged in wind power (19%), biomass energy (23%) or water (3%) is smaller compared to the share that is engaged in solar power production (70%), it is assumed that their influence on the distribution of wind power, biomass or water technology in Germany is smaller than 1.4%.

Table 6.3 Solar power capacity of renewable energy cooperatives and total installed solar power capacity in Germany, 2015

Solar power capacity	In megawatt	In percent
Total installed solar power capacity in Germany as of March 2015 (Deutsche Gesellschaft für Sonnenenergie e.V. 2015)	36,955	100
Total solar power capacity installed by all renewable energy production cooperatives as of May 2015 (own source)	515	1.4

---

[17]See Sect. 5.2.2 for further details on extrapolation.

6.2.4 Resource Interrelation

Since the resource power of energy cooperatives is based upon their capability to activate actors, capital, as well as technology, I evaluate here to what extent the mobilisation of high membership numbers, high amounts of capital and high levels of installed power capacity by this community are directly related to each other.

High equity ratios indicate that most energy cooperatives receive their capital from their members, so it could be assumed that large capital cooperatives also have many members. However, this is only true to some extent, as about 47%, that is, just less than half of those energy cooperatives that had more than two million Euros of capital in 2012, also had more than 200 members. In consequence, we can assume that renewable energy production cooperatives with high amounts of capital do not automatically have many members. Furthermore, only about 37% of those organisations that have more than 200 members and more than two million Euros of capital also achieved a positive return on investment in 2012. Hence, organisational size cannot be directly related to financial success.

We can also assess here whether large cooperatives can be associated with certain energy production types or to the application of more than one core business goal. The data reveals that 34% of the analysed cooperatives with more than 200 members and more than two million Euros of total capital invested in more than one energy resource, with some operating or investing in photovoltaic plants as well as in windmills or biomass energy plants, for example. Yet, it turns out that more cooperatives with many members and a high amount of total capital are focused on only one energy resource, such as the 45% of large organisations that solely focus on photovoltaic. This could mean that energy production cooperatives that have increased in size tend to produce solar energy. However, compared to all analysed solar energy production cooperatives, only 11% had more than 200 members and more than two million Euros capital in 2012, so it can be seen that solar production cooperatives are not automatically large. Meanwhile, 11% of large analysed renewable energy production cooperatives focus solely on wind. Thus, a direct relationship between large energy production cooperatives and a certain type of energy resource was not observed. There also does not seem to be a direct relationship between large membership numbers, a high amount of capital and number of business activities, as only 18% of analysed organisations with more than 200 members and two million Euros total capital have more than one business goal, such as producing as well as marketing and selling energy.

Organisations with more than 200 members and more than two million Euros total capital only make up 10% of all organisations. However, they are responsible for 30% of the installed solar power capacity. Hence, there does seem to be an interrelationship between large cooperatives and level of installed technology.

This joint evaluation of results regarding number of members, capital held and achieved technology levels among German energy cooperatives has revealed that other factors may exist that explicitly support mobilisation of their resources. Analysis of the actor network presented in the following section will seek to shed light on the influence of the actor-network composition of the enterprise community under study.

6.3 Associated Actors

In the following, actors and actor groups are identified which can be directly or indirectly associated with energy cooperatives, with the aim of assessing who belongs to the enterprise community around German energy cooperatives and who supports the group. Actors are differentiated into dominant collaborative partners (Sect. 6.3.1), intermediate actors (Sect. 6.3.2) and community outsiders (Sect. 6.3.3).

6.3.1 Dominant Collaborative Partners

Cooperatives have a strong tendency towards promoting collaboration due to their legal business structure, which includes an open membership approach. I conducted a basic actor network analysis in order to identify those collaborative partners who have particularly intense relationships with German energy cooperatives and strongly influence their organisational setups as well as their business activities. An added benefit of identifying dominant collaborative partners is that they can provide important insights into the network composition around energy cooperatives and existing power constellations within the enterprise community.

Collaborative partners, in general, and dominant collaborative partners, in particular, were identified by focussing on two groups (1) founding partner(s) of energy cooperatives and (2) members of cooperative steering boards. Both groups strongly shape the character of energy cooperatives, because they are directly engaged in developing and guiding these organisations.

As noted in Sect. 6.1.3, one of the goals of renewable energy cooperatives is to encourage and support citizen participation in renewable energy production. It is, thus, not surprising that many cooperatives are also founded and/or managed by private citizens. In this work, citizens are understood as people who primarily act out of their private motivations and not predominantly in the name of another association, institution or company. In almost all renewable energy production cooperatives, citizens were identified as founding partners or as members of steering boards. However, only 19% of all analysed organisations seemed to be exclusively founded or managed by citizens without the further involvement of other organisations or institutions. Hence, citizens are an important actor group for the cosmopolitan enterprise community, but they do not appear to be the only actor type guiding the business activities of energy cooperatives in Germany.

Table 6.4 provides an overview of all identified actor groups who directly collaborate with renewable energy production cooperatives that were registered by the end of December 2014, listing the shares of founding partnerships and steering

6.3 Associated Actors

Table 6.4 Institutional and organisational collaborative partners of German renewable energy production cooperatives; more than one category per cooperative is possible

Identified institutional and organisational collaborative partners of German renewable energy production cooperatives	Founding partner (n = 512) (%)	Member of supervisory board (n = 517) (%)	Member of executive board (n = 517) (%)
Communities			
Districts, municipalities, cities, villages	38	26	9
Banks			
Cooperative banks	31	16	19
Other banks (for example, savings banks)	3	3	1
Energy-related companies			
Municipal energy providers	11	6	7
Other energy cooperatives	5	1	1
Renewable energy advisory and project management companies	4	2	3
Renewable energy-related engineering companies, technical providers and technical installers	3	3	4
Renewable energy plant operators	3	0.4	1
Renewable energy related investment companies	2	1	1
Other energy providers	2	1	1
Renewable energy providers	0.6	0.2	0.4
Other energy-related companies	0.4	0.4	0.2
Regional associations			
Regional institutions and initiatives	10	6	4
Regional climate and renewable energy institutions and initiatives	7	1	3
Municipal institutions	4	3	2
National associations and research institutions			
National operating associations	0.4	0.6	0.4
Research institutions	0.4	2.3	0.8
Other larger companies			
Regional operating companies	1	0.8	0.4
National operating companies	0.8	0.2	0.6

board memberships for each actor type. More than one category per cooperative is possible. For example, an energy cooperative can be founded by a cooperative bank as well as by a community. At the same time, a bank can be a founding partner or may in addition be represented in the steering board.

As can be seen, renewable energy production cooperatives are founded and managed by many different organisational and institutional actor types, which are grouped here into six main categories. The group with the strongest presence are communities, which includes districts, municipalities, cities and villages. When this group is involved, the mayor or the district's chief administrative officer becomes a member of the cooperative's steering board.

The second group represents banks, with almost all involved financial institutions here being cooperative banks. Their own directors and supervisory board members are generally the ones who guide the foundation process of a cooperative or become a member of its steering board. A few savings banks (*Sparkassen*) are also involved as founding partners or management supporters. However, large internationally operating banks, such as the Commerzbank or the Deutsche Bank or state-run banks, such as the HSH Nordbank (Hamburg), could not be identified as founding partners of any renewable energy production cooperatives or as members of their steering boards.

The third group is composed of energy-related companies—a group involving a number of different actors, all of which operate along the energy value chain (see Fig. 2.1). Municipal energy providers are the strongest actor group in this category. Here it is the managing director who is predominantly active in guiding a cooperative's foundation process and becomes a member of its steering board. Compared to the share of municipal energy providers, only a few conventional or renewable energy providers could be identified as founding partners or members of steering boards. Other actor types within this category include other energy cooperatives, renewable energy advisory and project management companies, renewable energy-related engineering companies, technical providers and technical installers, renewable energy plant operators, renewable energy-related investment companies and other energy-related companies the concrete value chain position of which could not be readily identified.

The fourth category of cooperative partners represents regionally operating associations. Regional institutions and initiatives include churches, regional groups of a farmers' association (*Maschinenring*: farm machinery cooperative), as well as regional offices of larger foundations, such as the Heinrich Böll Foundation, Friends of the Earth Germany (BUND) or Nature And Biodiversity Conservation Union (NABU). Regional climate and renewable energy institutions or initiatives include bioenergy regions or official energy committees founded by citizens and local officials operating in networks with a focus on renewable energy. Municipal institutions involve amongst others climate protection agencies, agenda 21 offices, schools, or the municipal promotion of economic development.

The fifth category of cooperative partners represents nationally operating associations, such as Greenpeace, as well as research and development institutions, while the sixth category includes larger nationally (and internationally) operating companies, such as Unilever or Volkswagen, and large regionally operating companies, such as agricultural cooperatives or residential building cooperatives.

Identification of collaborative partners has revealed that energy cooperatives have been changing the actor composition of renewable energy supporters in Germany. Whereas energy-related companies already operated along the energy value chain, communities, banks, regional associations, national associations and initiatives, as well as other larger companies are primarily new actor groups that have become involved in developing and managing renewable energy projects. The classification

6.3 Associated Actors

presented here represents a significant improvement on previous analyses of founding partners or supporting actors, such as those by Volz (2012), Maron and Maron (2012) and the Research Institute for Cooperation and Cooperatives, Vienna University of Economics and Business (2012), because the empirical assessment of secondary data undertaken for the present work offers an extended and more detailed view of the whole network than the previous studies, and the data has been systematically classified into different actor groups in a more precise way as well.

A collaborative partner (e.g., bank, community) is defined as being dominant if it is present either as a founding partner or member of a cooperative's steering board in at least 30% of all analysed cooperatives (see Table 6.4), which implies that a considerably high number of energy cooperatives prefer to collaborate with this actor group in order to realise business activities. Dominant collaborative partners (1) have a particularly intense and influential relationship with energy cooperatives through direct interaction on a continuous and long-term basis and (2) are chosen by the majority of energy cooperatives as collaborative partners. Such dominant collaborative partners shed light upon dominant collaborative interaction models that exist within the niche community and, based upon which, energy cooperatives may primarily realise their business activities and structural change.

Table 6.5 summarises the overall presence of each actor category and points out which collaborative partner groups seem to be most dominant, meaning here groups that have founded and/or manage more than 30% of all renewable energy production cooperatives.

Next to private citizens, banks, communities and energy-related companies seem to play the most dominant role in this cosmopolitan niche community, as each of them is present in more than 30% of all analysed cooperatives. Communities, mostly represented by their mayors, are founding partners and/or members of the steering

Table 6.5 Overview of the six identified collaborative partner groups for German renewable energy production cooperatives, ordered according to their dominance; more than one category per cooperative is possible

Identified collaborative partner groups of German renewable energy production cooperatives	Founding partner (n = 512) (%)	Member of supervisory board (n = 517) (%)	Member of management board (n = 517) (%)	Overall presence (n = 566) (%)
Dominant collaborative partner groups (present in more than 72% of all analysed cooperatives)				
Communities	38	26	9	41
Banks, particularly cooperative ones	32	18	20	33
Energy-related companies, particularly municipal energy providers	29	11	17	31
Other collaborative partner groups				
Regional associations	20	9	8	23
National associations & research institutions	1	3	1	4
Other larger companies	2	1	1	2

150   6 German Energy Cooperatives: A Rising Cosmopolitan Enterprise Community

board at 41% of all analysed organisations. Banks are active as founding partners and/or as members of the steering board at 33% of cooperatives, while at least one energy-related company is founding and/or managing partner at 31%. In total, 72% of all analysed renewable energy production cooperatives have been founded and/or are managed by communities, banks or energy-related companies. Hence, the majority of energy cooperatives draw upon similar collaborative interaction models in order to realise their business goals:

1. Collaboration with communities;
2. collaboration with banks, particularly with cooperative banks; and
3. collaboration with energy related companies, particularly with municipal energy providers.

The presence of these three collaborative models is even higher among the largest energy cooperatives, meaning those that have more than 200 members and more than two million Euros in capital. About 81% of these have been founded and/or are managed by banks (mostly cooperatives banks), communities or energy related companies (mostly municipal energy providers), which underlines their important position as dominant collaborative partners. Table 6.6 displays an overview of their individual presence among the largest energy cooperatives.

As can be seen, almost 60% of analysed cooperatives have either been founded or are managed by banks, almost half collaborate with communities and about 30% with energy-related companies, of which almost all are municipal energy providers. Hence, there seems to be a direct relationship between mobilisation of resources and choice of collaborative partners.

In addition to identifying dominant collaborative partners, it is also important to achieve a better understanding of the steadiness of their presence over the long term. Figure 6.11 shows the share of annual energy cooperative registrations, differentiated according to their collaboration with banks, communities or energy-related companies.[18]

As displayed in the figure, the number of new collaborations between energy cooperatives and communities, banks or energy-related companies varies over the

Table 6.6 Dominant collaborative partners among the largest German renewable energy production cooperatives

Dominant collaborative partners among the largest German renewable energy production cooperatives	Founding partner (n = 35) (%)	Member of supervisory board (n = 35) (%)	Member of executive board (n = 35) (%)	Total presence (n = 35) (%)
Communities	34	40	3	49
Banks	57	37	46	59
Energy-related companies	26	9	14	27

[18]The data refers to registered German renewable energy production cooperatives for which the dominant organisational or institutional collaborative partners could be assessed (n = 554).

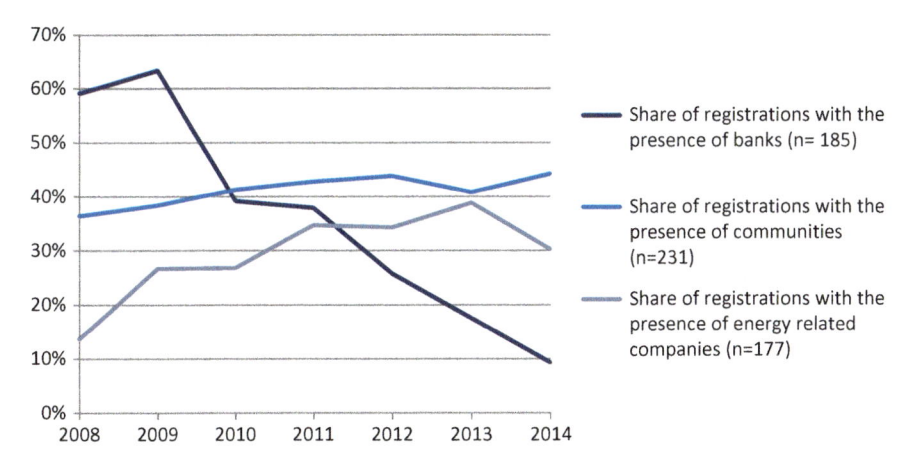

Fig. 6.11 Annual registration of German energy cooperatives collaborating with banks, communities or energy-related companies

years. The share of organisations in which communities function as foundation or management partners appears to have been stable since 2008, though it slightly increased in 2014. Meanwhile, the share of energy-related companies that have founded and/or managed a newly registered energy cooperative continuously increased until 2013. However, in 2014 their relative engagement seems to have decreased. Banks appear to have been the main collaborative partner type most involved in the founding and/or managing of energy cooperatives around the year 2009. But from then onwards, their relative presence strongly decreased, reaching its lowest level in 2014. It is assumed that one of the reasons for the decreased engagement of cooperatives banks has been the decreased profitability of renewable energy production projects. However, the reduction of their involvement already started in 2009 and, thus, does not directly correlate with the relevant amendments of the German Renewable Energy Act (EEG). Given this very general outline of dominant collaborative partner trends in Germany, it appears that further analysis is necessary in order to better understand the motivation of each key collaborative partner group towards founding and/or managing energy cooperatives as well as their positions and levels of influence on their business activities, given during the case study analysis.

6.3.2 Intermediate Actors

In 2012, Maron and Maron (2012, p. 241) claimed that the field of energy cooperatives needed stronger support from a network of secondary actors who can provide superior advice and know-how for energy cooperatives. As discussed in this section, my analysis of intermediate actors suggests that such a trans-local network within the energy cooperative community has experienced a substantial development in this regard.

6.3.2.1 Cooperative Associations

According to German cooperative law, all cooperatives have to become a member of a cooperative auditing association (*Prüfungsverband*). Such associations conduct obligatorily audits of their member cooperatives every 1–2 years, during which the economic conditions, compliance with defined business statutes and work of cooperative management is verified (§§ 53ff. GenG). These audits are intended to ensure that cooperatives are properly organised and managed in order to protect members and prevent insolvencies. Cooperative auditing associations not only support energy cooperatives in their foundation and registration processes but also offer juristic and economic advice whenever it is necessary. Many provide also additional education programs for their member organisations.

Cooperative-focused nationwide confederations also exist which seek to represent the interests of German cooperatives against political and administrative forces at the national level. One of the largest confederations, the German Cooperative and Raiffeisen Confederation, founded a special sub-agency in 2013 that exclusively focuses on supporting and fostering the interests of energy cooperatives. The agency was initiated in consequence of the strong growth dynamic among energy cooperatives.

This well-established network of superior associations and confederations is unique in the business sector, unifying all cooperatives in Germany and providing support as well as advice from the very beginning of a cooperative's foundation. Being embedded in such a strong network of supportive intermediate actors has played a crucial role for the strong growth dynamic of energy cooperatives in Germany, as further outlined during the case study analysis.

6.3.2.2 Associations

Next to the official cooperative-auditing associations and confederations, other intermediate actors exist who mainly focus on supporting energy cooperatives. The most important ones are the following:

- The Energiewende Now Network (*Netzwerk Energiewende jetzt e.V.*), which primarily concentrate on helping to found new energy cooperatives and it provides a platform for active networking among existing energy cooperatives. Amongst other activities, the initiative has developed an education course in which participants are trained to support the foundation of energy cooperatives with the goal of earning a qualification termed 'Project Developer for Energy Cooperatives'.

- The Union for Citizens Energy (*Bündnis Bürger Energie e.V.*) was founded in 2014 with the aim of unifying all actors who are active in the field of citizen-based energy production. Energy cooperatives represent a fundamental force of achieving citizen-organised energy due to their open, citizen-oriented membership approach and are, thus, well represented in the union. The initiative actively lobbies for the interests of citizen-based energy and tries to improve conditions in this field. It is in continual dialog with national politicians and organises regular conferences and meetings among its members.

Several associations for citizen-based energy cooperatives exist that have been founded in and are focused on providing support for energy cooperatives within certain Bundesländer. They connect energy cooperatives with each other, inform their supporters about important topics, offer advice and provide active support during a cooperative's foundation process. Furthermore, they aim to represent the interests of energy cooperatives. Currently existing associations include the following:

- The Association of Citizen-based Energy Cooperatives in Baden-Württemberg (*Verband Bürger Energiegenossenschaften in Baden-Württemberg e.V.*), which was founded in 2009.
- The Regional Network for Citizen-based Energy Cooperatives in Hessen (*Landesnetzwerk Bürger-Energiegenossenschaften Hessen e.V.*), which was founded in 2015.
- The Platform for Citizen-based Energy and Energy Cooperatives (*Plattform Bürgerenergie & Energiegenossenschaften*), which was founded by the energy agency of Nordrhein-Westfalia in 2016.

All in all, it can be said that energy cooperatives are supported by a growing network of intermediate actors. Of special note here is that energy cooperatives are, on the one hand, able to draw upon the long-existing and prevailing network of traditional cooperative associations while, on the other hand, also becoming embedded in a growing group of newly founded initiatives purely focused on the needs of energy cooperatives. Many of these initiatives have been founded from 2013 onwards, revealing a clear development and empowering of the overall network.

6.3.3 Community Outsiders

Energy cooperatives have also achieved considerable attention from outside their enterprise community, especially among governmental representatives and politicians as well as societal associations, which function as opinion leaders in society and thus have the potential to accelerate the transfer of new guiding principles and societal viewpoints into the wider public. The main interest of my analysis was, therefore, to observe whether and how actors belonging to these groups have started to notice and value the activities and development of energy cooperatives.

Communication is, according to Giddens (1984), a form of interaction dealing with the significance of structures. Governmental representatives and politicians, as well as members of societal associations, can increase the significance of certain communities and their ways of interaction by talking about their activities among the general public. In the following, examples are given of each group recognising energy cooperatives. It is not the aim here to provide a complete list of all group representatives who have done this but to demonstrate that recognition exists in general among these important intermediate actor groups.

6.3.3.1 Governmental Representatives and Politicians

Key national politicians have clearly started to notice energy cooperatives as important actors of the Energiewende and value their activities. Former Minister of the Federal Ministry for the Environment, Nature Conservation, Building and Nuclear Safety Norbert Röttgen put his appreciation in the following manner:

> "In dieser Konsequenz und mit den Erfolgen ist es das Projekt auch der Bundesregierung. Auch das ist ein Teil der Wahrheit. Ich glaube, wir können auch wechselseitig anerkennen, dass es diese Regierung ist, dass es die Länder sind, dass es Kommunen sind, dass es Energiegenossenschaften sind, dass es eben das Land ist, das vorangeht. Das ist auch gut" (Röttgen 2011).

> "In this respect and with these achievements, this is the federal government's project. But that is only partly true. I believe that we can also mutually recognise that it is this administration, that it is the federal states, that it is the municipalities, that it is the energy cooperatives, that it is the nation that is leading [this change]. And that is good" (Röttgen 2011).

Appreciation of politicians for the work of energy cooperatives was especially revealed in the course of the latest amendments of the Renewable Energy Act in 2014 and 2016. Many political representatives have announced that energy cooperatives and their contribution to the Energiewende are important. As the former Minister of the Federal Ministry for Economic Affairs and Energy, Sigmar Gabriel says, it must also be ensured that they are still able to operate under changing national regulations:

> "Niemand—darauf lege ich Wert—muss Angst davor haben, dass auf diesem Weg Bürgerwindparks oder Energiegenossenschaften keine Chance auf Teilnahme mehr erhalten. Im Gegenteil: Wir werden einen gesonderten Gesetzentwurf in den Bundestag einbringen, mit dem wir diese Beteiligung der Bürgerinnen und Bürger nachhaltig sichern werden" (Gabriel 2014).

> "Nobody—and I put great value in this—should be afraid that along this path citizen-based wind parks or energy cooperatives will no longer have a chance to take part. On the contrary, we will introduce a special draft bill to the Bundestag seeking to secure lasting participation of citizens [in this process]" (Gabriel 2014).

Also, Simone Peters, federal representative of the Green Party (BÜNDNIS 90/DIE GRÜNEN), has claimed that energy cooperatives should not be excluded from the energy market, because they are important for realising citizen participation (Peter and Glahr 2015).

6.3.3.2 Societal Associations

Key national associations, which function as societal opinion leaders and also greatly influence the work of the national government, have also started to explicitly value energy cooperatives, with several having begun to see their special mode of interaction as an important contribution to the development of a renewable energy system in Germany (interaction patterns are discussed in detail during the case study analysis in Chap. 7). The Ethics Commission for a Safe Energy Supply explicitly pointed towards energy cooperatives in its 2011 final report and argued that the participation of multiple actors, in particular of citizens, which is being fostered by energy cooperatives, is an important factor for achieving a change of Germany's energy system (Ethics Commission for a Safe Energy Supply 2011, pp. 30, 45, 96). The committee was appointed by Federal Chancellor Angela Merkel and functioned as a central voice in the process of finalising the decision to abandon the use of German nuclear power plants. The committee also greatly influenced the final political step of the former governmental position of the CDU and FDP to revise their initial decision for extending the operational life of German nuclear power plants. The German Advisory Council on Global Change explicitly names energy cooperatives in their final report on the energy transition, published in 2011, as actors who can make the change of energy system both technically and economically feasible (German Advisory Council on Global Change 2011, p. 182). Interesting is that the Council had not yet mentioned energy cooperatives in their final report on energy systems, published in 2003 (German Advisory Council on Global Change 2003). This more recent acknowledgment is thus likely to correlate with the growth dynamic of energy cooperatives which began in 2006 (see Fig. 6.1).

These brief examples seem sufficient to demonstrate that the work of energy cooperatives has begun to be acknowledged beyond the boundaries of their own community. The recognition of their work by central governmental representatives and politicians as well as by societal associations can help to unfold the potential for creating a wider institutionalisation of the particular manner of acting for which energy cooperatives stand.

6.4 Summary

The assessment of all registered German energy cooperatives, their business goals and locations, the development of their members and capital over a timeframe of 3 years, their installed power capacity, as well as the identification of their collaborative partners, associated actors and community outsiders has provided broad insights into the character and status quo of the enterprise community surrounding energy cooperatives.

Out of 1055 energy cooperatives that were registered through December 2015, 88% were active in the renewable energy sector and 73% produced renewable energy. The overwhelming majority—92% of all existing organisations—was registered from 2006 onwards, and they exist in each Bundesland across Germany.

By 2015, all registered renewable energy production cooperatives had

- Mobilised about 158,000 cooperative members;
- allocated about 1.3 billion Euros total capital, of which 400 million Euros came from its members in the form of member shares; and
- installed about 515 megawatts peak solar power capacity.

Yet the influence of German energy cooperatives on the distribution of renewable energy technology has ranged around or below 1.4% per technology and is, thus, still relatively small.

It was observed that renewable energy production cooperatives not only have technology-oriented business goals but have also articulated the following sociopolitical goals:

1. A special focus on their region when producing renewable energy (74% of all studied renewable energy cooperatives) and
2. the involvement of citizens in realising their business activities (71%).

Almost all energy cooperatives are founded and/or managed by citizens, meaning here people who are active within the energy cooperative out of their own private motivation. At the same time, 81% of German renewable energy production cooperatives have been founded and/or managed by organisational or institutional actors, which I have differentiated into six main categories: (1) communities, (2) banks, (3) energy-related companies, (4) regional associations, (5) national associations and research institutions, as well as (6) other larger companies. Except for energy-related companies, which already operate along the energy value chain, most of these are new actor groups in terms of developing and managing renewable energy projects. Out of the six categories, renewable energy production cooperatives seem to primarily prefer collaboration with the following three partner groups, based upon the aid of which they seek to realise their business goals:

1. Communities,
2. Banks, in particular cooperative ones; and
3. Energy-related companies, in particular municipal energy providers.

All in all, 72% of the analysed renewable energy production cooperatives were founded and/or are managed by communities (41%), banks (33%) or energy-related companies (31%), which can thus be understood as their dominant collaborative partners. They are particularly influential because they have shaped the character and aided the setup of most of the recently emergent energy cooperatives; hence, this cosmopolitan niche community is mainly guided by citizens, communities, banks

6.4 Summary

(particularly cooperative banks) and energy related companies (particularly municipal energy providers).

In addition, energy cooperatives receive support from a well-established and growing network of intermediate actors, and the existence and work of energy cooperatives has been officially recognised by important societal opinion leaders, most of all by central governmental representatives and politicians as well as societal associations. This attention has increased the significance of the particular modus operandi of energy cooperatives beyond the boundaries of their own community.

The energy cooperative community can grow by increasing the total number of organisations as well as by increasing the individual size of organisations. New company registrations reached their maximum of 203 organisations in 2011. Since then, annual registrations have strongly decreased, reaching 42 new organisations in 2015, the lowest level since 2008. The positive, as well as the negative, growth dynamic seems to be related to the creation or alteration of national regulations, in particular the introduction and later reduction of the feed-in tariff for renewable energy production in the German Renewable Energy Act and the inclusion and later exclusion of cooperatives in a more recently passed regulation for alternative investment funds.

In terms of size and assets, 80% of all renewable energy production cooperatives have less than 200 members or less than five million Euros total capital. It seems that most energy cooperatives stay within these ranges and are not becoming significantly larger. However, observed individual growth rates of organisations with more members and more capital reveal that renewable energy cooperatives could have the capacity to further increase their individual resource power for supporting the newly emerging renewable energy structure. For example, the share of organisations that had more than five million Euros total capital almost doubled: from 4% in 2010, to 7% in 2012. There seems to be a direct relationship between resource mobilisation and choice of collaborative partners, as 81% of the largest renewable energy production cooperatives—those possessing more than 200 members and more than two million Euros total capital—were founded and/or are managed by one of the three dominant collaborative partner groups. Hence the presence of communities, banks and energy-related companies among the largest energy cooperatives is higher compared to their presence among all renewable energy production cooperatives (72%). These largest organisations, which make up 10% of all renewable energy generating cooperatives, and their dominant collaborative partners are also responsible for almost 30% of the installed solar power capacity in Germany.

The growth potential of energy cooperatives and the accompanied increase of their influence thus seem to be strongly influenced by national regulations and the involvement of the three dominant collaborative partner groups of communities, banks and energy-related companies. The quantitative assessment of energy cooperatives undertaken here has provided detailed insights into the setup, character and the current status of the whole cosmopolitan community. In the following chapter, a case study analysis will provide detailed understanding of their interaction models as well as their aligned potentials, challenges, limits and pre-conditions.

Chapter 7
Analysis of Dominant Collaborative Interaction Models

Interaction of energy cooperatives is analysed based upon three different case studies. The case study analysis shall reveal whether and how energy cooperatives may support new patterns of rules and resources, which impact they have at the local level and how they achieved an influential position within their focus region. The chosen cases are relevant because they are typical and powerful examples of the whole enterprise community. Each of them stands for one of the three dominant collaborative interaction models that are preferred by 72% of renewable energy production cooperatives in order to do business. In addition each of the three cases belongs to the biggest German energy production cooperatives, regarding members and total capital. The typical case character ensures that results can be related to the whole organisational network. Their powerful character demonstrates how actor empowerment takes place within the community and sheds light on aligned potentials, challenges and pre-conditions. The cases were identified during the quantitative assessment. Figure 7.1 displays how each case study is analysed in the following.

Each case study begins with introducing the focus region of the energy cooperative. The geographical character and related regional energy transition goals being pursued in the region are described. The next section introduces the energy cooperative by explaining its foundation process and its business purpose. Aim is to comprehend why the energy cooperative emerged and how the organisation was formed. The next following section of each case study describes the collaborative concept. Main collaborative partners are identified and characterised, as well as their chosen collaborative interaction model. It is important to understand the existing position of the partners in the region, as well as their motivations for founding the energy cooperative and for collaborating. Then, detailed insights are provided in regard to the mobilisation of rules and resources. Aim is to understand in detail how the partners jointly mobilise crucial resources allowing them to pursue their business activities and whether and how they create new societal rules. In the end of each case study a comprehensive summary is given during which opportunities, difficulties, limits and pre-conditions of joint interaction are discussed. Each case study was analysed in great depths. In order to give the reader a comprehensive overview about

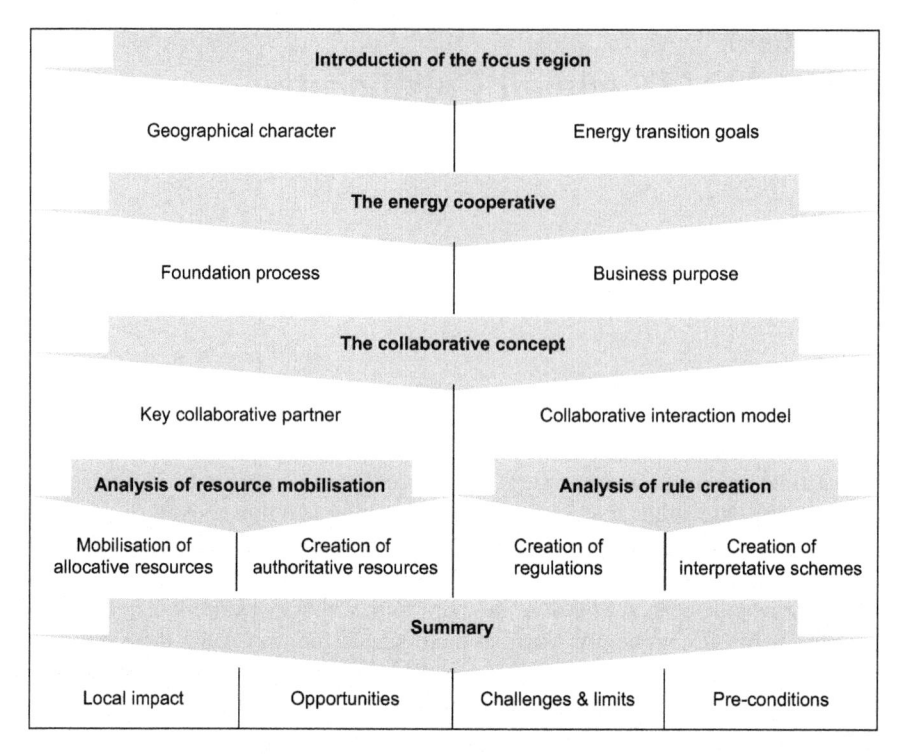

Fig. 7.1 Applied procedure for analysing the case studies

the results of each case a detailed summary is provided at the end of each case study section. I have underlined the argumentations and analysis in each case study with quotes from the interviews (a detailed description of how the interviews were conducted is provided in Sect. 5.3). All quotes used here are translated from German into English. The original German version of the used quotes can be found in annex 14.

Section 7.1 describes the collaborative interaction model between energy cooperatives and other energy related companies. Section 7.2 outlines the cooperation between energy cooperatives and banks. Section 7.3 describes the collaborative interaction between energy cooperatives and communities.

7.1 Collaboration Between Energy Cooperatives and Energy Related Companies

One of the three dominant collaborative interaction models favoured by most German energy cooperatives represents the collaboration with energy related companies in particular with municipal energy providers. As outlined in Sect. 6.3.1, 31%

of all analysed renewable energy production cooperatives were founded and/or are managed by energy-related companies.

The BürgerEnergiegenossesnschaft Wolfhagen eG (BEG Wolfhagen eG; The Citizens' Energy Cooperative of Wolfhagen) was chosen as a case for this study because it (1) has a strong and particularly innovative collaborative relationship with the Stadtwerke Wolfhagen, the municipal energy provider for the municipality of Wolfhagen; (2) has experienced strong organisational growth; and (3) is playing an important role in the development of a 100% renewable energy supply for Wolfhagen. The BEG Wolfhagen eG was founded by the Stadtwerke Wolfhagen and became, next to the city, the second shareholder of the municipal energy provider. Since 2012, the energy cooperative has held a 25% share of the Stadtwerke, and the two partners have been actively supporting Wolfhagen's goal of establishing a 100% renewable and regional energy structure. With 576 members and 3.1 million Euros total capital (as of September 2014), the energy cooperative belongs to the largest operating cooperatives of any kind in Germany.[1]

7.1.1 Introduction of the Focus Region

One of the core aims of the collaborative interaction between BEG Wolfhagen eG and Stadtwerke Wolfhagen is to support Wolfhagen in achieving a 100% renewable energy supply (Fraunhofer Institute for Building Physics IBP 2010a), making this municipality the focus region for this cooperative's business activities. In the following, the geographical character of Wolfhagen is described as well as its 100% renewable and regional energy goal.

7.1.1.1 Geographical Character

The municipality of Wolfhagen is located in the north of the Kassel district, which belongs to the Bundesland Hessen. Wolfhagen is situated about 25 kilometers to the west of the city of Kassel.

Wolfhagen's geographical structure, displayed in Table 7.1, is typical for the rural region of Hessen. The municipality includes an area of 112 square kilometres, which mainly consists of farmland (52%) and agricultural forest (34%), with living area and industrial real estate only making up 12% (Fraunhofer Institute for Building Physics IBP 2010a, p. 3).

Wolfhagen consists of the main city, also named Wolfhagen, and 11 smaller affiliated village districts (Fraunhofer Institute for Building Physics IBP 2010a, p. 3). As of 2014, the municipality had a population of 13,606 (Gemeinde Wolfhagen

---

[1]Parts of this case study have already been published by the author in (Debor 2017).

Table 7.1 Geographical character of the municipality of Wolfhagen as of 2010 (Fraunhofer Institute for Building Physics IBP 2010a, p. 3)

Geographical character of the region as of 2010	Municipality of Wolfhagen km²	Percent
Total size of region	112	100
Agricultural area	58	52
Forest area	38	34
Settlement and traffic infrastructure area	13	12
Other use areas	2	2

Table 7.2 The municipality of Wolhagen's population and its distribution as of 2014 (Gemeinde Wolfhagen 2014)

Overview of the municipality's population and its distribution as of 2014	Wolfhagen	Unit
Population	13,606	Number
Population per km²	121	Number/km²

2014), with around half of them—7532—living in the city Wolfhagen. As displayed in Table 7.2, Wolfhagen has about 121 citizens per square kilometre.

7.1.1.2 Energy Transition Goals

In April 2008, the assembly of city councillors of Wolfhagen (*Stadtverordnetenversammlung*) unanimously decided to supply 100% of the municipality's power demand with renewable and regional energy by 2015 (Stadtverordnetenversammlung Wolfhagen 2008). The 100% renewable energy goal was, amongst other reasons, initiated by the local network *Klimaoffensive Wolfhagen* (Climate Offensive of Wolfhagen), which was founded with the support of Stadtwerke Wolfhagen in 2007. This network, which consisted of citizens, politicians and the Stadtwerke itself, developed the first ideas for a more climate friendly municipality. In several meetings and discussions, they fostered a rethinking towards more climate protection among politicians and advocated the political resolution for a new renewable energy concept in Wolfhagen. After the parliament had passed the resolution, Wolfhagen started to participate in various national and regional climate and energy programmes. The programmes provided an important basis for developing a strategy through which Wolfhagen could realise its renewable and regional energy vision. Amongst others, it joined the Bundesland programme *Strategien von Kommunen zur Erreichung von Klimaneutralität* (strategies for communities in order to reach climate neutrality) (Kompetenznetzwerk Dezentrale Energietechnologien e.V. 2009). In 2011, it became part of the programme *100percent Erneuerbare-Energie-Region* (100% renewable energy regions), which is coordinated by the institute of decentralised energy technologies (Stadt Wolfhagen 2011a).[2] With the project *Wolfhagen 100% EE—Entwicklung einer nachhaltigen Energieversorgung für die Stadt Wolfhagen* (100% renewable energy—development of a sustainable energy supply for the city of Wolfhagen), the municipality participated in the national programme *Energy Efficient City*, which was funded by the Federal Ministry of Education and Research (Energie 2000 e.V. 2015). The municipal energy provider and the energy agency of the district Kassel Energie 2000 e.V. were the main partners of Wolfhagen in constructing a future energy concept. In addition Wolfhagen received support from expert institutions, such as the Fraunhofer Institute and the University of Kassel (for example, Fraunhofer Institute for Building Physics IBP 2010a).

Three main fields of action were identified on the way to a sustainable energy structure: (1) intelligent renewable energy supply, (2) energy efficiency and (3) electro-mobility. Hence, Wolfhagen's vision for a new energy structure includes most fields of action that are also identified by the national Government and leading national institutions as important development areas in order to achieve a renewable energy system in Germany (see Sect. 3.2). Before describing the goals, Wolfhagen's energy demand, the basis for any action-steps, are briefly described.

Energy Demand

In 2010, the Stadtwerke supplied Wolfhagen with about 66,000 megawatt hours[3] power (Fraunhofer Institute for Building Physics IBP 2010a, p. 13). As displayed in Table 7.3, about 59% were demanded by the industry, 33% by private households and 8% by others, presuming the public administration (Fraunhofer Institute for Building Physics IBP 2010b, p. 64).

Table 7.3 Power demand of Wolfhagen in 2010 (Fraunhofer Institute for Building Physics IBP 2010b, p. 64)

Power demand in 2010	MWh/a	Percent
Industry	38,684	59
Private households	21,922	33
Other	5394	8
Total	66,000	100

---

[2]The initiative *100ee Regions* identifies and supports regions in Germany with the aim to transform regional energy supply and demand towards 100% renewable energy in the long-term. The project is realised by the Institute for decentralised energy technologies and is funded by the Federal Ministry for the Environment, Nature Conservation and Nuclear Safety (Hoppenbrock and Fischer 2012).

[3]Most of the regional actors, including citizens and industry, receive their power from the Stadtwerke Wolfhagen. Wolfhagen exemplarily used their consumption figures in order to calculate the district's energy demand.

Table 7.4 Heat demand of Wolfhagen in 2009 (Fraunhofer Institute for Building Physics IBP 2010b, p. 68)

Head demand in 2009	MWh/a
Private households	160,760
Public administration	4432

As shown in Table 7.4, the heat demand of Wolfhagen's households laid around 161,000 megawatt hours in 2009. The public administration consumed about 4000 megawatt hours (Fraunhofer Institute for Building Physics IBP 2010b, p. 68). Data for the heat demand of the local industry was not provided.

Intelligent Renewable Energy Supply

The main goal being formulated in Wolfhagen's strategy for achieving a 100% renewable energy supply is to maximise the renewable energy production in the region (Fraunhofer Institute for Building Physics IBP 2010a). Two key renewable energy production projects were identified to cover most of the municipality's annual power demand:

- A solar power park with a capacity of 10 megawatt peak was constructed between Wolfhagen and Gasterfeld. Its operation started in 2012. Five megawatt peak are installed by the Stadtwerke Wolfhagen. The other five megawatt peak was installed by BLG-Project GmbH. The solar park shall produce 20,000 megawatt hours per year (BEG Wolfhagen eG 2015; Rühl 2013, p. 27).
- A wind park with four wind mills and a capacity of 12 megawatts was installed on top of the Rödeser Berg, a summit located north of the city Wolfhagen between Nothfelden and Niederelsungen. The wind park shall produce 30,000 megawatt hours per year (Rühl 2013, p. 27) and could thus cover about 50% of the municipality's power demand. The building licence was allocated in December 2013 (Müller 2013). The implementation of the wind park was finalised at the end of 2014.

The two lighthouse projects were supposed to complement already existing solar and biomass energy production (Fraunhofer Institute for Building Physics IBP 2010a, pp. 13, 28).

Wolfhagen's project choice is in line with the general calculated renewable power potential for the northern part of the Bundesland Hessen. Wind energy has with a presumed maximum supply of 3304 gigawatt hours per year the highest local potential in the district, followed by solar energy with 1760 gigawatt hours per year. The northern region of Kassel could produce 57% more renewable energy than needed, if the full potential was deployed (Fraunhofer Institute for Wind Energy and Energy System Technology & Stadtwerke Union Nordhessen (SUN) 2012, p. 28). However, as of 2012 only 15% of the region's energy demand was supplied by renewable energy produced in the north of Hessen (Fraunhofer Institute for Wind

Energy and Energy System Technology & Stadtwerke Union Nordhessen (SUN) 2012, p. 4). Only 9% of the wind power has been realised in the Bundesland so far (Ebert and Henke 2012, p. 14). It shows how challenging the installation of wind power is.

The aim of achieving a renewable supply by 2015 was focused on power. Nevertheless, heat was integrated in the energy efficiency plans of the municipality.

Energy Efficiency Through Retrofitting Old Buildings

Energy efficiency became the second main area on Wolfhagen's way towards attaining a sustainable energy supply (Fraunhofer Institute for Building Physics IBP 2010a, pp. 32ff.). The largest local potential for energy efficiency was identified in terms of heating, especially within the city of Wolfhagen, where most buildings are old and not well insulated. The aim then became to improve the local energy advisory service and increase the local insulation rate by providing financial support for citizens to retrofit old buildings. Amongst other measures, the municipality established an energy efficiency fund, and power demand was to be mainly reduced with the help of power reduction advisory services for citizens and through exchanging conventional light bulbs with LED lamps, for example, in street lights.

Electro-mobility

Wolfhagen's third main aim was to support electro-mobility in the region (Fraunhofer Institute for Building Physics IBP 2010a, pp. 39ff.) through pilot projects to help interested actors gain practical experience with the use of electric vehicles. Towards this end, electrical power-charging stations were installed in order to facilitate the use of electric vehicles in the region.

A 100% renewable energy goal means full coverage of overall annual energy demand through renewable energy. Accordingly, in order to estimate required supply, the accumulated energy demand for 1 whole year is compared to the accumulated renewable supply of the same year (overall annual energy balance). However, this annual energy balance approach ignores temporal differences between demand and supply levels. A precise 100% renewable energy supply would also need to integrate energy storage as well as energy distribution concepts in order to balance fluctuating renewable energy production and the resulting times of excess energy or insufficient energy supply at any given time (Federal Ministry for Economic Affairs and Energy 2014c).

7.1.2 The Energy Cooperative BürgerEnergiegenossenschaft Wolfhagen eG

In the following it is described how the BEG Wolfhagen eG was founded and which business purposes the organisation follows.

7.1.2.1 Foundation Process

The energy cooperative BEG Wolfhagen eG was founded on 28 March 2012 and became officially registered on 30 August 2012. As mentioned above, the idea to establish the energy cooperative had evolved within the context of realising a 100% renewable energy supply for the municipality of Wolfhagen.

Wolfhagen commissioned the municipal energy provider with development of the key renewable energy projects, in particular the wind park on top of the Rödeser Berg. However, the realisation of Wolfhagen's renewable energy concept and the planned wind park provoked a public conflict during which two groups with contradictory positions were formed. It began with the foundation of two citizens' initiatives. The citizens' initiative *ProWind Wolfhagen—Energiewende jetzt* (proWind Wolfhagen: Energy System Change Now) supported the wind park. According to them, the project was essential in order to be able to realise the goal of supplying 100% of Wolfhagen's power demand through regional and renewable energy. Meanwhile, the *Bürgerinitiative keine Windkraft in unseren Wäldern* (Citizen' Initiative: No Wind Power in Our Forests) was against the wind park. In their opinion, the summit of the Rödeser Berg—which is covered with forest—was not suitable for establishing wind mills, due to nature preservation concerns. Amongst other things, the initiative publically criticised the selection of the location and collected signatures against the park (for example, Bürgerinitiative (BI) Wolfhager Land—Keine Windkraft in unseren Wäldern 2013).

The conflict became a central political topic during the election of the assembly of city councillors (*Stadtverordnetenversammlung*) and the mayoral election of 2011. Three political parties—the SPD, CDU and *die Wolfhager Liste* (The Wolfhagen List)—as well as the sitting mayor of Wolfhagen had been supporting the wind energy project and the renewable energy strategy and, during the election campaign, they clearly committed themselves to the 100% renewable power goal and establishing the wind project. At the same time, they advocated direct participation of Wolfhagen's citizens in the wind park plans. Due to the conflicts around the wind park, they saw involvement of citizens as an important precondition in order to realise it. ProWind Wolfhagen—Energiewende jetzt supported the group during the election campaign by, for example, distributing posters on which they proposed that only those parties that were in favour of the wind park should be elected.

The Green Party, the *Bündnis Wolfhager Bürger* (The Citizens' Alliance of Wolfhagen), and the opponent of the sitting mayor had positioned themselves against the wind park during the election campaign. As stated by the Citizens'

Alliance, it was actually founded by members of the Bürgerinitiative keine Windkraft in unseren Wäldern prior to the elections in order to politically fight against the wind park project. Furthermore, the opponent of the sitting mayor was a member of this citizens' initiative.

In the end, however, the SPD, CDU and die Wolfhager Liste, who were all in favour of the wind energy plans, achieved 73.6% of the vote altogether and, thus, received the majority of the assembly seats of city councillors (Stadt Wolfhagen 2011b, p. 3). The sitting mayor also won re-election with 73.3% (Stadt Wolfhagen 2011b, p. 2). The winning actors interpreted their victory as a sign of political legitimation to continue with their plan to establish the wind park project, though with the precondition of involving the citizens in it.

However, the contradictory positions with respect to the wind park and the realisation of the renewable energy vision remained. The Bündnis Wolfhager Bürger and the Green Party, who were against the wind park, still achieved with 26.4% more than one fourth of the vote (Stadt Wolfhagen 2011b, p. 3), and they continued to argue against the wind project after the election (for example, Nord 2012). The election process and the result of the elections in 2011 illustrate how central renewable energy issues can become for a municipality and how challenging it can be to solve associated conflicts at a purely political level.

In the course of the elections, the sitting mayor and the Stadtwerke searched for the most applicable option for realising the wind park with the maximum of local acceptance and direct citizen participation. Several organisational concepts had been considered, including a limited company (GmbH Gesellschaft mit beschränkter Haftung) or a limited partnership with a limited company as general partner (GmbH & Co KG Gesellschaft mit beschränkter Haftung & Compagnie Kommanditgesellschaft). It was the proposition of the municipal energy provider's managing director to initiate an energy cooperative and to turn the organisation into a new shareholder of Stadtwerke Wolfhagen along with the city of Wolfhagen. The idea was born during a supervisory meeting between the Stadtwerke and the GLS Bank about applicable citizen participation models for renewable energy projects. The GLS Bank is a cooperative bank with strong social-ecological business principles.[4] As expressed in the following quotation, the shareholding concept between a new energy cooperative and the municipal energy provider was seen as the most effective form through which to involve citizens not only in the planned wind park but in all of the renewable energy projects which were supposed to be realised by Stadtwerke Wolfhagen:

> 1.1 "We have had very constructive support from the board of the GLS bank [...], who asked us, 'Have you ever thought about a cooperative?' Not a cooperative only for the wind park but, rather, one associated with the municipal energy provider of Wolfhagen. This is because we needed to consider the question, 'Do we want to found a separate subsidiary for each project that we start and for which we allow citizen participation and then only jump from [...] meeting to meeting, where the members of the solar park cooperative, wind park cooperative and biogas plant cooperative would meet?' That is why we came to the

---

[4]More details about the organisation are provided at www.gls.de

conclusion that the renewable energy projects that we plan in Wolfhagen should be founded as a direct subsidiary of the municipal energy provider [...] and that we would allow citizens to participate [therein]" (R1N1, 48–50).

Thus, this shareholder concept was supposed to implement direct citizen participation on a long-term basis within Stadtwerke Wolfhagen itself. The next quotation shows that it was also particularly important for political decision takers to provide participatory options for a broad group of actors, including low-income earners, as the cooperative model generally enables investments with small amounts of capital:

1.2 "Political stakeholders have said what I also personally thought was important: we want all citizens to have a chance to participate, not just those who have money" (R1N2, 32).

Furthermore, the municipal energy provider and the city of Wolfhagen aimed to collect additional financial capital through the energy cooperative in order to be able to realise investment-intensive projects, such as the wind park.

1.3 "The task of the BürgerEnergieGenossenschaft was ultimately to realise [citizen] participation and collect money, which it has done successfully" (R1N2, 198).

The benefits of the shareholding concept between the energy cooperative and the Stadtwerke were presented to the directory board of the Stadtwerke, the Magistrate (municipal authorities) and the assembly of city councillors. These committees agreed through a majority of votes to the planned project. Next to the managing director of the municipal energy provider, the sitting mayor of Wolfhagen, the local energy agency Energie2000 e.V and members of the citizen's alliance ProWind Wolfhagen—Energiewende jetzt played a key role in the process of achieving this positive decision, because they supported the concept from the very beginning. These actors also became founding members of the energy cooperative BEG Wolfhagen eG.

However, some actors who were against the wind park also criticized the foundation of the energy cooperative, which they saw as just a tool used by their counterparts in order to push forward implementation of the wind park. Hence, a transparent approach in establishing the energy cooperative and its tasks as well as public discussion about opportunities and challenges with respect of the shareholding concept were crucial building blocks in the cooperative foundation process. For example, the business structure of BEG Wolfhagen eG, including its statutes, was developed in open working groups. This way all interested actors, including those critical of the cooperative, were able to actively participate in its foundation and development. This approach stands in marked contrast to the one normally taken by the majority of German energy cooperatives, which are generally founded by small and exclusive groups of people who formulate and pass their statutes and elect the first members of their management and supervisory boards (German Cooperative and Raiffeisen Confederation – reg. assoc. 2013a, p. 6). Consequently, such cooperatives only first become open to further members once their organisational structure has already been established. The foundation of BEG Wolfhagen eG was governed by 'Genossenschaften-Gründen (founding cooperatives), a national wide operating intermediary association specialised in supporting the foundation of energy cooperatives,[5] as well as by the cooperative *Genossenschaftsverband e.V.* (Neu-Isenburg). Amongst other tasks, they helped to formulate the statutes, gave advice for legally setting-up the cooperative and helped in understanding cooperative law (GenG). After this process, BEG Wolfhagen eG was officially founded at a public inauguration meeting during which the supervisory board was elected.

7.1.2.2 Business Purpose

Main business goal of the BEG Wolfhagen eG is the ownership of a 25% minority share on the municipal energy provider. The purpose of the BEG Wolfhagen eG is to generally strengthen the Stadtwerke Wolfhagen and to support them in the implementation of a regional and renewable energy supply (BEG Wolfhagen eG 2013b, p. 2) by involving a broad group of citizens and by collecting additional capital for renewable energy projects.

Figure 7.2 illustrates the new ownership structure of the municipal energy provider and the shareholding concept of the BEG Wolfhagen eG on the Stadtwerke Wolfhagen. As can be seen, the BEG Wolfhagen eG holds two out of nine seats in the directory board of the municipal energy provider.

The second business goal of BEG Wolfhagen eG has been the development and support of local energy-efficiency projects (BEG Wolfhagen eG 2013b, p. 2). As illustrated in Fig. 7.2, the cooperative established a supervisory board for energy

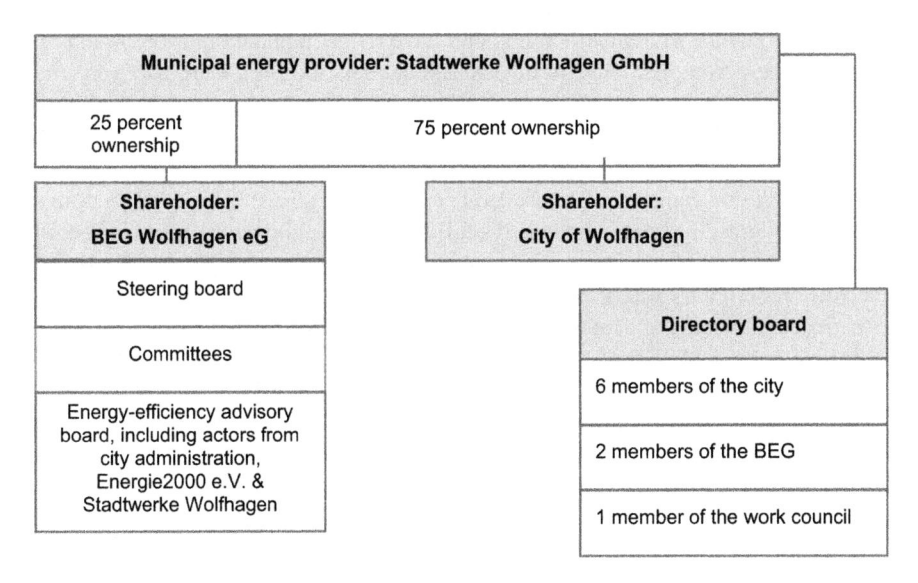

Fig. 7.2 Shareholding concept between BEG Wolfhagen eG and Stadtwerke Wolfhagen (Rühl 2013, p. 6)

---

[5]More information about the alliance is provided under www.energiegenossenschaften-gruenden.de

efficiency measures, consisting of between 9 and 12 members, with at least one member always being delegated by each of the following: the city of Wolfhagen, the local energy agency Energie 2000 e.V. and Stadtwerke Wolfhagen. Concrete topics for this board have included the creation of economic incentives, supply of technical energy-efficiency instruments, support during the construction of energy production projects, as well as the development of learning and information initiatives (BEG Wolfhagen eG 2013b, p. 7). In addition, BEG Wolfhagen eG is responsible for the administration of its own newly founded energy-efficiency fund (BEG Wolfhagen eG 2013b, p. 2), which is supplied through part of the cooperative's annual surplus, mainly created through the annual dividend distributed by Stadtwerke Wolfhagen.

A third aim of the energy cooperative has been to develop its own renewable energy projects as well as supporting the establishment of an environmentally friendly mobility concept (BEG Wolfhagen eG 2013b, p. 2).

The shareholding concept between BEG Wolfhagen eG and Stadtwerke Wolfhagen seemed to be at risk when the new Kapitalanlagegesetzbuch (KAGB)[6] was passed in 2013, intended to regulate alternative investment funds (Gesetz zur Umsetzung der Richtlinie 2011/61/EU über die Verwalter alternativer Investmentfonds 2013). Energy cooperatives which primarily follow a pure investment strategy, in the sense that they are a primarily shareholder in foreign projects and do not develop and operate their own, are at least partly covered by § 1 (1) KAGB. Since its main business purpose was to become a shareholder of the Stadtwerke, it was understood that the KAGB most likely applied to BEG Wolfhagen eG. In consequence, the energy cooperative was required to register itself under BaFin, without which it would have risked liquidation or been required to undertake a complete change of business model. As one of the two very first energy cooperatives to take this step, BEG Wolfhagen eG was able to successfully make it through the rigorous registration process for BaFin in December 2014 (BEG Wolfhagen eG 2014b). During this process, the cooperative was not allowed to collect additional capital, and interested actors were set on a waiting list. Amongst other procedures, the process included a full check of the cooperative steering board, which had to prove that its members had the expertise to manage the cooperative in the ways required by BaFin. For example, the board needed to present evidence of experience in managing investments which they were able to do, because one of the board members had worked at a bank. BaFin further required reformulation of the business approach set down in the cooperative's statutes. According to § 2 (4b) KAGB, relieving exception rules apply to an organisation if its assets are not greater than 100 million Euros and if it receives a secured minimum profit out of its projects.

Hence, BEG Wolfhagen eG changed its statutes and limited its maximum investments in assets to less than 100 million Euros. Furthermore, it narrowed its business activities to projects that generate at least the minimum amount of revenue guaranteed through national regulations, such as the feed-in tariff of the German

---

[6]Detailed information about the KAGB and the position of BaFin was provided in Sect. 6.1.1.

Renewable Energy Act (BEG Wolfhagen eG 2014a). The shareholding of the Stadtwerke was accepted by BaFin as an investment with secured profit because the municipal energy provider invested in photovoltaic and wind park projects covered by the German Renewable Energy Act. Moreover, it receives secured charges for the power distribution grid of Wolfhagen.

In the end, BEG Wolfhagen eG was able to keep its shareholding business model and to apply the requirements of the KAGB by slightly restructuring its approach. However, the change in their statutes has had an effect on the development of own projects, as further discussed in Sect. 7.1.4.

7.1.3 Collaborative Concept

7.1.3.1 Key Collaborating Partner: Stadtwerke Wolfhagen GmbH

Stadtwerke Wolfhagen GmbH supplies the municipality of Wolfhagen with energy and water (Fraunhofer Institute for Building Physics IBP 2010a, p. 1). As of February 2015, the Stadtwerke had about 11,000 customers, of which 6500 were located within Wolfhagen and 4500 located outside of it (Rühl 2015, p. 10). At that time, about 85% of the private citizens and businesses of Wolfhagen were supplied with energy by the municipal energy provider, which used to be a 100% owned and affiliated company of the city. As we have seen in the previous section, however, since 2012 it has been owned 25% by BEG Wolfhagen eG and 75% by Wolfhagen.

The municipal energy provider sees itself as a proactive organisation which supports the development of decentralised, regional and renewable energy. The organisation's self-understanding is especially evident in its interpretation of its own operative responsibility as a municipally owned energy provider, as demonstrated by the following quotation:

> 1.4 "We define 'provision of services' somewhat further, in the sense it should also interest us where our electricity comes from. And we should also care about how we consume this electricity, meaning here how much we consume [...]. And this also includes the fact that we also care about decentralised energy supply, that we build regional and local power plants based on renewable energy resources, our solar and wind parks" (R1N1, 68, 72).

This innovative understanding of their own organisational position was strongly influenced by the managing director of the Stadtwerke, who greatly contributed to forming its latest strategy. Stadtwerke Wolfhagen has been continually strengthening its regional position by purchasing regional energy grids and expanding their activities into the field of renewable energy. In 2006, the Stadtwerke took over the power distribution grid of Wolfhagen (Frey Martin, p. 42). One of the goals was to become more independent from non-regional energy providers, in this case E.ON Mitte, and to be able to better focus on regional energy services. Wolfhagen was the first municipality in Hessen where the distribution grid was returned to municipal ownership (Fraunhofer Institute for Building Physics IBP 2010a, p. 1). Further, Stadtwerke Wolfhagen also purchased the power distribution grid of the

Table 7.5 Key characteristics of Stadtwerke Wolfhagen

Key characteristics	Stadtwerke Wolfhagen eG
Position in the region	Well-established municipal energy provider, supplying the majority of customers in Wolfhagen
Self-understanding	Sees itself as a proactive and innovative organisation, with a strategy based upon the development of decentralised, regional and renewable energy
Motivation for founding the cooperative	Wanted to involve citizens in order to overcome rising public criticism with respect to municipal energy plans and establishment of the Rödeser Berg wind park

neighbouring municipality, Habichtswald, through which it increased its overall operative area. Since 2008, the Stadtwerke has been supplying private households via hydropower so that the majority of citizens of Wolfhagen have already been receiving renewable energy. However, this energy is not produced within the region but, rather, purchased from external providers (Fraunhofer Institute for Building Physics IBP 2010a, p. 13).

As described earlier, it was the managing director who started to mobilise regional actors in order to think about a renewable energy and climate-protection strategy for Wolfhagen (see Sect. 7.1.2). It was also his idea to found the BEG Wolfhagen eG energy cooperative. The core motivation here was to involve citizens in municipal energy plans and in the establishment of the Rödeser Berg wind park, in order to overcome rising public criticism.

Table 7.5 summarises the key characteristics of Stadtwerke Wolfhagen, including its position in the region, its self-understanding and its motivation for founding the cooperative.

7.1.3.2 Collaborative Interaction Model

Figure 7.3 displays the collaborative interaction model being employed between Stadtwerke Wolfhagen and BEG Wolfhagen eG.

As illustrated in the figure, the collaboration between the energy cooperative and the municipal energy provider has an internal and an external dimension. BEG Wolfhagen eG holds a minority share and has two seats in the supervisory board of the municipal energy provider (see also Fig. 7.2). Since the energy cooperative is part of Stadtwerke Wolfhagen, the two partners collaborate within the municipal energy provider itself. Of special note is that this internal collaborative relationship also involves important political decision takers, including the mayor, because the municipality is the main owner of the Stadtwerke, as well as private citizens, because they are members of BEG Wolfhagen eG. At the same time, BEG Wolfhagen eG and the Stadtwerke have, as separate entities, an external cooperative relationship as well and are, for example, both part of the new energy efficiency advisory board being implemented by BEG Wolfhagen eG.

7.1 Collaboration Between Energy Cooperatives and Energy Related Companies

Stadtwerke Wolfhagen GmbH		BEG Wolfhagen eG
Managing director of Stadtwerke **Supervisory board** Mayor of Wolfhagen Political representatives of Wolfhagen **Representatives of BEG Wolfhagen eG** Representatives of the work council Personel of the Stadtwerke	External collaboration ←——→	Executive board Civil representatives **Supervisory board** Civil representatives **Members** Energy-customers of the Stadtwerke Citizens of Wolfhagen Organisations located in Wolfhagen Municipal representiatives

(Internal collaboration arrows on both sides)

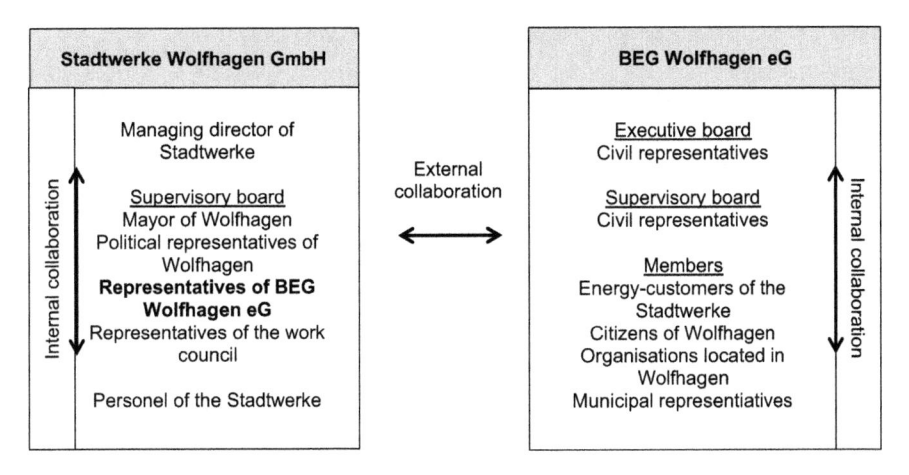

Fig. 7.3 Collaborative interaction model between Stadtwerke Wolfhagen and BEG Wolfhagen eG

In the following, I show how this particular actor constellation, the combination of an internal and external collaborative dimension, played a key role for the mobilisation of resources as well as for the creation of new energy-related norms and regulations. Furthermore, I also outline what kinds of challenges and boundaries are involved in this collaborative model.

7.1.4 Mobilisation of Allocative Resources

The shareholding concept between BEG Wolfhagen eG and Stadtwerke Wolfhagen has facilitated the allocation of crucial allocative resources, including, actors, financial capital, as well as technology projects, as detailed below.

7.1.4.1 Actors

The main actor groups that were jointly activated by collaboration between the energy cooperative and municipal energy provider were citizens, energy customers, cooperative managing personnel and experts. Each of these groups will be considered in the following sections.

Cooperative Members: Citizens and Energy Customers

Table 7.6 presents the membership development of BEG Wolfhagen eG, from September 2012 until September 2014.

Table 7.6 Membership development of BEG Wolfhagen eG 2011–2014[a]

BEG Wolfhagen eG	Sep 2012 (foundation year)	Sep 2013	Sep 2014
Members (number)	488	627	675
Member shares (number)	3385	5216	5997

Source: Annual business reports as explained in Sect. 5.2
[a]The official business year of the BEG Wolfhagen eG goes from 01 October until 30 September

As can be seen, the energy cooperative had already mobilised 488 members by the end of its foundation year. This is more than the overwhelming majority of renewable energy production cooperatives had been able to allocate throughout their complete business lives up until 2012. As revealed by the quantitative assessment of German energy cooperatives presented in Sect. 6.2, 80% of them have not had more than 200 members. Growing even further, however, BEG Wolfhagen increased the number of its members to 675 after its first 3 operating years.

According to the statutes of BEG Wolfhagen, actors who want to join the cooperative have to be or have to become an energy customer of the municipal energy provider. Energy customers are persons[7] or firms that are supplied with energy by Stadtwerke Wolfhagen[8] (BEG Wolfhagen eG 2013b, p. 2). Consequently, all members of BEG Wolfhagen eG are also clients of the municipal energy provider. As illustrated by the following quotation, the main intention for establishing a direct interdependency between cooperative members and Stadtwerke customers was to ensure that only those actors are able to benefit and exert direct power on the municipal energy provider (by holding a cooperative share) that are associated with how Stadtwerke Wolfhagen functions:

> 1.5 "Those who want to participate in the energy cooperative also want to benefit from it. But they should also be electricity customers of the Stadtwerke as well. They should not purchase their electricity from, for example, FlexStrom or some other low-cost provider and then still demand a share of the profit raised by the cooperative" (R1N4, 97).

This customer-member concept supported BEG Wolfhagen eG in acquiring new members—one of its most crucial resources. At the beginning of its foundation process, all customers of the municipal energy provider were directly informed about the new energy cooperative, its goals and about the possibility to become a cooperative member. Furthermore, the Stadtwerke invited its clients to information meetings and, later, to the cooperative's foundation meeting. The Stadtwerke organised these meetings, each of which was attended by around 300 interested parties. The invitation to the foundation meeting was accepted by 330 people, 265 of whom became founding members the same day (Frey Martin, p. 41). According to the following quotation, the expectation of becoming a shareholder of the municipal

---

[7]If the membership of a particular person is of special interest for BEG Wolfhagen eG, an exception can be made from this regulation (BEG Wolfhagen eG 2013b, p. 2).

[8]With respect to 'persons', an energy customer refers to a whole household, so that all private actors of a household which receives energy from Stadtwerke Wolfhagen can become member of the energy cooperative.

energy provider seemed to be a strong motivation for many actors to join the energy cooperative, even though it had just been founded.

For Stadtwerke Wolfhagen, the cooperative membership and client interdependency has represented a new and innovative approach for strengthening one of its most important resources—the loyalty of its customers. According to the following quotation, the cooperative member and Stadtwerke customer interrelationship is seen as a way to bind clients over the longer term:

> 1.6 "In the medium term, as far as I know, we have achieved a customer relationship as no other energy provider in Germany" (R1N1, 216).

> 1.7 "And, of course, it helps the Statdtwerke to keep its customer base over the long term" (R1N3a, 105).

As of February 2015, about 10% of the Stadtwerke's roughly 11,000 energy customers had already joined the energy cooperative.[9] Several actors even became new energy customers of the organisation, just in order to become members of BEG Wolfhagen eG. The binding effect is underlined by the fact that members of BEG Wolfhagen eG could be excluded from cooperative membership, if they change from Stadtwerke Wolfhagen to another energy provider (BEG Wolfhagen eG 2013b, p. 4).

However, the direct and required interdependency between cooperative membership and Stadtwerke client status has also been criticized, as social actors who are not energy customers of the municipal energy provider are excluded from becoming members of the energy cooperative. Yet, actors from outside of the region can become cooperative members, because the Statdwerke markets energy throughout Germany. As displayed in Table 7.7, about 67% of all members of BEG Wolfhagen eG also live in Wolfhagen, while about 33% come from other regions, though many of them still live within a radius of about 50–60 kilometres from the focus region.

Due to the fact that the cooperative is not open to all citizens of Wolfhagen while, at the same time, providing the opportunity for regional outsiders to become members, some actors seem to be questioning the status of BEG Wolfhagen eG as a true citizens' energy cooperative, as revealed by the following quotation:

> 1.8 "Only those who purchase energy from the municipal energy provider can become members. [...] So, from my point of view, one can no longer speak of a citizens' cooperative. At best, a 'customer energy cooperative' [...]. Customers come from all over Germany. And they can become members of the BEG, while other citizens of the city of

Table 7.7 Location of cooperative members as of September 2014 (information directly provided by BEG Wolfhagen eG)

Location of cooperative members as of September 2014	Numbers	Percent
Citizens of the municipalitiy of Wolfhagen	450	67
Citizens from other regions	225	33
Total	675	100

---

[9]The Stadtwerke had about 11,000 energy customers as Feb. 2015 (Rühl 2015, p. 10).

Wolfhagen are excluded because they aren't customers of the municipal energy provider" (R1N5, 55–59).

The focus on energy customers and, thus, a clear preference for a particular member type seems to be interpreted as a form of inequality. Furthermore, the circumstance that actors from outside of Wolfhagen can become cooperative members seems to stand in contradiction to the municipality's aim to especially integrating its own citizens into the development of a 100% renewable energy supply for the region.

Managing Personnel

As of September 2014, the supervisory and executive boards of the energy cooperative consisted of three and four persons, respectively, with different professional backgrounds, including financial and technical engineering expertise. All were civil representatives in so far as they were not in office in the name of an organisation or institution. Nevertheless, several board members were also active in regional politics. The executive board was elected during the cooperative's foundation meeting. The Stadtwerke had played an important role in finding people who were ready to and capable of taking a lead in the energy cooperative, amongst other ways, it had organised and moderated the foundation meeting and election.

All members of the steering board work on a voluntary basis but, as the following quotation indicates, the amount of work (and responsibility) such positions involve does not really seem to fit the fact that they are without any financial compensation:

> 1.9 "And when we don't have the motivation anymore [...] who will then take over? At least [no one] on a voluntary basis. Yes, it's always like this with the claim and the reality. Simply the question of time [spent for the position], which is already enormous" (R1N3b, 52).

It is feared that, due to such high voluntary commitment, it might be hard to find successors once the current leaders leave the steering board.

Experts

The Stadtwerke not only helped the energy cooperative in the acquisition of members and managing personnel but also in supplying its own legal experts to support its foundation process. It also provided marketing experts, who established a professional marketing concept for the cooperative which provided the basis for membership acquisition.

7.1.4.2 Financial Capital

BEG Wolfhagen eG was able to mobilise a substantial amount of financial capital within its first 3 operating years; Table 7.8 presents its economic development from September 2011 until September 2014.

As displayed in the figure, the energy cooperative had about 2.4 million Euros total capital at its disposal at the end of its foundation year of 2012 and was able to increase its total capital up to 3.1 million Euros by September 2014.

Nominal Capital

BEG Wolfhagen eG paid 2.3 million Euros for its 25% shareholder position to Stadtwerke Wolfhagen (BEG Wolfhagen eG 2012a, b, 2013a). The equity ratios, as displayed in Table 7.8, reveal that the energy cooperative was able to pay the full amount with equity capital, which was 100% mobilised from cooperative members. The payments to the Stadtwerke were divided into three tranches, in order to give the energy cooperative sufficient time for allocating members and capital. The last payment was finalised in March 2013 (BEG Wolfhagen eG 2013a). The almost 758,000 Euros of borrowed capital in 2012, as displayed in Table 7.8, do not represent a financial credit but, rather, the open payment that still had to be transferred to the Stadtwerke at that time.

A membership share of BEG Wolfhagen eG costs 500 Euros (BEG Wolfhagen eG 2013b, p. 2). By setting a relatively low entrance price, the cooperative's founding partners realised their first goal—a membership option for a broad group of citizens, including low-income earners.

The capital that was provided by the energy cooperative increased the nominal capital of the municipal energy provider and created a better investment basis for Stadtwerke Wolfhagen. In this way, the energy cooperative financially supported the development of the renewable energy technology mix needed to achieve a 100% renewable energy supply for Wolfhagen.

As revealed by the following quotation, neither the Stadtwerke nor its main owner—the municipality Wolfhagen—seemed to see themselves as being capable

Table 7.8 Financial development of BEG Wolfhagen eG, 2011–2014

BEG Wolfhagen eG	Sep 2012 (foundation year)	Sep 2013	Sep 2014
Total capital (Euros)	2,446,974	2,742,923	3,148,699
Equity capital (Euros)	1,685,740	2,737,315	3,139,916
Member shares (Euros)	1,692,500	2,628,100	3,026,850
Borrowed capital (Euros)	757,924	579	644
Equity ratio (%)	69	100	100
Total revenues/losses (Euros)	−6760	115,975	68,583
Distributed dividends (%)		3.00	2.75

of financing the envisioned renewable energy production projects, the Rödeser Berg wind park and the solar park between Wolfhagen and Gasterfeld, on their own:

> 1.10 "We realised that we also care about decentralised energy supply. That we [want to] build regional and local power plants based on renewable energy resources: our solar park, the wind park. And that we cannot do this alone and that this is not only the responsibility of the municipality alone" (R1N1, 72).

However, the exchange of financial resources between the partners was also criticised, especially by actors who were against the wind power project. The following quotation reveals that critical actors seemed to see the energy cooperative purely as a capital-acquisition tool for Stadtwerke Wolfhagen.

> 1.11 "The reason for this founding [of the cooperative] is that the Rödeser Berg or even these four plants, no matter where they are, must be financed. The Stadtwerke cannot do that out of its own strength, yes. For this purpose, then, the citizens' energy cooperative was established to act as a money laundering machine" (R1N5, 143).

By September 2013, 18 months after its foundation, BEG Wolfhagen eG had already collected 2.7 million Euros in nominal capital, which was in fact more than needed for realising the planned 25% ownership of Stadtwerke shares. As indicated by the following statement, the additional capital may not only have put the energy cooperative in the comfortable position of being able to realise more projects but may have also forced its leadership to search for new projects, in order to invest the money and generate dividends for new members. However, the speaker continues, the cooperative had difficulties in finding new technical and economically applicable investment opportunities in Wolfhagen:

> 1.12 "And then the question was: What do we do now with the other assets that are now coming in? I mean, the cake, [that is] last year's result from the Stadtwerke, is only there once. And it will be more and more divided [into smaller pieces]. And then this automatically reduces the dividends of the others. So one has to find new opportunities already in which we will be able to invest our assets, according to the statutes of course and such that they economically sensible and still generate a nice dividend. This is not so easy, as they need to also have a regional character. There is no concrete project somewhere in July [of this year] so that we cannot invest the next [... incoming Euros]" (R1N3a, 273).

Amongst other options, a plan to become partner of the Stadtwerke Union Nordhessen (in which Stadtwerke Wolfhagen is also a member) and to jointly implement a wind park project in the region could not be realised in the short term (BEG Wolfhagen eG 2013a). Due to such limited direct local renewable energy project options, BEG Wolfhagen eG consequently decided to become member of another energy cooperative, the BürgerWIND Westfalen eG, and to invest 200,000 Euros into it—about 7% of BEG Wolfhagen eG's capital at that time. The organisation is situated in Lichtenau, about 50 km to the north of Wolfhagen, in the neighbouring Bundesland of Nordrhein-Westfalia. The primary business goal of the cooperative is to establish wind energy projects throughout Nordrhein-Westfalia (BEG Wolfhagen eG 2013a). The high capital allocation that BEG Wolfhagen eG had achieved within a short amount of time, the goal of investing this capital in profitable projects, as well as a lack of short-term regional project options led to the

7.1 Collaboration Between Energy Cooperatives and Energy Related Companies   179

circumstance that it invested part of the mobilised capital outside its focus region of Wolfhagen.

Apparently due to this difficulty in finding feasible new investment options that were in accord with the cooperative's principles and goals, as of January 2014, the executive board of BEG Wolfhagen eG decided to limit the amount of capital that members are allowed to invest in the cooperative to 2500 Euros per member. This circumstance demonstrates that a low level of project-investment potential can strongly, and in this case quite artificially, decrease a cooperative's capability for financial capital mobilisation.

Profit and Distributed Dividend

Being a partial owner of a municipal energy provider appears to provide a solid income basis for an energy cooperative and its members. In proportion to its shareholder value, BEG Wolfhagen eG receives 25% of the Stadtwerke's annual distributed profit. Furthermore, as explained below, the Stadtwerke financially supported the energy cooperative in starting its business during its first 2 operating years. In consequence, BEG Wolfhagen eG had already made around 116,000 Euros in revenue during its second business year (see Table 7.8). Since 2013, BEG Wolfhagen eG has also been receiving income from its membership in the BürgerWIND Westfalen eG energy cooperative.

Due to the generated revenues, BEG Wolfhagen eG was already able to distribute a dividend of 3% to its members in its second business year and 2.75% in the third (see Table 7.8). The level of dividends is in line with the general cooperative business philosophy, based upon which it is not profit maximisation that is most important but support of members. Nevertheless, the following quotation of a leader of the cooperative reveals that the distributed dividends may still be an important motivating factor for actors to become members of the energy cooperative:

> 1.13 "There is already the fear that, at some point, when a high interest rate period returns, in which the banks can make quite different offers to investors, that then members may simply vanish. Because part of the[ir] motivation, of course, is also getting a dividend. I do not want to think about it yet" (R1N3a, 269).

BEG Wolfhagen eG assumes that actors evaluate their investments in the energy cooperative against the background of existing alternatives, many of which are currently economically less attractive due to low interest rates. It is thus possible that mobilisation of capital may become more challenging in the future, if interest rates rise and other investment options improve.

Foundation Capital for the Energy Cooperative

As mentioned above regarding personnel, the Stadtwerke financially supported the energy cooperative in the period of its foundation, providing financially relevant resources for free, such as meeting rooms, office space, personnel expertise, administrative support, and marketing material. In addition, BEG Wolfhagen eG received a fixed fee from the Stadtwerke in the first 2 years for each new cooperative member. The fee was doubled if it was not only a new cooperative member but also a new energy customer for the Stadtwerke. The money was primarily used for realising the first energy efficiency projects and for paying operational costs. It seems that this financial start-up support enabled BEG Wolfhagen eG to demonstrate its capability to pursue its business goals from the very start of its operations. As the following quotation implies, the professional image of the energy cooperative attracted members and, thus, affected the number of member shares and the amount of mobilised financial capital:

> 1.14 "The members simply realise: Something is really happening here. And even if we do not yet receive any dividends, something substantial is happening" (R1N3a, 73).

However, as with other dimensions of this process, the support of the Stadtwerke was also criticised. As revealed by the following quotation, some actors interpreted the financial aid as a hidden distribution of profit to the energy cooperative which, in their opinion, negatively affected the dividend distributed to the city of Wolfhagen, the main owner of the municipal energy provider:

> 1.15 "But in the long run it is not good, because political opponents always ask: What kind of shareholder is the city? Is it not like being taken to the cleaners, if the other partner, the cooperative, suddenly receives services free of charge? It amounts to concealed profit distribution through municipal employees working for the cooperative and so on" (R1N3a, 95).

7.1.4.3 Technology Mix

Installed Projects

The municipality of Wolfhagen planned to produce the maximum amount of power that it was able to with renewable resources in the region. In addition, it also wanted to increase energy efficiency, especially in the private dwelling sector, and to introduce electro-mobility (Fraunhofer Institute for Building Physics IBP 2010a, p. 43). Stadtwerke Wolfhagen has been the main executor of most of the municipality's associated projects and has, among other things, implemented the two key power production projects, which eventually enabled a 100% renewable power supply for the municipality: the solar park and the Rödeser Berg wind park. Table 7.9 provides an overview of these two core renewable energy power projects.

Both projects have already been installed and, with 17 megawatts and an expected power yield of 40,000 megawatt hours, cover about 61% of the 66,000 megawatt hours of annual power demand for Wolfhagen. The rest is covered by other already existing plants owned by private operators or other project developers. But the energy cooperative has installed more capacity than the overwhelming majority of other renewable energy production cooperatives in the region, only 5%

7.1 Collaboration Between Energy Cooperatives and Energy Related Companies   181

Table 7.9 Core renewable energy power projects in Wolfhagen (Rühl 2013, p. 27)

Technology type	Number of plants	Installed capacity (Megawatts)	Expected power yield (MWh/a)
Photovoltaic power	1	5	10,000
Wind power	1	12	30,000
Total		17	40,000

of which had installed more than five megawatts peak photovoltaic capacity by May 2015.

Being a new shareholder, the energy cooperative especially strengthened the position of the municipal energy provider as the main executor of renewable energy projects in the region. As one of the Stadtwerke owners, the cooperative is part of all projects realised by the energy provider. The financial resources mobilised by BEG Wolfhagen eG and transferred to the Stadtwerke were mainly used for financing the wind and solar parks. Together, the energy cooperative and the municipal energy provider directly influenced the technical power production mix of Wolfhagen and was able to finally reach the goal of a 100% renewable power supply.

Another transitional goal formulated by the municipality was the development of innovation projects in order to foster new technology fields, such as electro-mobility. Thus, the Stadtwerke initiated an electro-mobility project where it lends an electric vehicle to citizens at no cost. The aim here is to support private actors in becoming acquainted with this new technology, as test trips help them to gain concrete experience with electric vehicles (Stadtwerke Wolfhagen). BEG Wolfhagen eG indirectly supports this innovation project as shareholder of the organisation.

In addition, BEG Wolfhagen eG also supports renewable power production projects outside Wolfhagen. It became member of another energy cooperative—the BürgerWIND Westfalen eG—which develops wind parks in the neighbouring Bundesland of Nordrhein-Westfalia (BEG Wolfhagen eG 2013a). The main reasons for this investment of 200,000 Euros outside their focus region were a lack of investment opportunities within Wolfhagen and, at the same time, having accumulated a high amount of additional capital that had to be invested in order to maintain an acceptable rate of member dividends.

Future Goals

Thus far, as a Stadtwerke Wolfhagen shareholder, BEG Wolfhagen eG has only been indirectly aligned to the projects realised by the municipal energy provider. But the cooperative aims to realise its own projects in the near future and, as revealed by the following quotation, wishes to become proactive in a more direct sense:

> 1.16 "Oh, we would like for our concept to be successful over the long term. That there are not so many difficulties, as there are now with the EEG, that at least there will still be alternatives for us to realise our own projects. And I really hope that we do not finally

Table 7.10 Planned and installed projects of BEG Wolfhagen in relation to the municipality of Wolfhagen's energy transition goals

Technologies that were part of Wolfhagen's energy-transition goals	BEG Wolfhagen eG project status (as of December 2014)
Photovoltaic power technology	One project installed
Wind power technology	One project within and One project outside of Wolfhagen installed
Heat technology	Not engaged
Energy efficiency	Strongly pursued by the energy cooperative
Electro-mobility/energy storage	Not engaged

degenerate into a pure affiliate cooperative. That is, 'degenerate' in quotes! But that we can have both. That, on the one hand, we can have a broader portfolio, where we have a share of larger projects, which are in the end probably much more efficient, of course, in terms of energy generation and the like. And, on the other hand, that we have our own projects through which we can show our members that we ourselves can do something worthwhile too" (R1N3b, 52).

However, as of May 2015, the energy cooperative had not yet realised any projects of its own. The above quotation reveals that one of the reasons for this were the amendments in the German Renewable Energy Act. At the time of the interviews I conducted, planned amendments were still uncertain, and the energy cooperative assumed that they would most likely reduce the potential for its own future projects. Since July 2014, it became clear that the feed-in tariff was indeed to be reduced for several project types, including photovoltaic projects (Erneuerbare-Energien-Gesetz—EEG 2014). Furthermore, BEG Wolfhagen eG limited possible future activities to projects with a guaranteed minimum profit, in order to complete its BaFin registration and apply to the KAGB (BEG Wolfhagen eG 2014a). Consequently, it may be difficult for BEG Wolfhagen eG to realise projects that are not covered by the guaranteed feed-in tariff, such as energy self-consumption projects or innovation projects such as electro-mobility and energy storage. It seems like the required BaFin registration has reduced the innovation potential of BEG Wolfhagen eG. All in all, it can be said that national regulations seem to be the main factors that have been hampering the energy cooperative in realising its own projects.

Table 7.10 compares the project statues of BEG Wolfhagen eG with the energy transition goals that were planned by the municipality of Wolfhagen in order to achieve a 100% renewable and regional energy supply.

7.1.5 Creation of Authoritative Resources

In the following, I describe how the collaborative model employed between BEG Wolfhagen eG and Stadtwerke Wolfhagen has influenced authoritative resources in terms of (1) organisational identity, (2) image and publicity, (3) trust between involved actors and trust in the cooperative business model, (4) acceptance for the wind park, (5) credibility for energy efficiency projects, (6) access to potential projects, (7) access to information and new ideas, (8) transparency, as well as (9) expertise and know-how.

7.1.5.1 Organisational Identity

The shareholding concept has functioned as a strong source of organisational identity for BEG Wolfhagen eG. With being a large shareholder of Stadtwerke Wolfhagen as its main business purpose, BEG Wolfhagen has been able to present this shareholding concept as its unique selling point to the public (for example, Degenhardt-Meister 2012). In this sense, the energy cooperative mainly defines itself through its partial ownership of the municipal energy provider.

7.1.5.2 Image and Publicity

Both the Stadtwerke and BEG Wolfhagen eG have been able to raise their image and increase publicity since the start of their collaborative relationship and have even received national and international attention for their innovative shareholding concept. Numerous articles and documentaries have described the approach of Wolfhagen as a best practise example in a local realisation of the Energiewende (for a selection of articles see http://www.beg-wolfhagen.de/index.php/presse). Managing personnel from both organisations presented the concept at conferences and expert meetings throughout Germany (for example, Rühl 2013). The strong membership growth to more than 675 actors within the short timeframe of 3 years underlines this successful creation of a positive image and high publicity.

For this shareholding concept, Stadtwerke Wolfhagen won second place in the Municipal Utility (*Stadtwerke*) Award competition for 2013, an award given out by a major sector-specific national municipal utility conference. In making its decisions, the jury concluded that the municipal provider was taking

> 1.17 "a very courageous step, because citizens are really brought into the company. At the same time, this is a substantial contribution towards the financing activities of smaller cities. Its integration into the overall strategy has also been successful" (Stadtwerke Wolfhagen 2013).

Stadtwerke Wolfhagen stated that, even though it had previously been owned 100% by the municipality, citizens would not have automatically characterised it as an organisation predominantly working for the needs of the community. In this respect, some actors seem to have compared it with mainly commercially oriented national and international operating energy providers, such as e.on or Vattenfall, as illustrated by this quotation:

> 1.18 "It is really difficult for a small municipal energy provider to represent its own identity. One notices in conversations with members of the public that, again and again, they actually no longer distinguish between municipal energy providers and larger energy providers. They

are seen as all being the same. So it happens from time to time that one hears this: 'Yes, oh, E.ON and municipal energy provider, that is anyway all one unit'. [...] But this is different now, due to the energy cooperative, because it gives us something to think about. And many people do have 500 Euros available [to invest in it]" (R1N4, 79).

Thus, in the end, collaboration with the energy cooperative provided a new unique selling point for the Stadtwerke and helped both organisations in positioning themselves as an energy organisation with a mainly citizen-focused orientation.

7.1.5.3 Trust

Trust Between Involved Actors

The creation of trust between involved actors was crucial for initiating BEG Wolfhagen eG and for actually realising the shareholding concept. As indicated by the following quotation, the actors within the newly formed cooperative steering board only partly knew each other beforehand, so trust between them needed to be built, providing the basis for a harmonious working atmosphere.

> 1.19 "You really almost did not know anyone at all. [...] And then I did not know what kind of a person this could be. Who knows what kind of interests they have [...]. Could I work with them at all, and how do they think? What are their motives? [...] And that it has been harmonious so far for almost 2 years is, of course, optimal" (R1N3b, 38).

At the same time, a certain level of trust appeared to exist from the beginning between some actors of the young energy cooperative and the municipal energy provider. As one example, the energy cooperative trusted the municipal energy provider in setting a fair price for its new shareholding position, as it was the Stadtwerke and not the energy cooperative that commissioned and paid the external company that evaluated the Stadtwerke in order to define the price.

It is assumed that previous joint activities between some actors of both organisations were the starting point for building a trusting relationship. Several people who eventually became member of the cooperative's steering board had, for example, already been engaged in the ProWind Wolfhagen—Energiewende jetzt citizens' initiative, which had supported the wind park project of the Stadtwerke at an early stage. In this way, some members of the cooperative steering board and the managing director of the Stadtwerke already knew each other beforehand.

Trust in the Cooperative Business Model

Out in the public, it seems clear that Stadtwerke Wolfhagen greatly contributed towards creating trust in the business model proposed by BEG Wolfhagen eG, as 265 people already joined the energy cooperative as a member and provided capital of 853,000 Euros on the day of its foundation. It is assumed that such a great number of actors willing to provide a high amount of capital was mainly motivated to join the energy cooperative because it planned to become a part owner of the Stadtwerke. Actors seemed to perceive this shareholder position as a sustainable business approach for the energy cooperative. In fact, Stadtwerke Wolfhagen had already been economically well-established, as it made about 560,000 Euros in profit in 2012 (Stadtwerke Wolfhagen 2014). To become shareholder of such a strong partner would seem to provide business security for a young energy cooperative and, in consequence, investment security for its members. The next quotation reveals that the cooperative itself appears to see the achievement of business security through its collaboration with the Stadtwerke as an important resource, even though it may reduce its potential profit overall, because it will not be the one to realise great profit-generating, but sometimes risky, projects:

> 1.20 "Well, there are a lot of risks for municipal energy providers, especially when it comes to larger projects. Of course, we have less income—less dividends—compared to if we had become the operator as investor but, in turn, we have also achieved a broader spread and greater security, which is also something valuable in itself" (R1N3b, 44–46).

All in all, the general trust in the business model of BEG Wolfhagen eG was crucial for attracting many members and capital so as to empower the energy cooperative for exerting influence on the existing energy structure of the Stadtwerke.

7.1.5.4 Citizen Acceptance of Technology Projects

Acceptance is an important authoritative resource for changing a local technical resource mix. People that accept a certain technology project may support its implementation or may at least not impede it. In turn, actors who do not accept a certain technology may start to position themselves against its implementation by, for example, activating their right to public criticism and resistance, their right to demonstrate or their general right to take legal action.

In Wolfhagen, acceptance became a central resource for implementing the Rödeser Berg wind park. Especially the foundation of the two citizen initiatives Pro Wind Wolfhagen—Energiewende jetzt and Bürgerinitiative keine Windkraft in unseren Wäldern, of which one was in favour of the wind park and the other was against it, demonstrated the parallel existence of actors who accepted implementation of the wind project and those who did not. Both initiatives made use of their rights to publically position themselves. For example, members of Bürgerinitiative keine Windkraft in unseren Wäldern conducted public signature-gathering campaigns and initiated a petition with the EU parliament against the project[10] (Bürgerinitiative (BI) Wolfhager Land—Keine Windkraft in unseren Wäldern 2012; uli 2014). They assumed that the clearance of trees and the noise of the windmills would have a negative effect on animals living around the summit (Bürgerinitiative (BI) Wolfhager Land—Keine Windkraft in unseren Wäldern 2011; Bürgerinitiative (BI) Wolfhager Land—Keine Windkraft in unseren Wäldern 2013). Meanwhile, among their other activities, members of Pro Wind Wolfhagen—Energiewende jetzt organised events in which they underlined the importance of renewable energy (Pro Wind Wolfhagen—Energiewende jetzt 2010). Such activities demonstrated the strong will of the opposed groups to enforce or hinder the realisation of the wind park project. However, the size of the two groups was never transparent. For example, no publically available indications could be found regarding the number of people who were actively involved or supported the positions of either one of the citizen's initiatives.

In contrast to these two citizen's initiatives, BEG Wolfhagen eG represents a clear and tractable group of actors, because it is based upon a structured membership approach. Since the energy cooperative has always been a strong supporter of the wind park project, all cooperative members also automatically became supporters of the wind park, independent of their motivation for joining the energy cooperative. Consequently, BEG Wolfhagen helped to unify many more actors around this issue than the Pro Wind Wolfhagen—Energiewende jetzt.[11] The reason for this could be the combination of socio-political goals and a clear business approach, as the cooperative provides the opportunity for its members to economically benefit from the wind park in proportion to their cooperative member shares. Receiving a dividend seems to be one motivation for joining the energy cooperative. Furthermore, BEG Wolfhagen increased the visibility of wind park supporters, since the number of cooperative members has been publically known throughout the process (for example, BEG Wolfhagen eG 2013a). The following quotation underlines the unique role of BEG Wolfhagen eG in this respect:

> 1.21 "And so there were both advocates and opponents of the wind park. And also a large silent majority on both sides. And from there, I would say, we have proactively brought a completely new perspective into this very difficult situation. From having only a pro-wind park citizens' initiative and a counter-initiative against it, we brought the whole thing to a totally different level" (R1N1, 58).

However, the engagement of the newly founded BEG Wolfhagen eG was not able to fully resolve the conflict, as Bürgerinitiative keine Windkraft in unseren Wäldern, the Green Party and the political party BWB (Bündnis Wolfhager Bürger) continued to fight against the wind project by planning, amongst other measures, to go to court against the wind park (Müller 2014). As one opponent of the wind park said:

---

[10]The EU Commission suspended the petition (uli 2014).
[11]Concrete numbers of actors that participated in the citizens' initiative were offered by one of the organisers confidentiality, but I was asked not to publicise them.

7.1 Collaboration Between Energy Cooperatives and Energy Related Companies 187

1.22 "We are not against the use of wind power in the region, but we are still clearly against this planned location" (R1N5, 19).

According to the following quotation, the late involvement of local actors in developing the wind park and the early focus on only one wind park location may have been the main sources for the conflict and the reason why it was so difficult to resolve:

1.23 "We have said from the beginning: We want to go there, on top of the Rödeser Berg. And that was the goal from the beginning. There have been information events, but only after we asked for them. Before, citizens were not involved at all in these plans. Then there were information events, and there were five alternative locations under debate. Within a short time, I believe the whole thing took about a quarter of a year, and several meetings and such, selection of the Rödeser Berg was solidified, with all possible arguments [being heard]" (R1N5, 45).

According to a leading person of the Stadtwerke:

1.24 "Based on the example of the wind park, it became clear that, if we had founded the BEG earlier ... if we had had it from the beginning, then we, I believe, would have managed the project more smoothly" (R1N1, 172).

It is stated here that the foundation of the cooperative at an earlier stage, for example right at the start of planning the park, could have increased its positive effect on the project. However, the energy cooperative was founded at a time when the wind park plans had already been well developed and after the conflict had already emerged. It is probably much more challenging to convince actors about a project once they have already positioned themselves against it.

In the following two statements, we can see that both actor groups—supporters and opponents of the wind park—positively mentioned a future congress which was organised by the city and the Stadtwerke to inform about and discuss the wind park:

1.25 "In the course of the process, we also took a whole new path and conducted workshops with the citizens and so on. But it was all very late. That was very good. But today I would do it earlier. [...] It was, I think, a good way to clarify and explain the wind park project, to convey the information [to the public]" (R1N2, 432–448).

1.26 "Early last year, yes, this congress about the future [of the project] ... I was there, yes. So, basically, I participated in several discussion sessions there as well. The people who were there were all very interested in the subject and also brought in some good ideas themselves [...] As I've said, however, it actually came a couple of years too late here. Such an event for information and participation should have taken place much earlier" (R1N5, 471–479).

As hinted here, a central aspect of the congress was that it gave both groups the room to discuss their positions and the reasons for them. Further, it was moderated by a neutral person. However, in the opinion of both sides, the congress was organised too late. At an earlier stage, such mediation may have ameliorated the conflict-ridden situation surrounding the wind park.

All in all, it can be said that BEG Wolfhagen may actually have strengthened acceptance for the Rödeser Berg wind park by politically and economically empowering its supporters. At bottom, the energy cooperative activated more people than the pro-wind citizens' initiative, creating stronger group visibility and combining this actor unification with mobilisation of other important allocative resources, such as financial capital. However, the involvement of local actors seemed to come too late, at a time when the conflict had already emerged, and it is questionable whether the energy cooperative actually did help to convince any actors to support the wind park among those who had already clearly positioned themselves against it.

7.1.5.5 Credibility of Energy Efficiency Projects

The shareholding concept between the energy cooperative and the municipal energy provider seems to have increased both their levels of credibility in business activities. This was especially revealed with regard to the energy efficiency projects initiated by the energy cooperative. Credibility may increase motivation to act and can, thus, have a direct impact on the consumption patterns of measures such as energy efficiency ones. The efficiency projects in this case included the exchange of conventional light bulbs with LED lights and a thermal imaging procedure through which an energy inventory could be generated for diagnosing the insulation levels of private buildings. Crucial was the new approach through which the projects were developed and marketed. The measures were developed by the cooperative's energy supervisory board, which consisted of experienced actors from the municipality, the municipal energy provider, the energy agency Energie 2000 e.V. and citizens of Wolfhagen (cooperative members). The energy efficiency measures were mainly marketed by the energy cooperative, and some projects were technically supported by the municipal energy provider. The projects were open to all citizens, but cooperative members could participate to a reduced price. According to the local energy agency, Energie 2000 e.V., these energy efficiency measures were more strongly in demand than previously offered ones, as illustrated by this quotation:

> 1.27 "And I also noticed that the energy cooperative had started in spring, or actually it was still in winter, a new programme for conducting thermographic inspection on residential buildings. And I was very surprised about the positive response, because I know from other projects, which even banks have already attempted, that the response used to be much more restrained" (R1N4, 128).

It seems that the collaborative work between the new energy efficiency supervisory board, the energy cooperative and the municipal energy provider made the difference in this case. For example, the Stadtwerke on its own may not have been used to being perceived as a credible supporter of energy efficiency, since they mainly generate profit through the consumption of energy, as indicated by the following quotation:

> 1.28 "We have always heard or have been asked—How can a municipal utility really seriously mean that they want to drive forward energy efficiency and energy savings" (R1N1, 170)?

7.1 Collaboration Between Energy Cooperatives and Energy Related Companies 189

As outlined in the next quotation, the strong citizen-based approach of the cooperative may have made energy efficiency projects more compelling to local actors:

> 1.29 "But when it comes to the issue of energy savings, it is much more difficult to address individual citizens and convince them to do something. And it is even difficult because it is part of their personal lifestyle. So here, I think, the energy cooperative has had a great advantage, because as an institution it is considered in a different way. Citizens are members. There is another trust base and ... yes, what comes from the citizens' energy cooperative is maybe more convincing in terms of people thinking: They do not want to sell us anything, since we ourselves are actually the cooperative, and we therefore believe them" (R1N4, 122–124).

According to this logic, private citizens may identify themselves more with these projects, because they are also co-developed and marketed by people like them. This may have changed the perception pattern in relation to energy efficiency from a neutral sales approach, such as may have been fostered by other companies, such as the Stadtwerke, to a more personnel, member value-creation approach, as promoted by the energy cooperative.

7.1.5.6 Access to Potential Projects

For BEG Wolfhagen eG, the shareholding concept creates indirect access to renewable energy production projects, and it benefits from each project realised by the Stadtwerke. In its position as a partial shareholder, the energy cooperative receives part of the dividends generated through, for example, the photovoltaic and wind parks. At the same time, the Stadtwerke has access to energy efficiency projects that have been developed by the energy efficiency advisory committee of the energy cooperative.

However, the circumstance that BEG Wolfhagen eG invested a major part of its available capital in another energy cooperative outside of Wolfhagen (see Sect. 7.1.4.3 for more details) suggests that project access solely through the municipal energy provider is also relatively limited. One could have assumed that both actors may use their collaborative relationship in order to invest the complete amount of capital made available from Stadtwerke customers either in projects in which the Stadtwerke was participating or at least in projects located within the municipality of Wolfhagen. However, a lack of time for an intensive project search in both organisations combined with upcoming and uncertain amendments in the EEG and the need to invest the money in the short term, in order to generate a dividend, motivated the energy cooperative to look for options that were not aligned to the Stadtwerke.

7.1.5.7 Access to Information and New Ideas

The collaborative relationship between the energy cooperative and the Stadtwerke created innovative spaces for developing new ideas. A good example here is the energy efficiency committee, where a newly formed set of actors—including citizens, the municipality, the municipal energy provider, the energy agency Energie 2000 e.V. and the cooperative—searched for new options in order to save energy in Wolfhagen. The committee was accompanied by various working groups in which any interested member of BEG Wolfhagen eG could participate. This way, the energy cooperative and the Stadtwerke have had direct access to the expertise and innovation potential of cooperative members who are at the same time Stadtwerke customers. The voluntary work of the committee and its working groups complements the activities of other energy initiatives, such as the public trust Energieoffensive Wolfhagen. Not only citizens but also regional businesses have already benefited from the projects that were planned and conducted by the committee and the energy cooperative. For example, local energy advisors were involved in the LED-lamp project.

7.1.5.8 Transparency

The municipal energy provider is able to present its activities in a more transparent manner through its cooperative relationship with BEG Wolfhagen eG, which may create greater transparency because it continually informs its members, who as we know are simultaneously customers of the municipal energy provider and citizens of Wolfhagen, in a very detailed and open manner. Information provided includes facts about the projects being realised by the Stadtwerke, news about cooperative activities and internal developments (for example, BEG Wolfhagen eG 2013a). The energy cooperative followed a transparent business philosophy from the beginning of its foundation process, and the development of its business structure, including the formulation of statutes, was organised in an open process in which anyone could participate. Even the foundation meeting, during which the steering board was elected, was open to everyone. Protocols of all general assemblies are also publically accessible via its website.

7.1.5.9 Expertise and Know-How

The Stadtwerke has offered important expertise, which BEG Wolfhagen eG does not have, including technical, marketing and legal expertise. The utility's employees have, for example, key knowledge about the existing technical energy infrastructure in Wolfhagen. Meanwhile, the high rate of participation in and consequent success of the foundation of BEG Wolfhagen eG was to some degree a product of the professional marketing measures mainly organised by the Stadtwerke (out of 330 interested parties, 265 became members during the inauguration meeting). The BEG Wolfhagen eG website was also developed by personnel from the Stadtwerke.

7.1.6 Creation of New Regulations

Based upon their shareholding concept, Stadtwerke Wolfhagen and BEG Wolfhagen eG introduced new societal rules, which they used for structuring achievement of their joint goal: the development of renewable energy. New official regulations include the (1) direct ownership of private citizens and energy customers of the municipal energy provider, (2) new co-decision and participation rights for citizens with respect to municipal energy issues, (3) a general limitation of individual member investment volumes and profit distribution for the sake of the community, as well as (4) new forms a regional value creation. In the following, I describe how these regulations were created, how they influence the existing energy structure and what kinds of challenges they have involved.

7.1.6.1 Direct Ownership by Citizens and Energy Customers of Municipal Energy Provider

BEG Wolfhagen eG's 25% shareholding of Stadtwerke Wolfhagen created direct ownership by Wolfhagen's citizens, who are at the same time energy customers, over their own municipal energy provider. This new rule—the direct ownership of citizens and energy customers of their energy provider—is the heart of the energy-related structural change being triggered by BEG Wolfhagen eG and Stadtwerke Wolfhagen and, I hold, the main reason for the achieved 100% renewable energy supply[12] in their focus region of Wolfhagen. The established energy provider and the city reinforced their position as structuring actors by factually giving up (ownership) power to a young organisation with a completely different organisational self-understanding, based upon an open membership approach and democratic decision-taking structures.

The new, immediate, collaborative relationship between politicians (who represent the municipality as Stadtwerke shareholders), citizens and the Stadtwerke provided the basis for the mobilisation of great allocative and authoritative resources. Amongst other things, it triggered the allocation of 2.3 million Euros of new equity capital for the municipal energy provider and helped to mobilise 675 cooperative members by September 2014, who became new shareholding owners of Stadtwerke Wolfhagen. This collaboration further strengthened gaining acceptance for the Rödeser Berg wind park, increased the credibility of the energy efficiency measures being offered, and led to new renewable energy-related ideas being created in new innovative spaces, such as the energy efficiency advisory committee. The next quotation demonstrates that the goal of achieving 100% renewable energy in Wolfhagen and the involvement of the BEG Wolfhagen eG energy cooperative was one of the core factors that, from the outset, shaped the energy provider as an organisation.

---

[12]Using the annual energy balance approach.

1.30 "This means the task has been changed; it has not stayed the same. One hundred percent renewable energy has been added; then the citizens' energy cooperative has been added. It is already a process that has become recognizable" (R1N2, 190).

21 of 50 articles written in the dominant regional newspaper Hessische/ Niedersächsische Allgemeine—District Kassel und Wolfhagen about the energy cooperative between 2012 and 2014 emphasised the value of being a partial owner of Stadtwerke Wolfhagen. The new rule was not only discussed within the interaction model between the two partners but also within the wider regional public. However, there were also critical voices from actors who questioned the collaborative concept and interpreted the 25% shareholding of BEG Wolfhagen eG over the Stadtwerke as a reduction of municipal property. Critical actors assumed that the profit, which is annually distributed by Stadtwerke Wolfhagen to the city of Wolfhagen, may be reduced because it has to be shared with BEG Wolfhagen eG. The money represents part of the municipal budget and is, thus, an important capital source for the municipality. But, as the following quotation implies, that was not the only way to see the situation:

1.31 "So you say, 'Oh, now we must give up 25% of the cake'. [...] But we also needed the money to build the facilities. We would not have been able to handle it solely based on the municipality. And from there you have to ask, ultimately, 'Do we want to eat a cupcake all by ourselves, or do we want a fat pie and give up a piece?' That is the question. And I mean, then make the pie really thick and keep three-quarters of it. So, in terms of the bottom line, the cake for the city is also bigger. And from there it becomes a win-win situation, as the city profits from it too." (R1N1, 322–326).

As the quotation also reveals, the opening up towards the energy cooperative not only strengthened the Stadtwerke but also, potentially, the position of the city of Wolfhagen. In fact, since the shareholding concept was realised through an ordinary increase of the Stadtwerke's equity capital, the involvement of BEG Wolfhagen eG actually increased the value and reliability of the municipal energy provider. Also, the Stadtwerke itself assumed that it would be increasing its profit through new renewable energy production projects, receiving a guaranteed feed-in tariff for its produced power under the Germany Renewable Energy Act. Accordingly, the Stadtwerke expected that the profit share for the city would also be likely to rise in the future. This needed to be approved, however.

Yet, other actors seem to perceive a new exclusivity in the shareholder composition of the Stadtwerke, because the energy cooperative only represents its members and not all citizens who live in Wolfhagen. Only those actors that are able to afford the minimum cooperative member share of 500 Euros can join the cooperative and can, thus, become direct owners of Stadtwerke Wolfhagen, as expressed in the following quotation:

1.32 "The municipal energy provider was a one hundred percent daughter company of the city. Every citizen was involved. Now it is an exclusive circle, for those who can afford to be a member in the BEG, who can pay the share. Now, they are always involved, up to 25%" (R1N5, 29).

In general, the main owner of a municipal organisation is the municipality itself. Since the municipality represents its citizens, municipal organisations can be interpreted as publically owned enterprises (see for example, Wagner and Berlo 2011), though the great majority of citizens may not really have the opportunity to directly exert influence on these organisations, because an immediate relationship between them and the organisational management usually does not exist. Citizens may indirectly exert influence during political elections, where they choose the political parties and the mayor, who also represent the municipality as a shareholder in municipally owned organisations. However, elections often only take place every 4–5 years. As outlined in the following, the direct partial ownership of BEG Wolfhagen eG over the Stadtwerke created new decision rights for citizens regarding municipal energy services.

BEG Wolfhagen eG and the Stadtwerke tried to counteract critical voices by developing the cooperative foundation and their own collaboration in an open and transparent manner. Anyone, including critics, could participate in elaborating the business approach of BEG Wolfhagen eG. Open working groups and information meetings gave room for discussing advantages and challenges regarding the shareholding concept at a very early stage. The successful realisation of their envisioned collaboration reveals that critical actors seem to have remained in the minority.

7.1.6.2 New Co-Decision and Participation Rights for Citizens

Creating new co-decision and participation rights for citizens was one of the core reasons for founding BEG Wolfhagen eG. The following quotations underline this aim:

> 1.33 "Yes, it makes sense that you [the citizens] should be able to co-decide and participate, so that you are also present within these renewable projects" (R1N1, 134).

> 1.34 "It was said that we definitely want wide public participation" (R1N2, 210).

New Co-Decision Rights for Citizens Over Municipal Energy Strategy

Stadtwerke Wolfhagen is one of the main executors of energy-related political goals, since it is owned by the city of Wolfhagen. As we have seen, prior to the foundation of the energy cooperative, the city used to be the single shareholder and used to have full decision rights over the municipal energy provider. The new shareholding concept changed this decision-making authority, and the energy cooperative gained co-decision rights over the strategic direction of the Stadtwerke by becoming partial owner. The content and scope of this ownership are officially formulated in the shareholders' agreement, in the statutes of BEG Wolfhagen eG and in the cooperation contract between the partners. The new co-decision rights of BEG Wolfhagen eG are predominantly exerted during annual shareholder meetings and within the

```
┌─────────────────────────────────────────────────────────┐
│     Supervisory board of Stadtwerke Wolfhagen GmbH      │
│            Chairman: Mayor of Wolfhagen                  │
└─────────────────────────────────────────────────────────┘
        │                    │                    │
   ┌─────────┐          ┌─────────┐          ┌─────────┐
   │ 6 seats │          │ 2 seats │          │ 1 seat  │
   ├─────────┤          ├─────────┤          ├─────────┤
   │City of  │          │  BEG    │          │  Works  │
   │Wolfhagen│          │Wolfhagen│          │ council │
   │         │          │   eG    │          │         │
   └─────────┘          └─────────┘          └─────────┘
```

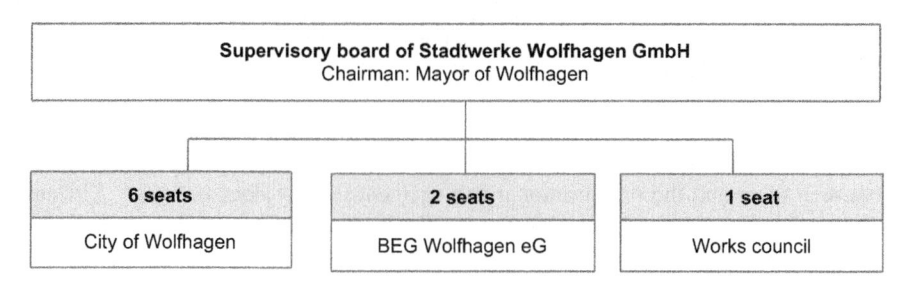

Fig. 7.4 Supervisory board of Stadtwerke Wolfhagen (Rühl 2013, p. 6)

supervisory board of Stadtwerke Wolfhagen, in which the energy cooperative holds two out of nine seats (see Fig. 7.2).

The shareholder group set up a preemption right for the existing shareholders—the city and the energy cooperative—as well as a veto right for the energy cooperative, in case the Stadtwerke ownership structure should be changed. These rights are intended to secure, over the long term, that main owner of Stadtwerke Wolfhagen remains the city of Wolfhagen and/or BEG Wolfhagen eG. Both, the city and the cooperative, are institutions that do not focus on the maximisation of profit but on the creation of value for its citizens and members. In this sense, the new preemption and veto right help to prevent a sale of Stadtwerke Wolfhagen to private shareholders who predominantly have their own economic interests in mind.

Figure 7.4 displays the structure of the supervisory board of the Stadtwerke Wolfhagen.

As can be seen, citizens, being represented by BEG Wolfhagen eG, have become part of this important decision making body, next to political representatives and representatives from the Stadtwerke.[13] Now, local energy issues can be directly discussed and agreed upon between citizens, politicians and the municipal energy provider on a regular basis. The topics discussed here include new project and investment plans in renewable energy as well as the setting of energy prices. Furthermore, the energy cooperative can co-decide over the use of the Stadtwerke's annual earnings, discussed at the annual shareholder meeting. For example, the energy cooperative has a veto right if less than 70% of the Stadtwerke's annual profit is to be distributed. The shareholding concept empowers citizens to co-decide over future renewable energy projects being realised by the Stadtwerke, assess its energy prices and discuss the co-financing of other municipal projects, such as the local swimming pool, using the Stadtwerke's profits. The following quotation underlines how being a shareholder of the Stadtwerke entails stronger co-decision rights for citizens than being shareholders of single energy projects:

> 1.35 "The opponents have always said with regard to the wind park that you only do your own thing and do not let anyone participate [. . . . But] with the cooperative, we can actively

---

[13] One of the six seats of the city of Wolfhagen used to be given to a private person, who always had to been chosen by the political representatives.

counteract this argument in the sense of: Citizens, you can co-determine what happens regarding energy topics on a day-to-day basis, by becoming a shareholder of the municipal energy provider and, thereby, gain influence [...]. So it is a much more comprehensive shareholding concept than only taking part in a single renewable energy plant" (R1N1, 50–54).

However, citizen possibilities for exerting influence on local energy issues and on the municipal energy provider through the energy cooperative are also limited. The cooperative's veto right is focused on the distribution of the Stadtwerke's profit and on changing the ownership structure of the energy provider. The city Wolfhagen still holds the majority of seats in the supervisory board of Stadtwerke Wolfhagen and can, thus, outvote the cooperative regarding other issues. The cooperative has no direct influence on the Stadtwerke's operative activities, including the implementation and operation of planned renewable energy projects, as it is only a shareholder of the Stadtwerke and not a direct project partner. Thus, participation of citizens as renewable energy project owners is missing from the collaborative relationship. As indicated by the next quotation, the cooperative wishes to gain more decision rights in the operative realisation of renewable energy projects:

1.36 "Of course, there already exists a certain attitude among members of the cooperative: Yes, actually, we would like to be able to participate more in nuts and bolts decisions—to have even more direct influence on, for example, business policy. But it is quite clearly defined by the agreements between us that the cooperative only has membership in the supervisory board of the municipal energy provider and that the cooperative has to exert its influence through this channel" (R1N4, 364).

General Participation Right for Citizens in the Energy Cooperative

All actors that become members of BEG Wolfhagen eG have a general right to participate in the energy cooperative and financially participate by holding member shares. By September 2014, 675 members had made use of this right and acquired 5997 member shares, with a financial value of more than three million Euros (see Tables 7.6 and 7.8). Furthermore, each member has the right to participate in decisions about cooperative activities and strategy. This right can primarily be exercised through attending general assemblies and making use of a member's voting right. Table 7.11 displays how many members attended the general assemblies of BEG Wolfhagen eG between 2012 and 2014.

As can be seen, between 27% and 50% of all members attended the assemblies of the energy cooperative in 2012–2014. The first assembly, it should be noted, was the

Table 7.11 Share of members who attended the general assemblies of BEG Wolfhagen eG, 2012–2014 (source: Information directly provided by BEG Wolfhagen eG)

General assemblies	28 March 2012	29 Jan 2013	23 Nov 2013	22 Nov 2014
Participants	265	244	249	182
Total members at that time	265	488	627	675
Share of members	100%	50%	40%	27%

foundation meeting of the cooperative, during which all attendees also became the first cooperative members. Almost half of all members came to the second general assembly, but the third was only attended by 40%, decreasing to 27% by the fourth assembly. Thus, we can see that only a minority of members took part in the general assemblies of BEG Wolfhagen eG. More members seem to be interested in financial participation rather than in participating in organisational decisions. This assumption is underlined by the following quotation:

> 1.37 "Actually, I really have to say that [membership] is in order to have a financial investment option, I think. So to a great extent—that is unfortunately the case—I think, when we have discussions during our consultation hours, nobody asks: And how does the municipal energy provider exactly use the money? Rather, they ask: How do the dividends look?" (R1N3a, 131–133)?

This circumstance that only a minority of members attend assemblies could also mean that most members generally agree with the overall organisational direction and support, or at least go along with, the work of the cooperative steering board. The continual decrease of member participation over the years can have several reasons. Actors may, for example, have had a greater interest in attending the meetings after the organisational foundation in order to get to know the organisation. However, one would have to conduct a member survey in order to fully understand the motivation of members to attend the assemblies or not.

Next to exercising their voting rights, members can also participate in, and potentially exert influence on, general assemblies of the energy cooperative by asking questions, making comments, giving their own opinion or announcing their own ideas for future projects. Yet, as indicated by the following statement, members seem to be rather passive during these meetings. Only few seem to get actively involved, while most attendees seem to remain silent:

> 1.38 "It's really only a few who actually ask questions. And these questions almost never really tackle fundamental issues or are not really seriously discussed, such as when we talked about power prices. [...] I think most participants are simply overwhelmed that there are so many issues on the agenda at these meetings" (R1N3c, 51).

The quotation also points towards one of the possible reasons for this lack of active participation. General assemblies are very formal meetings, during which the annual results are presented, elections are conducted and the steering board is reconfigured. Such a formal and dense meeting structure may prevent members from being creative and wanting to express and discuss their own ideas or opinions.

However, in addition to the general assemblies, BEG Wolfhagen eG offers its members additional concrete options for actively participating in developing organisational projects and bringing in their own ideas. The energy efficiency advisory committee, which is part of the cooperative, is a new group that meets regularly. It is further accompanied by working groups, in which all members are welcome to participate. The committee and the groups have a more informal interaction structure and are focused on developing projects and new ideas. The committee has already realised several projects. This concept can, thus, perhaps be seen as a more successful tool for active and concrete citizen participation in cooperative project development than the assemblies.

All in all, it can be said that BEG Wolfhagen eG created new rights for its members to co-decide and participate in the operation of municipal energy services. Different levels of participation and co-decision have been identified here, including financial and company ownership participation, participation in terms of guiding the municipal energy provider, active participation in developing the business of the energy cooperative, as well as participation in concrete project development. The new participation rights created by BEG Wolfhagen eG were also mentioned in 22 of 50 articles written about the energy cooperative in the dominant regional newspaper *Hessische/Niedersächsische Allgemeine—District Kassel und Wolfhagen* between 2012 and 2014, seeming to demonstrate that this rule has been transported into the wider public sphere, beyond the boundaries of BEG Wolfhagen eG.

7.1.6.3 General Limitation of Individual Member Investment Volume and Profit Distribution for the Sake of the Community

In line with typical cooperative principles, it is not the aim of BEG Wolfhagen eG to maximise revenues but, rather, to maximise the benefit of its members. This approach has led to the creation of several investment regulations, which can be summarised under the following formula: a general limitation of individual member investment volume and profit distribution for the sake of the community. The following quotations reveal that the energy cooperative did not want to attract actors with a lot of money who only had a sole interest in dividends:

> 1.39 "So we do not want people who say Ah, I've been investing in the wind park because I can receive a good return on my investment there" (R1N2, 126).

> 1.40 "We would have had a few rich people who would have put their money on the table and then would have been happy and received a nice return that they would not have gotten at any bank" (R1N2, 34–36).

In consequence, BEG Wolfhagen defined in its statutes:

1. A limitation on investment volume from individual members (BEG Wolfhagen eG 2013b, p. 2), as the maximum number of cooperative shares that can be held by any one member is limited to 40 (20,000 Euros).[14] This capital constraint is intended to particularly support the cooperative's goal of attracting many actors with little capital instead of few actors with a lot of capital. The regulation thus has a direct influence on the composition of actors who invest in the energy cooperative and can exert influence on municipal energy services by becoming shareholders of the municipal energy provider.

---

[14]In January 2014, the general investment limitation for individual members was further reduced to 2500 Euros per member, because projects seemed to be lacking at that time in which BEG Wolfhagen eG could have invested its newly mobilised member capital.

2. A limitation of profit distribution to its members to 6% of the cooperative's nominal capital (BEG Wolfhagen eG 2013b, p. 5). Any additional profit generated by BEG Wolfhagen eG is, amongst other possibilities, transferred to the cooperative's energy efficiency fund. As outlined in the following quotation, this approach is called *Gierbremse* (gear braking).

> 1.41 "We call it gear braking, which means that not more than 6% of the capital is distributed to the municipal energy provider or to the cooperative. So, in a monetary sense, it is limited to 6%. At a maximum investment of 10,000 Euros that would be 600 Euros. And the surplus beyond 6%, if we have a very successful year, is distributed to the energy fund of the cooperative in order to, of course, bring the issue of energy efficiency forward in a different way. I also call it [...] the Wolfhager energy-saving group—become a member, and there's always something social involved" (R1N1, 160).

Especially of note here is that BEG Wolfhagen eG dividends represent annual earnings from the Stadtwerke. Hence, this regulation initiated a change in the application of portions of the municipal energy provider's annual profit, with a special focus on fostering a more efficient renewable energy structure. By introducing this investment regulation, BEG Wolfhagen eG is directing its member focus towards generating social value for the community.

7.1.6.4 New Forms of Regional Value Creation

By becoming a shareholder of the Stadtwerke, BEG Wolfhagen eG turned the creation of regional value through citizen investment into a new official rule. Being the municipal energy provider of Wolfhagen, the Stadtwerke does its main business in Wolfhagen: most of its customers live there, it holds the municipality's power distribution grid and it is the project developer of the core renewable energy projects through which it was able to achieve a 100% renewable energy supply for the region.[15] By becoming a member of the energy cooperative, actors invest in the Stadtwerke's business activities and economically strengthen this regionally embedded organisation. The following quotations underline the strong alignment between the business purposes of BEG Wolfhagen eG and new regional value creation:

> 1.42 "Yes, there are two motives for membership. [...] One is to invest money in the region" (R1N3a, 131).

> 1.43 "To know that the money we spend on energy is not distributed to any big companies or to any old shareholders who simply receive a nice dividend at the end of the year [...]. Rather, it stays in the region. It stays here. We ourselves benefit" (R1N2, 133).

The Stadtwerke, the city and its citizens financially benefit from collaboration between BEG Wolfhagen eG and Stadtwerke Wolfhagen. The city receives business taxes and rent for the realised renewable energy projects, such as the wind and solar parks. The municipal energy provider invested the additional equity capital from the energy cooperative in renewable energy projects and, consequently, earns additional

---

[15] Annual energy balance approach.

profits therefrom. Citizens of Wolfhagen who are members of the cooperative also benefit from these earnings, because part of the Stadtwerke's profit is annually distributed to the cooperative and, thus, to each member. The solar park is situated on land that belongs, amongst others, to several local farmers who receive rent from its use. Local businesses also financially benefit from maintenance and advisory work that needs to be done at the wind and solar parks.

14 of 50 articles written about the energy cooperative between 2012 and 2014 by the most dominant regional newspaper emphasised the importance of regional value creation through the energy cooperative and Stadtwerke Wolfhagen. Hence, a wider public has been able to take notice of this new rule.

However, about 33% of its members do not live in Wolfhagen, so that part of the dividend being distributed by the energy cooperative is actually transferred to other regions. Furthermore, BEG Wolfhagen eG, during the study period, did not exclusively invest its capital within the region, due to difficulties with finding profitable renewable energy projects there that were commensurate with the capital capacity that needed to be allocated. The following quotation regarding the realisation of the solar park underlines a potential conflict between the exclusive creation of regional value, on the one hand, and the focus on being economically profitable, on the other:

> 1.44 "I mean, you always have the feeling that we are producing a regional value added, which is outstanding. But, naturally, this cannot take precedence over all considerations of profitability. It has to be balanced by asking: How much regionality can we afford? [...] And then you have to find such a balance, which unfortunately means having to make a lot of compromises" (R1N3a, 209).

The quotation indicates that being a shareholder of the Stadtwerke and being a citizen-focused company can lead to contradictory expectations for which compromise solutions have to be found. In terms of the cooperative's investment in the neighbouring region of Nordrhein-Westfalia, profitability was evaluated as being more important than waiting for an applicable regional project. Since it is assumed that financial participation plays a major role for members (see Sect. 7.1.4), the investment decision of the energy cooperative was most likely in line with the preferences of most members.

7.1.7 Drawing Upon New Interpretative Schemes

Based upon their shareholding concept, BEG Wolfhagen eG and Stadtwerke Wolfhagen jointly created new interpretative schemes for their envisioned energy structure in Wolfhagen. These societal guiding principles included (1) taking matters into one's own hands, (2) regular democratic feedback for municipal energy services, (3) citizen identification with change process, (4) new co-responsibility of citizens for regional energy issues, (5) new responsibility of municipal energy provider for citizens and energy customers and (5) symbiotic relationship between citizens and municipal energy provider. In the following, the meanings, influence, potentials and challenges entailed by these principles are described. Here I seek to demonstrate how they not only facilitated the development of a new energy infrastructure but also may have restrained certain developments or made them more difficult and complex.

7.1.7.1 Taking Matters into One's Own Hands

The shareholding concept between the energy cooperative and the municipal energy provider enforces a guiding principle which can be formulated as 'We take matters into our own hands by collaborating with each other'. The following quotations reveal that the collaboration between the two partners has been perceived as one of the most important steps towards actually achieving the goal of developing a 100% renewable power supply for the region. The integration of the energy cooperative into the municipal utility is seen as meaning 'We are now going in the direction into which we really wanted to go' and 'We are finding a way that allows us to become independent'.

> 1.45 "We are now going the way that we are going. I think this is absolutely the most important idea for the future. If we only say what we do not want, then we also have to go our own ways and make clear that it works" (R1N2, 86).

> 1.46 "We are finding a way in Wolfhagen, where one can make clear and show that we are making ourselves a bit more independent. We can actually produce our own energy" (R1N2, 123).

Wolfhagen wanted to shape its own energy future, and the energy cooperative has been seen as the final important step that was missing to do so. As outlined by the following statements, key regional actors have believed in the strength of the collaboration between regional actors. The shareholding concept between the Stadtwerke and the energy cooperative made clear how such collaboration can look and what role it can play in mobilising capital, supporters, trust, credibility and acceptance.

> 1.47 "It is not about the fact that an actor takes over everything in the region but, rather, that those who are active in the region work together in its interests" (R1N4, 392).

> 1.48 "It is important that the renewable energy potentials of a region are jointly planned strategically, in a holistic sense [...] that we are, so to speak, organising broad majorities in the municipalities" (R1N1, 274, 290).

Taking matters into one's own hands has a direct effect on regional value creation. It seems that through the collaboration between Stadtwerke Wolfhagen and BEG Wolfhagen, regional actors have become able to determine for themselves what values are created with energy in their region.

7.1.7.2 Regular Democratic Feedback for Municipal Energy Goals

The concept of BEG Wolfhagen eG holding shares in Stadtwerke Wolfhagen created a new regular and democratic feedback structure for municipal energy services. As outlined by the following statements, this new guiding principle is based upon a new discussion culture between citizens, politicians and the municipal energy provider regarding coordination of local energy affairs. Discussions mainly take place in the supervisory board of the Stadtwerke, in which the energy cooperative is a new member.

> 1.49 "The management says we have a project here and we want to go in a particular direction. And then the issue goes to a supervisory board, occupied by our associates—in brackets, now representatives of 600 citizens. To me, this is a kind of democratic feedback back into society" (R1N1, 60).

> 1.50 "And when we talk about it in the supervisory board, then it is also reflected by the supervisory board of the citizens' energy cooperative. This means, again, that it is carried into a group of citizens from Wolfhagen. So I [as a member of the city Wolfhagen] receive a completely different kind of feedback" (R1N2, 420).

An example of this democratic feedback process was joint reflection of advantages and disadvantages of installation of more small photovoltaic plants on private rooftops. The increase of these projects was favoured by the energy cooperative but was seen as being problematic by the Stadtwerke. As the municipal energy provider is the local grid operator, it is responsible for connecting power production plants to the distribution grid. Since the rate of implementation of small photovoltaic plants in Wolfhagen already lay above the national average by that time (Fraunhofer Institute for Building Physics IBP 2010a, p. 13), certain parts of the local distribution grid had reached their capacity limit. In order to satisfy both sides, the partners decided to partly support the installation of additional small photovoltaic plants in selected areas.

This example shows that regular and democratic feedback can influence the activities of the Stadtwerke and the energy cooperative in two ways. On the one hand, the viewpoints of citizens could be better reflected in strategic decisions regarding the development of new energy structures, as suggested by this quotation:

> 1.51 "And so BEG is also always helpful in the sense that the organisation also pays attention [...] and takes care that we are, so to speak, sensitive to the interests of the citizens, even when it comes to critical issues" (R1N1, 294).

On the other hand, it also seems to be improving the understanding of citizens, represented by the cooperative, regarding the activities and viewpoints of the municipal energy provider. Otherwise, it would not have been possible to find a compromise in the discussion around installation of new small photovoltaic plants. The next quotation indicates that the Stadtwerke seems to clearly point out difficulties once they arise:

> 1.52 "I have to say that, where we think investment in power grid expansion will not result in the necessary amount of value added in terms of electricity income, and where we are within our legal rights, we seek to protect ourselves by coming out against them" (R1N2, 296).

According to the following statements, the new direct exchange between the different actors has led to more rapid reflection on certain topics from different, and even unfamiliar, perspectives. Representatives of BEG Wolfhagen eG expressed several viewpoints and ideas at an earlier stage and in a different way in the supervisory board of the Stadtwerke than probably been would have done by established members of the committee.

> 1.53 "The citizens' energy cooperative has been actively involved in the project and also perceives problems at an early stage while also bringing along its own concerns. [...] So one is much closer to the concerns of citizens through the energy cooperative than otherwise" (R1N2, 296–300).

> 1.54 "The style in which decisions are made [has changed]. Particularly in the supervisory board, where the discussion culture that has been aroused may also come from the fact that others are questioning differently, looking at things differently" (R1N3a, 231).

However, the new democratic feedback structure also involves several challenges. Due to the new personnel composition in the supervisory board of Stadtwerke Wolfhagen, additional viewpoints needed to be acknowledged and reflected. The following quotation reveals that this can make decision-taking processes more complex and time intensive:

> 1.55 "It is simply that two more representatives sit at the table of the supervisory board who, of course, bring in their own thoughts, their concerns, their suggestions, their goals, and the like. And of course that makes it—not more difficult, that's the wrong word—but it makes it even more demanding" (R1N2, 270–272).

According to the next statement, the cooperative feels more involved in discussions than it is normally expected to, but this is not always positively seen by others:

> 1.56 "And I have the feeling that we think a lot more than we should. So, and this is maybe again something that [...] is annoying" (R1N3a, 187–191).

Furthermore, the cooperative feels limited in framing its activities and standpoints around topics discussed in the board in a transparent manner and through an interactive process with its members:

> 1.57 "They are actually not supposed to reveal what is discussed and voted on in the supervisory board, but how is that supposed to work with democratisation?" (R1N3a, 151).

A general professional secrecy for members of the Stadtwerke's supervisory board regarding the content of their meetings impedes establishment of public communication with respect to decision processes that take place within the board. But this professional secrecy for members of supervisory boards written into various legal clauses of the GmbH-Gesetz (GmbHG), the Aktiengesetz (AktG) und the Genossenschaftsgesetz (GenG).

7.1.7.3 Citizen Identification with the Change Process

The new guiding principle of 'Taking matters into one's own hands' has been accompanied by a new desire to create an emotional relationship between citizens and the fostered changes within the regional energy structure. As illustrated by the following quotation, the Stadtwerke, being responsible for realising the 100% renewable power supply, had wished for citizens to become able to identify themselves with the planned wind park:

> 1.58 "Enabling citizens to identify themselves with these changes, and from here we have a direction to say to the citizens [...] How are they to take place and why; Why are we only building windmills and not something else; And why on this mountain? In this way, we are also involving them in this process" (R1N1, 72).

In this sense, regional value creation is not only focused on creating an economic and technical value but also on creating new social values, such as the identification of regional citizens with the envisioned energy structure. According to the above presented quotation, identification is supposed to be created by directly involving citizens and by actively explaining to them how and why decisions for particular measures were taken.

Two developments may reveal that the shareholding concept may actually increase the identification of citizens with new energy projects, such as the Rödeser Berg wind park: (1) All members who became members of BEG Wolfhagen eG financially supported the wind park. (2) The steering board of the energy cooperative reported that, at times when the wind mills are not working, they have received phone calls from their members, wanting to inform the energy cooperative that the wind mills are not operating and want to know why. Hence, cooperative members seem to feel responsible for the wellbeing of the wind mills, with the result that they seem to regularly observe them. The wind mills are located on top of a summit, so they can generally be easily seen by citizens that live in Wolfhagen. Both investing in the wind park and feeling responsible for the wind mills indicate that members seem to have made the project into one of their own issues, identifying themselves with it.

7.1.7.4 New Co-Responsibility of Citizens for Regional Energy Issues

As we have seen, in this process citizens have turned from 'passive' customers to shareholders of their own municipal energy provider. According to the next statement, this new influence of local actors in the supervisory board of the Stadtwerke and in the expert committee for energy efficiency requires a new 'co-responsibility' of citizens regarding regional energy issues:

> 1.59 "[To be] powerless or co-responsible. The powerless just watch and lament. The co-responsible also bear responsibility" (R1N1, 344).

The energy cooperative perceives this responsibility in several ways. On the one hand, it feels responsible for the Stadtwerke's operations, because of being a new shareholder:

1.60 "This is now our municipal energy provider, and we are also a bit responsible for making sure that nothing goes wrong" (R1N3a, 191).

On the other hand, the energy cooperative also feels responsible for its members. This responsibility seems to be particularly influenced by personal relationships between the managing board of the cooperative and cooperative members:

1.61 "We're hanging in the middle of it. We must also hold our faces and our heads towards the people who know us here. And they say: Did you not know that?" (R1N3a, 191)

One example that illustrates how this new citizens' energy responsibility turned into a new guiding principle is the previously described Gierbremse. Profit distribution to members of the cooperative is limited to 6%, so any additional profit is transferred to the energy efficiency fund in order to reduce local energy demand, for the sake of the community.

7.1.7.5 New Responsibility of the Municipal Energy Provider for Citizens and Energy Customers

The shareholding concept between BEG Wolfhagen eG and Stadtwerke Wolfhagen and the accompanying capital exchange led to a new business relationship between the municipal energy provider and its customers, most of whom also live in Wolfhagen. Hence Stadtwerke Wolfhagen has become responsible for the investment capital of both citizens and energy customers. This kind of responsibility can be described as a new guiding principle, because it influences the business activities of the Stadtwerke, as it now has to take into account the expectations of its new shareholders when planning new projects. One of the consequences of this new responsibility is indicated by the next statement, which describes how the steering board of the municipal energy provider may be more risk averse in order to protect the capital, citizens and customers who have invested in the organisation:

1.62 "We also pay attention that not just any risks are taken [...] because their own money is involved [...] and that everything possible will be done not to take risks with it. That would otherwise not be done [to such an extent], but you can already see that the supervisory board is looking after that, which is basically also quite good" (R1N2, 326–328).

In this sense, the new responsibility of Stadtwerke Wolfhagen for the capital of citizens and its own energy customers may limit its engagement in innovative projects, which can often be risky and capital intensive. This rule can thus have a direct effect on how the Stadtwerke may be able to further support the energy technology mix of Wolfhagen in the future.

7.1.7.6 Symbiotic Relationship Between a Citizens' Cooperative and a Municipal Energy Provider

The Stadtwerke and the energy cooperative, which mainly represents citizens, see themselves in a symbiotic relationship with each other. The municipal energy provider appears to be happy about the development of BEG Wolfhagen eG, because every new member provides new capital resources for potential energy projects. At the same time, the energy cooperative is interested in the well-being of the Stadtwerke, as the cooperative's annual earnings directly depend upon the annual profit of the Stadtwerke. These sentiments seem to be indicated by the following quotations:

> 1.63 "I see us being in symbiosis. I am happy that the BEG is growing and thriving, because every new member invests money [...] in the municipal energy provider [...] and provides fresh equity" (R1N1, 216).

> 1.64 "If the municipal energy provider is doing well, we are doing well. And vice versa" (R1N3a, 73).

The symbiotic relationship between the Stadtwerke and BEG Wolfhagen eG has led to an intense relationship between the city and the energy cooperative that goes beyond the work of the Stadtwerke's supervisory board. Other projects which are planned by the city of Wolfhagen are now being coordinated with the energy cooperative, including the Energieoffensive project, which works together with the cooperative's energy efficiency advisory committee and even shares the same office.

However, in a symbiotic relationship it can also be challenging to stick to initially defined roles and act within agreed boundaries. For example, this relationship has made it more difficult for the energy cooperative to decide independently about own future plans, as indicated here:

> 1.65 "At first we felt like we were, all of us, together. But we are not the municipal energy provider, and the municipal energy provider is not the cooperative" (R1N3a, 77).

> 1.66 "Actually, we were specially created for this symbiosis, so there is little free room [for own development]. We have to see how this develops in reality" (R1N3b, 42).

Thus, positions and roles may have to be redefined throughout the collaboration process.

7.1.8 Summary

In 2012, BEG Wolfhagen eG became a new partial shareholder of Stadtwerke Wolfhagen, and it is now—next to the city of Wolfhagen—the second owner of the municipal energy provider. The core aim of this move has been to (1) strengthen the municipal organisation in realising renewable energy projects and (2) directly involve citizens of Wolfhagen in achieving a 100% renewable energy supply[16] for

---

[16] Annual energy balance approach.

the region. The shareholding concept has helped BEG Wolfhagen eG and Stadtwerke Wolfhagen in mobilising allocative and authoritative resources (summarised in Table 7.12), empowering them to pursue their goals.

Table 7.12 Allocative and authoritative resources mobilised through collaboration between BEG Wolfhagen eG and Stadtwerke Wolfhagen

Mobilised resources between BEG Wolfhagen eG and Stadtwerke Wolfhagen	
Allocative resources	Authoritative resources
Actors – Activation of 675 cooperative members by 2014, who became new shareholders of Stadtwerke Wolfhagen. All members are, per the cooperative's statutes, also energy customers of the municipal energy provider. – Cooperative steering board is staffed by private persons, elected in a publically open process. – Activation of personnel from the municipal energy provider, who professionally supported the complete foundation process of the energy cooperative. *Financial capital* – Mobilisation of 3.1 million Euros total capital by 2014 for becoming a 25% shareholder of Stadtwerke Wolfhagen and for renewable energy projects realised by the energy cooperative. – Transfer of 2.3 million Euros nominal capital to Stadtwerke Wolfhagen, which strengthened it in realising renewable energy projects, such as a wind park. – Generation up to 116,000 Euros annual profit, resulting in up to 3% dividend for members. – Early mobilisation of foundation capital for initiating BEG Wolfhagen eG. *Technology mix* – Installation of one wind park and one solar park by 2014. – With an expected power yield of 40,000 megawatt hours, projects cover about 61% of the annual power demand of Wolfhagen.	– *Organisational identify*: BEG Wolfhagen eG mainly defines itself as a new partial shareholder of Stadtwerke Wolfhagen. – *Image and publicity* related to the shareholding concept. Collaboration with the energy cooperative provided a new unique selling point for the Stadtwerke. – *Trust* of 675 members and Stadtwerke customers (as of 2014) in the business model of BEG Wolfhagen eG. – *Trust* between involved representatives of the cooperative, the Stadtwerke and the city of Wolfhagen—provided the basis for a harmonious work atmosphere. – *Citizen acceptance of energy projects:* BEG Wolfhagen eG strengthened acceptance for the wind park by strengthening its supporters. – *Credibility of energy efficiency projects:* Involvement of BEG Wolfhagen eG and its energy efficiency advisory board made energy efficiency projects more compelling to local actors. – *Access to potential projects:* As a shareholder of Stadtwerke Wolfhagen, the energy cooperative is automatically part of all projects realised by the municipal energy provider. – *Access to information and new ideas:* New organisational spaces, such as the energy efficiency advisory committee, create new project ideas for the energy cooperative and Stadtwerke Wolfhagen. – *Transparency*: Direct information flow between the energy cooperative and its members, who are at the same time customers of the municipal energy provider and citizens of Wolfhagen, which increases project transparency for the Stadtwerke. – *Expertise and know-how:* Stadtwerke Wolfhagen offers key legal and marketing expertise to BEG Wolfhagen eG.

The shareholding concept between BEG Wolfhagen eG and Stadtwerke Wolfhagen has also created new regulations and interpretative schemes. These new societal rules (summarised in Table 7.13) have provided the basis for their envisioned renewable energy structure.

The new rules created through and within the collaborative relationship between BEG Wolfhagen eG and Stadtwerke Wolfhagen have been discussed in public newspapers, demonstrating that they have been diffused into the wider regional public, beyond the direct interaction boundaries of the two partners. Table 7.14 provides an overview of articles about BEG Wolfhagen eG published in the Hessische/Niedersächsische Allgemeine—District Kassel und Wolfhagen between 2012 and 2014 that explicitly mention rules used during collaboration between the energy cooperative and Stadtwerke Wolfhagen.

The collaborative shareholding concept between BEG Wolfhagen eG and Stadtwerke Wolfhagen has influenced the energy structure of the municipality of Wolfhagen in the following ways:

Table 7.13 Rules jointly created through cooperation between BEG Wolfhagen eG and Stadtwerke Wolfhagen

Rules jointly created by BEG Wolfhagen eG and Stadtwerke Wolfhagen	
Regulations	Interpretative schemes
– Direct ownership of citizens and energy customers over their own municipal energy provider. The energy cooperative holds a 25% share of Stadtwerke Wolfhagen. – New co-decision and participation rights for citizens and energy customers with respect to municipal energy services, especially because the energy cooperative holds two of nine seats in the supervisory board of Stadtwerke Wolfhagen and itself has open membership and democratic decision-taking structures. – General limitation of individual member investment volume and of profit distribution for the sake of the community, by defining maximum investment levels for members and by defining a limit for distributed dividends to members in the statutes of BEG Wolfhagen eG. – Strengthening the ability of the Stadtwerke and the municipality of Wolfhagen to create regional value through involving citizen investments and by developing their own renewable energy projects.	– Together with citizens, Stadtwerke Wolfhagen and the municipality of Wolfhagen 'take matters into their own hands' in order to develop a 100% renewable power supply for their region. – New regular democratic feedback for municipal energy goals by allowing citizens to become shareholders of Stadtwerke Wolfhagen. – New identification of citizens and energy customers with change process fostered by their municipal energy provider, as they become joint owners of the Stadtwerke when joining BEG Wolfhagen eG. – New energy co-responsibility of citizens for regional energy issues. Since citizens invest their capital in the municipal energy provider, they are also co-responsible for the projects that are implemented by it. – New co-responsibility of municipal energy provider for capital from citizens and customers. – New symbiotic relationship between citizens and municipal energy provider in coordinating energy issues in the region, as they depend upon each other in developing renewable energy projects.

Table 7.14 Articles about BEG Wolfhagen eG published in the *Hessische/Niedersächsische Allgemeine—District Kassel und Wolfhagen* in 2012–2014 that mention new rules created between the cooperative and Stadtwerke Wolfhagen

New societal rules created between Stadtwerke Wolfhagen and BEG Wolfhagen eG	Percentage of articles that mention and reflect the respective rule (n = 50)
New co-decision and participation rights for citizens	44
Direct ownership of citizens and energy customers over their own municipal energy provider	42
Regional value creation	28
New regular democratic feedback	20
Taking matters into one's own hands	18
General limitation of individual member investment volume and community-oriented profit distribution	16
New identification of citizens with energy change process	6
New symbiotic relationship between citizens and municipal energy provider	4
New energy co-responsibility of citizens	2
New co-responsibility of municipal energy provider	2

1. *Joint empowerment of a municipal energy provider and a citizen-focused energy cooperative in developing renewable energy in their region:* By 2014, BEG Wolfhagen eG had mobilised 3.1 million Euros total capital, mainly from its 675 members, 2.3 million Euros of which were directly transferred to the Stadtwerke as new equity capital. These financial and personnel resources empowered the two partners to realise two core renewable power production projects, through which about 61% of Wolfhagen's power demand can be covered. The Stadtwerke and the energy cooperative positioned themselves as capable organisations by taking advantage of each other's unique business backgrounds. BEG Wolfhagen eG had direct access to the Stadtwerke's expertise, which particularly helped them in their foundation and registration process. Meanwhile, Stadtwerke Wolfhagen was able to increase the loyalty of its customers, because cooperative membership is aligned to the need to also be energy customer of the municipal energy provider. At the same time, the member-customer alignment represented a valuable tool for the energy cooperative to attract new members. BEG Wolfhagen eG mainly identifies itself as being a shareholder of Stadtwerke Wolfhagen. By becoming an owner of the well-established municipal energy provider, the newly initiated energy cooperative quickly evolved into a trustworthy and well-known organisation right from the start of its business operations. These authoritative resources, including organisational identity, image and publicity, played a key role for gaining cooperative members and capital. As a result, both BEG Wolfhagen eG and Stadtwerke Wolfhagen turned into the two most influential actors in developing

renewable power production within the region as well as putting them in a position to improve energy efficiency structures in the municipality of Wolfhagen.

2. *Reaching long-term legitimacy for municipal energy goals on a more direct and continual basis:* The new form of direct ownership of citizens and energy customers—represented via the energy cooperative—over their own municipal energy provider supports legitimation of municipal energy goals, such as the development of a 100% renewable power supply and implementation of the Rödeser Berg wind park, for two primary reasons. First, the new presence of the energy cooperative on the supervisory board and at shareholder meetings of the Stadtwerke has created an immediate and regular democratic feedback structure between political representatives, citizens and the municipal energy provider. Second, the new co-decision rights of citizens with respect to the business strategy of the Stadtwerke—which they have gained through the shareholder position of the energy cooperative—requires a joint decision making process between all involved actors. This has been accompanied by direct exchange regarding municipal energy issues that go beyond the boundaries of single energy projects, which increases understanding about each other's position. This is important, as the viewpoint of each side must be taken into account when deciding about energy pricing, the Stadtwerke's profit distribution or the selection of new projects that are supposed to be realised by the Stadtwerke in the future. In this sense, the collaborative shareholder concept has increased the significance of creating a joint will between citizens, politicians and the municipal energy provider with respect to municipal energy goals. Of special note is that Stadtwerke Wolfhagen and the city of Wolfhagen have strengthened public legitimation for their actions by actually giving up part of their established power to a new shareholder: the energy cooperative, with its open membership and democratic decision-taking structures. The election process for the mayor and city council in 2011, with its hotly contested debate over the wind park that was being planned by the Stadtwerke, illustrate how central renewable energy issues can become for a municipality and how challenging it can be to solve associated conflicts at a purely political level. Through the involvement of the energy cooperative, the Stadtwerke appears to have achieved a better image as a citizen-focused and mainly regionally situated energy provider. In turn, the municipal energy provider further increased transparency regarding its own business activities, as an important pre-condition for integrating the energy cooperative as a new shareholder was building a trustful relationship between politicians, representatives of the Stadtwerke and those citizens that eventually became members of the cooperative steering board.

3. *Giving citizen participation higher value through the creation of various interaction spaces and co-decision rights:* The energy cooperative and its shareholding position within the Stadtwerke has provided citizens direct participation and

co-decision rights at a number of levels regarding municipal energy issues. First, citizens financially participate in the business activities of the Stadtwerke when they become members of BEG Wolfhagen eG. By 2014, 675 members made use of this right by buying almost 6000 member shares, with a total value of about three million Euros. Second, citizen-members can participate in strategic decisions regarding the positioning of the Stadtwerke, because they are represented by the energy cooperative at its supervisory board and shareholder meetings. Third, citizens can participate in the development of energy efficiency projects in the newly created energy efficiency advisory committee and its aligned working groups, in which any interested cooperative member can join. Fourth, newly created interaction spaces between politicians, the municipal energy provider and citizens, where direct participation can actually take place on a regular and frequent basis, play a key role. Examples here are the just-mentioned supervisory board meetings and shareholder meetings of the Stadtwerke as well as the energy efficiency advisory committee meetings. This direct citizen participation has effected the creation of several important authoritative resources. Through direct interaction, Stadtwerke Wolfhagen, politicians and the energy cooperative have been receiving immediate access to the ideas, opinions and expertise of citizens. As outlined above, the credibility of energy efficiency measures seemed to increase when citizen participated in developing these projects. Hence, participation has influenced the energy technology mix that was fostered by the municipal energy provider and the energy cooperative. At the same time, direct participation entails a new sense of citizen co-responsibility with regard to municipal energy issues. To some extent, BEG Wolfhagen eG has embedded this responsibility by introducing new rules, such as limiting individual member investment volume, in order to demonstrate that private profit maximisation was not supposed to be the main reason for becoming a cooperative member. Furthermore, part of the Stadtwerke's annual distributed profit is not allocated to cooperative members but, rather, directed into a new energy efficiency fund for the sake of the cooperative community.

4. *Strengthening citizen acceptance of renewable energy projects and identification of citizens with the energy change process:* The shareholding concept between BEG Wolfhagen eG and Stadtwerke Wolfhagen strengthened acceptance for the planned wind park. Citizen acceptance was created through increasing their engagement, in particular their financial engagement when becoming members of BEG Wolfhagen eG. All cooperative members automatically supported the wind park financially, because most of the cooperative member shares were transferred to the Stadtwerke in order to implement it. The achievement of 675 cooperative members by 2014 also increased the number of actors that publically positioned themselves in favour of the wind power project compared to previously existing informal citizen's initiatives. At the same time, the energy cooperative increased the power of this group by mobilising resources, such as

capital, expertise and publicity, enabling them to support the targeted project. In this sense, the energy cooperative provided a tool for the activation and coordination of wind park acceptors. Investing in the wind park and feeling a sense of responsibility for the wind mills after their installation indicate that members seem to have made the project into their own issue, identifying themselves with the change process.

5. *Giving regional value creation a new meaning by taking matters into one's own hands:* The aim of developing a 100% renewable power supply in Wolfhagen was based upon a new guiding principal which can be described as taking matters into one's own hands. The energy cooperative was founded in order to support the municipal energy provider and the city of Wolfhagen in realising their energy vision out of own combined strength. The idea of taking matters into one's own hands reinforced the desire to create value for one's own region. Special here is that the involved partners have sought to combine the creation of economic values—such as generating a dividend for cooperative members, project rents and business taxes for the city or profit for the Stadtwerke—with the creation of social values. The partners have had the desire of helping citizens in identifying themselves with the fostered changes. BEG Wolfhagen, for example, limited the options for individual profit generation for the sake of the community by putting a cap on total shares one investor can own and transferring part of the dividend to the energy efficiency fund. Regional value creation by taking renewable energy production into one's own hands increases the significance of the community, because it makes the region into a new centre of action, initiated by regionally situated actors, such as the municipal energy provider and the citizen-focused energy cooperative, two central activists, who now determine what kinds of values are created through energy investment in the region. Such value is created through a new symbiotic relationship between the municipal energy provider and citizens, who build upon each other in developing regional energy projects.

However, the shareholding concept between BEG Wolfhagen eG and Stadtwerke Wolfhagen has also run into a number of challenges and boundaries.

1. *The regional focus of BEG Wolfhagen eG has been challenged by strong membership and capital growth as well as a lack of applicable projects:* The energy cooperative did not need to fully use its allocated financial resources in order to support the development of Wolfhagen's energy structure. Subsequently, about 7% of its member share-generated capital was invested in another trans-locally operating energy cooperative, which establishes wind parks throughout the neighbouring Bundesland of Nordrhein-Westfalia. From its founding day, BEG Wolfhagen eG had experienced strong growth in terms of members and capital within a relatively short amount of time, and it felt pressured to invest this accumulated capital in the short term in order to generate at least a small dividend for all members. At the same time, it had difficulties in finding

profitable projects in Wolfhagen. National amendments to the German Renewable Energy Act that were passed during this time further decreased the profitability of several project types, making the project search even more difficult. As a consequence, the cooperative's leadership looked for alternatives outside its focus region, with the profitability of a project being evaluated as more important than its regional location. Furthermore, cooperative membership is not strictly limited to citizens of Wolfhagen but to energy customers of the Stadtwerke. Because of this, about 33% of cooperative members were mobilised outside of Wolfhagen, though many of them live within a radius of 50–60 kilometers to the focus region.

2. *Influence of citizens on strategic business decisions and on project development is limited*: Citizens participate in strategic decision-taking processes of the energy cooperative by making use of their voting rights during general assemblies. Due to this possibility to influence the energy cooperative and its activities, they also have a new responsibility regarding energy issues in the region. However, only 27–50% of all members, that is generally the minority, have actually made use of this right. Hence more members seem to be interested solely in financial participation rather than in influencing the direction of cooperative business. At the same time, the level of citizen influence on the municipal energy provider is legally limited. The energy cooperative, representing the citizens in the supervisory board of the Stadtwerke, can be outvoted by the city of Wolfhagen, the main Stadtwerke owner, on most issues, as the city holds the majority of board seats. Furthermore, BEG Wolfhagen eG has no direct influence on the Stadtwerke's operational activities, because it is shareholder of the Stadtwerke and not a direct project partner. The influence of citizens on its implementation processes is, thus, also limited.

3. *Strengthening of acceptance for renewable energy but no overall resolution of local conflict regarding the Rödeser Berg wind park:* The close work between the energy cooperative and the Stadtwerke strengthened acceptance for the Rödeser Berg wind park. However, the collaboration did not help to fully resolve the conflicts between wind park supporters and opponents, which had already begun before the energy cooperative was founded. Opponents and supporters had had positive experiences coming together for a conference around the topics of energy and the wind park, which was organised by the city and which was open to all interested citizens. However, according to both sides, the conference had been organised too late, providing evidence that conflict management needs the early integration of citizens, a well-organised and open exchange of positions, as well as serious consideration of alternatives.

4. *New responsibility for citizens' capital as well as national investment regulations tend to foster low-risk investment options instead of innovative pilot projects:* BEG Wolfhagen eG receives its financial resources from Stadtwerke customers, most of whom are citizens of Wolfhagen. Both BEG Wolfhagen eG

7.1 Collaboration Between Energy Cooperatives and Energy Related Companies

and the Stadtwerke feel a strong responsibility for the private capital of their members and customers, preferring to invest citizens' capital in projects that involve low investment risks. The focus on 'safe' investments was further increased through the registration of BEG Wolfhagen eG under BaFin. In order to be in accordance with the rules of the KAGB (Kapitalanlagegesetz), the cooperative has needed to limit its activities to those projects that receive a secured minimum return on investment, such as a guaranteed feed-in tariff under the German Renewable Energy Act. Consequently, other important technologies that have been envisioned by the city in order to achieve a 100% renewable energy structure, such as power storage, smart energy distribution and electro-mobility, have not been fostered by BEG Wolfhagen eG. Hence, the energy cooperative has been limited in its ability to support innovative technologies.

5. *Organisational growth may be limited in the future:* The mobilisation of citizens' capital and cooperative members provided the main basis for all joint activities between the Stadtwerke and BEG Wolfhagen eG. On the one hand, the activation of members and capital was strongly supported by internal aspects of the collaborative effort, such as the creation of an organisational identity, trust in the business model of the energy cooperative and publicity for joint goals. On the other hand, their successful allocation was facilitated through external aspects of the collaboration structure, including a more or less secured renewable energy project profitability through the German Renewable Energy Act as well as through the circumstance that cooperative membership represents a valuable investment alternative for citizens. In the future, however, it could be challenging for BEG Wolfhagen eG to grow further, due to several reasons. Perhaps most importantly, the latest amendments to the German Renewable Energy Act have reduced renewable energy project profitability. Hence, it has already become more challenging to find applicable projects, especially considering the circumstance that the steering board of the energy cooperative works on a voluntary basis. As a consequence, the maximum amount of individual cooperative member investment volume was already reduced from 20,000 Euros to 2500 Euros in January 2014. Yet, a change of the general investment landscape may make other investment opportunities more attractive for citizens in the future.

Table 7.15 summarises the influence of the shareholding concept between BEG Wolfhagen eG and Stadtwerke Wolfhagen on Wolfhagen's energy structure, ordered in terms of realised potential and challenges or limits.

Having looked in detail at the benefits and constraints of collaboration between a cooperative and an energy-related company, we now move on to investigating the second dominant interaction model pursued by German energy cooperatives these days.

Table 7.15 Influence of collaboration between BEG Wolfhagen eG and Stadtwerke Wolfhagen on the local energy structure

Impacts of collaboration between BEG Wolfhagen eG and Stadtwerke Wolfhagen on Wolfhagen's energy structure	
Realised potential	Challenges and limits of influence
Empowerment of the municipal energy provider through citizen support – The joint mobilisation of 3.1 million Euros total capital and 675 cooperative members (2014) empowered the partners to realise two core renewable power production projects: solar and wind parks, with a capacity of 17 megawatts, covering 61% of Wolfhagen's power demand. The collaboration also became the basis for successfully initiating energy efficiency projects. – Jointly created authoritative resources—including organisational identity, trust in the business model of the cooperative, a positive public image, publicity and access to potential projects—played a key role in attracting cooperative members and capital. Reaching long term legitimacy for municipal energy goals on a more direct and regular basis – Stadtwerke Wolfhagen and the city of Wolfhagen strengthened the legitimacy of their actions by giving up part of their established power to a new shareholder: the democratic citizen and customer managed energy cooperative. – The new presence of the energy cooperative on the supervisory board and at the shareholder meetings of the Stadtwerke created an immediate and regular democratic feedback structure between political representatives, citizens and the municipal energy provider. – The accompanying direct exchange about municipal energy issues increased understanding for each other's positions and helped to create a joint will. Giving citizen participation a higher value – The energy cooperative and its shareholding position at the Stadtwerke give direct citizens participation in municipal energy issues a significant value through the creation of various interaction spaces and co-decisional rights. – Direct citizen participation in the form of involvement in strategic organisational decisions and in project development created immediate access to the ideas, opinions and the expertise of citizens.	Difficult to keep regional focus – Short-term investment pressure due to rapid capital growth, amendments in national regulations that reduced returns on project investment and lack of regional and profitable project options motivated BEG Wolfhagen eG to invest part of its capital in the neighbouring Bundesland of Nordrhein-Westfalia. – Profitability of a project was, thus, evaluated as being more important than its regional location. – Cooperative membership is not strictly limited to citizens of Wolfhagen but to energy customers of the Stadtwerke. In consequence, 33% of cooperative members live outside of Wolfhagen. Influence of citizens on strategic business decisions and project development is limited – Citizen participation is aligned to a new sense of citizen responsibility, though more members seem to be interested in solely financial participation than in co-deciding strategic business issues, as only between 27% and 50% of all members attend annual assemblies of the cooperative. – The level of direct citizen influence in the supervisory board of the Stadtwerke is limited because the energy cooperative, representing citizens, holds only two of nine seats. It can thus be outvoted by the city of Wolfhagen, the main Stadtwerke owner, on most issues. – Citizens, as represented by BEG Wolfhagen eG, have no direct influence on the Stadtwerke's operations, because the energy cooperative is not a direct project partner and, thus, not involved in project implementation processes. No overall resolution of local conflict regarding the Rödeser wind park – Collaboration between BEG Wolfhagen eG and Stadtwerke did not help to fully resolve existing conflicts between wind park supporters and opponents. – Experience from Wolfhagen suggest that conflict management needs early integration of citizens into a well organised and open

(continued)

7.1 Collaboration Between Energy Cooperatives and Energy Related Companies 215

Table 7.15 (continued)

Impacts of collaboration between BEG Wolfhagen eG and Stadtwerke Wolfhagen on Wolfhagen's energy structure	
Realised potential	Challenges and limits of influence
– Citizen participation in developing energy efficiency measures has seemed to increase the credibility of these projects.	exchange of positions, including serious consideration of alternatives.
Strengthening citizen acceptance of renewable energy projects and citizens identification with change process	Limited support for technology innovations
– New participation rights, in particular regarding financial participation, strengthened citizen acceptance for renewable energy projects and helped citizens to identify themselves with the change process.	– BEG Wolfhagen eG and the Stadtwerke prefer to invest citizens' capital in projects that involve low investment risks, because they feel highly responsible for the capital of their members and customers.
– BEG Wolfhagen eG strengthened project acceptance by increasing the number of actors who publically positioned themselves in favour of the wind park and by helping this group in mobilising capital, expertise and publicity.	– In order to comply with the rules of the KAGB laws, BEG Wolfhagen eG has limited its activities to those projects that can receive a secured minimum return on investment.
– Members seem to have made the project as their own issue and, thus, identify themselves with it.	– Since technology pilot projects often involve higher investment risks and do not receive secured return on investments, the energy cooperative may be constrained from supporting the development or distribution of innovative energy technologies, such as storage or e-mobility.
Giving regional value creation a new meaning by taking matters into one's own hands	Cooperative growth may be limited in the future
– The shareholding concept and accompanying renewable energy project activities have been creating new regional value and increasing the significance of the municipality in developing renewable energy, as it has become a new reference point for interaction.	– The latest amendments to the German Renewable Energy Act have reduced project profitability and made it more challenging for a voluntary working cooperative steering board to find further applicable projects.
– Stadtwerke Wolfhagen and BEG Wolfhagen eG, and the citizens supporting them, are taking matters into their own hands and, thus, determine what kinds of values are created through energy in the region.	– The maximum amount of individual member investment volume was thus reduced from 20,000 Euros to 2500 Euros.
– The partners combine economic value creation, including dividends, business taxes, and rent, with the creation of social values, such as citizen identification with the whole change process and a regular democratic feedback structure.	– A change of the general investment landscape, for example, if general interest rates rise, may make other investment opportunities more attractive for current cooperative members in the future.
– These values have been created through the new symbiotic relationship between the municipal energy provider and primarily local citizens.	

7.2 Collaboration Between Energy Cooperatives and Banks: Energie + Umwelt eG

The second dominant interaction model favoured by recently founded German energy cooperatives involves collaboration with banks, particularly with cooperative ones. The Energie + Umwelt eG energy cooperative was chosen for my second case study for the following reasons: (1) The organisation was founded by three cooperative banks and is also fully managed by them. In addition it closely works together with eight other cooperative banks. (2) The energy cooperative has the aim of being a structuring actor regarding energy in the districts Main-Tauber and Neckar-Odenwald and (3) it has experienced a strong organisational growth. With almost 1525 members and 11.9 million Euros total capital (as of December 2014), Energie + Umwelt eG belongs amongst the largest renewable energy producing cooperatives in Germany.

7.2.1 Introduction of the Focus Region

Energie + Umwelt eG has been seeking to support the two districts of Landdistrict Neckar-Odenwald and Landdistrict Main-Tauber in becoming a zero-emission region (for example, Energie + Umwelt eG 2014c). These districts can thus be described as the preferred focus region of the energy cooperative.

7.2.1.1 Geographical Character

The districts Neckar-Odenwald and Main-Tauber, together covering 2431 square kilometres, are situated in the northern part of the Bundesland of Baden-Württemberg, representing about 7% of its overall area. Settlement areas, industrial real estate and infrastructure for mobility make up 11% of these districts, whereas 52% is used for farmland and 35% is covered with forest. Neckar-Odenwald and Main-Tauber can thus be characterised as a rural region. Table 7.16 summarises key data of these districts from 2012.

In 2012, both districts together had around 272,000 people, representing about 2.6% of the population of Baden-Württemberg (Statistisches Landesamt Baden-Württemberg 2014). Table 7.17 summarises the districts' total population and population per square kilometre in 2012.

With about 126 people per square kilometre, Neckar-Odenwald has a slightly higher population density than Main-Tauber, which has about 100 per square kilometre, resulting in an average of 113 people per square kilometre for both districts in 2012.

Table 7.16 Geographical character of the study region of Neckar-Odenwald and Main-Tauber (Statistisches Landesamt Baden-Württemberg 2014)

Geographical character of the study region through 2012	Neckar-Odenwald	Main-Tauber	Total	Total
	Sq km	Sq km	Sq km	Percent
Total size of the region	1126	1304	2431	100
Agriculture	518	754	1271	52
Forest	474	386	861	35
Settlement and traffic infrastructure	120	141	261	11
Watter	8	11	18	1
Other use	7	13	20	1

Table 7.17 Population of Neckar-Odenwald and Main-Tauber and its distribution in 2012 (Statistisches Landesamt Baden-Württemberg 2014)

Overview of the Neckar-Odenwald and Main-Tauber population and distribution in 2012	Neckar-Odenwald	Main-Tauber	Total	Unit
Population	141,847	129,842	271,689	Number
Population per sq. km	126	100	113	Number/sq. km

7.2.1.2 Energy Transition Goals

The three neighbouring districts of Main-Tauber, Neckar-Odenwald and Hohenlohe set themselves the goal of jointly becoming zero-emission districts, wanting 100% of their power and heat demand to be supplied by renewable energy (Bioenergie-Region Hohenlohe-Odenwald-Tauber GmbH 2015; Landratsamt Neckar-Odenwald-Kreis 2014, p. 2). This goal evolved out of the districts' prior activities, which had been undertaken in the energy sector over the previous years. The three districts initiated several projects together, including becoming an official bio-energy region—*Bioenergieregion H-O-T*—funded by the Federal Ministry of Food and Agriculture (Bioenergie-Region Hohenlohe-Odenwald-Tauber GmbH 2014) and aiming to support the districts in realising their bioenergy potential. All three districts established their own energy agencies—Neckar-Odenwald in 2007 and Main-Tauber in 2008—which help to coordinate and realise projects in the areas of renewable energy and climate protection (Energieagentur Main-Tauber-Kreis GmbH; EnergieAgentur Neckar-Odenwald-Kreis 2014). All of these projects have had regional political support as well.

Thus far, however, Neckar-Odenwald is the only one of the three districts to have officially developed a concrete climate and energy strategy. Its climate-oriented investment program includes an analysis of the district's actual energy demand, the renewable energy potential in the region, as well as planned action steps (Landratsamt Neckar-Odenwald-Kreis 2014). These will be outlined in the following, in order to demonstrate how both districts, Neckar-Odenwald and Main-Tauber, could achieve zero emissions. A concrete master plan for climate protection is also being developed by Main-Tauber and was supposed to have been finalised by 2016. Before describing the main steps of Neckar-Odenwald's strategy, the energy demand of the two districts is outlined below.

Energy Demand[17]

As displayed in Table 7.18, the two districts consumed about two million megawatt hours power in 2011, of which 75% was demanded by industry, meaning about 684,000 megawatt hours in Main-Tauber and approximately 748,000 megawatt hours in Neckar-Odenwald. Household demand, at 25%, was about 224,000 megawatt hours in Main-Tauber and 244,000 megawatt hours in Neckar-Odenwald. Public administration consumed approximately 2000 megawatt hours in each district.

As displayed in Table 7.19, the two districts consumed about three million megawatt hours of heat in 2011. Household demand, at 73%, consumed most of the heat, meaning about 1.1 million megawatt hours in Main-Tauber and about 1.2 million megawatt hours in Neckar-Odenwald. Industry consumed about 27% of the heat, representing around 416,000 megawatt hours in Main-Tauber and about 454,000 megawatt hours in Neckar-Odenwald. Public administration consumed,

Table 7.18 Power demand of the Main-Tauber and Neckar-Odenwald districts in 2011 (based upon Landratsamt Neckar-Odenwald-Kreis 2014, p. 11)

Power demand in 2011	Main-Tauber MWh/a	Neckar-Odenwald MWh/a	Total MWh/a	Total Percent
Industry	684,328	747,600	1,431,928	75
Private households	223,716	244,400	468,116	25
Public administration	2014	2200	4214	0.2
Total	910,057	994,200	1,904,257	100

Table 7.19 Heat demand of the Main-Tauber and Neckar-Odenwald districts in 2011 (based upon Landratsamt Neckar-Odenwald-Kreis 2014, p. 13)

Heat demand in 2011	Main-Tauber MWh/a	Neckar-Odenwald MWh/a	Total MWh/a	Total Percent
Private households	1,141,462	1,247,000	2,388,462	73
Industry	415,576	454,000	869,576	27
Public administration	10,527	11,500	22,027	1
Total	1,567,565	1,712,500	3,280,065	100

---

[17]Energy demand for the Neckar-Odenwald district was directly drawn from its climate-oriented investment program (Landratsamt Neckar-Odenwald-Kreis 2014, pp. 11ff.), whereas energy demand for the Main-Tauber district was calculated by extrapolating the per-person power and heat demand of Neckar-Odenwald to the population of Main-Tauber.

1%, approximately 11,000 megawatt hours in Main-Tauber and about 12,000 megawatt hours in Neckar-Odenwald.

Development of a Renewable Energy Supply

According to the climate-oriented investment program of Neckar-Odenwald (Landratsamt Neckar-Odenwald-Kreis 2014, p. 115), the first planned measure is to increase employment of renewable energy production technology. Table 7.20 shows the renewable power potential for both districts.

As can be seen, the maximum estimated renewable power potential, calculated by the districts themselves, is with almost 8.6 million megawatt hours per year quite high. In such a maximum scenario, Main-Tauber and Neckar-Odenwald could produce more than four times as much renewable power as needed for covering their actual power demand (see Table 7.18). By November 2014, both Main-Tauber and Neckar-Odenwald had already become able to cover a high proportion of their power demand through renewable energy, with about 457,609 megawatt hours of renewable power being produced in Main-Tauber and about 560,046 megawatt hours in Neckar-Odenwald (Deutsche Gesellschaft für Sonnenenergie e.V. 2014a, b). Thus, compared to the power demand displayed in Table 7.18, Neckar-Odenwald was able at that time to cover 56% and Main-Tauber 50% of power demand with renewable power.

Neckar-Odenwald estimates that it can only cover 45% of its heat demand with renewable and regionally produced heat. Hence, the district has been prioritising reduction of heat demand on the way towards becoming a zero-emission region. Furthermore, it aims to support small district heating systems and small combined heat and power projects (Landratsamt Neckar-Odenwald-Kreis 2014, p. 131), which is in line with goals that were also expressed by the energy agency for the Main-Tauber district during interviews I conducted there (R2N5).

The 100% renewable energy goal refers to an overall annual balance between energy demand and renewable energy supply; management of time delays between actual demand and actual renewable energy supply is not included. A complete 100% renewable energy strategy would also have to involve energy storage and

Table 7.20 Potential renewable power supply in the Main-Tauber and Neckar-Odenwald districts (Energieagentur Main-Tauber-Kreis GmbH 2011; Landratsamt Neckar-Odenwald-Kreis 2014, p. 110)

Overview of the potential renewable power supply in Main-Tauber and Neckar-Odenwald	Main-Tauber MWh/a	Neckar-Odenwald MWh/a	Total MWh/a
Wind	2,725,470	3,826,000	6,551,470
Solar	576,884	415,967	992,851
Biomass	444,549	218,320	662,869
Water	319,175	68,939	388,114
Total	4,066,078	4,529,226	8,595,304

distribution concepts, through which excess or insufficient energy supply could be balanced at any time (Federal Ministry for Economic Affairs and Energy 2014c). Such approaches are, however, not yet being discussed within the existing political programmes for these districts (for example, Landratsamt Neckar-Odenwald-Kreis 2014).

Energy Efficiency

A second step towards realising a 100% renewable energy supply is reduction of energy demand from private households. According to their climate-oriented investment program, Neckar-Odenwald aims to reduce 26% of power and 54% of heat demand from private households by 2050 (Landratsamt Neckar-Odenwald-Kreis 2014, p. 40). With respect to heat, the largest energy efficiency potential lies in the improvement of housing insulation and the exchange of heat pumps (Landratsamt Neckar-Odenwald-Kreis 2014, p. 35). The highest energy efficiency potential with respect to power was identified as using more efficient domestic appliances and light bulbs. Neckar-Odenwald wants to improve the energy efficiency of its street lights as well as lighting for public buildings, especially by using LED technology (Landratsamt Neckar-Odenwald-Kreis 2014, pp. 123ff.).

Other Measures

In addition, Neckar-Odenwald aims to develop innovation projects and wants to activate electro-mobility options (Landratsamt Neckar-Odenwald-Kreis 2014, pp. 123ff.), such as supporting the establishment of charging infrastructure for electric vehicles and increasing car-sharing possibilities (Landratsamt Neckar-Odenwald-Kreis 2014, p. 129).

With respect to coordinating all of its measures, Neckar-Odenwald wants to improve communication, as well as to support the exchange of know-how and experience, between all involved actors (Landratsamt Neckar-Odenwald-Kreis 2014, pp. 130ff.).

The climate-investment programme of Neckar-Odenwald shows in an exemplary manner that the two districts of Neckar-Odenwald and Main-Tauber could achieve a 100% renewable energy structure (annual balance approach), provided that:

- Wind and solar power is strongly expanded. Especially wind plays a crucial future role, since the potential wind power supply alone could be three times higher than the annual power demand;
- energy demand is reduced. Especially heat demand would have to be decreased, since only 45% of it can presumably be supplied through renewable resources from the region; and
- innovation projects are activated, including electro-mobility.

7.2.2 The Energy Cooperative Energie + Umwelt eG

In the following, the foundation process of the energy cooperative Energie + Umwelt eG and its business approach are described.

7.2.2.1 Foundation Process

Energie + Umwelt eG was founded on 19 November 2010 and officially registered on 17 August 2011. The energy cooperative was initiated by three cooperative banks situated within the north of Baden-Württemberg: Volksbank Mosbach, Volksbank Main-Tauber and Volksbank Franken. The idea for it was born between the three chairmen of the banks' executive boards. The financial institutions had noticed among their customers an increasing demand for investment support and investment advisory services in the renewable energy sector. The banks planned to better position themselves in this potential new business field and began to search for options to increase their know-how and expertise with respect to the implementation and operation of renewable energy. Their aim was to become expert banks in financing renewable energy projects and to be noticed as such in the region. On the one hand, they wanted to be able to offer professional advice to customers in financing renewable energy projects. On the other hand, they planned to increase their capabilities in evaluating renewable energy project risks in order to better assess their own potential advantages and disadvantages in providing credits for such projects.

A second goal of the banks was socially oriented. According to the following quotation, they wanted to 'create a small movement in order to achieve a larger vision' and aimed to involve many people in order to develop renewable energy:

> 2.1 "So we simply wanted to figure out: How could we set up a concept in the business field of renewable energies? In order to create something, let's say a small citizen's movement—many small contributions to achieve something great" (R2N3, 18).

Another cooperative bank situated in a neighbouring region, Volksbank Odenwald eG, had already founded a large energy cooperative. The three other banks heard of this initiative and were impressed by the success of the concept, which motivated them to become operationally active themselves in this area. The strong position of the Main-Tauber and Neckar-Odenwald districts towards fostering renewable energy motivated the banks to become engaged in renewable energy project development in the region.

A working group, including the chairmen of the banks' executive boards and other bank personnel, was formed in order to analyse potential concepts. They sought advice from the *Energiegenossenschaft Odenwald eG* and the *baden-württembergischen Genossenschaftsverband*. The Genossenschaftsverband helped throughout the whole founding process and provided important know-how about the legal structure of cooperatives.

The banks compared the cooperative form with other possible legal constructions being applied for realising renewable energy projects, including GmbH & Co. KG and GbR. However, as revealed by the following quotation, the fact that the banks are organised themselves as cooperatives and are, thus, directly involved in this organisational form led to the final choice of founding an energy cooperative:

> 2.2 "Then we came to the opinion that we could do it best working together, that we would set this up and get it going in a cooperative" (R2N1, 7).

From the beginning, the three partners tended to think in larger business dimensions. According to the next statement, it was clear to them that they would found the energy cooperative together and that they would also involve other banks, as they wanted to link their capacities in order to be able to operate at a large scale within the two districts of Main-Tauber and Neckar-Odenwald:

> 2.3 "We just wanted to show that, okay, cooperatives are strong, yes. And when we bundle ourselves together, that we are even stronger" (R2N3, 108).

In their opinion, a large scale of action would also lead to a larger scale of value creation:

> 2.4 "I really mean that, I am convinced. I mean how we established that, with several institutes, a larger space, not just a bank by itself—although we discussed that as well. [...] And I believe that this is also positive for the E + U and its members and also for the other banks, that we were not thinking in small-small terms but, rather, that we said: Come, let's begin with a bigger vision, then it will also become a bigger thing" (R2N1, 90)!

For such a foundational approach, it was crucial that all of the banks (1) shared the same goals, since all three wanted to participate in the new business field renewable energy; (2) not stand in competition with each other but, rather, focus on different regional areas in which they do business; and (3) already knew each other well beforehand, as they had been working together prior to the foundation of the energy cooperative. Amongst other ways, they are members of a regional working group of banks (Districtarbeitsgemeinschaft) that meets regularly in order to discuss how to support the development of the region. In this sense, the energy cooperative is a continuation of their previous collaborations, as expressed here:

> 2.5 "We are all peoples banks (Volksbanken); we belong to the same family" (R2N1, 15).

> 2.6 "The banks had already worked [...] together in one or another regional project. And the collaboration for the E + U, was already given strength due to this prior collaboration" (R2N1, 34).

The business structure of Energie + Umwelt eG, including its statutes, was developed internally in monthly working groups, which exclusively consisted of personnel from the three founding banks. This business development process prior to the cooperative's registration lasted about a year. The inauguration meeting of Energie + Umwelt eG was held internally between Volksbank Mosbach, Volksbank Main-Tauber and Volksbank Franken, as well as between the newly assigned executive and supervisory boards of the energy cooperative. After its official registration, the energy cooperative went public and started to acquire members.

7.2.2.2 Business Purpose

According to its statutes, the main business goals of Energie + Umwelt eG are the

- initiation of renewable energy production projects,
- shareholding of renewable energy production projects,
- initiation of projects that support renewable energy and climate protection in the region, as well as
- support of energy efficiency in private households (Energie + Umwelt eG 2013, p. 3).

The business activities of the energy cooperative are predominantly focused on the two districts of Neckar-Odenwald und Main-Tauber, where the founding banks are situated. The energy cooperative aims to support Neckar-Odenwald und Main-Tauber in realising their vision of full energy independence through joint interaction between citizens, the municipality and regional organisations (Energie + Umwelt eG 2014c). Thus, its business activities should follow three main approaches:

1. contribution towards the development of a regional energy future,
2. participation of regional actors in joint projects, and
3. creation of regional value through renewable energy production (Energie + Umwelt eG 2014d).

7.2.3 Collaborative Concept

In the following, the main founding partners of Energie + Umwelt eG and their chosen collaborative interaction model are described in detail.

7.2.3.1 Key Collaborative Partners: Cooperative Banks

Volksbank Main-Tauber, Volksbank Mosbach and Volksbank Franken can be seen as the key collaborating partners of Energie + Umwelt eG, because they not only founded the energy cooperative but also fully manage it, as explained in the following sections. However, the three banks also encouraged other regional banks to become members of the energy cooperative. Consequently, as of 2015, Energie + Umwelt eG has been able to draw upon a network of 11 Volks- and Raiffeisenbanken that are all situated in Neckar-Odenwald and Main-Tauber. Table 7.21 presents key data, including the number of their customers, members, personnel and branches, as well as amounts of total capital from 2013.

As displayed, altogether the banks that are aligned with the energy cooperative had about 239,000 customers, 169,000 members and 1481 personnel at their disposal and have established a well distributed net of 181 office branches in the region. Their total capital lay at around 7.5 billion Euros in 2013. As shown in the following

Table 7.21 Key data from the collaborative banks owning shares in Energie + Umwelt eG as of 2013 (Companies' annual reports from 2013)

Data regarding cooperative banks associated with Energie + Umwelt eG energy cooperative, from 2013	Customers	Members	Personnel	Branches in the region	Total capital
Units	Number	Number	Number	Number	Million €
Founding banks					
Volksbank Main-Tauber eG	75,000	39,300	395	38	3040
Volksbank Mosbach eG	55,173	29,916	244	18	762
Volksbank Franken eG	46,000	24,197	231	29	813
Further associated banks					
Volksbank Neckartal eG	NA	44,000	374	42	1720
Raiffeisenbank Kraichgau eG	16,703	8333	77	9	343
Volksbank Vorbach-Tauber eG	15,277	7847	NA	24	286
Volksbank Krautheim eG	6935	4135	39	3	118
Volksbank Kirnau eG	9048	4032	52	11	150
Raiffeisenbank eG Elztal	4855	3054	29	3	111
Raiffeisenbank Neudenau-Stein-Herbolzheim eG	5433	2366	17	3	88
Volksbank Limbach eG	4300	1722	23	1	76
Total	238,724	168,902	1481	181	7507

Volksbank Main-Tauber eG (2013), Volksbank Mosbach eG (2013), Volksbank Franken eG (2013), Volksbank Neckartal eG (2013), Raiffeisenbank Kraichgau eG (2013), Volksbank Vorbach-Tauber eG (2013), Volksbank Krautheim eG (2013), Volksbank Kirnau eG (2013), Raiffeisenbank eG Elztal (2013), Raiffeisenbank Neudenau-Stein-Herbolzheim eG (2013), Volksbank Limbach eG (2013)

sections, these resources played a key role in empowering Energie + Umwelt eG to rise to becoming one of the largest renewable energy-producing cooperative (in terms of members) operating in Germany (as of 2012).

As can be seen in Table 7.21, the three founding banks are the strongest banks within the network. Volksbank Main-Tauber is situated in the Main-Tauber district, whereas Volksbank Franken and Volksbank Mosbach are situated in the district of Neckar-Odenwald. Their core business activities include classical financing and investment services as well as services related to insurance and old-age provision. A strong and important part of the banks' organisational self-understanding is their connection to the region and their focus on regional business, which is expressly stated, amongst other places, in their general business principles:

> 2.7 "As an active and reliable partner, we are committed to our regional responsibility" (Volksbank Franken eG 2014).

> 2.8 "Our drive has to do with our vision to be the first go-to partner for the people of the region—[...]" (Volksbank Mosbach eG 2014).

> 2.9 "As a cooperative bank situated within the region, we have a close relationship with the region" (Volksbank Main-Tauber eG 2014).

7.2 Collaboration Between Energy Cooperatives and Banks: Energie + Umwelt eG 225

Table 7.22 Key characteristics of Volksbank Main-Tauber, Volksbank Mosbach and Volksbank Franken

Key characteristics	Volksbank Main-Tauber, Volksbank Mosbach and Volksbank Franken
Position in the region	They belong to the largest cooperative banks in the districts of Main-Tauber and Neckar-Odenwald.
Self-understanding	Their general guiding principles are to do business within the region and serve the regional public.
Motivation for founding the cooperative	They want to position themselves as experts and potential partners in the new business field around renewable energy in the region and have also involved private actors with smaller amounts of money in renewable energy project investments.

The regional self-understanding of these banks goes beyond just offering financial services for the region. As can be seen in the above statements, it also involves the aim to take "regional responsibility" and to "be the initial go-to partner for people of the region". The foundation and management of an energy cooperative with an open membership approach and a regional preference would then seem to fit well into their regional business approach. Table 7.22 summarises the key characteristics of the three banks.

7.2.3.2 Collaborative Interaction Model

Figure 7.5 visualises the collaborative interaction model between the Energie + Umwelt eG energy cooperative and the banks that founded it.

As can be seen, the collaborative interaction model chosen by the energy cooperative and the banks involves both internal and external dimensions. Of special note here is that not only Energie + Umwelt eG but also the banks are legally organised as cooperatives. Due to their membership approach, cooperatives generally have a strong internal focus with regard to collaborative actions. Since all the banks are members of Energie + Umwelt eG, they also cooperate with each other within the energy cooperative (internal collaboration). At the same time, Energie + Umwelt eG and the banks hold, as separate organisational entities, an external collaborative relationship. For example, the energy cooperative receives support from the banks in achieving financial credits.

Figure 7.5 shows that the collaborative concept also involves a number of other actors. Apart from the banks, members of Energie + Umwelt eG include regional citizens, regional businesses, as well as important regional decision makers.

As shown in the following, this particular mix of actors, as well as the particular combination of an internal and external cooperative relationship, have played a key role for the development of Energie + Umwelt eG and its eventual impact on the local energy structure.

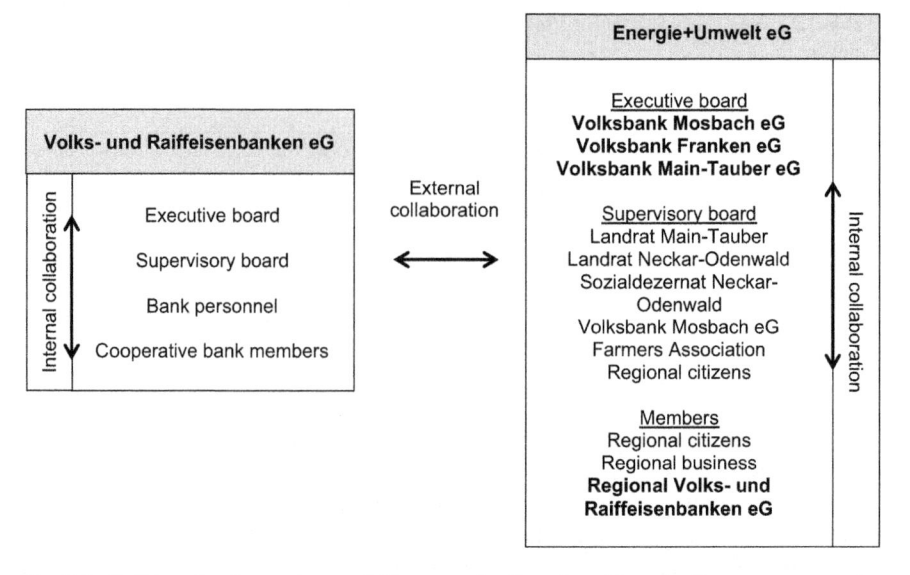

Fig. 7.5 Collaborative interaction model between the Energie + Umwelt eG energy cooperative and cooperative banks

7.2.4 Mobilisation of Allocative Resources

The close interrelationships between Energie + Umwelt eG and the cooperative banks led to the mobilisation of important allocative resources, including actors, financial capital and technology projects.

7.2.4.1 Actors

Citizens, bank customers, bank experts and regional decision takers represent the main actor groups being activated through this collaboration between the energy cooperative and the banks.

Cooperative Members: Citizens and Bank Customers

According to its statutes, any natural, legal or organisational actor that is officially situated in one of the two districts of Neckar-Odenwald or Main-Tauber can become a member of Energie + Umwelt eG. Actors that live or are situated in one of the districts directly bordering Neckar-Odenwald or Main-Tauber can also become cooperative members (Energie + Umwelt eG 2013, p. 3). Membership is thus almost exclusively limited to the focus region of the energy cooperative. Despite this restriction to a certain region, the energy cooperative experienced strong member growth; Table 7.23 presents membership development from its foundation year until 2014.

7.2 Collaboration Between Energy Cooperatives and Banks: Energie + Umwelt eG 227

Table 7.23 Membership development of Energie + Umwelt eG 2011–2014

Energie + Umwelt eG	Dec 2011 (foundation year)	Dec 2012	Dec 2013	Dec 2014
Members (number)	882	1290	1482	1525
Member shares (number)	43,629	63,739	65,407	65,784

Source: Annual business reports, as explained in Sect. 5.2

As displayed, Energie + Umwelt eG attracted almost 1525 members within 4 years. In 2012, the organisation became the fourth largest renewable energy producing cooperative (in terms of members) operating in Germany.[18]

As illustrated by the following quotation, the large network of 11 associated cooperative banks, with their 239,000 bank customers and 169,000 cooperative bank members (see Table 7.21), provided the basis for this high membership growth within a relatively short amount of time:

> 2.10 "Without this sales and distribution network, they would never have grown to be so gigantic" (R2N3, 59).

Of all the members mobilised by December 2014, 97% (1475) were either a cooperative member of one of the banks or a bank customer. The banks in the cooperative network actively marketed its concept all over the region. Amongst other ways, informational material about the energy cooperative was offered in many of the banks' 181 regional branches, where the application for becoming a cooperative member could also be submitted. The three founding banks actively used their sales departments in order to directly contact their bank customers and inform them about the energy cooperative. Right after Energie + Umwelt eG became registered, the founding banks organised events in their respective cities of Mosbach, Franken and Tauberbischofsheim in order to present the new energy cooperative to the general public. These events were advertised in the local press and in the banks' branches. The events were well visited and already acquired the first 650 members. Furthermore, members of the cooperative banks were personally informed about the foundation of Energie + Umwelt eG during the general assemblies of several associated banks.

The banks have benefitted from the strong mobilisation of members for the energy cooperative. As illustrated by the following quotation, they see membership in the energy cooperative as value added for their own customers and members and also market it this way:

> 2.11 "This is good, it will be successful. I can offer something to my people, my members in the bank, if they also become members of the energy cooperative" (R2N1, 19).

According to the next statement, the energy cooperative has even supported the banks in acquiring new customers:

---

[18]Their position was assessed during the quantitative analysis of energy cooperatives presented in Sect. 6.2.

2.12 "We have also won customers for the bank through the energy cooperative. There have also been customers who became interested in us through the homepage of the energy cooperative, not through the normal marketing of the Volks- und Raiffeisenbanken. Then they called via our info hotline, which is automatically forwarded to me. [...] And then they raised questions, saying that they might be very interested in becoming a customer at the cooperative bank. And then I said to them: Alright, so let's try it together" (R2N3, 310–318).

Collaboration with an energy cooperative can thus be described as a new and innovative tool for increasing the loyalty of cooperative bank customers and for acquiring new ones.

Cooperative Managing Personnel: Bank Experts and Regional Decision Takers

The executive board of Energie + Umwelt eG is exclusively staffed by bank personnel (as of 2014), and each cooperative founding bank—Volksbank Mosbach, Volksbank, Main-Tauber and Volksbank Franken—is represented by one person on the board who was already part of coordinating the cooperative's foundation process and thus knows it very well. Since the cooperative is officially registered, they are also responsible for the planning and coordination of all energy projects as well as for all marketing activities, member acquisition and administrative tasks. These actors work on a voluntary basis, in addition to their work at the bank, and have the full support of their bank directors. Two of these managers have additional support through administrative bank personnel. As displayed by the following quotation, it was the aim of the founding banks to establish professional management for the energy cooperative at no cost:

2.13 "To have professionalism that, in administrative terms, does not cost anything" (R2N1, 17).

According to the next statement, this strong and professional commitment of the executive board members—who are also bank personnel—was a crucial resource for the successful development of Energie + Umwelt eG:

2.14 "I think we have a very good executive board that really does this on a voluntary basis and invests a lot of time and energy there. And I think—as you can see from the last 2 years—that this is going very well" (R2N2, 128).

However, due to the strong growth of Energie + Umwelt eG, it became more and more challenging to manage all of their tasks on a voluntary basis. As illustrated by the following quotation, the work load of the managerial staff reached the limits of their personnel capabilities. As a consequence, it is under discussion whether to turn the cooperative's managing positions into full time jobs or to appoint an additional managing director.

2.15 "We have to restructure. [...] We can no longer manage it this way. [...] It takes a lot of time away—tons. And at some point—I'll tell you quite openly—it is enough, you do not want to do that anymore. Because you also have to live. And I'll say it quite clearly: None of us three goes home after less than 60 work hours a week—no one! No one. No one. I'd even say more like 70 [hours]. And then the pain threshold is now reached" (R2N3, 278–282).

7.2 Collaboration Between Energy Cooperatives and Banks: Energie + Umwelt eG

Not only the members of the executive board but also those of the energy cooperative's supervisory board have been exclusively selected by the three founding banks. As illustrated by the following quotation, one of the aims behind the establishment of a first-class supervisory board was to increase the positive publicity and image of the energy cooperative:

> 2.16 "Then we were thinking: What [local] personalities could help us with marketing? [...] And then, of course, we came to the following conclusion: Okay, we'll take two heads of the district authorities, [...], we'll take a member of the executive board from each of the founding banks and we'll take [... one person] from the Green Party" (R2N3, 26).

Since the directors of the banks are well connected and well known in the region, they were able to personally contact and assign regional leaders (as of 2014):

- the chairman of the executive board of Volksbank Mosbach, who is also chairman of the supervisory board of the energy cooperative;
- The two *Landräte* of the districts of Neckar-Odenwald and Main-Tauber. A *Landrat* is the administrative head of a district in Germany, democratically elected during political district elections;
- the *Sozialdezernentin* of the Neckar-Odenwald district. The *Dezernat* is an important part of the regional public administration; and
- the chairman of the farmer's association of Neckar-Odenwald (*Districtbauernverband*). The association is the local section of the larger farmer's association of the Bundesland of Baden-Württemberg. Their aim is to support local farmers, provide them with special advice and represent their needs in the political sphere.

7.2.4.2 Financial Capital

Table 7.24 presents the financial development of Energie + Umwelt eG from its foundation year of 2011 through 2014.

Table 7.24 Financial development of Energie + Umwelt eG, 2011–2014

Energie + Umwelt eG	Dec 2011 (foundation year)	Dec 2012	Dec 2013	Dec 2014
Total capital (Euros)	4,546,243	11,235,576	10,571,477	11,861,846
Equity capital (Euros)	3,833,908	6,599,744	6,864,640	6,974,075
Member shares (Euros)	3,800,900	6,373,900	6,556,700	6,578,400
Borrowed capital (Euros)	693,393	4,531,567	3,662,936	4,821,270
Equity ratio (%)	84	59	65	59
Total revenues (Euros)	33,008	192,837	231,437	264,542
Distributed dividend (%)		Almost 3	2.75	3

Source: Annual business reports, as explained in Sect. 5.2 and information directly provided by the energy cooperative

As can be seen, Energie + Umwelt eG already had about 4.5 million Euros total capital at its disposal by the end of its foundation year of 2011. By the end of 2014, however, it increased its capital to almost 12 million Euros.

Nominal Capital

Table 7.24 illustrates that most of the cooperative's financial resources represented equity in the form of member shares. Similarly to BEG Wolfhagen eG, membership in Energie + Umwelt eG requires a minimum investment of 500 Euros (Energie + Umwelt eG 2013, p. 3)—a low entrance fee that is in line with one of the core goals of the three banks that founded the energy cooperative, who wanted to especially provide investment opportunities for private people with relatively small amounts of capital at their disposal. Despite this focus on small individual capital shares, members provided 3.8 million Euros in shares in the foundation year of the organisation and almost doubled this amount to 6.6 million Euros by the end of 2014.

Volksbank Main-Tauber, Volksbank Mosbach and Volksbank Franken belonged to the first actors that provided equity capital to the cooperative. Right after the cooperative's foundation, each bank invested 100,000 Euros, the maximum capital amount that an individual member was allowed to invest at that time.[19] Due to this early financial support from its founding partners, Energie + Umwelt eG was well positioned to start realising its business goals from the day of its foundation. In 2011, eight other cooperative banks became members of Energie + Umwelt eG and provided an additional 500,000 Euros. In total, then, cooperative banks invested 800,000 Euros in equity capital, representing 21% of the total amount of 3.8 million Euros member shares which had been mobilised by Energie + Umwelt eG in its first year. Hence, the collaborative relationship with cooperative banks strongly influenced the equity basis of Energie + Umwelt eG.

Borrowed Capital

As displayed in Table 7.24, the energy cooperative mobilised about 4.8 million Euros of borrowed capital through 2014, which is a considerably high amount against the background that 80% of all analysed renewable energy production cooperatives did not have more than 200,000 Euros total capital at their disposal in 2012 (see Sect. 6.2 for more details). Energie + Umwelt eG has been able to receive borrowed capital from its associated cooperative banks, for whom the cooperative's demand for borrowed capital has created a new source of credit business. As suggested by the following quotation, the banks perceive the energy cooperative as a new option for generating profit:

---

[19]The maximum amount for individual member investment was later reduced to 1000 Euros, as will be explained below.

7.2 Collaboration Between Energy Cooperatives and Banks: Energie + Umwelt eG

2.17 "It makes sense, firstly, to fill these energy cooperatives with life but also, secondly, to help our banks—our Volks- und Raiffeisenbanken—to generate income. [...] The dependency is simply there" (R2N3, 402).

It was also an aim of the banks to construct novel business options in the field of renewable energy through founding Energie + Umwelt eG. However, not only the banks but also the energy cooperative have benefitted, as its close collaborative relationship with these financial institutions has created direct and rapid access for borrowing capital. The following quotation reveals that Energie + Umwelt eG may not have to undergo the banks' general, sometimes long-lasting, procedures for acquiring borrowed capital because it is already directly aligned with the lending institutions:

2.18 "And normally there would be a terrible surprise [...] if you as a smaller cooperative needed 100,000—what you would have to go through in order to receive this money. And that was all not relevant for us! In this respect, it was simple" (R2N3, 59).

As displayed in Table 7.24, the cooperative's equity ratio has been equal to or higher than 60% from the beginning of its foundation year. Despite the banks' aim to strengthen their credit business, they have focused the project investment into Energie + Umwelt eG on equity, as it is not borrowed capital but member shares that provide the basis for all investment activities of the energy cooperative. In this sense, the founding banks appear to mainly be taking into account the interests of the energy cooperative and its members.

Foundation Capital for the Cooperative

The three banks—Volksbank Mosbach, Volksbank, Main-Tauber and Volksbank Franken—coordinated the complete foundation process of the energy cooperative and carried all foundation costs, including for the obligatory registration at one of the national associations for cooperative organisations. The following quotation underlines that it was the aim of the banks to financially care for the energy cooperative until it had developed its own solid economic basis:

2.19 "So all costs were taken over by the respective banks. Whether it was marketing brochures or other things, they were all paid for by the bank. The banks paid for all the marketing events. So they have made life easier for us and also provided all the experts [we needed]" (R2N3, 65).

In addition, bank personnel developed the cooperative's business concept and took care of all administrative tasks.

Profit and Distributed Dividends

Table 7.24 indicates that Energie + Umwelt eG had a positive return on equity from the start of its operations, as the organisation had already made about 33,000 Euros in revenues by the end of its foundation year of 2011. But it strongly increased its revenues to about 265,000 Euros by the end of 2014. Accordingly, during this period it performed better than almost all other renewable energy production cooperatives in Germany, as only 8% of analysed energy cooperatives generated more than 50,000 Euros in revenues 2 years after registration, while 80% of them suffered losses in their registration year (see Sect. 6.2 for more details). Due to its financial position, Energie + Umwelt eG was able to annually distribute a dividend of around 2.75–3% to its members (see Table 7.24).

The collaborative relationship with cooperative banks played a key role for the strong business performance of Energie + Umwelt eG, as the banks not only provided its first equity capital and ensured quick and easy access to borrowed capital but also followed a profit-oriented business approach when managing it. As illustrated by the following quotation, profitability represents one of the main selection criteria for all the cooperative's projects:

> 2.20 "And we are always thinking in economic terms, we always fit every activity into an economic program. And if a particular activity doesn't give us a corresponding return on equity, then—except for the windmills—it become a dead end for us and we don't consider it anymore" (R2N3, 380).

Economic strength can influence a number of other important resources, such as the technology resource mix. The following statement shows that Energie + Umwelt eG perceives itself as an organisation that has become financially strong enough to manage larger and more capital intensive energy projects completely on its own.

> 2.21 "Theoretically, we could take over a wind park, we could build a wind park—and we wouldn't break down even if it went badly" (R2N3, 361).

However, the strong mobilisation of capital, especially in form of member shares, also challenged Energie + Umwelt eG, which faced difficulties in continually needing to find new projects in the region that it evaluated as being applicable. In order to reduce its capital flow, the investment volume an individual member could invest was reduced to 1000 Euros in October 2012, demonstrating that Energie + Umwelt eG did not have any problems to attract equity capital and members but, rather, to find worthwhile investment options.

As indicated by the following quotation, actors seem to evaluate investment in Energie + Umwelt eG against the background of missing investment alternatives. The decrease of the European key interest rate from around 3% in 2008 to 0.05% at the end of 2014 (European Central Bank) decreased the dividends of other investment options. For example, actors received a higher dividend for member shares at Energie + Umwelt eG than for capital in bank saving accounts.

> 2.22 "So everyone also gets a small economic advantage ... that is just now, because the normal interest rates are very low, it's very good, in terms of the distribution of dividends, what the energy cooperative can do there—compared to current interest rates or a savings account or something like that" (R2N2, 130).

7.2 Collaboration Between Energy Cooperatives and Banks: Energie + Umwelt eG   233

Hence, it could become more challenging for the energy cooperative to mobilise members and capital if other investment options improve in the future.

7.2.4.3 Technology Mix

Installed Projects

By December 2014, Energie + Umwelt eG had realised 36 photovoltaic production projects, with a total capacity of about 5.81 megawatts peak. It is shareholder of one wind park located in Hettingen, which involves five wind mills and has a capacity of 16 megawatts. The energy cooperative also operates a biomass plant that produces power with liquid manure, with a capacity of 0.06 megawatts. Together, the projects have an expected power yield of about 7478 megawatt hours per year. Table 7.25 provides an overview of all projects that were realised by Energie + Umwelt eG up to December 2014.

Most of the installed capacity is located in the focus region of the energy cooperative: the two districts of Main-Tauber and Neckar-Odenwald. Only two photovoltaic projects, with a capacity of 476 kilowatts peak, are located in a neighbouring district. Many photovoltaic plants are installed on top of buildings that belong to the involved cooperative banks or on top of public buildings, such as a sports facility or fire station. As we have seen, the energy cooperative supports the vision of developing a 100% renewable energy supply in Neckar-Odenwald and Main-Tauber. Based upon the energy demand of the districts, as outlined in Sect. 7.2.1.2, the energy cooperative can supply about 1.6% of household power demand, which could be interpreted as a small amount. However, with two districts, the chosen focus region of the energy cooperatives is quite large. Compared to other energy cooperatives, Energie + Umwelt eG has actually achieved a very high level of installed power capacity, as only 5% of all analysed renewable energy cooperatives had installed more than five megawatts of solar power by May 2015 (see Sect. 6.2).

Table 7.25 Projects realised by Energie + Umwelt eG as of December 2014 (based upon Energie + Umwelt eG 2014a, 2015)

Technology type	Number of plants	Installed capacity (MW)	Expected power yield (MWh/a)
Photovoltaic	36	5.81	4864
Wind	7% share of one 16 MW wind park	1.12	2014
Biomass	1	0.06	600
Total	38	7.0	7478

Future Goals

Of the cooperative's 38 existing projects as of 2014, 33 were implemented between July 2009 and August 2013. Hence, it can be seen that the cooperative had strongly reduced its project development activities after 2013. The amendments of the German Renewable Energy Act that were passed in August 2014 (Erneuerbare-Energien-Gesetz—EEG 2014) have been cited as the main reason for this decrease, particularly the reduction of the feed-in tariff for several renewable energy technology types, especially for photovoltaic projects. This directly affected Energie + Umwelt eG, which had been predominantly focused on solar power. The following quotations clearly express the effects that the changes to the Renewable Energy Act had on the cooperative:

> 2.23 "And if the federal government had not just now reduced the feed-in tariff to an unacceptable rate, we would have grown even more than we have so far" (R2N1, 62).
>
> 2.24 "Yes, the biggest challenge lies in complying with the EEG—we do not need to keep that a secret. [In a situation] where we ourselves do not yet know everything, how can we actually do everything properly? So that's the primary thing that's putting the brakes on [our progress], the new EEG" (R2N3, 378).

New concepts that are supported by the revised German Renewable Energy Act, including direct marketing and self-consumption of one's own produced energy, were discussed and concrete plans to become engaged in this direction have existed since 2015 (Energie + Umwelt eG 2015). However, it is taking time to analyse the potential of new project concepts which are now being fostered by the national government.

As outlined in Sect. 7.2.1.2, wind power has the highest renewable energy potential in the region, but only 5% of this potential has been installed yet. Thus, the deployment of wind power is likely to play a crucial role in achieving a 100% renewable power supply in the districts. Energie + Umwelt eG has participated in one regional wind power project so far and has been interested in realising more wind power, reviewing several other wind parks in 2014. One of the main problems that prevented the cooperative from becoming engaged in more wind park projects has been the new regulation for alternative investment funds (KAGB). Energie + Umwelt eG was mainly looking for options to become a shareholder in wind parks being planned by other project developers, as there are lower project risks when becoming a partial shareholder and not fully responsible for the insecure and capital-intensive project preparation and project development phases involved. However, by March 2015 it was uncertain whether and to what extent energy cooperatives would be subject to the KAGB, which would mean higher personal liabilities for the organisational executive board and stronger regulations. A statement from BaFin, according to which energy cooperatives are generally not to be regulated by the KAGB, was only first made public in March 2015, almost 2 years after the regulation was officially passed (see Sect. 6.1.1).

As a consequence of the increasing regulative complexity, Energie + Umwelt eG changed its organisational priorities for the near future. Instead of focussing on the

7.2 Collaboration Between Energy Cooperatives and Banks: Energie + Umwelt eG   235

search for new projects, the energy cooperative has rather been concentrating on restructuring the organisation and developing a position which takes into account all the new amendments in national regulations, the majority of which were still in a state of uncertainty at time of conducting my interviews around mid-2014.

Energy efficiency represents an important task for achieving a 100% regional renewable energy supply in Neckar-Odenwald and Main-Tauber, and the need for it especially applies to heat, as it is assumed that only 45% of the districts' heat demand can be covered with renewable resources. Consequently, Energie + Umwelt eG offers energy efficiency advice to its customers, although energy efficiency has not played a major role in the organisation's activities to date nor has its plan to become engaged in producing renewable heat in the near or midterm future.

A third transition goal of the two districts is to support innovation projects. The energy cooperative is aware of the need for further innovations and has, amongst other options, discussed potential energy storage concepts. However, no concrete plans for becoming engaged in such innovation areas exist for the near future. Table 7.26 compares the project statues of Energie + Umwelt eG with the energy transition steps that are planned by the two districts in order to achieve a 100% renewable and regional energy supply.

All in all, Energie + Umwelt eG has predominantly affected the power production technology mix in the region by increasing the photovoltaic power production, where it has installed a higher capacity than the overwhelming majority of all other renewable energy production cooperatives in Germany. In its focus region, it is one of many other actors that have strongly increased the renewable power capacity. But national regulations, particularly the amendments of the German Renewable Energy Act and the unsure position of energy cooperatives under the new KAGB, reduced the cooperative's project activities from 2013 onwards. Innovative concepts, such as direct marketing or self-consumption of renewable energy, which have been supported by the German Renewable Energy Act since 2014, as well as innovative projects, including storage, are not planned for the near future.

Table 7.26 Planned and installed projects of Energie + Umwelt eG in relation to the energy transition goals of the Neckar-Odenwald and Main-Tauber districts, as of December 2014

Technologies that are part of the districts' energy transition goal	Energie + Umwelt eG project status (as of December 2014)
Photovoltaic power	36 projects installed
Biomass energy	1 project installed
Wind power	Engaged as partial shareholder in one park
Heat	Not engaged
Energy efficiency	Generally included in cooperative goals but not seriously activated yet
Electro-mobility/energy storage	Not planned

7.2.5 *Creation of Authoritative Resources*

The collaboration between cooperative banks—in particular, the founding banks Volksbank Mosbach, Volksbank Main-Tauber and Volksbank Franken—and the energy cooperative has affected important authoritative resources, including their (1) organisational identity, (2) degree of publicity, (3) public image, (4) trust between involved actors and trust for the cooperative business model, (5) business transparency, (6) access to potential projects, (7) access to information and new ideas, as well as (8) level of expertise and know-how. In the following, I seek to demonstrate that these authoritative resources had a strong influence on the involved organisations' capabilities for mobilising allocative resources, such as members, capital and technology projects. Authoritative resources appeared to be fundamental for empowering the two partners in order to influence the regional energy structure in the districts of Main-Tauber and Neckar-Odenwald. Yet, as outlined below, the collaboration also reached certain limits in terms of strengthening authoritative resources, such as regarding internal transparency in the steering board and access for new project ideas from members.

7.2.5.1 Organisational Identity

The organisational identity of Energie + Umwelt eG is primarily drawn from its collaborative relationship with cooperative banks. As expressed by the following quotation, the energy cooperative even describes itself as a "bank cooperative":

> 2.25 "Today we have presented it from a different angle. Yes, actually a bank cooperative. [...]" (R2N3, 418).

As we have seen, Energie + Umwelt eG has been formed and founded and is now fully managed and represented by cooperative banks. The three founding banks constitute the executive board and select the supervisory board of Energie + Umwelt eG—the organisational heart of the cooperative. The financial institutions have formulated its business strategy. They seek out and choose its projects. The energy cooperative's public presence is closely aligned to the presence of the cooperative banks, and almost all public meetings held by Energie + Umwelt eG have been accompanied or co-organised by one of the associated financial institutions. This close interrelationship has been pointed out in many newspaper articles (for example, Fränkische Nachrichten (Main-Tauber) 2010; Fränkische Nachrichten (Neckar-Odenwald) 2010; Rhein-Neckar-Zeitung 2010; Stadtanzeiger Mosbach 2013). The banks are not only visually present on the homepage of the energy cooperative but are also named in several of its business reports (for example, Energie + Umwelt eG 2014b). The official office address of Energie + Umwelt eG is even the same as that of Volksbank Franken.

In turn, Energie + Umwelt eG functions as a tool through which the cooperative banks have been able to strengthen their own organisational identity. As the following quotation indicates, the banks want to be directly associated with the energy cooperative in public:

> 2.26 "One must, of course, know one thing: the cooperative Energy + Environment is always associated with banks. And it is also perceived this way in the public. But this is the way we represent the cooperative in public as well" (R2N3, 57–59).

The banks also use their position as the core founding and managing partners of the energy cooperative in order to spotlight their organisational values and their own organisational roots as cooperatives. By initiating the energy cooperative, the banks wanted to demonstrate that they do actively follow cooperative principles, such as self-help, self-responsibility and self-administration, and that they are entities that can listen to their members, who appear to be demanding more services in the renewable energy sector. The following quotation makes this logic clear:

> 2.27 "The cooperative's basic ideas are self-help, self-management and self-responsibility. And in paragraph one it says: Support of the members. So, that is to say, this requirement must also be guaranteed. And when the voice of a member calls and says: People, we have to change a bit, we have to see that we are creating something which was, until now, only purely bank-specific, then we can only realise such a change if we follow the basic idea of the cooperative: Thus founding [an energy cooperative] and not drawing upon [external energy expertise]" (R2N3, 98–100).

7.2.5.2 Publicity

The collaborative relationship between cooperative banks and Energie + Umwelt eG has helped to increase the publicity of all involved organisations. The banks have used their large customer network and well-distributed business branches in order to make Energie + Umwelt eG popular. Furthermore, the banks have been using their political connections—many mayors in the districts are, for example, members of the banks' supervisory boards—in order to increase publicity for the energy cooperative. One of the most important consequences of such publicity activities was the rapid and strong membership growth of the energy cooperative.

Likewise, the banks have been heightening their public profile through their activities around Energie + Umwelt eG, often being named, for example, in articles about the energy cooperative (for example, Rhein-Neckar-Zeitung 2010). Furthermore, the banks present themselves along with the energy cooperative during events, such as the fourth Taubertäler Climate Protection Forum. As demonstrated by the following statement, support for each other's organisational self-marketing has been seen as a clear win-win situation:

> 2.28 "So the there is potential here for mutual public promotion of each side. It does not hurt the bank if it represents itself at E + U events among other bankers who aren't from the cooperative. Meanwhile, if we also publicise the E + U at other events, congresses or meetings it is also, of course [...] a form of mutual fertilisation, which is good for us too" (R2N1, 86).

7.2.5.3 Image

Furthermore, the collaborating partners—the banks and the energy cooperative—have helped each other in creating or maintaining their organisational image. By supporting Energie + Umwelt eG, the banks not only wanted to gain new experience in financing and managing renewable energy projects but also, as articulated in the following quotation, used the energy cooperative in order to establish a new expert image in the field of renewable energy:

> 2.29 "As regional cooperative banks, we had to position ourselves in the investment field of energy and environment, anyway, as they say for purely financial reasons. But we also had to offer know-how so that we can say that we can offer more than just money and interest [...] and that is a lot. The combination of an energy cooperative and, thus, competence in the field of renewable energy" (R2N3, 78, 90).

Newspaper articles provide evidence that the banks were publicly noticed as being new experts in renewable energy (for example, Fränkische Nachrichten (Main-Tauber) 2010; Fränkische Nachrichten (Neckar-Odenwald) 2010; Rhein-Neckar-Zeitung 2010; Stadtanzeiger Mosbach 2013).

At the same time, the founding banks set up a strategy in order to introduce the young Energie + Umwelt eG as a new player in realising renewable and regional energy projects. As an example, the following statement demonstrates that the founding banks used the cooperative supervisory board in order to select prominent people who could help in forming a positive image for the energy cooperative:

> 2.30 "We have the head of the district administration on board—it was also very important for us to involve him in the build-up phase, simply because it provides us with an excellent reputation [...]" (R2N1, 46).

However, the intertwined image of the energy cooperative and the banks can also lead to difficulties. Amongst other concerns, the banks see the risk that possible business problems of the energy cooperative could also negatively affect their image. Since a clear differentiation between the organisations has not been fostered in public, members of the energy cooperative, of which many are at the same time bank customers, could make both the energy cooperative as well as the banks responsible for possible negative business developments within Energie + Umwelt eG. The banks' concerns are expressed in the following statement:

> 2.31 "If something went wrong—in quotes—we would be burdening the image and the reputation of the banking houses" (R2N1, 36).

In some cases, the lack of differentiation between the banks and the energy cooperative has also been challenging for Energie + Umwelt eG, as actors seem demand the same business management standard from it that they are accustomed to receiving from the cooperative banks. For example, members of the cooperative's supervisory board have demanded that it achieve the same performance levels with respect to project risk management that are usually associated with financial institutions. From the cooperative's perspective, however, this seems to be setting the bar too high, as explained here:

7.2 Collaboration Between Energy Cooperatives and Banks: Energie + Umwelt eG

2.32 "And we always say: People, we are not a bank, we are a cooperative, an energy cooperative, and this is a meeting of the supervisory board of the energy cooperative. We cannot always prepare everything to perfection; we cannot make a complete check of all risks. We do not have a top-rung lawyer sitting next to us every day, yes, who removes all legal concerns ... But this kind of thinking is so widespread" (R2N3, 218).

As indicated by this quotation, the capacities of the young energy cooperative, in which work is predominantly done on a voluntary basis by a small executive board of three persons, cannot be compared to the capacities of well-established cooperative banks with upwards of almost 400 employees (see Table 7.21). The lack of clear differentiation between the banks and the energy cooperative can, thus, lead to expectations in terms of how to run the business that may lie beyond the actual capabilities of the energy cooperative.

7.2.5.4 Trust

Trust Between Involved Actors

Trust between the three founding cooperative banks played a crucial role in the foundation process of Energie + Umwelt eG. Leaders within the banks knew each other beforehand and had already been working together in other regional alliances. According to the next quotation, such prior network activities helped the banks to trust each other and jointly initiate the energy cooperative:

2.33 "That's something you take for granted, and the way we thought about it [...] during the start-up phase, we simply knew that we could rely on each other, and could not imagine something that we cannot carry out" (R2N1, 62).

Trust in the Business Model of the Energy Cooperative

The energy cooperative was registered in August 2011, and by December 2011—only 4 months later—it had already attracted 882 members and 3.8 million Euros in member shares (see Tables 7.23 and 7.24). The strong membership growth suggests that many actors, especially those who became early members of the cooperative, may have evaluated Energie + Umwelt eG as a trustworthy organisation right from the beginning of its operation. As shown by the following quotation, it is also the opinion of Energie + Umwelt eG that trust in its business model was one of the main motivations for so many actors to become members within a short amount of time:

2.34 "And in the short time it was revealed that we have got so many members and so many projects and so many available resources. So it quickly became apparent that members could put their entire trust in this cooperative, because they know there are professionals at work" (R2N3, 358).

Based upon the measures publicly taken to promote the cooperative, it is most likely that the three founding banks played a key role in creating trust in the business model of the cooperative. The banks represent experts in managing capital and investments, and investing in renewable energy projects is one of the core activities of the energy cooperative, guided by the banks because they form the executive board. The cooperative founding banks made sure that Energie + Umwelt eG was able to present itself as a professional and active organisation from the day of its registration. Each founding bank had bought one photovoltaic plant for the energy cooperative prior to its foundation. As shown by the following quotation, the three banks wanted to create trust in the cooperative's business model by presenting well-functioning energy projects at a very early stage and demonstrating that the energy cooperative was already making a profit:

> 2.35 "What was also important was that they just saw: Okay, they already have projects, they are now progressing rapidly, here we can invest money and trust is simply there" (R2N3, 76).

As a result, about 650 people had already decided to become members after the first three public marketing events. In this sense, the concrete trust building strategy of the banks seems to have been very successful.

7.2.5.5 Transparency

Energie + Umwelt eG practices a transparent business philosophy. Its organisational website (www.epueg.de) provides detailed information about the energy cooperative and its cooperative partners. Each installed project is listed with detailed project information. The website lists key figures, such as number of members, allocated capital, installed capacity and invested capital, and figures are continually updated. Protocols and presentations for each general assembly are also publically accessible. This way, interested actors are able to create for themselves a very detailed picture about the organisation and its activities.

7.2.5.6 Access to Potential Projects

Access to potential projects is crucial for influencing the energy technology mix, and ease of project access guides an organisation's business focus in terms of what kinds of technology projects can be implemented, how many projects can be realised and within what kinds of timeframe. In the case being presented here, collaboration with cooperative banks has been providing an important means of access to new energy production projects for the young energy cooperative Energie + Umwelt eG. The banks used every possible internal project source to set the energy cooperative in motion. For example, they instructed their business customer consultants to list all firms with project opportunities, such as firms with applicable rooftops for installing photovoltaic plants, as explained here:

> 2.36 "As I said, we activated our relationship management. So we are in the big banks—you always have to see the three big ones; [...]—and said: We have gathered all consultants—corporate in-house consultants—said: So, please write us a list about where

7.2 Collaboration Between Energy Cooperatives and Banks: Energie + Umwelt eG   241

perhaps interesting projects could be. Where can we establish ourselves? Which roofs must be renovated? Where could we deal with one-time roof rents appropriately? And and and. So, and because of that, we really quickly had—I do not want to lie now, but—about 70, 80 projects that were carried to us. [...] So we have taken all feasible sales and distribution paths available to us" (R2N3, 136–142).

In this way, Energie + Umwelt eG was able to identify a large amount of potential projects within a very short timeframe at the very beginning of its business activities, became well known in the region, and consequently achieved a considerable size in terms of members, projects and capital. The cooperative's strong position further increased its project access, as illustrated by the following quotations:

2.37 "And now, as I said, in the meantime Energy + Environment has become well known. And so it gets many offers" (R2N1, 152).

2.38 "So, as I said, we also see that, due to our size, of course, a lot is brought to us out of insolvencies. Yes, so insolvency administrators take care of these projects and then call us: Are you able to take over this project" (R2N3, 361)?

Meanwhile, the energy cooperative has improved the banks' access to new business. As shown by the following quotation, projects being offered to the energy cooperative are at the same time being offered to the banks—mainly to the three founding banks:

2.39 "And therefore, because of the size that we have [as an energy cooperative], project requests to these three [bank] houses are as common as sliced bread" (R2N3, 90).

Each project realised by the energy cooperative can also potentially involve new business for the banks, such as for borrowing capital.

7.2.5.7 Access to Information and New Ideas

Access to information and new ideas has complemented access to projects and facilitated the work of the energy cooperative. The energy cooperative collaborates with important regional decision authorities, such as banks, heads of the administrative districts or the farmers association. As explained by the following quotation, access to information is closely aligned with access to such decision takers:

2.40 "That you get information at an early stage: this is a bank, this is an administrative site, some district administrative offices here. This allows us to exert influence with regard to, for example, the regional plan for wind parks and so on, where we simply have a better and more direct information channel" (R2N2, 150).

However, the energy cooperative does not seem to have fully used the potential of its many members in order to improve its access to new ideas. The business approach of cooperatives is based upon joint realisation of projects, which may not only involve allocation of financial capital but also joint development of projects and new ideas. For Energie + Umwelt eG, project development, choice of future goals, as well as formulation of cooperative strategy is exclusively in the hands of the cooperative management and supervisory board. As explained in detail in

Sect. 7.2.6.2, cooperative members have been rather passive, apart from providing capital, and the only direct exchange between them seems to happen during the annual general assembly. Additional informal meetings or working groups in which members could become engaged in project development or which would allow them to bring up their own business ideas have not been evident.

7.2.5.8 Expertise and Know-How

At the beginning of this process, the banks had wanted to learn more about the potentials and risks involved in renewable energy investments, so gathering know-how in this new business field was one of their main reasons for founding and operating the energy cooperative. The experience that they have gained through their work within Energie + Umwelt eG have helped them to evaluate credit demand and provide professional advice regarding renewable energy investments.

As we have seen, the founding banks also provided important expertise for establishing the new energy cooperative, offering their in-house knowledge in order to develop a business concept for it as well as helping it carry out its administrative and marketing tasks. As demonstrated by the following quotation, the banks have used their financial expertise in order to make sure that all projects planned by Energie + Umwelt eG are based upon proven investment plans:

> 2.41 "The bankers stand for a financially balanced orientation and resilience and can examine the concepts" (R2N1, 46).

The political heads of the two districts, who are also members of the cooperative's supervisory board, have provided additional expertise, including help in evaluating new projects:

> 2.42 "Both heads of the district administrations are members who can also evaluate the legal situation quite well" (R1N2, 144).

These examples show that collaboration has enabled both partners to draw upon each other's particular organisational background, in order to improve each other's expertise and know-how.

7.2.6 Creation of New Regulations

The collaborative relationship between Energie + Umwelt eG and the cooperative banks has led to the creation of several new societal rules that have provided the basis for developing renewable energy projects together. These new official regulations, described further below, include (1) official accreditation of cooperative banks for guiding citizen-focused regional energy projects, (2) new participation and co-decision rights for citizens in renewable energy projects, (3) new co-decision rights for politicians in the execution of renewable energy projects, (4) limitation of individual member investment volumes in renewable energy projects and (5) new forms of regional value creation.

7.2.6.1 Official Accreditation of Cooperative Banks for Guiding Citizen-Focused Renewable Energy Projects

Cooperative banks gave Energie + Umwelt eG its legal basis for realising its business goals. The founding banks defined the statutes of the energy cooperative (Energie + Umwelt eG 2013) and decided to primarily realise renewable energy projects in the region and limit membership to local citizens living in the districts of Main-Tauber and Neckar-Odenwald (Energie + Umwelt eG 2013, p. 3); (see also Fränkische Nachrichten (Main-Tauber) 2010; Fränkische Nachrichten (Neckar-Odenwald) 2010; Rhein-Neckar-Zeitung 2010).

By founding and managing Energie + Umwelt eG, the cooperative banks officially accredited themselves for guiding citizen-focused renewable energy projects in the region. As the next quotation indicates, the banks feel well positioned for taking the lead in founding energy cooperatives in order to unify regional actors in the development of renewable energy projects:

> 2.43 "I need people here who [...] want to develop something together. And there is no one more predestined than the people who have that in their blood: the Volks- und Raiffeisen banks. [...] Then one can also say that, if there is no cooperative in a region, no initiative for an energy community, then it is utterly appropriate that it should come from a Volks- or Raiffeisen bank" (R2N1, 94).

The circumstance that the banks and the energy cooperative share similar values has been pointed out as the main reason for a generally well-functioning interrelationship between these two institutions—energy cooperatives on the one hand and cooperative banks on the other. Here, financial institutions have turned into new initiators of citizen-based renewable and regional energy projects. The official guidance of such projects through cooperative banks has not only helped to mobilise important resources, such as members, capital, technology, image and trust, but also helped to create other rules, such as the new citizens' right to become engaged in energy projects. This new official accreditation of the founding cooperative banks for guiding renewable energy projects has also been transported into the wider regional public, as evidenced by the fact that 81% of all analysed newspaper articles that wrote about the energy cooperative in one of the most dominant regional newspapers between 2010 and 2014 pointed out the leading position of these banks in the energy cooperative.

7.2.6.2 New Participation and Co-Decision Rights for Citizens

Just like any other cooperative, Energie + Umwelt eG has a democratic decision-taking structure and is based upon an open membership approach. All actors who become members of Energie + Umwelt eG receive the right to financially participate and invest private capital in the cooperative's energy projects. By the end of 2014, 1525 members had joined the cooperative and made use of this right by acquiring almost 65,800 member shares, with a total value of about 6.6 million Euros (see Tables 7.23 and 7.24). Since Energie + Umwelt eG operates almost all of its projects itself, members also act as direct project owners.

Member rights further include the privilege to co-decide with respect to strategic business decisions. Member participation during the cooperative's assemblies can provide us with initial insights into how members have actually made use of their new co-decision right. Strong member participation in the assemblies of Energie + Umwelt eG was perceived by the following interviewee:

2.44 "The general assemblies are really crowded" (R2N3, 292).

Table 7.27 lists the number of members that visited the assemblies in 2012, 2013 and 2014 in relation to the total number of members the cooperative had for each year.

As can be seen, between about 170 and 190 members visited the assemblies during these years, which seems to reveal the interest of a large actor group. However, considering that these figures only represented a share of 11–13% of all cooperative members, it can be concluded that only a minority of members have actually made use of their right to co-decide the direction of the cooperative's business activities.

Once at the assemblies, members can further participate by actively commenting or expressing their opinions about the work of the energy cooperative or by articulating their own ideas and standpoints. However, although members seemed to be interested during the assemblies, they remained rather passive. According to the cooperative management, presented future plans have normally been accepted without any discussions, and critical questions are usually not posed during these meetings.

The passive manner of participation during the assemblies can have several reasons. It could indicate that most members are satisfied with the work of the executive board and, thus, may not feel the need to intervene. As outlined in Sect. 7.2.5.4, Energie + Umwelt eG seems to enjoy a high level of trust among its members. Furthermore, it could mean that more members are interested in participating financially than in participating in strategic business decisions, an assumption that is underlined by the following quotation:

2.45 "It is always important, I'll say it again, what comes out in the end" (R2N3, 304–306).

Table 7.27 Share of members that have visited the general assemblies of Energie + Umwelt eG, 2012–2014

General assemblies	2012	2013	2014
Participants	174	169	187
Total members at that time	1290	1482	1525
Share of members (%)	13	11	12

Source: Information directly provided by the energy cooperative

But, as the next quotation indicates, not all members may feel comfortable with speaking in front of a large group of people, such as a well-visited annual assembly:

> 2.46 "So yes, at the general assembly of the E & U one or another question is raised, but no more, and that is then more or less answered in detail. What may also be related to [this lack of participation], I mean, not everyone stands up and says something if 400 people are looking at you. Of course that also plays a role" (R2N4, 97).

In this sense, large assemblies may not be the adequate place for members to openly evaluate future business plans or to communicate their own ideas with respect to business activities. Perhaps, then, additional exchange opportunities for members and the cooperative steering board may be necessary in order to foster more active member participation regarding such matters. However, other informal meetings in which members would have an opportunity to discuss their own ideas have not existed, mostly because the management does not seem to have additional time for organising and coordinating such informal meetings. This missing space for member participation in project development also decreases the cooperative's access to innovative ideas from members and limits its access to additional work that might be performed by members.

At the same time, it seems as if members expect the energy cooperative to be successful without their pro-active engagement in guiding it. Energie + Umwelt eG seems to be perceived as a well-positioned organisation, because it already receives strong support by well-established cooperative banks, as expressed at length in real and metaphorical terms here:

> 2.47 "We also present what we intend to implement in the future. But—and I have now participated in three assemblies—it never comes to classical discussions. [...] And if they see how cost-efficiently we work, if we look at the labor or other costs, then they already say: Oh, great! Then they are already more than satisfied. And also with regard to projects. But it does not happen that they say: Okay, please look at this or another topic. This is missing but also wanted. Also wanted! That is why ... the sap from the root is missing. This is so because we have begun at this level. We already have a standard. I say, a forest was expanded. But we already had a really solid tree where everybody could participate, because we knew very well that it was producing fruit. So, in such a case, you don't have ideas from below" (R2N3, 292, 306).

The above quotation also reveals that active influence from members on the business of the energy cooperative may not really be desired by the cooperative steering boards either. This standpoint is underlined by the fact that the founding banks have exclusively chosen actors for the management and supervisory board who either have a bank background or another official position. Private individuals without some kind of official background have not yet been represented in the legally legitimised steering groups of Energie + Umwelt eG.

Yet, it appears that Energie + Umwelt eG would welcome member support for administrative and operational tasks. However, only one person responded to an open call by the cooperative for operational help and was willing to assist in administrative tasks. The low interest of members in supporting the energy cooperative beyond providing capital seems to underline the assumption that Energie + Umwelt eG is perceived as an organisation that does not need further member help or guidance, due to the strong support of well-established cooperative banks. This contradictory situation is summed up by the following interviewee:

> 2.48 "It's like this, the people want to participate: Yes, they are with the program. But they still say: "OK, but in the background is a large apparatus which ensures that we can receive a good dividend in the front". It is just that way, nothing wrong about that" (R2N3, 254).

Nevertheless, members have actively used their direct connection to the energy cooperative in order to realise individual energy production projects on their own private real estate. Farmers, for example, have leased their land and the rooftops of their sheds to Energie + Umwelt eG for implementing a biogas plant and a photovoltaic plant. The energy cooperative became the owner of the technology, provided the financial capital and coordinated the construction, while the farmers have been taking care of daily plant operation. As revealed by the next statement, this approach has allowed farmers, who are at the same time members of the cooperative, to realise projects on their own land without taking the overall investment risks and project responsibility:

> 2.49 "They could have done that by themselves, but their business is so big, let's say, in terms of space and animals, that they only ... wanted to provide the land, and they did not want to burden themselves with the whole thing. And also the credit lines from the companies and so, which are huge amounts of money that need to be invested here, are something that they don't want to be additionally burdened by" (R2N2, 124).

All in all, it can be said that Energie + Umwelt eG has created a right for its members to co-decide and participate in its renewable energy activities. There are, however, different levels of participation and co-decision observable, including financial and project ownership participation, participation in strategic voting, active participation in developing business and company strategy as well as participation in project realisation. The intensity of actual member participation in any of these different areas seems to be strongly dependent upon the set-up of the cooperative, how it was founded and to what extent it fosters active exchange with and between its own members. Participation as a new value with regard to developing renewable energy projects seems to have also been transported into the public sphere, as 77% of the 26 articles that were published about the energy cooperative in the newspaper Fränkische Nachrichten between 2010 and 2014 described the new participation options it offers.

7.2.6.3 New Co-Decision Right for Politicians in the Execution of Renewable Energy Projects

The foundation of Energie + Umwelt eG also created new co-decision rights for politicians with respect to realising regional energy projects. On a regular basis, the executive board of Energie + Umwelt eG presents planned projects to the supervisory board and seeks their approval. Since the administrative heads of the two districts of Neckar-Odenwald and Main-Tauber became members of the cooperative's supervisory board, they have been authorised to co-decide which projects are to be realised or not by the energy cooperative.

7.2 Collaboration Between Energy Cooperatives and Banks: Energie + Umwelt eG   247

The involvement of these key political actors has had a direct effect on the development of the energy technology mix in the region. As revealed by the following quotation, not all projects that were evaluated as economically satisfactory by the cooperative's executive board were accepted by supervisory board members with a political background:

> 2.50 "That depends upon certain political restrictions. From an economic point of view, well, there are concerns. Because—I'll say it again—there are two heads of the district administration, who know a lot about life, what is in it, but do not communicate everything, yes, and suddenly impede projects. We also wanted to participate in a project in their area. [...] But it was thrown out of the supervisory board, because the relevant political bodies had vetoed it. So this happens when you have top leaders sitting on the board. If we had only six [members of the supervisory board], who were not politically organized, the thing would have gone through" (R2N3, 196–204).

This new co-decision right for politicians has, thus, led to a situation where projects realised by Energie + Umwelt eG need to comply with reigning political viewpoints; otherwise they may not be realised.

7.2.6.4 Limitation of Individual Member Investment Volume for Renewable Energy Projects

As mentioned above, the banks that founded the energy cooperative primarily wanted to provide investment possibilities for actors with smaller amounts of capital at their disposal. Therefore, and similar to BEG Wofhagen eG, they defined a low minimum entry share of 500 Euros and initially limited the highest amount of member investment to 100,000 Euros. The limit was later reduced to 1000 Euros per member at the end of 2012 due to a lack of feasible project options. The legal capping of individual member investment volume for energy projects, which was introduced by the banks and defined in the statues of the energy cooperative (Energie + Umwelt eG 2013, pp. 19f.), has influenced the composition of actors supporting renewable energy projects in the region, as those who want to invest large amounts of capital cannot become members of Energie + Umwelt eG.

The focus on actors with small capital assets was not welcomed by everyone. For example, one bank customer threatened to withdraw his capital from the bank if no exception was made for him so that he could invest more money in the cooperative than the maximum amount of 100,000 Euros, as explained here:

> 2.51 "A situation which has touched me in a particularly negative way that was what you could call an attempt at "extortion" from someone who wanted to invest money above the accepted limit and then threatened me: if we did not do that, then he would withdraw his money completely from the bank [...]. And this was an experience that was not so much fun, though we were been able to, let's say, amicably resolve the situation in two talks" (R2N1, 56).

Due to the bank's management position in the energy cooperative, we can assume that this bank customer expected to have the same rights as a member of the energy cooperative that he has as a bank customer. Normally, bank customers are not limited in terms of the size of the investment goals that they seek to realise with a particular bank. Thus, this example illustrates that it can be challenging for established actors, such as banks, to introduce rules that deviate from their usual operational approach.

7.2.6.5 New Forms of Regional Value Creation

Energie + Umwelt eG can be described as a concrete tool for creating regional value, as 36 of its 38 realised projects (through 2014) are situated in Neckar-Odenwald and Main-Tauber. The two districts economically benefit from the projects because they receive business taxes for all projects located within their boundaries as well as rent for all projects located on their properties. Meanwhile, the generated profits are exclusively distributed to members that live in or are situated in the region, because cooperative membership has been limited by its statutes to those living in Neckar-Odenwald and Main-Tauber or at least in a neighbouring district (Energie + Umwelt eG 2013, p. 3). This circumstance, that the cooperative's regional focus is to some extent even legally defined, thus turns regional value creation into a new official rule.

A regional focus is also part of the founding banks' business philosophy. By founding and managing the energy cooperative, the banks have sought to transfer their existing guiding principle of 'creating value for the region' into the energy sector. Of special note here is that regional value creation is not limited to a purely economic perspective but also involves a strong social perspective. In fact, leaders of the banks reflected upon whether developing renewable energy projects really 'fits' the region—in other words, whether the business plans fit to the self-understanding and identity of the region and the people that live there.

However, creating a direct connection between consumers and projects while trying to maintain the regional focus has been challenging. So far, Energie + Umwelt eG has decided to sell its produced power to only one larger energy provider. A direct connection between energy demand and supply does not yet exist. The executive board of the energy cooperative has indicated that limited time has been making it difficult to start innovative projects, such as direct marketing of its own produced energy to its members.

It is becoming more difficult for the energy cooperative to find profitable projects in the region. Profitability depends to a great extent upon geographical aspects, including wind and solar radiation intensity. As outlined by the following quotation, the focus on profitable as well as regional projects has been limiting the organisational growth of the cooperative:

> 2.52 "I now see boundaries in our normal [investment] area, indicating that we have limited ourselves a bit by only wanting to invest in our region. And, yes, [sighs] the economic situation for photovoltaics is simply getting worse, and wind is not possible everywhere. And that is the reason why we are slowly reaching our limits" (R2N1, 132).

In public, Energie + Umwelt eG represents an organisation fostering regional value creation, and about 73% of the 26 articles that were published about the energy cooperative in the newspaper Fränkische Nachrichten between 2010 and 2014 point out the value that it has been bringing into the region.

7.2.7 Drawing Upon New Interpretative Schemes

The three founding cooperative banks and Energie + Umwelt eG have jointly introduced new interpretative schemes that act as guidelines for their engagement in renewable energy development, including (1) the idea of taking matters into one's own hands, (2) citizen acceptance as a pre-condition for energy project involvement, (3) a new co-responsibility of citizens for regional energy issues, (4) a new co-responsibility of cooperative banks for renewable energy development, as well as (5) a novel symbiotic relationship between citizens and cooperative banks in the field of renewable energy.

7.2.7.1 Taking Matters into One's Own Hands

Based upon its experience managing the energy cooperative, the cooperative banks activated a guiding principle that can be described as 'We take matters into our own hands' and have fostered the development of renewable energy in their region. According to the next statement, the cooperative banks wanted to demonstrate that one can actively support the region in becoming more independent from fossil energy and give citizens the opportunity to be able to say "I helped too":

> 2.53 "Or we ourselves, we do not just talk, we also put it into practice [...]. And with regard to E + U, the idea was also to help to give citizens the opportunity to participate with a small amount of money and to be able to say: I have also helped here for our region to become more self-sufficient" (R2N1, 9).

As indicated by the following quotation, cooperative banks feel predestined to take matters into their own hands:

> 2.54 "The cooperative banks have existed for more than 150 years. Raiffeisen founded them so that the people in the regions could do something for themselves that cannot be done solely by an individual. This is how we live, the Volks- und Raiffeisen banks. [...] And therefore, in my opinion, there is no other kind of bank more prepared than Volks- und Raiffeisen banks to approach the subject of energy" (R2N1, 92).

Cooperative banks understand themselves as organisations that take a lead in the region and may not only aim to create value for themselves. This orientation towards the community is in line with the identified business philosophy of the founding banks Volksbank Main-Tauber, Volksbank Mosbach and Volksbank Franken, which has included creating business opportunities in the region and serving the regional public. In public, Energie + Umwelt eG strongly stands for this new value of 'taking matters into one's own hands', and about 81% of all 26 articles published about the cooperative in the newspaper Fränkische Nachrichten between 2010 and 2014 emphasised this attitude.

7.2.7.2 Citizen Acceptance as a Pre-condition for Energy Project Involvement

Energie + Umwelt eG has sought to strengthen regional acceptance for renewable energy projects by unifying and coordinating a great number of supporters as well as by making the cooperative and its activities visible to the public. Energie + Umwelt eG had 1525 members by December 2014 who have publicly demonstrated their support for renewable energy projects in particular by having provided 6.6 million Euros capital for concrete renewable energy project development. The energy cooperative has coordinated the mobilisation of financial resources and organised project implementation on a professional basis over the long term, and the public has been regularly informed about its activities. Key information, such as number of members, amount of allocated capital and number of realised projects are easy accessible and actively presented to the public via the continually updated homepage of the cooperative's website and as reported in several local newspaper articles (for example, Stadtanzeiger Mosbach 2013). In this sense 'project acceptance' can be seen as an authoritative resource which may have been increased by collaboration between the energy cooperative and cooperative banks.

However, Energie + Umwelt eG has taken this a step further by perceiving acceptance for energy projects to be a primary guiding principle for its activities and only undertaking projects under the precondition of full citizen support. In fact, clear public acceptance is one of the main selection criteria for new projects, as indicated by this quotation:

> 2.55 "So, to make it very short, we're just getting involved in one project where citizens can stand behind it. [...] We are not really a movement in the classical sense, which, despite enormous resistance, realises projects. Rather, we listen to the voice of the citizens" (R2N3, 330–332).

Consequently, before pursuing a potential project, the cooperative analyses general local perspectives on it in detail. Amongst other ways of doing this, cooperative leaders speak with responsible mayors and accompany them, if possible, to events where respective projects are discussed in public. If there are any signs of local conflict or activities against the project, Energie + Umwelt eG does not become engaged. The following two examples illustrate this organisational approach:

1. Energie + Umwelt eG had considered becoming project partner in a wind power project which was supposed to be realised by the large wind project developer juwi AG. However, the project developer wanted to modify its building permission in order to increase the hub height of the planned wind mills from 100 meters to 148 meters. Actors that lived near the planned wind park began to position themselves against it, especially due to the planned increase of hub height and the small distance between their houses and the wind mills. Energie + Umwelt eG

7.2 Collaboration Between Energy Cooperatives and Banks: Energie + Umwelt eG   251

withdrew from their planned project engagement after recognising that full support for the project had seemed to vanish.

2. In 2013, Energie + Umwelt eG became engaged in its first wind park in Hettingen, which already had several energy cooperatives as partners. The project developer informed local citizens about the background of the project from the beginning, and no conflicts seemed to arise, so Energie + Umwelt eG continued with its project engagement.

This philosophy of being careful to heed public opinion regarding the projects it becomes involved in has been put in the following manner by one interviewee:

> 2.56 "But we really only participate in projects where the citizens say: "Yes". We do not go beyond that" (R2N3, 356).

The cooperative's business approach of avoiding a business involvement in projects with potential conflicts is strongly influenced by the founding banks. As indicated by the next statement, the financial institutions fear that engagement of the energy cooperative in projects without full local acceptance may have a negative effect on their own image:

> 2.57 "Not a simple topic, but one must not conceal that there may be certain conflicts of interest. And we would surely want to take into account the perspective of the bank, that is, in a region where we are represented, with a high market share of [...] for example, 50% in an area. And to do projects in such a region with E + U, against the will of the citizens, could hurt us badly in the banking business" (R2N1, 40).

As a consequence, Energie + Umwelt eG has not yet been engaged in any projects where acceptance had to be created or increased in order to realise them. Energie + Umwelt eG accepts local viewpoints about renewable energy without having the goal of changing them in favour of a certain project. Nevertheless, at some stage it may become important to convince critical actors about the benefit of local renewable energy projects. Otherwise, it may not be possible to achieve the project density that is necessary for a 100% renewable energy supply in the districts of Main-Tauber and Neckar-Odenwald.

7.2.7.3 New Co-Responsibility of Citizens Regarding Regional Energy Issues

The activities of Energie + Umwelt eG and its particular business approach has created a new sense of co-responsibility amongst citizens in the development of regional energy in two ways:

1. First, the energy cooperative entails responsibilities for its members. Since they have the right to participate and co-decide with respect to the cooperative's business activities, they are also co-responsible for all of its undertakings as well as for the impacts that come along with them.
2. Second, the energy cooperative creates new co-responsibilities for all actors that are directly affected by renewable energy projects considered or planned by the

organisation. Since Energie + Umwelt eG only becomes engaged in projects for which full citizen acceptance exists, the decision of whether a project is eventually realised or not can also be influenced by non-members – that is, those who are potentially affected by a future project. As underlined by the following quotations, this influence also makes such actors to some extent potentially co-responsible for the eventually realised project mix of Energie + Umwelt eG:

2.58 "If we notice that citizens are greatly opposed, then we'll totally and immediately withdraw from a project" (R2N3, 340, 352).

2.59 "People, if we are ready and we have the wish to do something together, then we'll do it now and make something out of it—we'll be available. But we would never push anything without acceptance" (R2N3, 340).

The new energy co-responsibility of cooperative members, as well as of non-members in project development, has had two consequences. On the one hand, the energy cooperative seems to be trying to avoid the risk of overburdening the regional society in the development of a new energy structure, because citizens have the right and the responsibility to express their viewpoints. On the other hand, due to the potential influence of oppositional perspectives, not all potential projects that may be necessary in order to achieve 100% renewable energy supply in the region might be realisable (see also Sect. 7.2.1.2).

7.2.7.4 New Co-Responsibility of Banks Regarding Renewable Energy Development

Energie + Umwelt eG has also created a new sense of energy co-responsibility for the cooperative banks it is associated with. The banks are the ones that primarily decide how many projects and which kinds of projects are realised. Hence, the financial institutions are also responsible for the business activities for Energie + Umwelt eG and for any project consequences resulting from them. The way in which the banks interpret their newly achieved energy responsibility can be seen through their standpoint regarding the investment philosophy of Energie + Umwelt eG. While managing the energy cooperative, the banks follow a business code which could be described as 'No cooperative bank megalomania in the renewable energy sector!' The founding banks clearly want to differentiate themselves from other specialised and nationally operating renewable energy project developers. As the following interviewee puts it, such organisations can be compared to 'plundering knights'—that is, actors who only strive to realise their own business advantages as well as maximise the number of their nationally implemented projects in order to generate a maximum of profit without sharing the benefits with the respective regions:

2.60 "There are robbers around here—large-scale project implementers, sorry" (R2N3, 330)!

7.2 Collaboration Between Energy Cooperatives and Banks: Energie + Umwelt eG

In contrast to large renewable energy project developers, cooperative banks claim to put social goals at the centre of their business intensions, especially by supporting small private households to benefit from the Energiewende:

> 2.61 "We want the Energiewende for the everyday citizen. Because Volks banks do not stand for megalomania, but for the ordinary person [...]" (R2N3, 116).

In this vein, they further point out that it is not their aim to primarily maximise profit. Thus, the banks seek to distinguish themselves from a 'megalomania' which—in their opinion—reigns over some other organisations in the energy sector. Their standpoint, as expressed in the following quotation, is bolstered by the fact that Energie + Umwelt eG financed almost all of its realised projects up to 60% with its own equity capital and by the fact that individual member investments are capped (see previous section):

> 2.62 "For us, dividends don't stand in the foreground. But we have always said: A lot of equity, little debt—we do not want to have to leverage" (R2N3, 24).

That the kind of 'megalomania' that the cooperative banks are trying to avoid has already intruded on the renewable energy sector can be seen via the concrete example of the company PROKON. This organisation invested heavily in German renewable energy projects, especially in wind parks, and collected 1.4 billion Euros of capital from 75,000 mostly small private investors. Its biggest selling point was the promise of high dividends. However, the company had to announce its insolvency in 2014 (dpa 2015; Hielscher 2014).[20]

The banks' interpretation of their responsibility for renewable energy development is primarily influenced by the norms and guiding principles that they follow in the banking business. Business values may not only include doing business in the region but also serving the regional public. Furthermore, their sense of a new energy responsibility is influenced by what they think members of the energy cooperative demand from them. The banks assume that members require manageable and calculable business risks and fear that abuse of member confidence could have a negative effect on important resources, such as the number of their bank customers, invested capital and image:

> 2.63 "We are not approaching it with the eyes of the bank and in order to get a higher rate of return, but we have a responsibility to the 1400 members who trust us as persons who are bankers [...]. They rely on us as persons, that we conduct serious, manageable, and predictable business in E + U" (R2N1, 27).

So far, Energie + Umwelt eG has invested in 36 small to medium-sized photovoltaic projects and in one biomass plant. In addition, it became shareholder of one wind park. Photovoltaic plants tend to cover most of their investments. They have a

---

[20]In 2015, the remaining part of PROKON was turned into an energy cooperative (see entry in the German cooperative register) and thus follows a completely different approach. PROKON Regenerative Energien eG represents with more than 37,000 members and about 847 million Euros total capital the largest energy cooperative in Germany as of 2015 (PROKON regenerative Energien eG 2016).

proven technology design and receive a rather secure return on investment through the guaranteed feed-in tariff under the German Renewable Energy Act. Hence, the new responsibility in developing energy projects with citizen capital and the accompanying fear of image loss may have led to a primarily conservative business approach. Therefore, it seems, motivation for supporting innovative technology projects, such as electro-mobility or energy storage, has thus far been limited.

7.2.7.5 Symbiotic Relationship Between Citizens and Cooperative Banks

Citizens—represented by the energy cooperative—and cooperative banks have created a symbiotic relationship, building and depending upon each other in the development of renewable energy in their region. Cooperative banks have taken it upon themselves to realise their own energy projects, but they have done so with the direct support of local citizens. In this process, leading bank personnel have stated that engagement in the energy cooperative has extended their job tasks. It is not only important to do a good job for the bank but also for Energie + Umwelt eG:

> 2.64 "Expansion of our task spectrum. To be brighter and more alert when there are any topics that you see and hear, to think about: Could this be something for us as E + U? Simply having the responsibility of the honorary mandate at the E + U, to be the supervisory board, to think together, to bring in information or other things about which you've heard" (R2N1, 76).

This new value can be described as a symbiotic relationship between the cooperative banks which founded Energie + Umwelt eG and all other actors engaged in the energy cooperative.

7.2.8 Summary

The energy cooperative Energie + Umwelt eG was founded and is managed by three cooperative banks from the study region. The primary aim of these financial institutions was to position themselves as experts and potential partners in the new business field of renewable energy as well as to involve citizens with small amounts of capital in renewable energy project development. The energy cooperative supports the goal of the districts of Main-Tauber and Neckar-Odenwald to become a zero-emission region and to achieve a 100% renewable energy supply. Together, Energie + Umwelt eG and the banks have mobilised several kinds of allocative and authoritative resources, summarised in Table 7.28, which have empowered them to act and concretely pursue their joint goals.

In addition, the close collaboration between Energie + Umwelt eG and the cooperative banks led to new regulations and interpretative schemes: rules that function as the basis for their activities. These are summarised in Table 7.29.

Table 7.28 Allocative and authoritative resources mobilised through collaboration between Energie + Umwelt eG and cooperative banks in the study region

Mobilised resources between Energie + Umwelt eG and cooperative banks	
Allocative resources	Authoritative resources
Actors – Activation of 1525 cooperative members through 2014, all citizens of Main-Tauber and Neckar-Odenwald or neighbouring districts. – Activation of new bank customers. – Activation and alignment of eleven cooperative banks with almost 239,000 bank customers and 169,000 cooperative bank members (as of 2013). – Selection of cooperative executive board, which is exclusively staffed with personnel of the three founding banks. – Selection of cooperative supervisory board, which is exclusively staffed with key decision takers from the region. *Financial capital* – Mobilisation of 11.9 million Euros total capital through 2014 for developing renewable energy projects in the Main-Tauber and Neckar-Odenwald districts. – Mobilisation of 6.6 million Euros equity capital from members until 2014. – Creation of direct and rapid access to 4.8 million Euros borrowed capital through cooperative banks until 2014. – Generation of up to 265,000 Euros annual revenues until 2014, which led to a distributed dividend of up to 3%. – Early mobilisation of foundation capital from cooperative banks for initiating Energie + Umwelt eG. *Technology mix* – Installation of 36 photovoltaic plants and one biomass plant in Main-Tauber and Neckar-Odenwald until 2014. – Shareholder of one wind park in Neckar-Odenwald until 2014. – All projects have an expected power yield of 7500 megawatt hours per year, which is equal to 1.6% of the annual household power demand that existed in Main-Tauber and Neckar-Odenwald in 2011.	– *Organisational identity:* Energie + Umwelt eG defines itself through the foundation and management of cooperative banks. In turn, the banks use the energy cooperative in order to strengthen their own cooperative identity. – *Publicity:* Large bank network made Energie + Umwelt eG well known in the region. In turn, banks market their newly gained knowledge about renewable energy through Energie + Umwelt eG. – *Image:* Banks established a new expert image in the field of renewable energy. In turn, banks introduced the energy cooperative as a new professional player for realising renewable energy projects in the region. – *Trust* of 1525 members (as of 2014) in the business model of Energie + Umwelt eG. – *Trust* between involved banks existed before through previous activities and was further increased within Energie + Umwelt eG. – *Transparency:* Energie + Umwelt eG created a transparent business philosophy. – *Access to potential projects:* Banks used their large customer network in order to acquire a high number of renewable energy project opportunities within a short amount of time. At the same time, the energy cooperative provided access to new credit business for the banks. – *Access to information and new ideas:* The energy cooperative collaborates with important regional decision authorities, such as the banks, heads of administrative districts or the farmers association. – *Expertise and Know-how:* Banks offer important financial, legal and marketing expertise supporting all business activities of Energie + Umwelt eG.

These new norms and guiding principles have also been mentioned in public newspapers, likely indicating that they have been further transferred into regional public beyond the boundaries of the energy cooperative that created them. Table 7.30 summarises the share of articles written about the energy cooperative

Table 7.29 New rules created through collaboration between Energie + Umwelt eG and the cooperative banks in the study region

Rules created between Energie + Umwelt eG and cooperative banks	
Regulations	Interpretative schemes
– Official accreditation of cooperative banks for guiding citizen-focused regional energy projects. By founding the energy cooperative, financial institutions turned into new initiators of citizen-based renewable and regional energy. – New co-decision and participation rights for citizens of the Main-Tauber and Neckar-Odenwald districts with respect to renewable energy projects, provided they become cooperative members. – New co-decision rights for politicians in the execution of renewable energy projects, since the administrative heads of Neckar-Odenwald and Main-Tauber became members of the cooperative's supervisory board. – General limitation of individual member investment volumes in renewable energy projects realised by the cooperative banks. – Strengthening the right of the cooperative banks to create regional value through renewable energy projects in Neckar-Odenwald and Main-Tauber by involving citizens of the region.	– Together with citizens, the cooperative banks are 'taking matters into their own hands' in order to develop renewable energy projects in the Main-Tauber and Neckar-Odenwald districts. – Full citizen acceptance is a new pre-condition for becoming involved in renewable energy projects in the first place and becomes the main project-selection criterion. – New co-responsibility of citizens for regional energy issues. Members invest their capital in energy projects and are, thus, co-responsible for them; non-members need to accept projects and are, thus, also responsible for their realisation. – New co-responsibility of cooperative banks for renewable energy project development and aligned citizens' capital. Banks follow the business philosophy 'no cooperative bank megalomania in the renewable energy sector'. – New symbiotic relationship between citizens and cooperative banks for coordinating energy issues in the region. They depend upon each other for developing renewable energy projects.

Table 7.30 Share of articles about the energy cooperative published in *Fränksiche Nachrichten* in 2010–2015 according to the respective new societal rules mentioned or discussed

New societal rules created through collaboration between the banks and energy cooperative	Share of articles that mention or discuss the respective rule (n = 26) (%)
Official accreditation of cooperative banks for guiding energy projects	81
Taking matters into one's own hands	81
New participation and co-decision rights for citizens	77
New forms of regional value creation	73
New symbiotic relationship	50
New co-decision right for politicians	35
Citizen acceptance as a pre-condition for energy project involvement	15
New co-responsibility of banks for renewable energy development	15
New co-responsibility of citizens for regional energy issues	12
Limitation of individual member investment volumes in renewable energy projects	4

in one of the most important regional newspapers, *Fränkische Nachrichten*, between 2010 and 2015 that mention or discuss the newly introduced rules.

The mobilisation of resources and the creation of rules through collaborative action between the cooperative banks and the energy cooperative have influenced the energy structure in the Main-Tauber and Neckar-Odenwald districts in several respects:

1. *Empowering cooperative banks through citizens for developing renewable energy projects:* A network of eleven cooperative banks and the energy cooperative Energie + Umwelt eG mutually reinforced each other in becoming new actors for developing renewable and regional energy projects by making use of their particular organisational backgrounds. Together they mobilised about 11.9 million Euros total capital and 1525 members by December 2014—only three and a half years after Energie + Umwelt eG was founded. In 2012, the energy cooperative was already amongst the largest renewable energy production cooperatives in Germany. By the end of 2014, it had installed 38 renewable energy production projects with a power production yield of about 7500 megawatt hours per year. Most of the installed power capacity is located in the partners' focus region of Main-Tauber and Neckar-Odenwald. Only two photovoltaic projects are located in a neighbouring region. In 2011, the energy cooperative was able to supply about 1.6% of the annual household power demand that existed in the two districts. The collaboration turned the partners into strong supporters of regional and renewable power production in the districts. The increase of renewable power is also one of the main goals of the districts on their way towards becoming a zero-emission region with a 100% renewable energy supply. Authoritative resources, including the creation of an organisational identity, a professional image and publicity for the energy cooperative, as well as trust in its business model, provided the most important basis for the mobilisation of members and capital. As established institutions with financial know-how, the three leading cooperative banks—Volksbank Mosbach, Volksbank Main-Tauber and Volksbank Franken—strongly supported the creating of these resources. Amongst other ways, they provided expert personnel, founding capital, initial equity capital, administrative and marketing competencies, as well as the first well-running projects. Energie + Umwelt eG was thus able to represent itself as a professional and capable organisation right from beginning of its business activities. The banks not only secured rapid access to borrowed capital but also, due to their large customer network and connections with important decision takers, became the main marketing partner of the cooperative (97% of the members of Energie + Umwelt eG are also customers or members of the banks). The banks also provided direct access to strategic information and potential projects. In turn, Energie + Umwelt eG helped the banks to position themselves as new experts in the renewable energy sector—a new unique selling point in their own competitive environment. The energy cooperative has supported the banks to acquire new customers and strengthen the loyalty of their existing customers. Projects planned

by the energy cooperative have also presented a new credit-business source for the banks. The partners have also supported each other by making their new competencies and business goals public. Together, the banks and the energy cooperative have created an image of citizen-focused organisations that can realise regional and renewable energy production projects on a highly professional basis.

2. *Legitimising cooperative banks as a new actor in the renewable energy sector*: The foundation and management of Energie + Umwelt eG publicly legitimised the cooperative banks as players that can pro-actively initiate, guide, and develop renewable energy projects in the region. The banks now feel mandated to 'take matters into their own hands' and see themselves as being the most capable actors for initiating energy cooperatives, because they both have the same legal business basis as cooperatives. The banks have also transferred own values into the energy sector, such as taking societal responsibility for the region. While managing Energie + Umwelt eG, Volksbank Mosbach, Volksbank Main-Tauber and Volksbank Franken have followed a norm-driven business approach in which profit maximisation does not represent their core business aim. They have pointed towards a new co-responsibility in developing renewable energy, which they express through the motto 'no cooperative bank megalomania in the renewable energy sector'. Accordingly, they have introduced an official limit on capital volumes from individual members in order to involve many citizens who have small amounts of capital, even though this investment cap was not positively evaluated by all of its bank customers. In this sense, the banks have influenced the composition of actors who invest in the region's energy projects. Nevertheless, banks still see the generation of profit as a crucial factor for all business activities of the energy cooperative. Consequently, they primarily select energy projects that are very likely to be profitable. A trustful relationship between the banks that founded the energy cooperative provided an important basis for their engagement in the pro-active development of renewable energy in their region.

3. *Creating greater significance for the participation of citizens and politicians in renewable energy projects:* By founding the energy cooperative, the cooperative banks created new decision and participation rights for citizens of Main-Tauber and Neckar-Odenwald with respect to developing renewable energy projects in their region. Several participation levels have been differentiated. By the end of 2014, 1525 citizens had made use of the right to financially participate by buying 66,000 member shares with a value of around 6.6 million Euros. Between 11 and 13% of all members made use of their right to co-decide by attending the cooperative's general assemblies, likely indicating that more members seem to be only interested in financial participation whereas a minority seems to be interested in co-deciding strategic business issues. Some members also made use of their right to become directly engaged in individual projects. The energy cooperative has also created new co-decision rights for key political actors in developing renewable energy projects, as membership of the administrative heads of the districts on the cooperative supervisory board authorises them to participate

in all strategic project decisions. It seems, though, that such projects need to comply with reigning political positions. Otherwise they are not passed. In this sense, not only cooperative banks but also politicians and citizens have turned into new developers of renewable energy projects in the study districts.

4. *Underlining the significance of regional value creation and combining economic and social values:* The cooperative banks have reinforced their position for creating economic value for their districts by founding and managing the energy cooperative. All except two projects that have been realised by Energie + Umwelt eG are located in the districts of Main-Tauber and Neckar-Odenwald. The cooperative's statutes mandate that all cooperative members must live within the two districts. Members of the energy cooperative economically benefit by receiving an annual dividend of around 3%, distributed by the energy cooperative based upon their regional project activities. The districts benefit because they receive business taxes for each project located within their boundaries as well as rent for those projects that are located on district property. The regional cooperative banks benefit by acquiring additional business potential, such as providing borrowed capital for the projects planned by the energy cooperative. In consequence, the energy cooperative appears to be underlining the significance of small renewable energy projects for economic value creation in the region. The cooperative banks have also been aligning the creation of regional economic value in the energy sector with the creation of social value. They point out that they want to take matters into their own hands and chose the cooperative form for developing renewable energy projects due to its democratic decision-taking structure. One of their main expressed aims has been to give regional citizens a participation right in developing renewable energy. Special here is that this value is being created through a new symbiotic relationship between the banks and citizens, who depend upon each other and also build upon each other in developing regional energy projects.

However, the collaboration between Energie + Umwelt eG and cooperative banks has also involved challenges and limitations with respect to influencing the energy structure of the region:

1. *Focus on proven technology investments limits support for innovative technology projects:* Energie + Umwelt eG has mainly fostered the distribution of proven renewable energy technologies. So far, it has not been actively involved in the development or distribution of innovative projects, including electro-mobility, storage concepts, energy self-consumption concepts or energy-efficiency measures. However, such projects are very likely to play an important role in achieving a 100% renewable energy structure in Main-Tauber and Neckar-Odenwald—a goal that is actively supported by the energy cooperative. The executive board of Energie + Umwelt eG, which is represented by the leading cooperative banks, has been following a risk-averse business approach: focussing on projects with low technology risks and with a more or less secure return on investment through the guaranteed feed-in tariff under the German Renewable

Energy Act. Consequently, 36 of its 38 already realised renewable energy projects are small to medium-sized photovoltaic plants. The main reason for the rather conservative business approach is their newly felt responsibility for citizen capital in renewable energy project development. Amongst other reasons, the banks fear risking their image if problems arise with renewable energy projects that are managed by them. At the same time, these cooperative banks are still economically oriented organisations. Even though profit maximisation may not be their main project goal, profitability is still one of the main project selection criteria. The strong interest of cooperative members in the economic development of the energy cooperative may indicate that they support the rather conservative investment approach of Energie + Umwelt eG, even though this may limit its technological innovation potential.

2. *The new symbiotic relationship is limited regarding activation of citizen participation in project development and strategic business decisions:* The symbiotic relationship between the cooperative banks and Energie + Umwelt eG influences the ways in which members may actually make use of their right to participate and to co-decide within the energy cooperative. The partners have triggered strong financial participation of citizens that has been accompanied by a rather low level of citizen participation regarding operational and strategic business or project issues. This is especially revealed by the passive position of members in the cooperative's general assemblies and their seemingly low interest in supporting the cooperative in its daily business activities. Lack of informal places, such as additional working groups, for regular exchange between members and the cooperative steering board also make it difficult for members to communicate their own viewpoints and ideas. The general assembly does not appear to be the right place to do so, due to its highly formal structure. At the same time, the energy cooperative seems to prefer to take strategic and important decisions between its management and supervisory board—a carefully chosen group of leaders from the region. It seems that a possible consequence here is that members also perceive Energie + Umwelt eG as an organisation that does not need further member help or guidance, because it has gained its strengths from its well-established founding partners: the banks. As a consequence of the low member participation regarding business issues, the cooperative seems to have limited access to innovative ideas and additional help from its members.

3. *Using full citizen acceptance as a rule can limit regional project activities and inhibit constructive societal discussion about projects:* The collaboration between Energie + Umwelt eG and cooperative banks has generally strengthened supporters of renewable energy projects and, thus, also strengthened the acceptance for these projects. Since all energy cooperative 1525 members have provided private capital of around 6.6 million Euros for realising renewable energy projects, they must also accept their implementation. Special here is that the energy cooperative has not only functioned as a tool for unifying and coordinating the renewable energy supporters but also turned them to a clearly defined and easy identifiable group by making key information, such as group size, project

activities and project results, publicly available. However, Energie + Umwelt eG turned project acceptance into a new rule and only realises projects under the precondition of full citizen support. In this sense, the energy cooperative may particularly help to mobilise and unify those actors that are already in favour of certain renewable energy projects or those that are at least not against them. Since the energy cooperative normally withdraws in cases where full citizen support is lacking, it is limited by its ability to convince actors that have actually positioned themselves against renewable energy projects about the importance of those projects. The selection criterion of full citizen support may ensure that the regional society does not become overburdened in the development of new renewable energy structures. Yet, at the same time, it also seems to inhibit constructive social exchange about conflictive aspects in the development of renewable energy. Such open discussion about difficult projects is crucial in order to fully realise the level of renewable energy production necessary for achieving a 100% renewable power supply in the districts of Main-Tauber and Neckar-Odenwald.

4. *In the future, organisational growth can be challenged by its regional focus and change of external factors:* The strong growth of Energie + Umwelt eG with respect to the cooperative members, financial capital and technology it has attracted within a relative short amount of time was, on the one hand, driven by the creation of a positive organisational image, trust in the business model and publicity. These authoritative resources were created through the collaboration with cooperative banks. On the other hand, organisational growth was also driven through external factors, such as the guaranteed feed-in tariff for renewable energy projects under the German Renewable Energy Act and by a general citizen interest in investing in the energy cooperative, as other investment alternatives seemed to be generally limited due to low interest rates. Thus, several internal and external factors can challenge the growth of Energie + Umwelt eG in the future. The cooperative's executive board has already reached the maximum work load they are able to manage on a voluntary basis. If the energy cooperative does not realise its goal to introduce professional employment structures, time for new project searches and implementation may be limited. The latest reduction of the government's feed-in tariff decreased new project opportunities in the region. As a direct consequence, the energy cooperative already reduced individual member investment volumes from 100,000 Euros to 1000 Euros. The unsure position of energy cooperatives under the KAGB legislation had prevented the energy cooperative from investing in several larger energy projects. Furthermore, Energie + Umwelt eG fears that future changes in the general investment landscape may make other investment alternatives more attractive for citizens in the future.

Table 7.31 summarises the impacts of the collaboration between Energie + Umwelt eG and the cooperative banks on the energy structure of Main-Tauber and Neckar-Odenwald.

Table 7.31 Impacts of collaboration between Energie + Umwelt eG and cooperative banks in the study region on local energy structure

Impacts of collaboration between Energie + Umwelt eG and the cooperative banks on the energy structure of Main-Tauber and Neckar-Odenwald	
Realised potential	Challenges and limitations
Empowering cooperative banks and citizens for developing renewable energy projects – The joint activation of about 11.9 million Euros empowered the partners to develop 38 renewable power production projects in Main-Tauber and Neckar-Odenwald with 6.6 million Euros provided by 1525 citizens of the region. – The partners have been able to supply about 1.6% of the regional household power demand with their seven megawatt installed power capacity. – Authoritative resources created in collaboration, including organisational identity, trust in the business model of the cooperative, image, publicity or access to potential projects, has played a key role for attracting cooperative members and capital. Legitimising cooperative banks as a new actor in the renewable energy sector – The energy cooperative legitimates the cooperative banks to pro-actively initiate and guide renewable energy projects. Financial institutions have become new players in concretely developing a renewable energy structure in their region. – Cooperative banks feel mandated to 'take matters into their own hands'. – Cooperative banks transfer their own values into the energy sector, such as taking societal responsibility for the region or the prevention of megalomania in investment activities. Creating greater significance for the participation of citizens and politicians – Politicians and citizens have turned into new developers of renewable energy projects in Main-Tauber and Neckar-Odenwald by receiving different participation and co-decision rights. – More members seem interested in financial participation than in co-deciding strategic business issues, as only between 11% and 13% of all members attend general assemblies. – Projects need to comply with reigning political positions. Underlining the significance of regional value	Limited support for innovative technology – A new responsibility for citizens' capital in renewable energy project development has led to a risk-averse business approach. – Even though profit maximisation may not be the main project goal, profitability is still one of the main project selection criteria. – The technological innovation potential of Energie + Umwelt eG may be limited, since projects, such as electro-mobility and energy storage, involve higher risks and less secure returns on investment. Symbiotic relationship is limited regarding activation of citizen participation in strategic decisions and project development – Energie + Umwelt eG is perceived as an organisation that does not need further member help or guidance, because it has gained its strengths from its well-established founding partners: the banks. – The energy cooperative prefers to take business decisions between its management and supervisory board: a carefully chosen group of leaders from the region. – Lack of informal places or work groups makes it difficult for members to communicate their own viewpoints and ideas and limits the cooperative's access to innovative ideas and additional help from its members. Using full citizen acceptance as a rule limits regional project activities and inhibits constructive societal project discussion – Energie + Umwelt eG only realises projects under the precondition of full citizen support. – Since the cooperative uses existing acceptance as a rule, it is limited by its ability to convince actors that have positioned themselves against renewable energy projects about their importance. – At the same time, the rule also inhibits constructive social exchange about conflictive aspects of the development of renewable energy. Organisational growth is challenged by internal and external factors – The executive board of Energie + Umwelt eG has reached the maximum work load its

(continued)

Table 7.31 (continued)

Impacts of collaboration between Energie + Umwelt eG and the cooperative banks on the energy structure of Main-Tauber and Neckar-Odenwald

Realised potential	Challenges and limitations

creation and combining economic and social values
  – Increased significance of small renewable energy projects for value creation in the region.
  – Economic value creation, such as business taxes, rent or dividends, is aligned with the creation of societal value, such as greater regional self-determination.
  – Regional value is created through a new symbiotic relationship between citizens and cooperative banks.
	members are capable of managing on a voluntary basis.
	  – The latest reduction of the government feed-in tariff decreased new project opportunities in the region.
	  – Changes in the general investment landscape may make other investment alternatives more attractive for citizens in the future.

7.3 Collaboration Between Energy Cooperatives and Communities: Neue Energien West eG and Bürgerenergiegenossenschaft West eG

Cooperation with communities represents the third dominant interaction model preferred by recently emerging energy cooperatives in Germany. As outlined in Sect. 6.3, 41% of analysed renewable energy production cooperatives, the largest group among energy cooperatives, are founded and/or are managed by communities. *Neue Energien West eG* (NEW eG) and *Bürgerenergiegenossenschaft West eG* (Bürger eG) were chosen for a third case study because they have had particularly intense collaboration with communities and their citizens. NEW eG unifies 17 municipalities and cities, as well as two municipal energy and water providers,[21] with the aim of jointly making their region independent from fossil energy resources by 2030 (Stadtwerke Grafenwöhr 2011, p. 31). Meanwhile, Bürger eG unifies all citizens who are interested in investing in the energy projects realised by NEW eG. Bürger eG has experienced strong organisational growth and, with 1303 members and 14.6 million Euros total capital, belongs among the largest renewable energy production cooperatives in Germany.

7.3.1 Introduction of the Focus Region

The inter-municipal energy cooperative NEW eG and the aligned citizens energy cooperative Bürger eG were founded by a group of communities that are located in

---

[21] In the following, I will speak of 18 involved cities and municipalities (or communities): 16 communities are represented by their mayors, one community is presented by its municipal utility (Floß) and one community is represented by its mayor and its municipal utility (Grafenwöhr).

the districts Neustadt an der Waldnaab, Amberg-Sulzbach and Tirschenreuth, near the Czech border in the Bundesland of Bavaria. Their goal is to become independent from fossil energy resources by 2030 (Stadtwerke Grafenwöhr 2011, p. 31). The municipalities and cities have defined their own administrative boundaries as the preferred business region for the two energy cooperatives.

7.3.1.1 Geographical Character

The administrative area of Oberpfalz, in which the districts Neustadt an der Waldnaab, Amberg-Sulzbach and Tirschenreuth are located, has a mainly rural character. As displayed in Table 7.32, 43% of the land is used for agricultural purposes, 40% is covered by forest, and only 10% is used for settlement and traffic infrastructure. Two percent is covered with water, and 4% is used for other purposes.

Of the 18 communities that are represented by NEW eG, 11 are situated in the western part of the district Neustadt an der Waldnaab. The city of Weiden,[22] as well as the municipalities of Altenstadt a. d. Waldnaab and Floß are located in the eastern part of the district. Three municipalities—Auberbach i.d. OPf. Kastl and Kemnath—belong to the neighbouring districts of Amberg-Sulzbach and Tirschenreuth and are situated on the border of the district of Neustadt an der Waldnaab. Figure 7.6 provides an overview of the municipalities that are represented by NEW eG and the region that they cover.

Table 7.33 lists all communities and cities involved in NEW eG, differentiated according to their population and area size.

As can be seen, the member communities of NEW eG have unified about 97,000 citizens (as of January 2013) and, taken together, they have a size of about 919 square kilometres (as of December 2013). The average population density ranges around 106 citizens per square kilometre, but the individual cities and municipalities exhibit a wide range of population numbers and land area sizes. The smallest municipality in

Table 7.32 Land use characteristics of the administrative region of Operpfalz, as of December 2012 (Bayerisches Landesamt für Statistik und Datenerhebung 2014, p. 12)

Land use characteristics of the region, as of December 2012	Oberpfalz Hectares	Percent
Size of region	969,014	100
Agriculture	419,838	43
Forest	389,164	40
Settlement and traffic infrastructure	99,835	10
Water	18,027	2
Other uses	42,150	4

---

[22]Weiden in der Oberpfalz is a municipality that is not associated with a district, meaning that it is self-governed.

7.3 Collaboration Between Energy Cooperatives and Communities: Neue... 265

Fig. 7.6 Location of the 18 municipalities represented by NEW eG (based upon Regierung der Oberpfalz 2015b)

Table 7.33 Population and size of all communities being represented NEW eG, as of 2013 (Regierung der Oberpfalz 2015a)

Communities represented by NEW eG	Population As of January 2013	Area in km^2 As of December 2013	Population per km^2
District of Neustadt a.d.Waldnaab			
Grafenwöhr, St	6470	216	30
Altenstadt a.d.Waldnaab	4742	22	216
Pressath, St	4394	66	66
Eschenbach i.d.OPf., St	3877	35	110
Weiherhammer	3875	40	97
Floß, M	3487	54	64
Kirchenthumbach, M	3264	67	48
Parkstein, M	2301	31	74
Trabitz	1295	27	49
Schwarzenbach	1179	12	99
Neustadt am Kulm, St	1170	20	58
Speinshart	1112	24	47
Vorbach	1039	14	77
Schlammersdorf	877	20	43
District of Amberg-Sulzbach			
Auerbach i.d.OPf., St	8859	78	113
Weiden i.d.OPf. (Krfr.St)			
Weiden i.d.OPf. (Krfr.St)	41,726	71	592
District of Tirschenreuth			
Kemnath, St	5317	57	94
Kastl, M	2395	65	37
Total	97,379	919	106

terms of citizens—Schlammersdorf—has a population of 877, whereas the largest community—the city of Weiden i.d.OPf.—has about 42,000 citizens. With 12 square kilometres, Schwarzenbach is the smallest municipality in terms of area; meanwhile, with 216 square kilometres, Grafenwöhr is the largest. Yet, all involved cities and municipalities are rather small, with less than 600 citizens per square kilometre.

7.3.1.2 Energy Transition Goals

The 18 municipalities that have come together through NEW eG have the aim of making their region independent from fossil energy resources by 2030 (Stadtwerke Grafenwöhr 2011, p. 30). In order to verify the feasibility of their goal and identify concrete tasks for its realisation, 11 municipalities, which form the western part of the district of Neustadt an der Waldnaab, commissioned a climate-protection concept for their region (Stadtwerke Grafenwöhr 2011). At the time the analysis was conducted, only these 11 communities and cities were members of NEW eG; the other seven joined the cooperative at a later stage. As a consequence, the concept focuses on the renewable energy and energy efficiency potential existing in the 11 municipalities which commissioned the report. However, the goal of establishing a 100% renewable energy structure is supported by all the municipalities represented by NEW eG. In the following, the results of the study are used in order to generally and specifically discuss the potential for the whole focus region of the two energy cooperatives to achieve such an energy transition goal.

Energy Demand

As displayed in Table 7.34, the western part of the district of Neustadt an der Waldnaab had a total power demand of about 158,000 megawatt hours in 2009. About 68% was consumed by companies and industry, about 27% by private households and about 5% by the public administration.

If the per person power consumption of the western part of Neutstadt an der Waldnaab is extrapolated to the population of all the municipalities that are represented by NEW eG, one can assume a total household power demand of about 147,089 megawatt hours per year for the complete focus region of NEW eG.

Table 7.34 Total power demand for the western part of the district of Neutstadt an der Waldnaab in 2009 (Stadtwerke Grafenwöhr 2011, pp. 20ff.)

Power demand in 2009	MWh/a	Percent
Industry	108,173	68
Private households	42,263	27
Public administration	7753	5
Total	158,189	100

7.3 Collaboration Between Energy Cooperatives and Communities: Neue... 267

Table 7.35 Total heat demand of the western part of the district of Neutstadt an der Waldnaab in 2010 (Stadtwerke Grafenwöhr 2011, pp. 20ff.)

Heat demand in 2010	MWh/a	Percent
Private households	235,770	60
Industry	145,959	37
Public administration	8388	2
Total	390,117	100

As shown in Table 7.35, the district had a total heat demand of about 390,000 megawatt hours in 2010. Private households consumed about 60%, companies and industry about 37% and the public administration consumed about 2% of the heat demand in 2010.

If the per person heat consumption of the western part of Neutstadt an der Waldnaab is extrapolated to the population of all communities that are represented by NEW eG, one can assume a total household heat demand of about 820,554 megawatt hours per year for the complete focus region of NEW eG.

Renewable Energy Supply

The first aim of the municipalities and cities has been to increase their renewable energy production (Stadtwerke Grafenwöhr 2011). Table 7.36 outlines the renewable energy potential of the region.

As can be seen, it is assumed that about 188,000 megawatt hours of renewable power could be produced annually in the western part of Neutstadt an der Waldnaab. This would cover 119% of its needed power and is, thus, more renewable power than needed. Solar and wind power have the highest potential. Solar energy could cover about 44% of annual power demand, whereas wind energy could produce about 68,400 megawatt hours and could, thus, cover about 43% (see Tables 7.34 and 7.36). In order to achieve a 100% renewable energy supply in the region, the establishment of wind parks and an increase of solar power would have to be fostered. The climate protection plan proposes to install at least 12 wind mills in the district by 2030 and, further, proposes to define new locations for solar parks in the land development plan (Stadtwerke Grafenwöhr 2011, p. 102).

Meanwhile, it seems to be challenging to establish a totally renewable heat supply for the region. As can be seen in Table 7.36, only 141,000 megawatt hours, 36% of the total heat demand, may be potentially supplied by renewable resources from the region. The reduction of heat demand has thus been prioritised (Stadtwerke Grafenwöhr 2011, p. 37). Further analysis is planned to see whether additional biomass plants make sense against the background of nature conversation as well as economic and regulative considerations, and excess heat from existing biomass plants is to be better activated (Stadtwerke Grafenwöhr 2011, pp. 101ff.).

The aim of achieving a 100% renewable energy supply is mainly focused on an overall annual balance between energy demand and renewable energy supply.

Table 7.36 Renewable energy potential of the western part of Neustadt an der Waldnaab (Stadtwerke Grafenwöhr 2011, p. 70)

Overview of renewable energy sources in the western part of Neutstadt an der Waldnaab		Potential renewable energy supply	
		Energy electric MWh/a	Energy thermal MWh/a
Photovoltaic	50% of applicable rooftops	41,965	–
Photovoltaic	Solar parks	27,470	–
Solarthermal		–	5422
Biomass	Forestproducts	–	80,765
Biomass	Combined heat and power	no inf.	no inf.
Biogas	Agricultural products	49,120	55,260
Wind power		68,400	–
Hydropower		855	–
Total		187,810	141,447

Consequently, temporal differences between real demand and renewable energy supply have not been included. Energy storage concepts, as well as distribution concepts through which excess energy or insufficient energy supply can be balanced at any time, would have to be integrated into a full 100% renewable energy strategy (Federal Ministry for Economic Affairs and Energy 2014c).

Energy Efficiency

A second aim of the municipalities on their way towards achieving a 100% renewable energy supply is to increase their energy efficiency rate. According to the integrated climate protection concept, the current heat demand could be reduced by almost 50%, to 123,318 megawatt hours per year, if all dwellings were to be renovated according to the EnEV Standard[23] (an energy efficiency decree) by 2030 (Stadtwerke Grafenwöhr 2011, p. 37). The potential for reducing the power demand of private households is calculated at around 30%. Needed actions here include replacing incandescent light bulbs by LED lamps and replacing old electronic devices and appliances, such as washing machines, with more efficient ones (Stadtwerke Grafenwöhr 2011, p. 86). Furthermore, all public administration buildings are to be renovated and optimised. Street lights are to be changed to LED technology, and photovoltaic production on public rooftops will be maximised (Stadtwerke Grafenwöhr 2011, pp. 90f.).

---

[23]*Energieeinspeiseverordnung* (EnEV) (Zweite Verordnung zur Änderung der Energieeinsparverordnung 2013).

Electro-mobility

A third aim is to change the mobility infrastructure in the region. The climate concept foresees that about 10% of the vehicles should be electric-powered in the district by 2030 (Stadtwerke Grafenwöhr 2011, pp. 99f.). Thus, charging stations need to be distributed and the purchase of electric vehicles prioritised by the public administration.

The climate protection concept indicates that the municipalities of the western part of the district Neustadt an der Waldnaab, represented by NEW eG, most likely have the capability to become independent from fossil energy resources (annual balance approach), provided that:

- Wind and solar power is strongly expanded, as 87% of power demand could be met;
- energy demand is reduced. Especially heat demand must be strongly reduced, as only 36% could be potentially supplied by renewable resources from the region; and provided that
- innovation projects, including electro-mobility, are fostered.

7.3.2 The Energy Cooperatives Neue Energien West eG and Bürgerenergiegenossenschaft West eG

In the following, the two energy cooperatives NEW eG and Bürger eG are introduced in detail.

7.3.2.1 The Foundation Process

The inter-municipal energy cooperative NEW eG was founded on 27 February 2009 and registered on 27 January 2010. Meanwhile, the Bürger eG energy cooperative was founded shortly afterwards, on 08 June 2009, and was registered on 12 August 2010.

The plan to found a new energy cooperative in order to unify several communities for the development of a 100% renewable energy supply in the region was initiated by the director of Stadtwerke Grafenwöhr (the municipal utility of Grafenwöhr) and by the former mayor of Grafenwöhr, both of whom had become more and more dissatisfied with increasing energy prices. They felt dependent upon large energy providers. In their opinion, these companies had achieved such strong negotiation status that they could dictate their prices to the municipality. In consequence, the mayor and the Stadtwerke director started to think about options for how to change their own position. Subsequently, the idea evolved of producing their own energy in collaboration with other neighbouring cities and municipalities.

The idea of founding a new energy cooperative in order to realise this goal was born during a meeting between the regional cooperative bank (*Raiffeisenbank*) and the director of the municipal energy provider, where the former mayor of Grafenwöhr and the director of Stadtwerke Grafenwöhr successfully convinced the administrative council of Stadtwerke Grafenwöhr. Then they presented the project to all communities belonging to the EU LEADER region *VierStädtedreieck* (Four City Triangle), which is a collaborative initiative between 10 communities and cities of the western part of the district of Neutstadt an der Waldnaab. The initiative had evolved out of a long-lasting collaborative relationship between the cities of Grafenwöhr, Pressath and Eschenback, which had already been realising joint projects together for many years. Together with the city of Kirchenthumbach, they had formed VierStädtedreieck in 2004, and the other six municipalities joined the group during the initiative's application for the EU LEADER programme in 2004 (Landratsamt Neustadt a.d.Waldnaab 2015), which supports local actors in realising the long-term potential of their region (European Commission 2015). Since 2008, VierStädtedreieck has been an official LEADER partner (Landratsamt Neustadt a.d. Waldnaab 2015).

Being the closest collaborative partners of Grafenwöhr, the mayors of Pressath and Eschenbach were the first to be personally contacted by the former mayor of Grafenwöhr and the director of Stadtwerke Grafenwöhr. They were quickly convinced by the idea of founding an inter-municipal energy cooperative in order to jointly realise renewable energy projects. Together, they presented the plan to the mayors of the other communities and cities comprising VierStädtedreieck as well as to the respective community and city councils.

Eventually, all 10 municipalities that belong to the LEADER region, as well as the community of Parkstein, became founding members of the NEW eG energy cooperative. During interviews that I conducted in the region, two key reasons were explicitly cited to explain the great interest of the mayors and councils in joining the energy cooperative. First, as indicated by the following quotation, the perception of a high and undesired degree of dependency upon large trans-regional energy providers and their price policies not only existed in Grafenwöhr but also among many of the involved cities and communities, and the option to become more independent by producing their own energy was appealing to them:

> 3.1 "I just said that, ultimately, this inter-communal cooperation will enable us to shape the future of energy policy. I just mentioned how much we pay as a community to the energy companies, what kind of problems we have with public street lighting. We are really treated dictatorially from above by the big energy companies. And that this is simply an opportunity to move away from it, at least the first step" (R3N3, 27).

Second, the energy cooperative was presented as a small economic stimulus plan for the district, from which especially smaller cities and communities were supposed to benefit (Neue Energien West eG 2009c). This argument convinced many involved cities and communities as the majority are small in size (see Table 7.33). Involved actors did not appear to see any major risks in organising themselves in a cooperative in order to develop their own renewable energy projects.

The Bürger eG citizens cooperative was initiated by the same group of municipalities and founded closely after NEW eG. As illustrated by the following quotation, the involved cities and communities wanted to integrate their citizens into their renewable energy development activities by establishing this second energy cooperative:

> 3.2 "We said that we do not want to have any special people or capital shareholders or anything of the kind; rather we want to win over our own citizens" (R3N5, 31).

The purpose of NEW eG was to unify municipalities in order to jointly realise renewable energy projects, whereas Bürger eG was meant to unify and coordinate all citizens of the region who are interested in investing in projects implemented by NEW eG. The two energy cooperatives are directly interrelated with each other, in so far as Bürger eG became a member of NEW eG.

The business structures of Bürger eG and NEW eG, including their statutes, were primarily developed by the director of Stadtwerke Grafenwöhr with support from the cooperative association of Bavaria. The actual foundation meeting of the two energy cooperatives took place within a closed group, including the 10 founding municipalities and Stadtwerke Grafenwöhr. The foundation process of both cooperatives was also strongly supported by a central cooperative interest association as well as by a cooperative auditing association. Both provided important advice with regard to how to legally set up a cooperative, define cooperative statues and stay in line with cooperative law (GenG).

7.3.2.2 Business Purpose

Both energy cooperatives—NEW eG and Bürger eG—aim to support the development of regional and renewable energy structures in their focus region (Bürgerenergiegenossenschaft West eG 2015; Neue Energien West eG 2009b; Stadtwerke Grafenwöhr 2011). According to its statutes, main business goal of NEW eG is the planning, development and operation of renewable energy production projects that support an environmentally friendly, sustainable and innovative energy supply (Neue Energien West eG 2009a, p. 1). According to the statues of Bürger eG, its core business goal is financial investment in renewable energy projects realised by NEW eG. Other business goals of Bürger eG include:

- providing renewable energy advisory services, such as advice regarding the use of renewable energy through regional experts or by carrying out thermal imaging, through which the insulation of private buildings can be checked;
- bundling of purchasing activities for members as well as offering group contracts for insurance or cleaning services of photovoltaic plants; and
- developing and operating of its own renewable energy projects (Bürgerenergiegenossenschaft West eG 2013, p. 1; Bürgerenergiegenossenschaft West eG 2015).

Another business goal is the integration of local businesses, in order to support them and ensure local jobs (Bürgerenergiegenossenschaft West eG 2015).

7.3.3 Collaborative Concept

7.3.3.1 Key Collaborative Partners: Municipalities

NEW eG and Bürger eG were founded by municipalities, which are the key collaborative partners of both energy cooperatives. Their core motivation for founding the two energy cooperatives was to become more independent from fossil fuels and from large energy providers by jointly supporting a 100% renewable energy supply for their region. Furthermore, they wanted to involve citizens in their activities by allowing them to become members of Bürger eG. Apart from having the same goals, however, the involved cities and communities are quite different. As outlined in Sect. 7.3.1, their geographical set-up varies in terms of population, ranging between 900 and 42,000 citizens, as well as in size, lying between 12 and 216 square kilometres. The municipalities also have different political self-understandings. Table 7.37 presents the distribution of the political affiliations of the mayors who were in office in the municipalities represented by NEW eG as of 2015.

As can be seen, there is no clear tendency for one political party. Instead, mayors almost equally belong to the two most dominant German political parties the CDU and the SPD as well as to electoral coalitions. Thus, the aim of becoming more autonomous in terms of energy supply and the decision to realise energy projects with other municipalities through a cooperative do not seem to be exclusively aligned to a certain political self-understanding but, rather, appear to be compatible with a wide range of political approaches, such as conservative (CDU), social-democratic (SPD) or independent political lines.

The communities of Grafenwöhr and Floß incorporated their municipal utilities into NEW eG. Stadtwerke Grafenwöhr was founded in 2000 and belongs 100% to the city of Grafenwöhr, mainly operating the purification plant and water net for the city. Before its involvement in the energy cooperative, Stadtwerke Grafenwöhr had not been involved in any energy services (Stadtwerke Grafenwöhr 2015). The municipal utility of Floß belongs 100% to the community of Floß and is responsible

Table 7.37 Distribution of the political affiliations of encumbent mayors in the municipalities being represented by NEW eG as of 2015

Number of involved municipalities	Political affiliation of encumbent mayor (as of 2015)
7	CDU
6	SPD
5	Electoral coalitions

Source: Publically available information for the respective municipalities

7.3 Collaboration Between Energy Cooperatives and Communities: Neue... 273

Table 7.38 Key characteristics of the 18 cities and municipalities that are members of NEW eG

Key characteristics	18 Municipalities
Position in the region	Municipalities of up to 42,000 citizens and up to 216 square kilometres, located in the rural area of North Bavaria
Political self-understanding	Administrative entities with different political backgrounds, including mayors belonging to the CDU, the SPD and electoral coalitions.
Motivation for founding the cooperatives	They want to become independent from fossil fuels and large energy companies by producing their own energy. They also want to involve citizens in their renewable energy project development activities.

for its water supply and wastewater services. Together with the municipal utility of Flossenbürg, it also operates two photovoltaic plants with a capacity of 4.5 megawatts peak (Markt Floß 2015). Especially Stadtwerke Grafenwöhr played a key role in the collaborative process as, along with the former mayor of Grafenwöhr, the managing director of the Stadtwerke was the initiator of the whole idea of founding the two energy cooperatives.

Table 7.38 summarises the key characteristics of the 18 involved municipalities that are key collaborating partners of both energy cooperatives.

7.3.3.2 Collaborative Interaction Model

Figure 7.7 illustrates the particular collaborative interaction model between NEW eG and Bürger eG, as well as between involved actor groups.

As can be seen, the interaction model between NEW eG and Bürger eG has both internal and external collaborative dimensions. The 18 municipalities and cities, including the two associated municipal utilities, collaborate within NEW eG, as they are all cooperative members and represented on its managerial or supervisory board. Further, the communities have another kind of internal collaborative relationship with citizens within NEW eG and the citizens cooperative Bürger eG, which is also a member of NEW eG and represented on its supervisory board. At the same time, NEW eG is part of the supervisory board of Bürger eG and is also a member of the cooperative. Bürger eG is a place where all citizens, being members, can collaborate with each other. In addition, Bürger eG and NEW eG hold, as interrelated but autonomous entities, an external collaborative relationship, as Bürger eG and NEW eG are both direct project owners in some renewable energy projects.

In the following sections, I seek to illustrate how the special constellation created by these actors—including municipalities, associated municipal companies and citizens, on the one hand, and the combination of an internal and external collaborative relationship between them, on the other—has become the basis for mobilisation of resources and creation of new societal norms and guidelines for a new energy structure in their focus region.

```
┌─────────────────────────────────┐              ┌─────────────────────────────────┐
│          Bürger eG              │              │            NEW eG               │
├──┬──────────────────────────────┤              ├──────────────────────────────┬──┤
│  │   Executive board            │              │   Executive board            │  │
│  │   Civil representatives      │              │   Mayors of member communities│  │
│I │                              │              │                              │I │
│n │   Supervisory board          │   External   │   Supervisory board          │n │
│t │   Civil representatives      │ collaboration│   Mayors of member communities│t │
│e │   Municipal representatives  │              │   Director of municipal companies│e│
│r │   of NEW eG                  │   ←——→       │   Representatives of Bürger eG│r │
│n │                              │              │                              │n │
│a │   Members                    │              │   Members                    │a │
│l │   Citizens                   │              │   Communities                │l │
│  │                              │              │   Municipal companies        │  │
│  │                              │              │   Bürger eG                  │  │
└──┴──────────────────────────────┘              └──────────────────────────────┴──┘
```

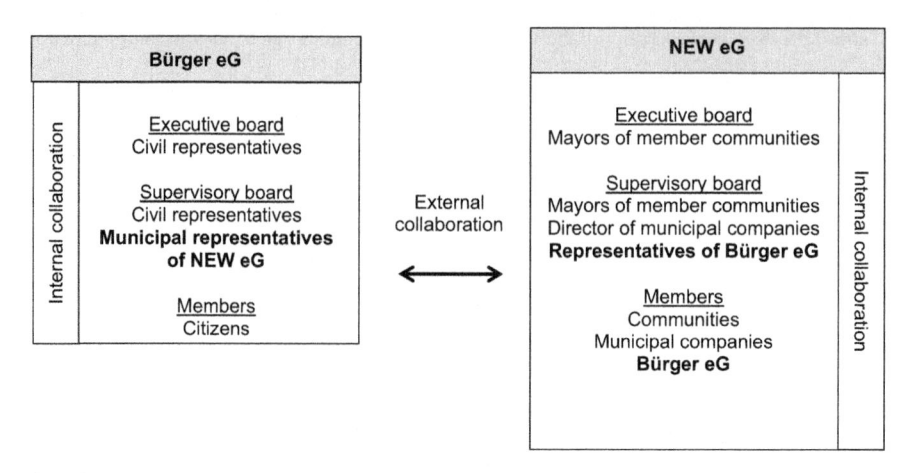

Fig. 7.7 Collaborative interaction model between NEW eG and Bürger eG

7.3.4 Mobilisation of Allocative Resources

Together, NEW eG and Bürger eG have mobilised important allocative resources, including actors, financial capital and technology projects.

7.3.4.1 Actors

Municipalities, citizens, experts, as well as personnel who manage and administrate the energy cooperatives represent the main actor groups that were activated through the joint work of Bürger eG and NEW eG. Further details about their roles and interrelationships are provided in the following.

Cooperative Members

NEW eG is a rather closed group of municipalities and municipal companies, unified in the aim of realising renewable energy projects. Meanwhile, Bürger eG seeks to unite all interested citizens of the region to become engaged in those projects implemented by NEW eG.

The statutes of NEW eG regulate that members of the energy cooperative need to primarily be legal agencies governed by public law, such as communities, cities or municipal utilities (Neue Energien West eG 2009a, p. 1). The only other actors that are, according to the statutes, allowed to join NEW eG are elected cooperative representatives of Bürger eG (Neue Energien West eG 2009a, p. 1). This ensures a direct business relationship between the two energy cooperatives. Table 7.39 displays the membership development of NEW eG, from its registration in 2010 through 2014.

7.3 Collaboration Between Energy Cooperatives and Communities: Neue... 275

Table 7.39 Membership development of NEW eG, 2010–2014

NEW eG	Dec 2010 (registration year)	Dec 2011	Dec 2012	Dec 2013	Dec 2014
Members	13	18	20	20	20
Total member shares	423	1035	1680	2260	2350

Source: Annual business reports, as explained in Sect. 5.2

By the end of 2014, NEW eG had 20 members, including 17 cities and communities, the municipal utility of Grafenwöhr, as well as the municipal energy and water provider Floß and Bürger eG. Compared to other energy cooperatives, NEW eG has only a few members and has exhibited almost no membership growth in 4 years. But here it is not the number of members that is important but the regional affiliation and close connection between them. It is not the aim of NEW eG to continually grow, even though other municipalities would like to join it. As outlined by the following quotation, the energy cooperative fears that the close relationship that has been developed between its municipalities could decrease if more communities were to join:

> 3.3 "The members know each other, are all neighbours. That is quite a different relationship, because, if the group broadens too much, the relationships will not be as tight. This is actually the main reason why the group shouldn't become too large" (R3N1, 215).

High entrance and cancellation barriers for becoming a member underline the closed character of NEW eG. A minimum member share costs 5000 Euros (Neue Energien West eG 2009a, p. 1), and the period of cancellation notice for members is, at 5 years, also quite long (Neue Energien West eG 2009a, p. 3).

In contrast to the closed approach of NEW eG, Bürger eG aims to unify and coordinate as many actors as possible. This is underlined by the following quotation:

> 3.4 "Actually everyone, yes, actually almost everyone can participate. This was the main intention" (R3N1, 9).

This open concept has made strong and continual membership growth possible, as seen in Table 7.40, which illustrates this development between 2010 and 2014.

As can be seen, the energy cooperative has already attracted 549 members by the end of its registration year of 2010, with this number more than doubling to 1303 by the end of 2014. In comparison, only 5% of all analysed renewable energy production cooperatives had more than 500 members in 2012 (see Sect. 6.2), so Bürger eG

Table 7.40 Membership development of Bürger eG, 2010–2014

Bürger eG	Dec 2010 (registration year)	Dec 2011	Dec 2012	Dec 2013	Dec 2014
Members	549	823	1132	1245	1303
Total member shares	6201	12,949	21,273	26,181	28,460

Source: Annual business reports as explained in Sect. 5.2

already had more members when it was founded than almost all other renewable production cooperatives in Germany.

According to its statutes, any natural or legal person can become a member of Bürger eG (Bürgerenergiegenossenschaft West eG 2013, p. 2), though membership is exclusively focused on individual citizens, as the communities and cities involved in NEW eG want to prevent their projects from being influenced by external companies, such as large energy providers. The core aim of the municipalities being members in NEW eG was to primarily integrate their own citizens into their plan to make themselves independent from fossil energy resources. Thus, they strongly supported Bürger eG in mobilising members. All involved mayors and the directors of the two municipal energy providers actively marketed the cooperatives' goals and visions in their region, presented planned projects in public meetings and talked to many people in the streets. According to the next statement, involved municipalities and representatives of Bürger eG did not expect such a strong level of affirmation from interested actors and have described the strong growth of Bürger eG within a relatively brief time frame as a "fast-selling item":

> 3.5 "And, as you can see, then we received so-called acceptance from our citizens in this region, and then this became a fast-selling item. Today one can say that it sells itself. At that time, we started with little and we said: "Yes, if we build one or two small plants, then we will be happy, then we will be satisfied". But we have really been overrun since then. If you can imagine, in the meantime we have collected 14 million Euros in 5 years" (R3N5, 75).

This perception underlines the high actor mobilisation potential of the collaborative interaction model created between NEW eG and Bürger eG.

Low entrance and cancellation barriers have further facilitated the increase of members and underline the cooperative's aim to activate as many actors as possible. The same as in the previous two case studies, the minimum member share of Bürger eG costs 500 Euros (Bürgerenergiegenossenschaft West eG 2013, p. 18), and members can withdraw from their membership at the end of each business year (Bürgerenergiegenossenschaft West eG 2013, p. 2). However, unlike the cooperatives of other two case studies, Bürger eG has no investment limit for individual members.

The representatives of the two energy cooperatives actively presented their work all over Germany, with the aim of disseminating the ideas and goals that stand behind NEW eG and Bürger eG. Due to this nationwide publicity for the energy cooperatives, new members were not only attracted from within the cooperatives' focus region but throughout Germany. As a consequence, about 20% of all members of Bürger eG live in regions throughout Germany. Thus, actors joined the cooperative independently of how far away they live from its focus region. As can be seen from the following quotation, Bürger eG and NEW eG have perceived this dynamic as a 'paradox', because their primary intention was to help other communities in setting up similar 'regional' concepts and not to acquire members from outside their focus region:

3.6 "The paradox has always been there, as we've realised that wherever we have appeared and have presented our model, people have said, "we are joining the cooperative, we will become members". This was a paradox. We were in Dresden, [...] we were in Upper Bavaria, Neuötting, we were in Regensburg. Wherever we were, we attracted members" (R3N5, 79).

The fact that the municipalities have mobilised a considerable amount of support from citizens outside their focus region can be viewed as an unintended consequence of their activities and may, further, stand in contradiction to their actual goal of almost exclusively integrating regional citizens in their project plans. However, during the interviews that I conducted, it became clear that this aspect of their growth was not perceived as a problem for the energy cooperatives in the end.

Managing Personnel

As of 2014, the managing board of Bürger eG consists of two private persons (private insofar as they do not represent any institution or other organisation within the energy cooperative). They both, however, have a political background. One of them is a member of the city council in Grafenwöhr, while the other is a member of the green party—Bündnis 90/die Grünen—and spokesman for the district Neustadt. The supervisory board of Bürger eG consists of four persons, one of whom is the mayor of the city of Pressath, who also represents the city in NEW eG. The other three are private persons and have no political background.

All cities and communities comprising the members of NEW eG are directly and exclusively represented by its mayors and/or their municipal utilities. Each member of NEW eG is also part of the cooperative's executive or supervisory board, which is a way of seeking to ensure that all involved communities and cities have the same level of influence within the energy cooperative. Accordingly, as of 2014, NEW eG is managed by the three mayors of the cities of Grafenwöhr, Eschenbach and Neustadt am Kuhn, and the supervisory board of NEW eG is formed by 19 representatives including:

- The mayors of the 14 remaining member cities and communities who are not part of the executive board of the energy cooperative;
- the directors of the municipal energy and water provider Floß and the municipal energy provider Grafenwöhr; and
- two members of the executive board and one member of the supervisory board of Bürger eG.

The strong and immediate presence of mayors shows that NEW eG has mobilised important political decision takers and turned their focus towards the development of renewable energy projects in their own municipalities. As illustrated by the following quotation, the mayors see themselves as an ideal group of actors for managing renewable energy projects because, in their opinion, they have greater potential for realising such projects than other actor groups:

> 3.7 "And I mean, a mayor has more possibilities than a private citizen. Hence, especially with regard to the initiation of projects such as large photovoltaic parks and so on, of course the mayor is the ideal contact person" (R3N3, 21).

In fact, as discussed below, mayors have relatively easy access to potential projects and information, which are important authoritative resources for becoming a structuring actor in the field of renewable energy.

However, the mobilisation of managing personnel with a focus on political decision takers has also involved challenges. As indicated by the next statement, it has been critically noted that one of the members of the cooperative's managing board stayed on even after he was no longer mayor:

> 3.8 "For me it was problematic that this person [...] is now second member of the executive board, even though he is no longer a mayor. I did not like that, because it has such a bad reputation such as [...] exists with regard to municipal energy providers. Once, there was an article in Spiegel magazine on how municipal energy providers have become—how do you say that?—a sideline area for politicians to obtain positions. I don't want such ideas to be associated with the energy cooperative" (R3N3, 127).

Yet, this personnel exchange would have led to a loss of important know-how and of a person who has been enthusiastically committed to the energy cooperative. Personal commitment is an important resource, especially since the limited available time of incumbent mayors became another challenge for accomplishing the work of NEW eG, as they have many political responsibilities and, thus, almost no time for additional work. Both energy cooperatives grew to a level in terms of members, capital and projects at which the executive boards could not handle the work anymore on a voluntary basis. This illustrates the strong interdependency between the mobilisation of capital, personnel-management, and technical project resources. The more capital and projects are activated, the more time and responsibilities are entailed for managerial work. As two interviewees remarked:

> 3.9 "Then we simply saw that the workload had become so extensive that the mayors could no longer manage it, that is impossible" (R3N2, 89).

> 3.10 "And now we have, I believe, 20 million Euros invested, and that cannot be managed as a side job" (R3N1, 27).

As illustrated by the following quotation, however, the limited managerial availability of involved persons was seen as the only limitation for the cooperatives' planned business activities:

> 3.11 "These boundaries are only momentary, so we say: "We don't have the staff to manage everything; otherwise we have no limits"" (R3N4, 115).

As a consequence, they have employed a paid managing director since January 2014.

7.3 Collaboration Between Energy Cooperatives and Communities: Neue... 279

Experts and Administrative Personnel

The municipal utilities of Grafenwöhr and Floß are experts in the field of providing municipal services. Especially Stadtwerke Grafenwöhr accompanied the foundation and business activities of the energy cooperatives with important know-how and, within their first 2 years, personnel from the utility took over the administrative work for both energy cooperatives, and part of the administrative work, such as member-related tasks, is still done by personnel from Stadtwerke Grafenwöhr. This administrative support was crucial in order to ensure a quick and professional business start for Bürger eG and NEW eG.

The involvement of experts seems to have had a positive effect on attracting cooperative members. As indicated by the following quotation, expert opinions played an important role in motivating actors to join Bürger eG:

> 3.12 "But when we presented the model, most of the time, the chairman of the city, was present at the meetings, and he explained everything and then we received the response, "Yes, that is a great thing, we will join"" (R3N5, 55).

7.3.4.2 Financial Capital

In terms of capital, NEW eG and Bürger eG belong among the largest renewable energy production cooperatives in Germany, as revealed by the quantitative assessment of energy cooperatives presented in Sect. 6.2. Tables 7.41 and 7.42 outline the economic development of NEW eG and Bürger eG from their registration year of 2010 through 2014.

As displayed in the tables, NEW eG had attracted almost 10 million Euros total capital, and Bürger eG had mobilised about three million Euros capital in 2010. The two cooperatives increased their total capital to 22.6 million Euros (NEW eG) and 14.6 million Euros (Bürger eG) within 4 years.

Table 7.41 Economic development of NEW eG, 2010–2014

NEW eG	Dec 2010 (registration year)	Dec 2011	Dec 2012	Dec 2013	Dec 2014
Total capital (Euros)	9,176,246	14,161,798	16,765,858	18,176,710	22,552,490
Equity capital (Euros)	2,136,689	5,277,544	8,731,585	11,525,892	11,931,308
Member shares (Euros)	2,115,000	5,175,000	8,400,000	11,300,000	11,750,000
Borrowed capital (Euros)	7,027,795	8,808,767	7,796,712	6,589,281	10,520,791
Equity ratio (%)	23	37	52	63	53
Net income (Euros)	20,305	95,420	270,515	156,580	98,118

Source: Annual business reports, as explained in Sect. 5.2

Table 7.42 Economic development of Bürger eG, 2010–2014

Bürger eG	Dec 2010 (registration year)	Dec 2011	Dec 2012	Dec 2013	Dec 2014
Total capital (Euros)	3,148,755	6,622,602	11,056,471	13,630,901	14,575,435
Equity capital (Euros)	3,142,607	6,597,894	11,021,414	13,573,712	14,545,140
Member shares (Euros)	3,127,000	6,484,500	10,733,000	13,399,000	14,439,000
Borrowed capital (Euros)	930	9883	23,031	46,989	17,595
Equity ratio (%)	100	100	100	100	100
Total revenues (Euros)	19,528	141,313	379,434	209,636	301,182
Distributed dividend (%)	3.8	4.3	3.8	3.0	2.5

Source: Annual business reports, as explained in Sect. 5.2 and information directly provided by the energy cooperative

Nominal Capital

By the end of 2010, Bürger eG members had already bought just over three million Euros in shares and, 4 years later, this figure had increased to about 14.4 million Euros. The following quotation points out that one of the main functions of Bürger eG, which transfers almost all of its equity capital to NEW eG, has been to attract nominal capital for the projects developed by the latter:

> 3.13 "The cooperative [Bürger eG] has been founded simply for collecting money" (R3N1, 9).

In this way, NEW eG can dispose of a high amount of equity capital, despite only having 20 members. Table 7.43 outlines the distribution of member shares within NEW eG, as of December 2014.

By the end of 2014, 93% of member shares of NEW eG—almost 11 million Euros—were provided by Bürger eG. By this point, involved cities and communities only invested a total 700,000 Euros, and 11 of the 18 member communities and cities did not provide more than 20,000 Euros. As revealed by the following statement, Bürger eG and its citizen members thus provided a new and innovative form of access to capital for the involved communities and cities, as many communities that have been engaged in NEW eG had been unable to dispose of enough nominal capital in order to develop renewable energy projects on their own:

> 3.14 "An advantage of the citizens cooperative is, I'll put it this way, the possibility for us to get more capital—in the long run. Each time municipalities start a new project, they have to raise capital. And that is, of course, not so easy in a municipality, especially because we also have a lot of small municipalities, which do not have so much of their own capital, in order to—let's say—immediately bring 50,000 Euros capital into the cooperative. And so the idea was that we practically organise our capital mainly through the citizens cooperative" (R3N2, 81).

As revealed by the next quotation, the strong increase of capacity to raise capital motivated involved municipalities to realise larger energy production projects:

> 3.15 "And that it developed so fast, we did not expect that. You start with a roof-top photovoltaic plant, and then it developed so fast that we had so much money that we dared to build a whole open land photovoltaic park" (R3N1, 9).

7.3 Collaboration Between Energy Cooperatives and Communities: Neue... 281

Table 7.43 Distribution of member shares attracted by NEW eG, as of December 2014 (Neue Energien West eG 2009a)

NEW eG members	Member shares	in Euros
Gemeinde Schwarzenbach	1	5000
Stadt Neustadt am Kulm	1	5000
Gemeinde Weiherhammer	1	5000
Gemeinde Schlammersdorf	2	10,000
Kommunalbetrieb Floss	2	10,000
Stadt Kemnath	2	10,000
Gemeinde Speinshart	3	15,000
Gemeinde Vorbach	3	15,000
Gemeinde Kirchenthumbach	4	20,000
Gemeinde Trabitz	4	20,000
Stadt Auerbach	4	20,000
Stadt Pressath	10	50,000
Stadt Eschenbach	10	50,000
Stadt Weiden	10	50,000
Stadtwerke Grafenwöhr	15	75,000
Gemeinde Altenstadt/WN	15	75,000
Markt Parkstein	20	100,000
Gemeinde Kastl	20	100,000
Stadt Grafenwöhr	27	135,000
Bürger eG	2196	10,980,000
Total	2350	11,750,000

Accordingly, NEW eG implemented its first large solar park, with a capacity of about two megawatts peak, in December 2010—that is, less than a year after its registration. This strong capital growth has, thus, had a direct impact on the mobilisation of technology projects.

Borrowed Capital

As can be seen from Table 7.41, NEW eG was already able to borrow around seven million Euros in its first business year, an amount that increased to 10.5 million Euros in 2014. The circumstance that the energy cooperative was founded and has been managed by municipalities facilitated access to borrowed capital, as underlined by the following statement:

> 3.16 "The municipalities already have relations there or loans with the banks. This is simple, if the municipalities stand behind [the cooperative], everything is much easier" (R3N1, 83).

As shown in Table 7.42, Bürger eG has not acquired any borrowed capital, because its primary aim is to provide equity capital to NEW eG.

Foundation Capital for the Cooperatives

Stadtwerke Grafenwöhr coordinated the complete foundation process of the two energy cooperatives. It paid for the founding costs and provided expertise, rooms and other important foundational resources at no charge. Along with Stadtwerke Grafenwöhr, several municipalities provided nominal capital before NEW eG and Bürger eG actually became registered. In this manner, three photovoltaic projects, with a capacity of 50 kilowatts peak, could already be installed by December 2009. The implementation of projects at such an early stage (the cooperatives were registered in January and August 2010) helped the two energy cooperatives to have a good business start.

Profit and Distributed Dividends

Based upon its membership in the project-developing NEW eG, Bürger eG generated revenues upwards of 379,000 Euros over the course of 4 years (see Table 7.42). As a result, since it became registered, Bürger eG was able to distribute an annual dividend to its members that ranged between 2.5 and 4.3%. As revealed by the following statement, NEW eG had officially announced offering at least 3% dividends to Bürger eG and its citizen members:

> 3.17 "So we had already made promises before we had actually decided internally that the minimum distribution would be 3% per year" (R3N5, 73).

The circumstance that NEW eG 'promised' a certain amount of dividend created a totally new business relationship between municipalities and citizens where citizens can commission municipalities to develop renewable energy while, at the same time, also expecting to receive at least a small dividend. In this sense, profit generation seems to have played a crucial rule for the functioning of the collaborative relationship between the two energy cooperatives, the importance of which is highlighted by the following quotation:

> 3.18 "This is not just based on good will but also the idea of making a bit of profit. It is of course good, if a certain yield is available; otherwise the whole project would not work. That one looks into things, that one is able to bring it down as far as possible, but in the end it must be profitable; otherwise the business will not work" (R3N1, 143).

However, the strong allocation of nominal capital from citizens and their expectation of receiving a dividend has also challenged both energy cooperatives. The continual growth of members and member shares within Bürger eG has obligated NEW eG to continually identify new projects in order to invest their money and generate a certain level of revenues. As indicated by the next interviewee, NEW eG feared that low or no return on investments could lead to cancellation of memberships in Bürger eG:

7.3 Collaboration Between Energy Cooperatives and Communities: Neue... 283

3.19 "Yes, we already met some kind of boundaries when we had a lot of money from the citizens or from the citizens cooperative but no projects that we could feasibly realise. [...] Then it will of course be a problem to distribute the 3% dividend. And then, when this financial model is turned backwards, it will lead to the circumstance that capital investors may vanish" (R3N2, 85).

As suggested by the next quotation, one motivating factor for actors to become members of the energy cooperative may have been that other investment options were unattractive due to low interest rates:

3.20 "Of course, everyone is currently convinced [to invest in us] by the current interest rate structure. But what happens in 5, 6, 7 years—or even in a shorter time—when interest rates rise again above our current rate of return? Then, of course, the question is: [...] will members nonetheless leave their money in the cooperative as a sort of green conscience" (R3N4, 151)?

Hence, as with the cooperatives examined in the previous two cases, not only reduced project profitability but also improvement of investment alternatives could make it more challenging for the two energy cooperatives to mobilise members and capital in the future.

7.3.4.3 Technology Mix

Installed Projects

NEW eG implemented 25 solar power projects between November 2009 and June 2014, including 18 rooftop photovoltaic plants with a total capacity of about 617 kilowatts peak and seven solar parks, with a total capacity of about 15 megawatts peak. Only a very small share of about 0.3% of all other analysed renewable energy production cooperatives had installed more than 10 megawatts peak solar capacity by May 2015. The expected power yield of the NEW eG parks is around 15,200 megawatts hours per year. Compared to the total household power demand of all involved communities (see Sect. 7.3.1 for more details), the cooperative was able to establish a renewable power supply for about 10% of the targeted citizens within four and a half years.

All rooftop photovoltaic plants except one—Grafenwöhr Megaplay—are installed on top of municipal buildings, such as fire stations, school buildings or buildings that belong to water purification plants. Three of the seven solar parks are installed in the municipalities of Pressath and Speinshart, and the respective city and community is also the owner of the land occupied by them. In this sense, the energy cooperative has functioned as a tool for its member municipalities to implement new technical energy resources on their public property.

Only three of the seven solar parks are located in the focus region of the energy cooperative. One of them is directly located on the border of the focus region, in the neighbouring districts to Tischenreuth and Neustadt an der Waldnaab, whereas three are installed in other districts of Bavaria, including Kelheim and Kitzingen, and they all lie within a radius of around 100 kilometres to Neustadt an der Waldnaab. One

Table 7.44 NEW eG project locations according to installed capacity, as of January 2015 (based upon Neue Energien West eG 2014)

Location of NEW eG's installed capacity	Capacity in kWp	Capcity in percent
Within or next to the focus region	9082	58
Outside the focus region	6687	42
Total installed capacity	15,769	100

park is located in the Kyffhähser district of the Bundesland of Thuringia, about 290 km to the north of Neustadt an der Waldnaab. Table 7.44 summarises the power capacity installed within and outside of the focus region, as of January 2015, according to which we can see that about 42% of the generated power is actually not produced in the focus region of two energy cooperatives.

That a considerable amount of power is produced outside the focus region of NEW eG and Bürger eG may be seen as standing in contradiction to their actual aim of supporting achievement of a 100% regionally produced and renewable energy supply. This goal requires a change of the energy technology mix within all involved municipalities. During the interviews I conducted for this study, it became clear that particular events and conditions led to this development. The large solar parks that are located in other districts were all installed between November 2013 and May 2014. During this time, NEW eG faced the challenge to invest a great amount of capital that had been collected by Bürger eG in a relatively short amount of time. The aim of the NEW eG to distribute at least a three-percent dividend to its members annually had pressured the cooperative to invest its available money sooner rather than later in applicable projects. Otherwise, it may not have been able reach this goal. At the same time, the energy cooperative was confronted by planned, but as yet uncertain, legal changes to the German Renewable Energy Act (EEG) (Erneuerbare-Energien-Gesetz—EEG 2014). According to the following interviewee, NEW eG tried everything it could to invest its available capital before the expected regulatory amendments would likely reduce returns on investment for renewable power projects:

> 3.21 "Yes. This became a problem, especially in 2014, because of the fact that the EEG feed-in tariff intended just for large power plants would melt away so quickly, and we were given so-called time steps, [which led to questions]: When can I sign the contract at the latest? When must everything be finalised and tested so that it can still be built by the last day of the month when the EEG runs out? And that led to at least a few sleepless nights" (R3N4, 67).

In 2012, the energy cooperative had the option to buy a large solar park with an investment volume of about 31 million Euros, located in the municipality of Grafenwöhr (Neue Energien West eG 2013). However, the supervisory board members of NEW eG decided against purchasing the park, because they questioned its forecasted profitability. Consequently, NEW eG ended up searching for other profitable project alternatives that would be available in the short term and found them outside their own region, such as in Thuringia. The location and geographical character of such projects, including available space and incidental solar radiation, have a high impact on their eventual power yield and possible revenues. Such geographical selection criteria are, thus, strongly aligned to project feasibility, as confirmed by the following interviewee:

> 3.22 "So buying parks in those respective communities was only possible because the right spatial conditions existed there" (R3N2, 123).

Future Goals

As outlined in Sect. 7.3.1, solar and wind power are playing a key role in the process of attaining 100% regional and renewable energy supply for regions in Germany. For example, 44% of the total power demand of the western part of the district of Neutstadt an der Waldnaab, involving 11 of the 18 member municipalities of NEW eG, could be covered by solar power, and 43% could be covered with wind power. Thus far, the energy cooperative has exclusively focused its project activities on photovoltaic technology but, as indicated by the following quotation, is aware of the importance of wind power and has been planning to become engaged in wind-power projects:

> 3.23 "Yes, the future depends on wind power. If we can realise wind power plants, that is really for the moment the most important goal. And this is the next major project: to speed-up our involvement with wind power" (R3N1, 231).

Two reasons were mainly given to explain why the cooperative has not yet realised a wind park. First, the development of wind parks involves higher project risks than solar or other possible projects. The intensive preparation time prior to project start requires more capital and more time than other alternatives. Second, the realisation of wind park projects in the region has been challenged by an obstructive district office representative. As illustrated by the following quotation and also stated in local newspaper articles (for example, Raab 2015), the director of the rural district office of Neustadt an der Waldnaab appears to have been against wind projects and has been trying as much as possible to prevent implementation of wind parks in the district:

> 3.24 "The hurdles, the regulatory hurdles, are also quite high here. And our regional district office is very restrictive, in my opinion. So we do not have a wind-friendly district office—that's my personal opinion. This will be confirmed by the other mayors, too. [...] I was also part of the meetings at the regional association, and I simply realised that the chairman—the former chairman – head of the regional district office [...] was simply employing delaying tactics. He wanted to remove it [from the agenda ...]. And then it would be put aside" (R3N3, 53).

Consequently, by 2009 no wind mills had yet been installed in the western part of the district (Stadtwerke Grafenwöhr 2011, p. 70). Delaying tactics on the part of the district office and additional restrictions passed by the Bundesland of Bavaria also made the search for applicable project locations more challenging and more time consuming than initially expected.

In addition to an uncooperative regional policy, national regulations have further impeded the activities of the two energy cooperatives. As demonstrated by the following quotation, and similar to the situation faced by the cooperatives examined in the other two case studies presented above, the potential coverage of energy cooperatives under the KAGB and potential amendments to the EEG were perceived as major challenges at the time of conducting my interviews:

> 3.25 "So, on to the current legal situation. Especially the EEG, especially the KAGB, these are gigantic topics for us where we say: How can we do this? How can we make this conform to our situation" (R3N4, 103)?

In general, Bürger eG is not operationally active and functions almost exclusviely as a capital provider by being a member of NEW eG. Hence, the cooperative steering boards decided to register Bürger eG under BaFin, which was a very time-consuming and complex process for all involved actors. Amendments to the EEG reduced the profitability of certain project types (Erneuerbare-Energien-Gesetz—EEG 2014). Due to all of these regulative challenges, NEW eG now sees its planned engagement in wind power development as a midterm goal.

For the future, the two energy cooperatives envision becoming engaged in several other technology-related project fields. According to the following interviewee, one of their concrete aims is to take over their regional power distribution grids:

> 3.26 "Of course, to take over our own grid structure would be a dream of mine. To take over the grid and then really offer a cooperative tariff for all and really try to promote energy self-sufficiency here at least in Northern Bavaria or in the counties, in these three counties: Tirschenreuth, Neustadt/Kulm, Bayreuth. That would be great" (R3N3, 121).

Several of the member municipalities have already informed themselves about how to achieve the acquisition of their power grids. However, as this is also a highly complex process and the needed know-how does not yet exist among them, the municipalities have defined the purchase of their power grids as a long-term goal.

Electro-mobility has been strongly pushed by one of the involved mayors, who has already founded a special initiative for electromobility in which 13 municipalities around Neustadt am Kulm are working together to foster the topic. Two thirds of the involved municipalities are also members of NEW eG. He aims to purchase the required power in direct cooperation with NEW eG, which may not be the main coordinator of the project but seems to be strongly facilitating the foundation and functioning of this new alliance around electro-mobility. As one interviewee explains:

> 3.27 "Promoting electromobility, for example. [...] This is another form of intercommunal cooperation, but they want to work together with NEW: the network wants to strengthen electromobility and may need NEW as an electricity supplier" (R3N3, 11–15).

The energy cooperative would like to become engaged in the development of small district-level heating systems. However, so far no applicable projects have been identified:

7.3 Collaboration Between Energy Cooperatives and Communities: Neue... 287

3.28 "Heat and heating grids is at the moment a big topic in Bavaria. We have taken a look at it here. We tried it because of the spatial distances involved, but I have yet not found a project where we can implement it" (R3N4, 111).

Energy efficiency does not seem to be a concrete topic yet, even though it as been listed as a business goal in the cooperative statutes of Bürger eG (Bürgerenergiegenossenschaft West eG 2013, p. 1). Table 7.45 lists the installed and planned projects of NEW eG in relation to the municipal energy transition goals outlined in Sect. 7.3.1.

In sum, the inter-municipal energy cooperative NEW eG, in cooperation with the citizens energy cooperative Bürger eG, had by the end of December 2014 installed more solar power capacity than 99% of all other renewable energy production cooperatives in Germany and were already able to supply 10% of their private household power demand with their own produced power. In order to further support their goal of developing a 100% renewable energy supply in all member municipalities, they would have to maximise power production within their region and become engaged in wind-power projects. However, the rapid accumulation of a large amount of capital that had to be invested within a short amount of time, impending amendments to national regulations that would reduce returns on relevant project investments in the near future, and a lack of realisable and profitable projects within their region motivated the energy cooperatives to establish their largest projects outside their defined focus region. The complexity of wind power and a resistant district administration has prevented NEW eG from developing wind parks so far. Other innovative project ideas, such as small district-level heating systems or energy efficiency measures, do exist but have been problematic to realise due to missing know-how, limited time and limited concrete project options.

Table 7.45 Planned and installed projects of NEW eB and Bürger eG in relation to the municipal energy transition goals

Technologies that are part of the municipal energy transition goal	NEW eG project status (as of December 2014)
Photovoltaic power projects	18 projects installed
Solar power parks	7 projects installed
Wind power	Planned for the midterm
Heat technology	Small district-level heating systems are planned for the long term
Energy efficiency	Generally included among the cooperatives' business goals but not activated yet
Electro-mobility	Planned by another coalition that is directly linked to both energy cooperatives

7.3.5 Mobilisation of Authoritative Resources

The relationships between municipalities, municipal utilities and citizens within the energy cooperatives NEW eG and Bürger eG have created several kinds of authoritative resources, including (1) an organisational identity for both energy cooperatives; (2) a high level of publicity for them; (3) a generally positive image; (4) their courage in project development; (5) trust and respect between involved actors, and trust in the business model of the energy cooperatives; (6) internal and external transparency; (7) generally high acceptance for their energy projects; (8) access to potential projects, information and decision takers; (9) access to new ideas; (10) as well as the creation or exchange of expertise and know-how. These resources are detailed below.

7.3.5.1 Organisational Identity

The organisational identity of both energy cooperatives is closely aligned to the 18 involved communities and two municipal utilities. Most of them are cooperative founding partners, and all of them are represented on the steering board of NEW eG. The communities are responsible for project management and organisational strategy and form the focus region in which both energy cooperatives aim to do their core business. In this sense, NEW eG and Bürger eG mainly define themselves through these 18 cities and municipalities as well as the two municipal utilities.

7.3.5.2 Publicity

The strong membership growth of Bürger eG—1303 members within four and a half years by 2014 (see Table 7.40)—implies that the two energy cooperatives have achieved a high level of publicity. As vividly expressed by the following interviewee, the successful marketing of Bürger eG and NEW eG has been a result of the personal dedication and enthusiasm of the cooperatives' managers:

> 3.29 "This is madness: how much they are involved, with how much passion they market our cooperation" (R3N4, 125)!

From the beginning, they presented the concept of the two energy cooperatives and their aim of achieving a regional energy autonomy to the local and national public. Mayors and municipal utility directors played a key role here, as they not only organised information events within their municipalities but attended congresses, workshops and meetings in many other regions. Involved actors described their personal dedication as 'literally creating an information wave', as described by one interviewee:

> 3.30 "For me this was one of the triggers, through this information wave that we have set off. We went from community to community, held presentations several times a year about our

concept, and then the registrations came rolling in, the shares, it was a stream without end. So I was totally surprised when I heard the numbers [...]" (R3N3, 37).

In addition to public meetings, the two energy cooperatives achieved a strong presence in the local press. A review of the archives of the newspaper Der neue Tag—Oberpfälzischer Kurier—one of the most dominant newspapers of the region—has revealed that NEW eG was mentioned in 194 articles between 1 January 2009 and 31 December 2013. In most of these articles, one of the mayors or directors of the two municipal utilities were also mentioned. I thus assume that the involvement of representatives of the municipalities was an important criterion for the press to publish articles about the energy cooperatives. In turn, the mayors and utility directors appear to have used NEW eG and Bürger eG as a tool for publicly presenting and disseminating their political goal of actively fostering renewable energy, as 63% of all analysed articles describe the production of renewable energy as a municipal goal. These marketing activities can be seen as a win–win situation for the communities and the energy cooperatives.

7.3.5.3 Image

Together, the two energy cooperatives have established an image of pioneering but seriously operating organisations. According to their own organisational descriptions (for example, Neue Energien West eG 2009b), they understand themselves to be a collaborative project with a pioneering character. And they also seem to be perceived this way as, for example, newspaper articles have described them as trend-setting (for example, bjp 2009). At the same time, both energy cooperatives have sought to create a serious image, as illustrated by the following quotation:

3.31 "Our goal and public image is absolute seriousness" (R3N3, 45).

Being serious here means being trustworthy and operating on a solid basis (Dudenverlag 2015b). As outlined in Sect. 7.3.5.5, the energy cooperatives have established trust in and credibility for their business model, with the fact that they have mobilised the support of more than 1300 citizen members seeming to confirm this assumption.

7.3.5.4 Courage

The realisation of renewable energy technology projects can involve several risks, such as technical implementation, financial or regulative ones. Courage—based upon a strong confidence in one's own capabilities—is thus an authoritative resource that can help in starting or continuing projects that involve such risks (Dudenverlag 2015a). As revealed by the following interviewee, courage was seen as one of the main advantages of the two energy cooperatives in attracting support:

3.33 "Yes, the citizens simply trusted us [...]" (R3N1, 139).

It is assumed that the courage of the members of the executive boards of the cooperatives helped in implementing a high number of projects—in total 25—and for focusing on larger more investment-intensive projects, such as solar parks, at a very early stage. Their first solar park, with a capacity of two megawatts peak, had already been implemented 1 year after cooperative registration. Bürger eG had the courage to attract about 14.4 million Euros capital from members by the end of 2014. The courage of the two energy cooperatives appears to be especially based upon two other authoritative resources: trust and respect between involved actors.

7.3.5.5 Trust and Respect

Trust and respect between involved communities, cities, municipal companies and civil representatives of Bürger eG provided an important basis for the successful work of the two energy cooperatives. Two kinds of trust can be differentiated: (1) Trust in the business activities of the two energy cooperatives, which involves a high degree of credibility for their work. (2) Trust between involved actors and respect for each other.

Trust in the Business Model of the Energy Cooperatives

In the opinion of representatives from NEW eG and Bürger eG, they were able to win the trust of citizens. More specifically, they suppose that citizens who support them have confidence that the energy cooperatives are capable of actually realising the business goals that they proposed to the public. As one interviewee put it,

3.32 "We simply trust ourselves. That is our greatest advantage" (R3N1, 119).

The assumption that citizens have developed trust in the business model of the two energy cooperatives is underlined by the strong membership growth of Bürger eG. With its 1303 members by December 2014, the cooperative attracted almost six times more actors than the overwhelming majority of all renewable energy production cooperatives in Germany, 80% of which only have up to 200 members (see Sect. 6.2). These members then confided about 14.4 million Euros of their private capital to the energy cooperative. The strong capital flow is described by one cooperative manager as if each day citizens want to provide money to it, underlining the level of trust that they have won:

3.34 "I can only say that every day people come to us and want to deposit money—really" (R3N5, 133).

NEW eG and Bürger eG assume that many citizens particularly perceive the two energy cooperatives as trustworthy and credible organisations because the mayors of the involved municipalities are actually the ones who in part manage and coordinate the renewable energy projects. The 17 involved mayors seem to enjoy a high degree

7.3 Collaboration Between Energy Cooperatives and Communities: Neue... 291

of credibility among the public, which is further transferred to the activities that they support, as explained by one interviewee:

> 3.35 "If my mayor is active in this, then it must be good. So this function of the NEW being an ideal for the citizens cooperative, yes, that is based so much on just people. We have, I'll say, the 50+ generation as members who have invested large amounts of money, and they really trust the word of their mayor. And because of that, you really have to say, here a great level of trust was developed; these are not just any people, but the mayors. The municipalities are behind it" (R3N4, 73).

The assumption that regional mayors have played a key role in building trust in the activities of NEW eG and Bürger eG is in line with the results of a survey of the Bertelsmann association from 2008, according to which credibility, assertiveness and being close to citizens are cited by German citizens as the most important characteristics of a mayor (Bertelsmann Stiftung et al. 2008, p. 61). Furthermore, 78% of German citizens say that they are pleased with the work of their mayors (Bertelsmann Stiftung et al. 2008, p. 63) and, thus, seem to be of the opinion that their mayors also possess such authoritative resources. For the general trust-building process it was an advantage that those mayors participating in NEW eG had different political backgrounds, involving mayors from the SPD, CDU and politically independent coalitions (see Table 7.37). This political diversity helped to build a trustful relationship with a broader group of citizens.

Trust and Respect Between Involved Actors

Trust, respect, and having a similar attitude were named as authoritative resources that have enabled close collaboration between all member municipalities, despite their different traits, including size, population, political composition of councils and political backgrounds of current mayors.

> 3.36 "This is a very colorful group, politically. However, they are on the same wavelength with regard to certain points, which is actually the secret why this works, I think. And also hold this necessary trust for each other" (R3N3, 99).

This degree of trust between the involved actors is especially based upon the fact that many of the mayors already knew each other personally. In the past, 10 of the 18 municipalities had been continually working together in the context of the VierStädtedreieck initiative, which aimed towards realising local potential and has been supported by the EU LEADER programme. The experiences that they went through with each other while working together in the past benefitted them later in being able to rely on each other's credibility and capabilities, as outlined by the following interviewee:

> 3.37 "The people know each other. It's almost a blind belief [in each other], as we rely on each other. [...] So really these agreements, this reliability, is actually a huge advantage" (R3N4, 91).

According to the following quotation, those member municipalities that joined the cooperative at a later stage seemed to further strengthen the trust-filled working atmosphere, such that the resource of trust was actually even improved through the activities of NEW eG:

> 3.38 "In the case of a new mayor, he/she has to become acquainted with it. And you can see that the one who has already been involved for 5 years, since the founding, knows how it works. And there is often a new one who exercises the right to ask critical questions, who says: "How does it work? Why is it this way?"" (R3N4, 95).

One important consequence of the resource of trust between the municipalities is fast and efficient work within the cooperatives' executive boards. As revealed by the next statement, the 14 mayors that form the supervisory board of NEW eG allow the three mayors that form the executive board of NEW eG to coordinate the investment budget and planned projects in the most self-contained manner possible:

> 3.39 "Often, the three members of the board are given a free hand in decisions that are made over seven-digit amounts. There is already a certain level of trust now, one really has to say" (R3N3, 57).

As an example for their fast work, NEW eG installed 65% of their total capacity, involving six of their rooftop photovoltaic plants and four of their photovoltaic plants, within a timeframe of only 10 months: between September 2013 and June 2014. This way NEW eG was able to establish an additional 10 megawatts peak capacity before national amendments to the German renewable energy act reduced the feed-in tariffs and returns on investment for several renewable energy project types (Erneuerbare-Energien-Gesetz – EEG 2014).

Respect was especially built between the political representatives of NEW eG and the civil representatives of Bürger eG and was mentioned by that latter as an important resource for smooth communication between the two energy cooperatives:

> 3.40 "Yes, in the beginning it was strange, because you just think that the mayors are a somewhat far off beyond you. But, in the end, they were the same as everyone else. [...] So we are taken seriously, we can also talk and everything. We often make suggestions." (R3N1, 95)

However, one situation shows that trust and respect can also be endangered when cooperative members pursue individual interests that stand in competition with or in contradiction to the interests of the energy cooperative. In 2012, a large solar park of 36 square kilometres and an investment volume of about 31 million Euros was planned in the member municipality Grafenwöhr. The park was implemented by the project developer WIRSOL Solar, which planned to sell it to others. With 15 megawatts peak, the park has a capacity that is as large as the complete capacity that had been installed by NEW eG by June 2014 (Neue Energien West eG 2013). NEW eG was interested in buying the park, since it was located in one of its municipalities. But members of the supervisory board questioned the forecasted profitability of the park, so ongoing negotiations with WIRSOL Solar were supposed to be extended until the questions could be clarified. However, in the meantime, the park was already bought by another inter-municipal energy-oriented initiative in which two municipalities—the city of Weiden and the municipal utility Floß—are involved that are also members of NEW eG. It is assumed that actors from Weiden and Floss received information about and initial access to the project through their membership in NEW eG. Such incidents can negatively affect the resource of trust in a long-lasting manner, as implied by the following interviewee's comment:

> 3.41 "So, in the end, they were playing a double game [...]. Yes, what have I learned? We should probably not have taken them as members at that time, both of them. Then we would not have had this problem" (R3N2, 105, 111).

Nevertheless, the conflict was resolved in the end and seems to be the only challenging situation the municipalities have experienced in their collaboration.

7.3.5.6 Transparency

Internal Transparency

The authoritative resource of transparency was mentioned as another important factor for creating a good working atmosphere in and between the two energy cooperatives:

> 3.42 "We always play with open cards, because I think: Is it not that one says the one cooperative and the other cooperative; rather, it is always together, with each other. And that is a great strength—yes" (R3N4, 101).

The organisations are coordinated by four boards: the executive and supervisory boards of NEW eG and Bürger eG. Internal transparency is sought through the holding of all board meetings together. This is intended to make sure that all coordinators are informed about and integrated into important discussions and decisions.

However, the negative experiences surrounding the negotiation process over the solar park mentioned in the previous section, which was eventually bought by two municipalities that are also members of NEW eG, led to changes in the work structure of the steering boards. After the incident, the group decided to try and prevent situations in which one or more of the mayors, while acting as supervisory board members of NEW eG, could use early project information in order to gain their own competitive advantage. In consequence, plans for projects under evaluation are now prepared by the executive board to such extent that the supervisory board is able to decide about their realisation during the same meeting at which they are presented for the first time. In this way, all steering board members still remain informed about all business activities, but they are now notified about new projects at a stage in which the project negotiation process is more advanced so that they just have to agree or disagree with buying or developing new projects. This concept is intended to reduce the possible window of opportunity that board members start becoming separately engaged in such negotiation processes on behalf of interests outside of the cooperatives. Here we see that a negative effect on the resource of trust has altered the information flow between the cooperative steering boards.

External Transparency

The two energy cooperatives have a business philosophy that is also built upon transparency. Especially NEW eG tries to present its business activities in a transparent manner, and each of its projects can be followed on a website—solar blog—which is publically accessible.[24] The blog provides daily project information, including the amount of power produced by each of its plants. The cooperative has further tried to increase its transparency by distributing its renewable energy projects throughout its member municipalities, with the aim to situating model projects near its members.

7.3.5.7 Acceptance of Projects

Bürger eG seems to have strengthened visibility and acceptance for renewable energy projects in the region and beyond, since it has unified a relatively high number of project supporters (1303 in 2014). Key data about the organisation and its activities, such as number of members, amount of allocated capital and number of realised projects, are easily accessible on its website. Between 2009 and 2013, 122 articles were published in one of the most dominant regional newspapers about the energy cooperative and its activities. Members of Bürger eG show their acceptance by having by the end of 2014 mobilised about 14.4 million Euros in capital for the concrete implementation of renewable energy projects.

Especially interesting in this context is the example of wind power, because it reveals the potential as well as the limits of the cooperative in creating the authoritative resource of project acceptance. Members of Bürger eG clearly communicated their general will to support wind power projects and actually forced the project developer NEW eG to become engaged in wind power as soon as possible. Hence, Bürger eG seems to have strengthened the acceptance of wind parks among its members. However, the regional administration of the district is strongly against wind power. Thus, general acceptance of wind power in this important political authority was completely lacking. NEW eG and Bürger eG were not able to solve the conflict or change the district's position, even though NEW eG is almost fully staffed by local mayors. The many restrictions placed on wind-power projects by the district were one of the core reasons why NEW eG has not yet become engaged in them.

The involvement of local mayors with the cooperatives has not automatically led to higher project acceptance among other political district authorities—especially those who are not members of the energy cooperative. Furthermore, the energy cooperative had difficulties in creating full acceptance among all citizens directly affected by their planned wind parks or solar parks. Concrete examples here are potential projects in the municipalities of Pressath and Kemnath. Even though both communities are part of NEW eG, citizen initiatives against the wind park plans were

---

[24]The blog is accessible under http://neumann-solar.solarlog-web.de/24974.html

formed, and it did not seem to make a difference to them that the wind parks are planned by regional energy cooperatives.

The installation process for the solar park in Döllmitz reveals that the ways in which affected actors are approached and communicated with can play a key role in creating acceptance. As stated in one of the newspaper articles about the project, citizens and regional politicians were not fully informed about all of this project's plans, which seemed to have had a negative effect on overall acceptance for it. Nevertheless, in the end the project was still able to be realised.

All in all, in can be said that the involvement of the two energy cooperatives has strengthened acceptance for renewable-energy projects by strengthening and coordinating a group of key supporters, including politicians and citizens. Nonetheless, the energy cooperatives have not been able to create full project acceptance among all citizens and politicians. This has created a dilemma, because each individual opponent can make use of their general right to fight against a project, independent of how large the group of supporters is. Thus, the overall communication approach for dealing with such critical actors can contribute strongly towards creating project acceptance.

7.3.5.8 Access to Potential Projects and Decision Takers

NEW eG and Bürger eG have immediate access to potential projects, mainly due to the direct involvement of the 18 municipalities. Due to their public presence and political position, mayors are well connected in the region and, thus, also have good access to and understanding of regional project potential. As the following interviewee also noted, the mayors themselves actively searched for potential projects or project locations:

> 3.43 "The mayors have all tried to realise projects in the municipalities, and everyone in their municipality looked around—there we can do a few roof projects or we can do this, we can do that" (R3N1, 91).

Collaboration between citizens and municipalities in NEW eG and Bürger eG also appears to support a general flow of important information and better connections with key decision takers in the energy sector. First of all, the mayors themselves possess crucial decision-making power and authority, but they also have direct access to other decision takers, such as actors in the municipal administration who can approve projects or confirm lease agreements for municipal land or buildings. One interviewee explained the connectedness of mayors as follows:

> 3.44 "Because he [the mayor] can also get his administration on board, he knows all the information, he has access to real estate, to people who are willing to provide areas [for projects]. He also has contacts in the industry or companies that ultimately build regenerative plants. And, of course, he has good contact with the regional district office, in order to obtain the permits" (R3N3, 23).

As a consequence, 20 of the 25 solar projects realised by NEW eG have been implemented either on top of municipal buildings or on land owned by involved municipalities. These projects were also able to be realised within a relatively short timeframe of four and a half years. In fact, the first projects had already been installed on top of public buildings in 2009, even before the energy cooperatives were officially registered in 2010.

However, this ease of access to projects or decision takers can also run into limits, despite the strong presence of municipal representatives on cooperative boards. As already mentioned, the plan of the two energy cooperatives to become engaged in wind-power projects in the district of Neustadt an der Waldnaab was especially rejected by the district's administrative director, whom the mayors were not able to convince about the importance of for achieving the region's power self-generation goals. Hence, beneficial access to decision takers may be limited to actors who are in favour of the plans of NEW eG and Bürger eG. The same can be assumed for information. As long as particular actors want to support the two energy cooperatives, they may also provide supportive information, but actors who are against their plans may hold back any information that could help the energy cooperatives.

Other potential entry points to projects do not seem to have thus far been activated. For example, the 1303 members of Bürger eG may offer interesting possibilities regarding the implementation of renewable energy production projects. However, so far none of the cooperatives' installed photovoltaic plants are directly associated with any private members. Hence, the relatively smoothly functioning project access of the municipality leaders may have led to overlooking other internal options towards approaching potential projects.

7.3.5.9 Access to New Ideas

New ideas seem to be particularly pushed by Bürger eG. According to one interviewee, the representatives from Bürger eG perceive themselves as being more 'active' compared to the representatives of NEW eG in terms of concrete and continual presentation of new ideas for potential future projects:

> 3.45 "We in the citizens cooperative are also quite active. [The things we do are] taken seriously and aren't just swept under the carpet. [...] We are definitely trying to realise projects or bring in new projects or introduce new ideas. And that is recognised and, then, also often works" (R3N1, 221–223).

Meanwhile, the representatives of NEW eG seem to respect the active participation of Bürger eG, as the latter's ideas have often ended up becoming concrete goals of NEW eG for the mid and long term. For example, it was Bürger eG that sought to place and push the issues of wind power and small district-level heating systems within NEW eG at a time when the latter did not yet seem have a clear opinion regarding these concepts, though they were subsequently incorporated into the cooperative's planning goals.

Despite the tendency for new project concepts to be brought into its collaboration with NEW eG, Bürger eG has not yet integrated its own members into developing such ideas. New plans pretty much come solely from the steering boards of Bürger eG. Additional working groups or meetings beside the annual general assembly, during which members may be able to communicate their own ideas or co-develop new projects, do not exist. Hence, access to ideas from non-managerial members is limited, a circumstance underlined by the fact that none of its projects is directly associated with any of them.

7.3.5.10 Expertise and Know-How

Exchange of experience is seen as an important advantage of the inter-municipal collaboration that takes place within NEW eG. As explained by one interviewee, the energy cooperative helped the participating mayors to overcome the limits of individual and uncoordinated activities and to learn from each other:

> 3.46 "This is also such an advantage: If a municipality can advance in a certain sector, because it simply has the opportunity to do so alone, then the others can observe what's happening. And the forerunners can let the others deliberately check what turned out positive. This is also such a forward step away from provincial thinking—another great advantage of this inter-municipal cooperative" (R3N3, 123).

As an example, one of the aims the member municipalities of NEW eG is re-communalisation of their power grids. However, none of them have thus far been able to successfully purchase or repurchase their grids. Important know-how for properly working through the complex grid-acquisition procedures does not exist in any of the cities and communities. Weiden, the largest municipality among the members of NEW eG, plans to purchase its grid in the near future and, due to its size and financial background, it is able to conduct such process on its own. The plan of the other municipalities has been to closely observe the process and to directly learn from the experiences made by Weiden.

According to a study of the Wuppertal Institute for Climate Environment and Energy, lack of know-how among municipalities has been one of the reasons for a very low re-communalisation rate of grids in Germany (Berlo and Wagner 2013). As an inter-municipal energy cooperative, NEW eG has been trying to help its member communities in building such know-how through joint effort.

7.3.6 Creation of New Regulations

Based upon their collaboration within the two energy cooperatives of NEW eG and Bürger eG, cities, communities, municipal utilities and citizens have created new regulations through which they frame their envisioned renewable energy concept for parts of the districts of Neustadt and der Waldnaab, Amberg-Sulzbach and Tirschenreuth. These new societal rules, explained in detail below, include (1) direct municipal guidance and ownership of renewable energy projects, (2) new co-decision and participation rights for citizens regarding municipal energy issues,

(3) a new direct energy consumer–producer relationship, and (4) new forms of regional value creation.

7.3.6.1 Municipal Guidance and Ownership of Renewable Energy Projects

The statutes of NEW eG primarily legitimize municipalities to become members of the energy cooperative (Neue Energien West eG 2009a, p. 1). So far, NEW eG has officially accredited 18 cities and communities to directly guide and own renewable energy projects. As one interviewee put it,

> 3.47 "Yes, the goal is actually that we put our energy supply in the hands of the citizens or local authorities" (R3N1, 65).

One can argue that Stadtwerke—municipal owned energy providers, mostly in form of GmbHs—already put municipalities in a legal position to do business along the energy value chain (for example, Berlo et al. 2008). Two of the 18 communities and cities participating in NEW eG already had a municipally owned utility beforehand. However, according to the following interviewee, the municipal company Grafenwöhr and the municipal company Floß did not feel capable of realising a large number of renewable energy projects on their own, mainly due to limited financial and personnel resources:

> 3.48 "Yes, it would not have worked to that extent with our municipal company. [...] All the others have no municipal companies, are mostly smaller municipalities, so in that area it is not quite so easy for us. That might work for a bigger city, but for us the structure is not there at all" (R3N1, 151).

In fact, the other 16 municipalities and cities—meaning the overwhelming majority—did not even have a municipally owned company, as most of them are even smaller in terms of size and number of citizens. For such municipalities, an energy cooperative can provide a legal and innovative alternative for becoming engaged in restructuring its renewable energy supply. In particular, the concept applies to communities that:

1. Do not have municipal energy utilities of a considerable size for taking a lead in renewable energy project management and
2. do not have the necessary financial resources.

The fact that the municipal utilities of Grafenwöhr and Floß also became members of NEW eG suggests that the energy cooperative has been able to help legally coordinate renewable energy project development between municipalities with and without existing municipal utilities. NEW eG can thus be described as a regulative bracket for inter-municipal collaboration in the field of renewable energy.

New here is that mayors representing the communities within NEW eG have been able to expand their mainly politically based decision-taking authority via an operationally business-focused decision-taking authority. Mayors not only form the controlling supervisory board of NEW eG but also form its executive board

7.3 Collaboration Between Energy Cooperatives and Communities: Neue... 299

and are responsible for its operational business activities, including searching for and implementing of renewable energy projects. As one supervisory board member explains,

> 3.49 "We are planning, we are looking for locations and we are realising projects" (R3N5, 107).

The mayors feel predestined to take such an operational business lead in developing renewable energy, due to their having direct access to projects, information and other decision takers.

This new norm of 'municipal guidance and ownership of renewable energy projects' has been further transported into the regional public. My archival research has revealed that at least 63% of all 120 articles written between 2009 and 2013 about the energy cooperatives in *der Neue Tag*, one of the most dominant regional newspapers, point out the leading position of involved communities and cities within them.

7.3.6.2 New Co-Decision and Participation Rights for Citizens

Civil participation in renewable energy projects was pointed out by the mayors who are members of NEW eG as one of the most important norms coming out of the collaboration between NEW eG and Bürger eG. Creating citizen participation was their main motivation for choosing a cooperative as the legal business form for realising renewable energy projects, because it offers an open-membership approach and basic democratic decision-taking structures, as explained in the following two quotations:

> 3.50 "Here is the possibility that every citizen of the county can participate" (R3N4, 77).
>
> 3.51 "So it's easy for us to get citizens on board [...] but also] I cannot see NEW eG without citizens [...]" (R3N3, 109,111).

The two energy cooperatives are also strongly associated with citizen participation among the regional public. About 40% of all 120 articles written between 2009 and 2013 about the energy cooperatives in the regional newspaper *der Neue Tag* referred to citizen participation in renewable energy as a new norm that is being created through NEW eG and Bürger eG.

It is, however, crucial here to better understand what is meant by 'citizen participation in renewable energy' and how the energy cooperatives actively involve citizens. During the interviews I conducted, next to 'citizen participation', a second crucial value mentioned for involved municipalities was that of 'being free and independent in decision-taking processes':

> 3.52 "We [NEW eG] can freely decide, and that is an advantage" (R3N5, 211).

To a certain extent, these two new principles—direct citizen participation, on the one hand, and autonomous decision-making authority for municipalities, on the other—appeared at first to stand in contradiction with each other. The initiators of the energy cooperatives saw a difficulty in coordinating their own plans with the individual interests that all of the new cooperative citizen members may have brought with them. Thus, direct, continual and immediate participation of many citizens in strategic project decisions was not really desired. As revealed by the following quotation, the idea of founding two separate energy cooperatives instead of one was primarily based upon the aim of exclusively giving mayors and municipalities direct strategic decision rights and giving citizens predominantly financial participation rights:

> 3.53 "And we wanted to deliberately separate this, so that one can say, the citizens are here as affiliates, so to speak,and on the other side there are the municipalities along with NEW, and we can have a free hand" (R3N5, 211).

Hence, NEW eG is legally defined as a project-developing energy cooperative and Bürger eG is legally defined as a project-financing energy cooperative (Bürgerenergiegenossenschaft West eG 2013; Neue Energien West eG 2009a). This concept of power distribution is underlined in the following statements:

> 3.54 "The citizens don't decide this; the citizens have no influence at all. We, as representatives of NEW and representatives of the citizens cooperative, we actually determine [what happens]. We don't have to ask the citizens. That would go too far, if we had to ask each individual what we had to do with the money. It is desired this way, as realised in form of the cooperative. So we are free. People take their money, invest it and we let it work" (R3N5, 193).

> 3.55 "Decision power, I would say, actually belongs to NEW" (R3N3, 105).

Against the background of this special organisational set-up, involved municipalities seem to generally understand citizen participation as the right to invest in the renewable energy projects that are implemented by them through NEW eG. By the end of 2014, 1303 citizens had made use of this right by becoming members of Bürger eG and acquiring 28,460 member shares with a total value of about 14.4 million Euros (see Tables 7.40 and 7.42). The circumstance that citizen participation does not seem to be understood as having the power to influence strategic decisions and project development is reinforced by the fact that, despite providing 93% of the project capital, Bürger eG holds only three of the 19 seats in the supervisory board in NEW eG.

However, in practice, the two energy cooperatives make sure that all decisions are actually taken together. All board meetings (advisory and executive board) of Bürger eG and NEW eG are held together so that desired project plans can be continually coordinated between both energy cooperatives. As a consequence, Bürger eG sees itself as being integrated in the project decision process of NEW eG, even though it may not have the same level of influence. This perception is articulated by the following interviewee:

> 3.56 "In effect, all of the work amounts to one thing in the end, because we have board meetings where the citizen and municipal board members [...]decide together in the end. So, at the moment, the cooperation works perfectly. There have been some quarrels here and there, but it works" (R3N1, 35).

7.3 Collaboration Between Energy Cooperatives and Communities: Neue... 301

Yet, as another interviewee also notes, Bürger eG is aware of the unequal power distribution between the two energy cooperatives and wants to change it:

3.57 "Yes, we are completely dependent, because we, yes, we have the shares, but we are not involved directly at the plants. That's how I'll put it" (R3N1, 105).

In order to increase its project influence and strengthen the position of citizen members, Bürger eG wishes to increase the number of its supervisory board seats in NEW eG and is striving towards direct project ownership. Bürger eG has already become project owner—next to NEW eG—in some of the previously realised renewable energy projects. This development shows that decision rights can be further improved over time through well-working collaboration. I assume that the authoritative resources of trust and respect have strongly shaped the rationale for transferring more decision-making authority to Bürger eG and its citizen members.

Regular members may still have limited rights to participate in decisions of NEW eG, but they do have full rights to directly participate in the decisions taken by Bürger eG, once they become members. To what extent members have been making use of this right, however, is revealed by the number and behaviour of general assembly attendees. Table 7.46 displays the share of members that have attended the assemblies of Bürger eG since its registration in 2010 up through 2014.

As displayed, between 15 and 30% have attended the annual general assemblies of Bürger eG between 2010 and 2014, indicating that only a minority of members actually exercise their co-decision right. Members can further influence company decisions by pro-actively involving themselves in topics discussed during the assemblies. Yet, according to one interviewee, only a few attendees seem to ask questions, make comments or mention their own ideas during these meetings, while most appear to be rather inactive, solely approving the plans of the executive board:

3.58 "At the annual general meeting, questions are raised and then topics are addressed, but the majority is actually inactive" (R3N1, 121).

It could be said that the majority of members seem to be satisfied with financial participation only and may not be interested in being directly involved in strategic business decisions. This assumption is underlined by the following quotation:

3.59 "Well, I think, of course, that the incentive was, of course, also the interest rate, especially in the area of the citizenship cooperative" (R3N2, 79).

As indicated by another interviewee, another reason could also be a high degree of satisfaction among regular members with the work of the cooperative's steering boards:

Table 7.46 Share of Bürger eG members who have attended its general assemblies, 2010–2014

General assemblies	15.06.2010	03.03.2011	02.03.2012	27.03.2013	06.05.2014
Participants	42	123	149	186	182
Total members at that time	141	549	823	1132	1245
Share of members	30%	22%	18%	16%	15%

Source: Information directly provided by Bürger eG

3.60 "Most, the vast majority, are very satisfied and very agreeable" (R3N4, 129).

According to this premise, members may not see any reason for actively influencing the cooperative management work to take another direction.

As mentioned above, other concrete participation options for regular members, such as provided by additional working groups in which they can meet in order to exchange ideas and opinions, do not yet exist. Members who may feel uncomfortable speaking in large formal meetings, such as the general assembly, or who may not have time to attend the annual assembly may, thus, have difficulties to personally participate beyond providing financial capital. This lack of opportunities for additional member exchange seems to have affected mobilisation of other important resources. Whereas the resource 'member capital' has been quite effectively mobilised, the resources 'member projects', 'member know-how' or 'member ideas' have not been activated yet. The fact that almost all projects are located on municipally owned rooftops or land indicates that the generation of ideas and projects is completely focused on access provided by involved communities, whereas the project potential that regular members may bring along with them remains unused.

All in all, Bürger eG and NEW eG have activated different grades of citizen participation with different levels of power, including financial participation in renewable energy projects realised by communities and participation in strategic decisions regarding the project-financing energy cooperative Bürger eG. Due to their particular organisational set-up, the energy cooperatives have mainly focused on the financial participation of citizen-members. At the same time, it seems that financial participation has been more desired by regular members than the right to co-decide.

7.3.6.3 New Direct Energy Consumer-Producer Relationship

In cooperation with the eco-energy provider *Grünstromwerk*,[25] the Bürger eG and NEW eG have created their own regional power tariff called *Regionalstrom Nordoberpfalz*,[26] with the aim of selling the power they produce in their focus region directly to their members or other citizens of the region. Hence, the power mix supplied by Regionalstrom Nordoberpfalz includes at least 25% of the power that is produced in the Speichersdorf open solar park, located in the district of Bayreuth, directly bordering the focus region. This solar power is mixed with hydropower, particularly at times during which the solar park is not productive, such as at night.

The possibility of eventually selling self-produced power to its own citizens strongly motivated the involved communities to become engaged in NEW eG, as explained by this interviewee:

---

[25]Since 2015, Grünstromwerk GmbH has become part of the eco-energy provider Naturstrom AG.

[26]More details are provided under http://www.gruenstromwerk.de/nordoberpfalz/

3.61 "Yes, yes, that was the goal that we practically produce our own electricity and then sell it to citizens, as members of the cooperative. And, I'll say that this enthusiasm, I think, then spread to other municipalities, which then more and more became members, yes" (R3N2, 29).

With their regional power tariff, the two energy cooperatives have created a direct power consumer–producer relationship. Here, the energy cooperatives connect the member municipalities that produce power with the regional citizens and cooperative members who consume it. As of April 2016, the regional power tariff has been chosen by 118 cooperative members. Hence, only about 8% of all cooperative members[27]—and thus the minority—have changed their existing power tariff and switched to the regional power mix of their own energy cooperative so far. The steering boards of NEW eG and Bürger eG have pointed out the strong challenge of acquiring regional power customers. The power tariff was introduced in 2014 through a professional marketing campaign; a second campaign was carried out in August 2015 and a third in October 2015. The campaign involved, amongst other means, the distribution of about 80,000 information flyers, personal information distributed during meetings, as well as advertisement over the regional radio station. Yet, despite such strong marketing activities, regional demand for the regional power tariff has not risen above 10%. Nonetheless, this experience seems to be in line with research regarding the power tariff-change rate of German consumers which indicates that, as of 2012, only 15% of all end-power customers in Germany have opted for eco-power (Bundesnetzagentur für Elektrizität 2013, p. 161). Whereas some empirical surveys have found out that power customers are generally willing to pay more for eco power than for conventional power, other surveys have shown that the price tolerance of private consumers is more heterogeneous than expected and is not always high enough in order to cover the price difference between an individual's current (conventional) power tariff and existing eco power tariffs (for example, Hasanov 2010; Rommel and Meyerhoff 2009). According to Rommel and Meyerhoff (2009), lack of trust in energy providers that offer eco power (for example, Vattenfall), lack of motivation to actually make the change, and lack of information about why making such a change should be seen as sensible are other reasons that may explain the mismatch between willingness to change and actual change. Willingness to change to eco power is, amongst other factors, positively affected by level of trust in energy providers, desire to support renewable energy in Germany and a positive attitude towards nature conversation and eco power (Rommel and Meyerhoff 2009; Sagebiel et al. 2014). Yet, more research appears to be necessary in order to better understand this phenomenon.

7.3.6.4 New Forms of Regional Value Creation

NEW eG and Bürger eG were explicitly founded under the motto "from the region—for the region" (Amschler 2014, p. 57), which became its guiding norm

---

[27] As of April 2016, Bürger eG had 1434 cooperative members.

and was further specified in the interviews I conducted as well as in press releases. According to these sources, it means that the energy cooperatives aim to create regional value by collecting capital from citizens that live in the member municipalities of NEW eG in order to invest it in renewable energy projects within those municipalities. This approach is outlined in the following quotations:

3.62 "The focus is on the idea of community and sustainability as well as the regional idea. This means that the members benefit from the work of the cooperative. [...] In addition, the money that is collected in the region is to benefit the citizens living here and the companies that are settled here" (New Energies West, 7.10.2009 #584).

3.63 "The money remains in the municipality or among the citizens. It remains with us, rather than God knows where... yes, taxes, everything, everything remains with us" (R3N1, 149).

As of 2014, 21 of the 25 realised photovoltaic projects have been located within the boundaries of the municipalities that are part of NEW eG; further, 20 of these projects are located on municipally owned rooftops or land. Communities and cities, which host these projects financially benefit because they receive business tax, as well as a rent for their used property. The majority of these renewable energy projects produce the kind of regional value envisioned by Bürger eG and NEW eG. As indicated by the following quotation, especially smaller cities and communities have economically benefited, since they usually host relatively small businesses and, thus, receive smaller amounts of business tax compared to municipalities with larger businesses:

3.64 "And projects can be realised that would not be possible because we are mostly smaller municipalities. And since we collect almost no commercial tax, a project is already good when it is in the municipality" (R3N1, 167).

The small community of Speinshart is a concrete example here. Due to its size of about 1000 citizens and 24 square kilometres (see Table 7.33), Speinshart has only a very few businesses and receives little business tax (Gemeinde Speinshart 2015). However, now it hosts two of the cooperatives' large solar parks and one rooftop photovoltaic plant, with a total capacity of almost four megawatt hours. Since the community is also owner of the land and the building upon which the plants are installed, Speinshart benefits by receiving business taxes and rent from the projects.

In public, the two energy cooperatives appear to stand for regional value creation. About 44% of all 120 articles written about NEW eG and Bürger eG in the newspaper *der neue Tag* between 2009 and 2013 associate the two energy cooperatives with the active creation of regional value. Of special note is that the energy cooperative has been seen as combining the creation of economic value with the creation of social value, including offering different levels of citizen participation and a new regional consumer–producer relationship in the power sector.

However, three of the five solar parks lay outside of the focus region of the energy cooperatives, comprising 42% of the total power capacity generated by NEW eG. Consequently, as a considerable amount of dividend is created outside the region, a considerable amount of business tax also goes to municipalities located in other parts of Germany. Furthermore, as 20% of the members of Bürger eG do not live in the region, part of its dividends are distributed to citizens who do not live in the districts of Neustadt an der Waldnaab, Amberg-Sulzbach or Tirschenreuth.

These developments can be seen as being in conflict with the cooperatives' guiding principles. However, project investments as well as attraction of members from outside the focus region were not initially part of the cooperatives' plan and can, thus, be described as an unintended consequence of their business activities. Citizens from all over Germany became members due to their successful publicity work and image building. Lack of applicable regional projects while, at the same time, having attracted a large amount of capital motivated NEW eG to search for projects outside of its member municipalities. Such trans-regional business activities were not pointed out as being a concrete problem during the interviews I conducted. The goal of mainly regional value creation in terms of the "from the region—for the region" motto seems to be understood as being the ultimate result of taking a longer path, during which value can also be created outside the region in order to eventually aid the focus region in achieving its renewable-energy aims.

7.3.7 Drawing Upon New Interpretative Schemes

The collaboration between Bürger eG and NEW eG has created new interpretative schemes which have functioned as societal guiding principles for their energy activities, including (1) taking matters into one's own hands, (2) regional self-confidence through greater self-determination, (3) the importance of achieving collective will between municipalities, (4) citizen identification with renewable energy projects, (5) a responsibility for politicians of handling citizen-members' capital, as well as (6) a symbiotic relationship between municipalities and citizens in developing renewable energy.

7.3.7.1 Taking Matters into One's Own Hands

NEW eG has been interpreted by the mayors involved in it as being a powerful means for realising political goals related to energy and climate protection. Two of the mayors explicitly referred to it as a new political and economic force that they have created to address energy and climate issues:

3.65 "I have just said that, ultimately, through this intercommunal cooperation, we have the possibility to shape the future in energy policy terms" (R3N3, 27).

3.66 "This is climate protection, and these are the costs that are there. And they simply force us to go in new directions. For me, these are the two decisive factors" (R3N2, 17).

The next quotation reveals that mayors see themselves as guiding actors who can pro-actively set best practice examples and standards for climate protection and renewable energy, due to their new management position in NEW eG:

3.67 "Yes, the energy cooperative is actually our medium for implementing our energy policy goals, our climate protection targets. This is what we have written on our banner, that we are actively contributing to climate protection. And if the municipality does not do anything, the citizens in the community will not do anything. We are the positive example" (R3N2, 39).

These mayors not only understand themselves as political decision takers but also as people who are actually beginning to make a difference through concrete business and project activities. Based upon their work in NEW eG they see themselves as creating a new attitude which, as with the cooperatives examined in the other two case studies above, can be described as 'taking matters into one's own hands'.

As implied by the following interviewee, this newly created force for achieving energy-related goals is also seen as a gamble that has been worthwhile, because it demonstrates that one can make a difference if one really wants to:

3.68 "We have entered into a venture. The venture has paid off, and I would do it again right away. I can really say that" (R3N5, 229).

This guiding principle of taking matters into one's own hands has also been noticed in public. For example, about 32% of the 120 articles about NEW eG and Bürger eG published in the regional newspaper *der neue Tag* between 2009 and 2013 point out that the municipalities are taking matters into their own hands regarding the development of renewable energy in their region.

7.3.7.2 New Regional Self-Confidence Through Greater Self-Determination

The attitude of taking matters into one's own hands has been accompanied by a new regional self-confidence. On their website, during my interviews and in published newspaper articles,[28] NEW eG and Bürger eG have criticised dependency on fossil fuels among "large international operating capital with monopolistic tendencies" and the associated "price dictation of large energy providers" (for example, Bürgerenergiegenossenschaft West eG 2015). They claim that they want to change this dependency by strengthening the "regional perspective", becoming "autonomous", and developing their own renewable energy supply (see also, Bürgerenergiegenossenschaft West eG 2015), as illustrated by the following quotations:

3.69 "Yes, if we can achieve our goals: That we build our plants where we need energy. That we become practically independent. That will all be for the region, in whichever form.... This will take some time, but the conditions are there" (R3N1, 245).

3.70 "We think regionally [...]" (R3N5, 225).

---

[28]About 23% of the 120 articles published about NEW eG and Bürger eG in *der neue Tag* between 2009 and 2013 critise a general dependency on the price dictation of large energy providers or mention becoming more energy independent as an important value.

7.3 Collaboration Between Energy Cooperatives and Communities: Neue... 307

A new sense of regional self-confidence has been created through belief in being able to achieve greater regional self-determination by organising regional energy supply through a self-owned energy cooperative, as underlined by the following interviewee:

> 3.71 "And even if the energy companies were to be producing clean energy now, the dependence would still exist. And that is also natural, as they can actually do what they want. And this is the reason why we want independence, so we can determine things for ourselves, and the money will remain with us in the region. And that dividends are not the primary focus, but rather having [...] our own energy at an affordable price. This is the real profit" (R3N1, 65).

Thus, the newly introduced guiding principle seems to have increased the significance of municipalities within the regional energy system, since they have now become the new reference points of interaction. However, greater regional self-confidence and self-determination has also had an impact on the evaluation of what is considered 'needed' energy technology resources. This is demonstrated by the following quotation, in which one mayor compares the options of regional wind power to a trans-regional power grid:

> 3.72 "We're just lucky now—in quotes, actually it is not luck—that the discussion about huge power transmission towers (Monstertrassen) is taking place in the region. Now such towers are being built. Why? Because then we will be able to transport wind power from the north to the south, yes, power that we could actually produce ourselves here. And then I always ask... If we had to do a survey here in Northern Bavaria, what would people prefer? Power towers or windmills? There would be a clear tendency towards wind power, I think" (R3N3, 49).

Even though wind power plants can affect the natural scenery of a region, the technology is positively connoted because it supports the desired regional self-determination with respect to energy matters. In contrast, a new trans-regional power grid, which may have a similar impact on the natural scenery, is negatively connoted and generally rejected, because it does not support regional self-determination with respect to energy. Consequently, the new guiding principal seems to be facilitating an increase of regionally produced renewable energy, while also appearing to lead to a decreased appreciation of trans-regional energy technology. However, in a fully functioning national renewable energy system, increased decentralised renewable energy production needs to be balanced by a well-coordinated trans-regional energy distribution infrastructure (50Hertz Transmission GmbH et al. 2014).

7.3.7.3 New Importance of a Collective Will Between Municipalities

The aim of strengthening the region by taking matters into one's own hands has also been accompanied by a third new guiding principle: 'The importance of finding collective will between several municipalities for developing renewable and regional energy'. As demonstrated by the following quotations, the success of the inter-municipal energy cooperative NEW eG has been defined, amongst other criteria, by its ability to enable the involved municipalities to act in concert and develop strategies together:

> 3.73 "But everyone comes along, because they know that. They say that our work will only be successful if we really develop strategies together [...]" (R3N5, 203).

> 3.74 "That we are all pulling together and continuing to push the work that we have begun, not losing sight of these goals" (R3N5, 283).

The approach of NEW eG is set in contrast to individual backyard politics, which seem to have existed at least to some extent before. Working within the inter-municipal energy cooperative has been pointed out as a sign of a reduction in individual and uncoordinated political thinking in all involved cities and communities. Expanding one's own limited political viewpoints has been described by the following interviewee as an important consequence of the collaboration that has taken place within NEW eG:

> 3.75 "I think it's a different way of thinking. People are now looking at the big picture, and one-sided provincial policy is no longer in demand" (R3N5, 227).

The aim of creating a space for joint collaboration was already set in motion by several municipalities when setting up VierStädtedreieck, which is a union of 10 municipalities that has been developing different regional projects for many years. Since these 10 municipalities are also members of NEW eG, in a sense this new inter-municipal energy cooperative can be seen as a further elaboration and intensification of their joint approach.

In comparison to other political committees that strive to create space for cooperation in order to achieve a common goal, NEW eG is special in five respects:

1. NEW eG is almost exclusively staffed by mayors—the key political decision takers in the region. The creation of collective will between the involved municipalities is ensured by the structure of the cooperative's steering boards, as each of the 20 cooperative members has one seat, either on the executive or supervisory board. Furthermore, each municipality is represented by its mayor or by the director of its municipal utility.
2. It understands cooperation as a form of striving for consensus, whereas other committees, such as the county council, may already define a collective will as the existence of only a slight majority of votes.
3. As members of the steering board, local mayors do not try to solely push the interests of their political party but to act in favour of the goals of the energy cooperatives.
4. The collective will created via NEW eG is aligned towards making large investments in renewable energy. The fact that the cooperative has had great amounts of capital at its disposal, as a means for implementing its decisions, was pointed out as one of the greatest differences and advantages it has in comparison to normal political committees:

> 3.76 "Advantages? We have more money" (R3N1, 81).

7.3 Collaboration Between Energy Cooperatives and Communities: Neue... 309

5. The collective will created via NEW eG is exclusively focused on a concrete issue: renewable energy.

The importance of having achieved a collective will between all involved municipalities while developing regional renewable energy projects has also been reflected in public. About 29% of the 120 newspaper articles about the energy cooperatives in the newspaper *der neue Tag* between 2009 and 2013 mentioned this collective will as an important guiding principle.

However, it may not always be so easy to find a common line, as there seems to be a tendency amongst board members to at least try to maximise advantages for their own communities. This is underlined by the circumstance that some involved municipalities benefit more from jointly realised projects than others by, for example, receiving business taxes or rent for hosting projects. Arguments especially arose due to the differing characteristics of the involved communities. Despite the cooperative principle of 'one member one vote', for example, larger municipalities appear to have more often demanded to have greater influence on the direction of the energy cooperative than smaller municipalities did. To successfully reach a collective will, such different viewpoints need to be harmonised, which can be exhausting. One example is the decision around the Hütten wind park, which triggered a great deal of controversy. Such problems, experienced in the process of trying to work together and develop a collective will, are illustrated by the following quotations:

> 3.77 "The one mayor wants this, the other that, depending on what suits their community. [...] The majority still follows the direction of the cooperative, yes. In the end, it still works. So far it has been shown that this works" (R3N1, 71–73).

> 3.78 "Sometimes small disputes arise: why, why not? Here, a mayor from a smaller municipality can relax, while those from larger municipalities are always pointing out their right and their strength and their size, and then they ask: What do you really want?" (R3N4, 62).

Nevertheless, as the first quotation suggests, in the end a joint position has always been found in favour of the whole energy cooperative. Having actors on the steering board who share concrete goals was cited as an important pre-condition for creating a collective will between the communities. Furthermore, trust and respect between the involved mayors appear to have been important authoritative resources for establishing the collective will as a new guiding principle between municipalities.

7.3.7.4 Citizen Identification with Renewable Energy Projects

The energy cooperatives see identification of citizens with their business activities and the region in which they prefer to operate as an important pre-condition for building a new sense of regional self-confidence and new forms of regional value creation. This is partly indicated by the following quotation:

> 3.79 "The citizens can identify themselves with it, and the whole thing is pretty straightforward" (R3N1, 259).

In fact, members of Bürger eG do seem to identify themselves with the organisation's goals and its projects. The quotation below illustrates in an exemplary manner how one member compared financial participation in the energy cooperative with 'becoming part of a good idea'. In order to become part of something, one needs to make the respective issue one's own's as well. Accordingly, one has to identify oneself with it.

> 3.80 "We happened to meet someone yesterday morning, I know him well, and he said, "Now you've convinced me, so I paid 10,000 Euros for myself and 10,000 Euros for my wife". He has not done anything yet, but he has said, "It is a good thing, now I understand it—I'm in""(R3N5, 135).

In the press, however, the importance of citizen identification with renewable energy projects does not really seem to have been discussed very much. In fact, only two of 120 articles about the energy cooperatives published in *der neue Tag* between 2009 and 2013 refer to citizen identification with renewable energy.

7.3.7.5 New Responsibility of Politicians for Citizens' Capital

The recently created and immediate financial relationship between mayors and citizens who are cooperative members has created a completely new responsibility for political decision takers who now handle citizen-members' capital. Bürger eG had mobilised 14.4 million Euros private capital by the end of 2014. This money was invested by NEW eG, which is almost fully managed by mayors. As illustrated by the following quotations, they also feel a strong sense of responsibility for those who have put their private money into the hands of public officials:

> 3.81 "So for 17 mayors this is a very large amount of money. And it is also the constellation here that is so special, in that we manage private wealth" (R3N3, 61).

> 3.82 "This was a bit strange at the beginning, because we were [suddenly] dealing with millions of Euros and, above all, with money from the citizens. And that is not so easy, because we cannot just muck it up, because we are all living in the same area, and they have given us their money" (R3N1, 9).

This perception of responsibility seems to be further intensified through the personnel relationships between cooperative members that live in the municipalities of NEW eG and the respective mayors coordinating the projects.

The quotation below illustrates how the mayors interpret their responsibility. They 'do not want to push their luck', clearly prefer solid projects, and strive to avoid profit losses and problematic projects:

> 3.83 "I cannot just play with the money of the citizens. We have to find very solid projects that are financially sound and ultimately yield the income we need to serve the interests of the citizens with regard to interest rates" (R3N2, 93).

It is this newly created responsibility of politicians for citizen-members' capital which has affected NEW eG's choices regarding potential project technology the most. Fulfilling their responsibilities means, to a great extent, making sure that member capital is not lost but, rather, increased through annual dividends. Hence, NEW eG is highly concentrated on making what it considers to be safe technology investments. Consequently, it has focused on well-proven technology designs, such as photovoltaic, and on projects that receive a guaranteed return on investment through the feed-in tariff under the German Renewable Energy Act (Erneuerbare-Energien-Gesetz—EEG 2014). This means primarily selecting projects due to their profitability. The mayors of NEW eG do not feel responsible for supporting innovative and risky pilot projects based upon new technology but, rather, prefer the image of a reliable and sincere energy cooperative. Since many citizens of the region did in fact become members of the energy cooperative, it seems that they agree with the business approach of their mayors.

7.3.7.6 Symbiotic Relationship Between Municipalities and Citizens in Developing Renewable Energy

Through the cooperatives, local mayors and citizens have entered into a symbiotic relationship to develop renewable energy for their region. The citizen-members finance renewable energy projects that are selected and implemented by the mayors. The following quotation points out the strong interdependence between the members and mayors, who have now been joined through the two energy cooperatives:

> 3.84 "Well, as I said, the inter-communal cooperative needs the donors of the citizens energy cooperative and the citizens energy cooperative needs the mayors who can implement the projects in the fastest possible time. This is precisely their dependence on each other" (R3N3, 103).

This new symbiotic relationship also seems to have been transported into the wider regional society as a new guiding principle. 18% of the 120 articles about the energy cooperatives written in *der neue Tag* between 2009 and 2013 positively reflect upon the new relationship between citizens and communities in developing renewable energy. As an example, one article points out that the more shares are acquired by citizens the more effectively renewable energy can be realised in their communities (lgc 2010). Another article names this symbiotic relationship between citizens and the mayors of their communities as the best option in order to become more independent and self-determined in the realisation of renewable energy in their own region (bjp 2009).

7.3.8 Summary

In the following, the impacts of the collaboration between Bürger eG and NEW eG revealed by my study are summarised. NEW eG consists of an exclusive and pre-selected group of 18 municipalities with the aim of jointly developing renewable energy projects. Meanwhile, Bürger eG has been created to bring together all citizens who are interested in investing in projects realised by NEW eG. The two energy cooperatives were founded together in order to support a 100% renewable energy supply in the municipalities and cities represented by NEW eG. Collaboration between the two energy cooperatives has helped the involved communities to mobilise allocative and authoritative resources which have empowered them to act, as summarised in Table 7.47.

Table 7.47 Allocative and authoritative resources mobilised through collaboration between Bürger eG and NEW eG

Resources mobilised by NEW eG and Bürger eG	
Allocative resources	Authoritative resources
Actors – Alliance of 18 communities and cities, two municipal companies as well as 1303 citizens (as of 2014) for the pro-active development of renewable energy in the districts of Neustadt an der Waldnaab, Amberg-Sulzbach and Tirschenreuth. – Seventeen mayors, two municipal company directors and five private persons manage the two energy cooperatives. – Technology experts and administrative personnel from involved municipal utilities provide support. – The energy cooperatives employed their own managing director in 2014. *Financial capital* – Mobilisation of 22.6 million Euros total capital for renewable energy project investments by the end of 2014. Citizens provided 14.4 million Euros of equity capital through 2014. – Generation of up to 379,000 Euros profit (Bürger eG), resulting in up to a 4.3% dividend for members. – Early mobilisation of foundation capital for establishing the two energy cooperatives. – Better access to borrowed capital through direct involvement of communities. *Technology mix* – Installation of 25 solar power production projects—18 roof top photovoltaic plants and seven solar parks by—June 2014. – Production of about 15,200 megawatt hours of renewable power per year (16 megawatts peak renewable power capacity). – The cooperatives was able to supply about 10% of the household power demand existing in the 18 involved municipalities within four and a half years.	– *Organisational identity*: NEW eG and Bürger eG mainly define themselves through their close interrelationships with the communities they serve. – *Publicity* for the collaborative approach of the two energy cooperatives was generated throughout Germany. – *Image* built of pioneering but serious operating organisations – *Courage* demonstrated by establishing 25 renewable energy projects within a relatively short time frame (4½ years) and by realising large solar parks only 10 months after cooperative registration. – *Trust* gained among 1303 citizens (as of 2014) in the business model of the two energy cooperatives. – *Trust* and respect strengthened between involved members of the cooperatives' steering boards (advisory and executive board), enabling a quick and effective business management. – *Internal transparency* created for the work of the cooperatives' steering boards. – *External project transparency* created to inform all interested actors. – *Acceptance* built among citizens regarding developed and future renewable energy projects. – *Access* to important information, political decision takers and installation of potential projects on public property. – *Access* to new project ideas, especially created by the steering board of the two energy cooperatives. – *Expertise and know-how* is shared between communities for developing renewable energy projects.

7.3 Collaboration Between Energy Cooperatives and Communities: Neue...

Furthermore, collaboration between Bürger eG and NEW eG became the starting point for the development of new societal regulations and interpretative schemes which frame the realisation of their renewable energy vision, as summarised in Table 7.48.

These new norms and guiding principles have been discussed publicly in the press, indicating that they have been further disseminated and diffused into the regional society beyond the boundaries of the energy cooperatives that created them. Table 7.49 summarises which of these newly introduced rules are mentioned or discussed in the 120 articles written about the energy cooperatives in one of the most important regional newspapers, *der neue Tag*, between 2009 and 2013.

Based upon the mobilisation of resources and creation of new rules described above, the inter-municipal energy cooperative NEW eG and the citizens energy cooperative Bürger eG have been able to influence the energy structure in parts of the districts of Neustadt an der Waldnaab, Amberg-Sulzbach and Tirschenreuth in several ways:

1. *Legitimation of municipalities to jointly pursue their own renewable energy goals:* NEW eG officially unifies 18 municipalities and cities as well as two

Table 7.48 New societal rules that have been created through collaboration between NEW eG and Bürger eG

New rules created between NEW eG and Bürger eG	
Regulations/Norms	Interpretative schemes/Guiding principles
– Municipal guidance and ownership of renewable energy projects—Legal approach, in particular for those communities that do not have utilities of considerable size and/or do not have the necessary financial resources for taking a lead in renewable energy development. – Co-decision and participation rights for citizens regarding energy projects realised by the 18 involved municipalities, with a special focus on their financial participation. – Direct energy consumer–producer relationship. – Right of municipalities to create regional value by installing renewable energy projects within the region.	– Mayors pro-actively 'take matters into their own hands' in order to increase climate protection and develop a 100% renewable energy supply for their region. – Regional self-confidence with respect to energy independence through greater regional self-determination in developing its own energy projects. – The collective will created among 18 municipalities located in the districts of Neustadt an der Waldnaab, Amberg-Sulzbach and Tirschenreuth became the basis for developing renewable energy within and outside their focus region. – Identification of citizens with energy projects realized by municipalities, as citizens become part of the projects when joining Bürger eG. – Co-responsibility of politicians for citizen-members' capital. Mayors involved in NEW eG have turned into managers of citizen capital. – Symbiotic relationship created between citizens and municipalities in coordinating and developing renewable energy projects in the region.

Table 7.49 Share of articles written about the energy cooperatives in *der neue Tag* (2009–2013) that mention or discuss new societal rules

New societal rules created through the two energy cooperatives	Share of articles about the energy cooperative that mention or discuss the respective rule (n = 120) (%)
Municipal guidance and ownership of renewable energy	63
Regional value creation	44
New participation rights for citizens	40
Taking matters into ones' own hands	32
Importance of achieving a collective will	29
New self-confidence through greater self-determination	23
Symbiotic relationship between citizens and communities	16
Responsibility of politicians for citizen-members capital	15
Identification of citizens with renewable energy projects	2
Direct energy consumer–producer relationship[a]	1

[a]The circumstance that the regional power tariff was introduced in 2014—that is, after the analysed timeframe—is most likely the reason why the direct consumer–producer relationship was not mentioned in more of the analysed articles

municipal utilities that have expressed a clear will to create a joint approach towards active development of a 100% renewable energy supply in parts of the districts of Neustadt an der Waldnaab, Amberg-Sulzbach and Tirschenreuth. The involved cities and municipalities have neither electric utilities of a considerable size nor the necessary financial resources for taking sole lead in renewable energy project development. The possibility of founding an energy cooperative presents a new regulative bracket and legally innovative alternative for communities to develop, guide and own renewable energy projects. The energy cooperative NEW eG is predominantly managed by 17 incumbent mayors and two directors of municipal utilities. In consequence, has expanded the politically based decision taking authority of the involved mayors with an operational business decision taking authority having an exclusive focus on renewable energy. Having developed a collective will has been the basis for their action. Within NEW eG, decisions have been understood as striving for consensus, detached from the agendas of particular political parties or municipalities. Authoritative resources, such as having trustful and respectful relationships as well as a transparent working atmosphere between all members of the steering boards, have provided the most important basis for rapid and effective business development. Joint work within the two energy cooperatives has resulted in 25 renewable energy projects within four and a half years. The newly created rule of 'municipal guidance and ownership of renewable energy projects' and the aligned 'importance of achieving a collective will among all involved communities in project development' are strongly associated with the energy cooperatives in the public sphere and belong to the most mentioned topics in articles concerned with the cooperatives in public newspapers.

2. *Financial empowerment of municipalities through citizen funding for developing renewable energy projects*: Citizens are the main capital provider for all renewable energy projects realised by NEW eG and are mobilised and brought together through the closely aligned citizen cooperative Bürger eG. Until the end of 2014, Bürger eG had allocated more than 14.4 million Euros from more than 1303 members. The municipalities that are members in NEW eG invested the capital in 25 solar renewable power projects with a total capacity of 16 megawatts peak power. The high mobilisation of citizen capital empowered the 18 involved cities and communities to establish a renewable energy supply for about 10% of their total household power demand (as of 2012) within four and a half years. Municipalities and citizens entered a new symbiotic relationship for developing renewable energy. They depend upon each other because citizens finance the projects that are implemented by their own municipalities. The mayors that coordinate NEW eG greatly contributed to the strong growth of Bürger eG in terms of capital and members. Their direct involvement in the project management increased trust in the business model of the two energy cooperatives, because citizens generally perceive mayors as being trustworthy and credible actors. The steering board members established an image of the cooperatives being pioneering but also seriously operating organisations. They even publicised the cooperatives all over Germany, creating a positive image for them through their great enthusiasm and personal dedication. Due to their strong presence and organisational engagement, the mayors played a large role in shaping the organisational identity of both energy cooperatives. At the same time, these politicians used the cooperatives in order to pursue their own energy goals.

3. *Increased significance of citizen financial participation and a continual feedback structure related to energy:* Creating citizen participation was one of the main drivers of all involved municipalities for founding the two energy cooperatives. By becoming members of Bürger eG, citizens receive a new right to financially participate in the renewable energy projects that are realised by the 18 municipalities engaged in NEW eG. Until the end of 2014, 1303 members made use this right. New here is that the private financial investment strategies of these citizens are now aligned to the municipal goal of developing a 100% renewable energy supply. Furthermore, direct collaboration between municipalities, municipal energy providers and citizens in Bürger eG and NEW eG creates a new continual feedback structure for reaching municipal energy goals. Based upon the will of its citizen members, NEW eG made the realisation of wind projects into one of its core future goals. This shows that key political decision takers of the region are in direct contact with their citizens and are able to jointly discover the kinds of projects that need to be realised as well as the way towards their realisation. In the public sphere, the two energy cooperatives are strongly associated with citizen participation, as 40% of all articles about them published in *der neue Tag* between

2009 and 2013 point out the (financial) participation options that they make available to their members.

4. *Strengthening project identification and project acceptance among citizens:* Financial participation on the part of normal citizens has seemed to increase their identification with as well as acceptance of regional energy projects. NEW eG and Bürger eG have strengthened acceptance for their envisioned projects by strengthening and coordinating groups of supporters, including politicians and citizens. By participating (mainly financially), citizens have made the cooperatives' projects into their own concern and seem to identify themselves with them. Even though wind power plants can affect the natural scenery of a region, and for that reason have been generally rejected by the district administration of Neustadt an der Waldnaab, members of Bürger eG have been strongly in favour of the implementation of wind parks and pushed NEW eG to become engaged in such projects sooner rather than later. The general project acceptance and project identification of citizens has also facilitated the implementation and planning of all projects, in so far as involved citizens did not activate their right to fight against them.

5. *Strengthening regional self-awareness through self-determined regional creation of economic and social values*: The joint business approach of 18 project-developing municipalities and 1303 project-financing citizens has increased the significance of municipalities in developing renewable energy, as they become the new reference point for interaction. All involved mayors and directors of municipal companies are literally 'taking matters into their own hands' in order to push their own energy vision. Since they form the cooperative steering board, they have turned into guiding forces that can pro-actively set best practice examples and standards for climate protection and renewable energy through concrete business actions. By founding the two energy cooperatives in order to pursue a 100% renewable and regional energy supply strategy, the municipalities and citizens have stressed their strong will to develop and finance renewable energy in their region in a self-determined manner. The result is a new regional self-confidence, exhibited by the strongly expressed aim of involved cities and communities to become more independent from fossil fuels and the perceived energy price determination of large internationally operating energy providers. Their new official norm 'from the region—for the region' is underlined by the circumstance that 20 of the 25 projects are built on the public property of involved communities. Municipalities benefit from this, as they receive rent and business taxes. Members of Bürger eG benefit from the projects, since they have received an annual dividend of up to 4%. By offering regional power tariff Regionalstrom Nordoperpfalz, the energy cooperatives have even been able to directly supply their citizens and members with their own produced power. In this way, they are increasing the significance of a more immediate energy consumer–producer relationship in their region. Special here is that the energy cooperatives are combining economic value creation with the creation of social values, including their focus on building a common political will and attracting the support of (financial) citizen participation in their projects. The direct project involvement of

mayors and municipal company directors has greatly helped to achieve this level of regional value creation. They had the courage to establish a high number of 25 renewable energy projects within a short timeframe of four and half years and to focus on larger solar parks only 10 months after official registration of the cooperatives. Their unique access to public buildings and land as well as to the public administration officials who eventually need to officially approve project clearance has fostered project development on public property. This new symbiotic relationship between citizens and municipalities shows that they depend upon each other and are building upon each other's strengths in the development of regional energy projects. Regional value creation and taking matters into one's own hands are strongly associated with the two energy cooperatives. As can be seen in Table 7.49, the majority of newspaper articles examined refer to these newly created norms when writing about NEW eG and Bürger eG.

However, the collaboration between NEW eG and Bürger eG has also experienced challenges and limits in influencing achievement of a renewable and regional energy supply:

1. *Challenge for municipalities to continually find a joint line:* It is not always easy for all involved actors to find a joint line. Arguments have especially arisen due to the different characteristics of the involved communities and cities. At one point, the existing trust between them was even endangered when some of the involved municipalities used cooperative project access in order to pursue their own extra-cooperative project interests. Due to these challenges, it seems to be important that the group of collaborating communities and cities does not become too large, otherwise it may be too difficult to find collective will and reach compromises as well as to maintain their trust, respect and internal transparency while working together.
2. *Strong capital and member growth, as well as lack of applicable projects challenges the regional approach*: Very near half of the generated renewable power capacity—42%—is located outside of the cooperatives' focus region and, consequently, neither creates regional value nor contributes towards achieving regional energy supply—two of the core aims of NEW eG and Bürger eG. Strong capital growth through a rising number of members and aligned expectations of citizens to receive a dividend has pressured the involved municipalities towards continually finding new projects and investment opportunities in order to generate at least a small profit. Amendments to Germany's Renewable Energy Act reduced the guaranteed feed-in tariff and, thus, the potential return on investment for several renewable power production project types. At the same time, it was challenging to find profitable projects in the region that could be realised in the short term. Furthermore, the energy cooperative was not able to create acceptance for regional wind power projects in the district administration. The combination of these factors motivated the energy cooperative to look for applicable renewable energy projects throughout Germany. The strong publicity for the cooperatives led to the circumstance that actors from all over Germany joined Bürger eG. Hence, as of 2014, 20% of the members did not live in one of the cities

and communities being represented by NEW eG. It has, further, been challenging to activate the new potential of the immediate energy consumer–producer relationship, with only about 8% of cooperative members having become power customers of the newly created regional power tariff by the end of 2014. The identification of cooperative members with cooperative projects, to the extent that they will also demand the resulting products (in this case power), seems to also be influenced by other issues, such as power prices. In this sense, self-determined regional value creation based upon the establishment of a 100% renewable energy supply—where energy is completely produced, financed and consumed within the region—seems to be understood as a long term goal, without a straightforward and exclusively regionally focused strategy.

3. *New responsibility for citizen-member capital limits support for technology innovations:* Mayors now have a new responsibility for citizen-member capital, since they are the ones managing investment capital within NEW eG, which is mainly provided by citizens. Accordingly, NEW eG has been following a conservative business strategy that is highly concentrated on making relatively safe investments in order to prevent capital loss. This led to the exclusive installation of rooftop and solar plants—this technology has a proven design and the projects have a guaranteed return on investment through a feed-in tariff under the German Renewable Energy Act. Over the long term, the energy cooperative may pursue the aim of supporting more innovative and possibly more risky technology-based approaches, such as electro-mobility or small district-level heating systems. It even aims to take over regional power-distribution grids. However, its conservative business approach as well as limited time resources have thus far prevented NEW eG from becoming engaged in innovative-technology pilot projects.

4. *Limited citizen participation in strategic project decisions and project development*: The involved municipalities have aimed to create direct citizen participation while, at the same time, remaining as autonomous as possible in steering the cooperative and developing renewable energy projects. This led to the foundation of the two separate energy cooperatives: one for citizen engagement through project financing and one for strategic project development, in which the municipalities are member. This means that the involved municipalities have guided citizen participation towards being predominantly financial. Direct participation of citizens in strategic project decisions has been only indirectly possible, and options to further involve them in project development have remained unused. Informal meetings or working groups in which members could exchange their own opinions and ideas on a regular basis do not exist. As a result, the involved municipalities have not had full access to their members' ideas, know-how or access to putting projects on their property. However, most of the members have not seemed to be demanding more direct participation either. Only 15–30% of the 1303 members have attended the general assemblies of Bürger eG in order to inform themselves about the cooperatives' activities. Most attendees seem to be rather passive and follow the line of the steering boards. It seems like citizens are satisfied with having the right to financially participate and with at least having the possibility to give feedback whenever they wish to. Even though the

municipalities kept most of the decision rights, they have also listened to the wishes of the members. Amongst other instances, it was Bürger eG that pushed NEW eG to become engaged in wind power. The municipalities have most of the decision power, then, but still have an important continual feedback channel with citizens through Bürger eG.

5. *Difficulties in dealing with renewable-energy project opponents:* The collaboration of the two energy cooperatives has generally seemed to strengthen acceptance for renewable energy projects, such as wind parks, especially among actors who are directly aligned to the energy cooperatives, such as citizens and political representatives who are cooperative members. However, Bürger eG and NEW eG have not been able to fully avoid problems with renewable-energy project opponents. The involvement of local mayors in the cooperatives has not led to greater project acceptance among respective political district authorities who have been strongly against the implementation of wind parks and have held firm in their position. Also, citizen initiatives were founded that positioned themselves against the wind parks being planned by NEW eG. The circumstance that every single opponent can make use of her or his general right to fight against a project, regardless of how large the group of its supporters is, creates an acceptance dilemma. It may not be enough to strengthen acceptance when, at the same time, active and vocal opponents still exist. The example the Döllmitz project, where residents initially became oppositional because they felt uninformed about the project, shows that transparent communication with critical actors can play a major role for creating project acceptance.

6. *New regional self-confidence decreases appreciation of trans-regional technologies*: The recently formulated aim of establishing a 100% regional energy supply combined with the greater self-confidence that involved municipalities have achieved through their joint work within NEW eG seem to some extent to have decreased their appreciation of trans-regional infrastructure technologies. Amongst other relevant issues, the establishment of a new trans-regional power grid for transporting renewable power from the north to the south of Germany has been criticized and rejected, even though it is generally needed to balance the increasingly decentralised and fluctuating renewable energy supply within a properly functioning national renewable energy system.

7. *National regulations and member expectations regarding dividends may limit further organisational growth:* The mobilisation of investment capital and cooperative members has provided the basis for all of NEW eG's project activities. With 1303 members and 22.6 million Euros total capital through 2014, NEW eG and Bürger eG were particularly successful in activating these important allocative resources. Their organisational growth not only depended upon internal factors, such as the creation of an organisational identity, trust in the business model, credibility and the achievement of organisational publicity but also upon project profitability and the circumstance that citizens seemed to view membership as the best option out of very few existing investment opportunities. However, reduced project profitability due to amendments in the German Renewable Energy Act, as well possible changes in the investment landscape could make it

more challenging for the two energy cooperatives to mobilise members and capital in the future. At the same time, the new KAGB passed in 2013 created greater business risks for the particular organisational set-up of the two energy cooperatives, wherein Bürger eG mainly functions as a financing actor without being operationally active itself. Hence, Bürger eG registered itself under BaFin. In 2015, BaFin announced that energy cooperatives are generally not covered under the KAGB. By then, however, Bürger eG had already registered.

Table 7.50 summarises the influence (and limitations) that collaboration between NEW eG and Bürger eG has had with regard to building a renewable and regional energy structure in parts of the districts of Neustadt an der Waldnaab, Amberg-Sulzbach and Tirschenreuth.

Table 7.50 Impacts and limitations of collaboration between NEW eG and Bürger eG regarding regional energy structure

Impacts and limitations of collaboration between NEW eG and Bürger eG on regional energy structure	
Realised potential	Challenges and limits of influence
Financial empowerment of municipalities through citizens – NEW eG and Bürger eG empowered municipalities to mobilise more than 22 million Euros of capital, based upon which 25 solar power projects were installed by 2014, 14.4 million Euros provided by 1303 citizens. – Involved municipalities were able to supply 10% of their household power demand with 16 megawatts of self-produced renewable energy within four and a half years of cooperative registration. – Authoritative resources, such as organisational identity, courage, trust in the business model of the cooperatives, positive image, publicity and access to decision takers, provided the basis for attracting cooperative members and capital. – Empowerment has been created through a new symbiotic relationship between citizens and their municipalities. Legitimation of municipalities to jointly pursue their own renewable energy goals – NEW eG unifies 18 different cities and communities, as well as two municipal utilities, behind a clear will to jointly develop a 100% renewable energy supply in parts of the districts of Neustadt an der Waldnaab, Amberg-Sulzbach and Tirschenreuth.	Challenge for municipalities to continually find a joint line – It is not always easy for involved mayors to find a joint line in choosing and developing projects. Arguments have especially arisen due to the different characteristics of involved communities and cities. – Now and then, regional decision takers have tried to maximise the advantages for their own community at the expense of the interests of the cooperative. Difficult to keep regional focus – Short-term investment pressure due to rapid capital growth, upcoming amendments in national regulations that would further reduce returns on project investment, and a lack of regional and profitable project options motivated NEW eG to install 42% of its power capacity outside of its defined focus region. – Publicity for NEW eG and Bürger eG all over Germany led to the circumstance that 20% of cooperative members do not live in the focus region. – It has been challenging to create a new direct energy consumer–producer relationship, because the motivation of cooperative members to change to the cooperative's own regional power tariff is limited and seems to be affected by factors such as power prices.

(continued)

7.3 Collaboration Between Energy Cooperatives and Communities: Neue... 321

Table 7.50 (continued)

Impacts and limitations of collaboration between NEW eG and Bürger eG on regional energy structure	
Realised potential	Challenges and limits of influence
– NEW eG has been particularly helpful for cities and communities that have neither electric utilities of a considerable size nor sufficient financial resources for taking a solo lead in renewable energy project development. Increased significance of citizen financial participation and a continual feedback structure related to energy – By becoming members of Bürger eG, citizens receive a new right to financially participate in renewable energy projects realised by their communities. – The close relationships between the steering boards of the two energy cooperatives creates a continual feedback structure between citizens, mayors and municipal energy providers with respect to energy issues. Strengthened project identification and acceptance among citizens – The financial participation of citizens in renewable energy projects realised by NEW eG has strengthened their identification with and acceptance of renewable energy projects in the region. Strengthened regional self-awareness through self-determined creation of social and economic values in the region – The two energy cooperatives have helped all involved municipalities to strengthen their self-awareness by turning the region into the new central reference point for all interaction, increasing the significance of communities and cities in developing renewable energy. – Based upon their management position in the energy cooperative, mayors take matters into their own hands and pro-actively set best practice examples for climate protection and renewable energy through concrete project action. – the cooperatives combine economic value creation, including business taxes, project rent, member dividends and their own regional power tariff, with social value creation, including development of a common political will and a democratic feedback structure.	Limited support for technology innovations – Due to its new responsibility for citizen capital, NEW eG is highly concentrated on making low-risk investments in projects with proven technology designs and a guaranteed return on investment. – Limited time resources of cooperative steering board members have made it challenging to become engaged in more complex technology-related projects. Limited activation of direct citizen participation in project development and strategic business decisions – Members of Bürger eG do not have direct co-decision rights with respect to strategic project decisions for most installed projects. – Access to citizens' ideas, know-how or to potentially putting projects on their property remains virtually untapped. Working groups or meetings in which members can interact with the steering boards of the two energy cooperatives on a regular basis do not exist. – Most members seem to be interested in financial participation rather than in co-deciding about strategic business issues; only between 15 and 30% of all members attend the annual assemblies of Bürger eG. Difficult to avoid renewable energy project opponents – The energy cooperatives have not been able to fully avoid, work around, or persuade some renewable energy project opponents obstructing their planned projects. Decreased appreciation of trans-regional technologies – The new aim of establishing a 100 regional energy supply and increased self-confidence of municipalities has decreased appreciation of a national power grid that transports renewable power from the north to the south. Organisational growth may be limited in the future – Reduced project profitability, due to amendments to the German Renewable Energy Act as well as possible changes in the investment landscape, could make it more challenging for the two energy cooperatives to mobilise members and capital in the future.

Chapter 8
Discussion of Results

In this section, the results of my quantitative assessment of secondary data and the results from the three detailed case studies are merged and jointly discussed. The aim here is to achieve a holistic understanding of the influence of energy cooperatives on Germany's on-going energy transition.

According to the IBAST framework which has guided my empirical analysis, conclusions are drawn here for the local (Sect. 8.1), trans-local (Sect. 8.2) and wider system level (Sect. 8.3). This differentiation into three analytically separated spatial perspectives systemisatises of my empirical data concerning energy cooperatives and in critically reflecting upon them in relation to a socio-technical energy system still undergoing transition. A final conclusion and suggestions for future research are provided in Chap. 9.

8.1 Influence of Energy Cooperatives from a Local Perspective

Exploring the influence of energy cooperatives on Germany's energy transition from a local perspective means analysing their concrete modes of interaction.

8.1.1 Turning Local Actors into Knowledgeable and Capable Renewable Energy Agents

My empirical assessment has revealed that energy cooperatives have become particularly powerful by breaking with prevailing local energy structures through collaborative interaction. Energy cooperatives can empower local actors in the development of concrete renewable energy projects by creating new collaborative interaction models.

Energy cooperatives can bring together (1) citizens; (2) local political decision takers, in particular mayors; (3) companies with a local business focus, in particular cooperative banks and municipally owned utilities; and (4) local associations, such as energy agencies, with the concrete goal of supporting their focus region in realising renewable energy goals. Energy cooperatives can not only activate local citizens, as has been done out by other studies (for example, Kayser 2014),[1] but can also draw together key local agents from civil society, politics and business around the common intention to fundamentally change the energy structures that they immediately draw upon.

In all three case studies, it was the process of initiating and managing the relevant energy cooperative(s) which led to conscious and discursive reflection regarding how prevailing local energy structures could actually be changed. During the foundation of the three energy cooperatives examined here, engaged actors intensely discussed options and, in the end, translated their abstract renewable energy goals into a new concrete agenda of interaction and collaboration. Helpful in prompting this turn towards finding concrete solutions was that energy cooperatives can only be officially registered if they present a realistic business model. Of special note here is that each energy cooperative examined in this study supports collaborative interaction that is tailored to the very particular local contexts (social, spatial and physical in kind) in which on-going and future interaction is to take place. Each energy cooperative brought together similar actor groups but did so via different internal and external collaborative relationships. For example, BEG Wolfhagen eG became partial owner of Stadtwerke Wolfhagen, whereas NEW eG is an energy cooperative in which only communities, as legal persons, can become members. In the end, the three energy cooperatives and their partners were able to realise various projects, including the implementation of renewable power-production projects through which between 1.6% and 61% of their respective region's power demand could be covered (see annex 4).

Energy cooperatives and their partners empower each other by taking advantage of each other's individual and quite different actor backgrounds. In this way, they become able to activate all of the important resources necessary for realising their plans (see annexes 7 and 8). Young energy cooperatives have strongly benefited from the well-established positions of their main founding partners. Cooperative banks, communities and municipally owned utilities have helped in mobilising members and capital, as well as in searching for potential renewable energy projects. Amongst other ways of achieving this, they have used their existing networks, for example, their customer networks, in order to market the goals of the respective energy cooperative and recruit new members. The engagement of local political decision takers and local companies has created trust in the cooperatives' business model, strengthened their image and increased their publicity. Founding partners

---

[1] See Sect. 2.2 for more details on existing studies about energy cooperatives.

8.1 Influence of Energy Cooperatives from a Local Perspective

helped the three energy cooperatives examined here to form their own organisational identity and self-understanding and to establish their working base. As a result, the three young energy cooperatives were already able to present themselves as successful and professionally oriented organisations from the day of their foundation. At the same time, it is the energy cooperative which have put existing local companies, such as municipal energy providers and cooperative banks, as well as communities, into the position of actually being able to concretely address renewable energy development, and their engagement within the energy cooperative has helped the latter to gain credibility and transparency with regard to supporting renewable energy activities.

At the local level, energy cooperatives function as something of an organisational roof, under which local actors commit themselves to engaging in a **new symbiotic relationship**, which means that local citizens, politicians, companies and associations directly depend upon each other in concretely implementing renewable energy projects (see Fig. 8.1).

My empirical results appear to support what Schneidewind (1998, pp. 286ff.), based upon Giddens (1984) has said regarding collaboration as a discursive and reflexive medium which can *make mechanisms of reproduction visible*. Collaboration not only helps actors to review and question their existing routine interaction patterns but, further, supports engaged actors in trying to adopt new ways of interaction in order to amend their prevailing routines. In collaboration with and within energy cooperatives, local key actors from civil society, politics and business have achieved the necessary level of discursivity in order to reflect upon their own interaction routines and have, in the process, achieved the capability to change the local energy structures that they immediately draw upon. My empirical assessment demonstrates how actor empowerment happens at the local level and originates in

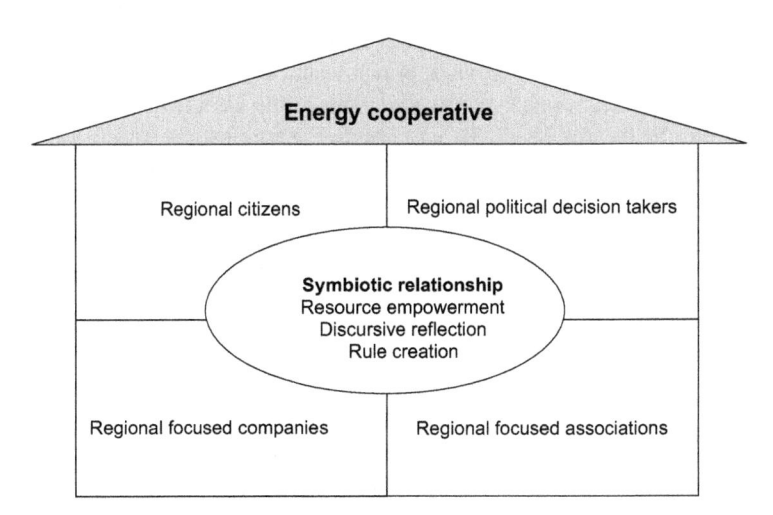

Fig. 8.1 Visualisation of the energy cooperative as a roof for local actor constellations and interaction in renewable energy project development

local interaction. My assessment further confirms earlier research on strategic niche management, which points out that the mobilisation of substantial resources is crucial for the development of niche structures (see Sect. 4.1.4.2). Last but not least, the detailed case studies that I have presented also shows how authoritative resources, such as publicity, organisational identity, trust and credibility, provide the fundamental basis for mobilising allocative resources, including money, members and technology projects.

8.1.2 Creating New Legitimation Pathways for Regional Renewable Energy

My empirical results demonstrate that energy cooperatives can help local peers from politics, business and society, on the one hand, as well as extended groups of citizens, on the other hand, to provide each other with the legal and normative permission to implement renewable energy projects in their focus region. In each case study, local actors used the cooperative as a basis in order to gain official legitimation for realising their own energy goals. One strength of energy cooperatives is that they can introduce new official co-decision and participation rights that have different intensity grades. By offering a high variety of participation options, energy cooperatives are able to activate concerted groups of multiple actors, including politicians, other companies, associations and citizens, for developing renewable energy projects, even though they all have different participation interests. I have observed that such a new participatory and co-decision culture provokes—independent of individual participation intensity—a joint decision-taking process within actor groups that goes beyond the boundaries of their individual projects. The search for a collective will at different participation levels makes taking a democratic feedback approach into a new guiding principle and a pathway towards legitimising the development of renewable energy structures at the local level.

According to the results of the case studies, participation within and through energy cooperatives is exercised in different ways and at various levels of intensity. I have identified four levels of participation: (1) financial, (2) consultative, (3) creative and (4) steering (see Fig. 8.2 and annex 10). These levels vary in depth (intensity of interaction) and breadth (number of actors that make use of respective participation and co-decision rights), as explained below.

The first participation level is *financial participation*. All actors who join an energy cooperative have the legal right to invest in its business activities (§1 GenG). In the three case studies, between 675 and 1303 actors made use of their right to financially participate by becoming cooperative members (as of 2014), acquiring member shares with a total value of between three and 14.4 million Euros. Financial participation is the most passive way to become engaged, since the only required activity is providing capital to the energy cooperative. Nevertheless, even those

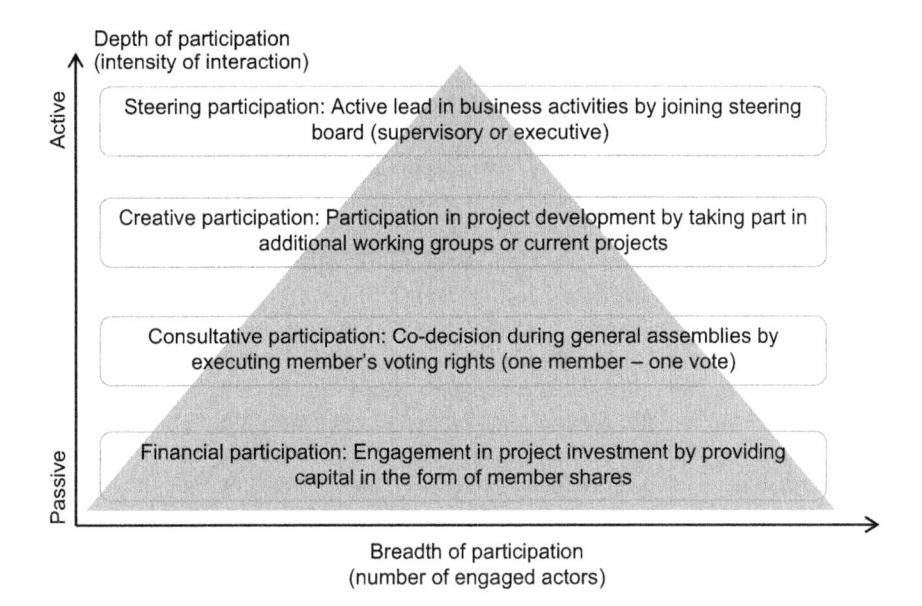

8.1 Influence of Energy Cooperatives from a Local Perspective

[Figure: Pyramid diagram with vertical axis labeled "Depth of participation (intensity of interaction)" ranging from Passive (bottom) to Active (top), and horizontal axis labeled "Breadth of participation (number of engaged actors)". From top to bottom:
- Steering participation: Active lead in business activities by joining steering board (supervisory or executive)
- Creative participation: Participation in project development by taking part in additional working groups or current projects
- Consultative participation: Co-decision during general assemblies by executing member's voting rights (one member – one vote)
- Financial participation: Engagement in project investment by providing capital in the form of member shares]

Fig. 8.2 Typology of participation levels that can be activated through energy cooperatives

members who 'just' participate financially must in some way agree with the cooperative's respective business goals; otherwise they would not invest.

The second participation level is *consultative participation*, where the overall direction for a cooperative is generally proposed by its steering board and agreed upon by a group of interested members. Cooperative members have the formal right to co-decide during general assemblies (§43 GenG), though only a minority of members seem to make use of their democratic voting right. In the three case studies, about 11–30% of members had attended their cooperative's annual assemblies since its foundation. The cases revealed that assembly attendees tend to be rather passive and mostly follow the line of the steering boards. Only seldom have members raised comments, criticisms or expressed their own ideas during meetings. On the one hand, members seem to be satisfied with the work of the steering board and agree with their management. Yet, on the other hand, a general assembly may not be an adequate place for members to actively share their opinions and ideas, as the assembly structure may be too formalised and members do not seem to feel comfortable speaking in front of a large group of assembly attendees.

Case study one demonstrated that creative and pro-active participation in renewable energy project development requires creation of additional spaces, such as working groups, through which members and the cooperative steering board can maintain a direct exchange on a regular basis. This leads us to the third level of participation, *creative participation*, which is characterised by greater depth of interaction but a lower degree of breadth, since working groups are usually only attended by interested members. Such additional interaction spaces affect several important authoritative resources, including access to member ideas, expertise, projects and the additional potential work force they represent. In case studies two and three, such interactive spaces have been lacking. In consequence, potential access to member ideas or additional work contributions for their own organisation has remained almost fully untapped.

The fourth and highest level of participation and co-decisional input is *steering participation*, which is aligned with the work of steering boards. Of special note here is that such work can be done by the steering board of an energy cooperative or within the steering board of an organisation in which the cooperative is a shareholder. In both cases, the board has a strong influence on all business activities undertaken. This participation level thus has the greatest degree of participation depth, although such boards usually consist of very few people compared to an organisation's number of total members (see annex 10). Therefore, it also has the lowest degree of breadth. Steering participation is generally only granted to local political and societal decision takers as well as those belonging to the municipal utilities that have helped to found an energy cooperative. Despite its apparent exclusivity, participation at the steering board level plays a fundamental role in creating a collective will for the organisation. In all three case studies, key local decision takers representing the steering board were able to enter into a new and direct exchange about renewable energy goals and projects. It appears that this kind of immediate communication helped to increase their understanding of each other's positions. Figure 8.2 provides an overview regarding the four identified participation levels.

The in-depth findings presented here regarding participation, joint decision-taking processes and effects on the structural dimension of cooperative legitimation represent a significant elaboration of and advance on results from previous research. Many studies have simply concluded that energy cooperatives influence the energy system because they provide participation options for citizens (for example, Hauser et al. 2015; Kayser 2014).[2] Meanwhile, Radtke (2016), who conducted a far-reaching analysis and reflection upon participation in citizens energy projects, comes to the conclusion that the level of actor engagement in citizens energy projects varies greatly. However, central is that energy cooperatives offer participation and co-decision rights with *different intensity levels and scopes*. This has the effect that multiple local actor groups, including citizens, local political decision takers, companies and associations, can find feasible ways to become engaged in renewable energy development, even though they all have different agendas and means. Participation at different intensity levels triggers a joint decision-taking process between a very diverse group of local actors and, therefore, helps them to give each other the legal and normative legitimation they need to act.

Nevertheless, the new importance of achieving a collective will can also be quite challenging, as it may not always be easy to find a joint line among the involved actors. In case study one, I observed that collective decision taking can become more complex and time-consuming than other modes, because the viewpoints of more

---

[2]See Sect. 2.2 for a detailed discussion of previous studies.

8.1 Influence of Energy Cooperatives from a Local Perspective

actors need to be taken into account. Meanwhile, case study three revealed that actors may tend to prioritise their own individual interests—beyond the cooperative—instead of the interest of the group. Case study three also shows us that arguments can especially arise due to the quite different characteristics of involved actors. Breadth and depth of participation is strongly influenced by an energy cooperative's main collaborative partners, because they partially control its organisational set-up and management as well as its self-understanding. Cooperative founding partners also greatly influence the selection of cooperative steering board members. First of all, it is a widespread practice that founding partners also become members of supervisory and/or executive boards. Second, they often have the right to pre-select actors for the remaining steering board seats. The three case studies indicate that cooperative steering boards do not always desire the pro-active and immediate participation of a large group of members in concrete project decisions and in day-to-day operational project management. As an example, in case studies one and three, founders of the energy cooperatives wanted to integrate citizens while also, at the same time, remaining as independent as possible in terms of final project development and management. In consequence, they created indirect co-decision rights for citizens regarding concrete projects by choosing a shareholder approach. The energy cooperative in which citizens can become members does not implement any of the projects itself. Instead, the cooperative itself becomes a shareholder of the organisation which is responsible for project realisation (the Stadtwerke in case study one and a second energy cooperative exclusively managed by mayors in case study three).

My empirical assessment has further revealed that creating legitimation needs to be differentiated from creating acceptance. Some speak of social acceptance by visualising a widely used mix of acceptance and legitimation (for example, Jakobsen 2008; Wüstenhagen et al. 2007). According to Giddens' (1984) understanding of structure consisting of rules and resources, legitimation represents a structural dimension—next to signification and domination. Achieving legitimation means to gain permission to do something based upon legal regulations and normative guidelines. In society, legitimation is defined by a group, for example, by 'the majority' or by a group of local peers (see Sect. 4.2.2.3). In contrast, acceptance must be understood as an authoritative resource which can be exchanged between actors. To accept something is a voluntary act of individuals and means that a person actively agrees with something or someone. If individuals do not accept a certain process or project, they can activate their right to public criticism and resistance, their right to demonstrate or their general right to go to court (depending upon the legal conditions surrounding the process or object). Single individuals can, therefore, actively oppose a process or a project, regardless of how many actors may support it.

According to the results of my assessment, energy cooperatives can become caught in an 'acceptance dilemma', meaning a situation where energy cooperatives can strengthen acceptance for envisioned energy change by strengthening groups of actors who support it without automatically being able to avoid problems from its opponents. Hence, energy cooperatives cannot always fully prevent conflict and rejection of their ideas or proposals in developing renewable energy projects at the local level.

In all three case studies, energy cooperatives were able to increase the number of actors who publically positioned themselves in favour of the implementation of renewable energy projects. The three energy cooperatives have brought together and unified their respective groups of renewable energy supporters over the long term and empowered them through the mobilisation of crucial resources, such as the collection of investment capital. In all three case studies, the strong unification and empowerment of renewable energy supporters led to the ultimate realisation of their projects and, thus, seemed to 'weaken' renewable energy opponents. At the same time, energy cooperatives and their main partners had to confront a number of opponents (citizens as well as politicians) who did not accept their projects. In case study one, BEG Wolfhagen eG was not able to fully solve the conflict it became embroiled in around the Rödeser Berg wind park, a conflict which had emerged long before the energy cooperative was founded. In case study three, NEW eG was not able to convince several key political leaders of the focus region to allow implementation of wind parks. Both cases demonstrate that energy cooperatives are hardly able to mobilise support from those actors who have already positioned themselves against their project plans. The argument of Hildebrand et al. (2012, p. 492), for example, according to which energy cooperatives can enable a congruence between cooperative members and actors who are directly affected by energy projects in a negative way—such as through smells, noise or unsightly change of immediate natural scenery—needs to be critically examined. According to my case study analysis, project opponents seem rather to perceive energy cooperatives as adversaries and not as actors that may provide a solution for their problem. Actors that have already positioned themselves against a certain project may, then, simply not be interested in becoming engaged in energy cooperatives in the first place, regardless of all the potential advantages a cooperative may have to offer.

Case study one demonstrates that conflict management between renewable energy project opponents and supporters not only requires early integration of and information for affected actors but also a well organised, overall and open exchange about existing opinions and positions as well as a serious consideration of all alternatives and compromises. These results are in line with, for example, Hildebrand et al. (2012, pp. 499f.) and Rau et al. (2012). The case reveals, however, that such process must be guided by a neutral actor and not those from an energy cooperative or others who have a clear position in favour of a certain project solution.

Case study two reveals that some energy cooperatives can choose to actively bypass projects that may present conflicts in terms of their acceptance. Energie + Umwelt eG views full project acceptance not simply as an authoritative resource that needs to be achieved but also as a pre-condition (rule) which must already exist before it becomes engaged in a project. In such cases, energy cooperatives and their partners are limited in strengthening acceptance, because they actively avoid potential conflicts instead of facing them. The main reason for such an approach can be a fear of image loss on the part of the energy cooperative's partners. My assessment shows clearly that pre-existing resources, such as the positive image of established partners, can turn into dominant but unquestioned conditions of innovative interaction (see annex 11).

The differentiation between the structural dimension of legitimation and the individual authoritative resource of acceptance that I am making here extends results from previous research. All in all, energy cooperatives are now creating new legitimation pathways for regional energy projects among broad and diverse groups of local key actors. Energy cooperatives have been greatly strengthening renewable energy project supporters and activating joint decision-taking processes allowing participation at various levels. Yet, it seems to be difficult for energy cooperatives to fully reduce conflicts and rejection in regional energy development, because individual actors can remain who simply will not accept project implementation. Perhaps because of that, some energy cooperatives may also try to bypass projects with a likely potential for conflict. It is, therefore, I believe misleading to abstractly claim that energy cooperatives generally increase acceptance for renewable energy projects without taking stubborn opposition into account (such as done by for example, Becker et al. 2012, p. 55; Hauser et al. 2015, pp. 79f.; Kaphengst and Velten 2014, pp. 56f.; Research Institute for Cooperation and Cooperatives & Vienna University of Economics and Business 2012, p. 17).

8.1.3 Increasing the Significance of Renewable Energy Within Regions

Energy cooperatives and their partners have been increasing the significance of renewable energy within particular regions, because they have made a regional focus into one of the main guiding principles for all their business activities. In all three case studies, energy cooperatives have been implementing most of their projects in a selected focus region, and most of their members, who finance the projects, live in the same region (see annex 4). The energy cooperatives have also been creating and regulating new direct ownership arrangements between regional actors and regionally installed energy projects.

The analysed energy cooperatives and their partners have made their selected regions into a new reference point for immediate interaction. A main strength of energy cooperatives is that they combine regional economic value creation with the creation of new regionally focused societal value. They can do this because they align the mobilisation of resources with the creation of new regulations and guiding principles. Due to their regional project activities, energy cooperatives pay business taxes to the respective community, pay rent to the owners of regional project locations and distribute dividends to citizens who have become cooperative members. Some energy cooperatives further intensify regional economic value creation by distributing part of their profit to community-specific funds, such as BEG Wolfhagen eG.

However, all three energy cooperatives clearly demonstrate that the generation of profit is not their main intention. Accordingly, the energy cooperatives in case studies one and two set an investment maximum for individual members in order to make clear that private profit maximisation should not be the main reason for actors to become members. Instead, energy cooperatives and their partners highly value being able to act in a more self-determined manner. 'Taking matters into one's own hands' has, consequently, become a new dominant guiding principle in developing renewable energy. In case study one, Stadtwerke Wolfhagen and BEG Wolfhagen eG are taking matters into their own hands in order to develop a 100% renewable power supply for Wolfhagen. In case study two, cooperative banks want to enable regional citizens to be able to say 'I also helped support the region in becoming more autonomous from fossil energy'. In case study three, NEW eG was the starting point for 17 mayors and two municipal utilities to officially announce an effort to make their region independent from fossil energy and large energy providers. As a result, energy cooperatives and their partners are expressing a new regional self-confidence and they desire that members feel the same. In case studies one and three, the energy cooperatives and their partners strongly wish for citizens to identify themselves with their renewable energy projects.

Special here is that the energy cooperatives and their partners have not defined a particular region merely according to geographical or administrative boundaries. In all three cases, they selected the boundaries of their focus region based upon:

(1) A collective vision regarding a future energy system and
(2) the interaction radius of their main collaborative partners: in particular, cooperative banks, regional political decision takers and municipal energy providers (see annex 5).

This assessment empirically illustrates the theoretical understanding of space applied in the IBAST framework, which is based upon the ideas of Anthony Giddens (1984, pp. 110ff.); (see also, Giddens 1989, p. 280), according to whom space is not a purely geographical dimension but must be understood as the intersection of the social, spatial and the physical. The results of the case studies also emphasize the approach of Hoopenbrock and Fischer (2012, p. 5), who propose that renewable energy regions should be defined as spaces of interaction and potential. However, in the course of my work it has become clear that an important pre-condition for cooperatives to unfold their structuring potential is a pre-existing (political) vision about the role of renewable energy in a particular region. In all three case studies, the energy cooperative was not the start but, rather, the advancement of an already-existing social and environmental movement in that direction. Furthermore, the cases show that the regional scope of energy cooperatives strongly depends upon the size of the selected region. The region of Wolfhagen is relatively small, so BEG Wolfhagen eG and Stadtwerke Wolfhagen eG have been able to cover 61% of total regional power demand with only two projects: a solar park with five megawatts peak and a wind park with 12 megawatts. In contrast, Energie + Umwelt eG covers the largest region among the three case studies and has only been able to supply 1.6% of regional household demand, even though it has installed a power capacity of seven megawatts (see annex 4 and annex 7). Hence, the larger the individually defined region, the lower the scope with regard to covering regional energy supply.

The new sense of regional empowerment and the aligned regionally oriented norms and interpretative schemes that have emerged from energy cooperatives operating in the case study regions have also involved some limits and challenges. In all three case studies, strong membership and capital growth as well as difficulties in finding new projects have challenged their ability to strictly maintain a regional focus. Energy cooperatives and their main partners still need to follow prevailing economic principles, even though profit maximisation is explicitly not their main business intention. Consequently, they feel pressured to invest their allocative capital in profitable projects as soon as possible, because they want to generate at least a small amount of profit (this aspect is further discussed from a trans-local perspective in the following Sect. 8.2). A lack of 'applicable' projects in their focus regions motivated the energy cooperatives in case studies two and three to implement a considerable amount of energy production capacity outside those regions (see annex 4). The financial profitability of renewable energy projects remains to a great extent an unquestioned condition of innovative action, an unintended consequence of which is that cooperatives must partly deviate from their regional focus when regional projects which fit their approach are lacking. One could assume that, the stronger an energy cooperative grows and the larger it cooperatives becomes, the less stable its regional focus will be.

As outlined in Sect. 3.3, a future renewable energy system in Germany requires a mixture of decentralised and centralised building blocks. The energy produced from offshore wind parks (see annex 2) needs to be transported from the north to the south, populated by large demand centres, via an extended transmission grid. However, greater regional self-determination and self-confidence—fostered by energy cooperatives—seems to be decreasing appreciation of centrally oriented energy infrastructures. In case study three, I observed that the energy cooperatives and their partners strongly reject the establishment of a new trans-regional power grid infrastructure, even though it is generally needed in order to achieve a fully renewable energy system. Energy cooperatives and aligned actors may have difficulties evaluating structures that are important but that may not fit into their own newly created norms and guiding principles.

8.2 Influence of Energy Cooperatives from a Trans-local Perspective

Exploring the influence of energy cooperatives on Germany's energy transition from a trans-local perspective means characterising and evaluating their collective action across time and space. According to the results of the assessment presented above, 1055 energy cooperatives were registered by December 2015 in Germany. They are located in all 16 Bundesländer and are, thus, well distributed throughout the whole country. About 92% of these organisations were registered from 2006 onwards. In this sense, energy cooperatives can be described as a cosmopolitan enterprise community which has recently emerged in Germany.

According to the IBAST framework elaborated in this work, innovative niche structures develop and become dominant if rules and resources are formed into a new structural property that is reproduced via similar interaction patterns beyond the boundaries of individual spaces. From a trans-local perspective, my main interest in this respect has been to examine how renewable energy cooperatives in Germany are providing transferable interaction models and what kinds of consequences their emergence has had on the energy cooperative community's contribution to structural change across time and space. The direction and scope of collective action is not only formed by the multiplication of consciously planned interaction models. Rather, the diffusion of new structures is always accompanied by the reproduction of prevailing structures. The co-evolution of niche and regime structures in time and space is best conceived as the rise of unintended consequences and unacknowledged conditions of action being unconsciously manifested by a socially dispersed but similarly acting group in multiple places. A second interest of my analysis has, therefore, been to reflect upon the ways in which the trans-local influence of energy cooperatives is being directed, challenged and limited by unintended consequences, as well as revealing the un-examined conditions of their own innovative action in time and space.

8.2.1 Changing Prevailing Actor-Sets Across Regions Through Active Networking

Energy cooperatives are influencing Germany's energy system structures by changing the overall actor composition of renewable energy supporters. This cosmopolitan enterprise community is mainly guided by citizens, communities, banks (in particular cooperative banks) and energy-related companies (in particular municipal energy providers).

My empirical assessment has revealed that the majority of energy cooperatives draw upon similar interaction patterns in order to realise their business activities. First and foremost, 72% of all renewable energy production cooperatives have been founded and/or are managed

(1) by communities, in particular by mayors;
(2) by banks, in particular by local cooperative banks; and/or
(3) by energy-related companies, in particular by local municipal energy providers.

Not as common, but still practiced, is the additional engagement of regional associations, such as energy agencies, in cooperative activities, with about 23% of all renewable energy production cooperatives having been founded and/or managed by regional associations. Citizens were identified as founding partners or as members of the steering boards in almost all German renewable energy production cooperatives. Of special note here is that energy cooperatives generally involve actors who are new to the renewable energy business. In each of the three cases that I examined, the minimum cooperative member share was set at 500 Euros in order to involve citizens in municipalities that may not be able to finance projects on their own. An extended group between 675 and 1303 citizens (as of 2014) became engaged in renewable energy projects by becoming cooperative members in the three different regional locations. Involved cooperative banks have been turned from pure financing partners to concrete business actors by initiating and managing energy cooperatives. Involved mayors have expanded their politically based decision taking authority with an operational business-focused decision taking authority in the field of renewable energy by founding and managing energy cooperatives. As revealed in all three case studies, regionally embedded actors, such as political decision takers, cooperative banks and regional municipal energy providers have already had a long tradition of taking responsibility for their regions. By founding energy cooperatives, they have now actively transferred this value into the energy sector, amending prevailing local energy structures in the process. Energy cooperatives are therefore much more than just citizen-focused organisations.

The results of my assessment appear to be in accord with Schneidewind (1998, pp. 201f.), who proposes that a change of actor composition is an important mechanism through which organisations unfold their structural change potential. At the same time, they also seem to be in agreement with Hegger et al. (2007) and Smith (2007), who argue that the inclusion of incumbent actors—those who have already achieved an established position in the same market or in other arenas—at an early innovation stage can increase possibilities to develop an innovation in a direction that has a realistic chance of being adopted by broader societal groups. Cooperative banks, municipal energy providers and politicians present such actor groups that are either already established in the energy sector or in other arenas.

Collective action arises when locally interacting actors join a community through active networking. The process of networking can be well observed throughout the enterprise community consisting of energy cooperatives and their partners. Two networking activities were identified by my analysis to be central:

(1) **Pre-existing local networking activities between initiators of energy cooperatives:** In all three case studies, the initiation of each energy cooperative was based upon pre-existing network activities between the actors who eventually founded it. Regional political decision takers, cooperative banks, municipal utilities, certain engaged citizens and regionally based associations had already collaborated with each other before founding their respective energy cooperative (see annex 12). The energy cooperatives and their particular collaborative interaction models thus did not suddenly come out of the blue but, rather, evolved out of a history of joint local action. Existing close and trusting relationships between founding partners, as well as their experience of exchanging ideas and know-how with each other, became the starting point for their joint initiation of an energy cooperative. This result complements the findings of Kaphengst and Velten (2014, pp. 32ff.) who found in a single

case study that the respective energy cooperative was initiated by an established network of local actors.

(2) **Trans-local networking activities between intermediates and between energy cooperatives:** My empirical assessment has revealed that already-existing national networks were a fundamental pre-condition for the rise of a new cosmopolitan enterprise community. The diffusion of energy cooperatives through many regions has been strongly supported by a well-established group of central cooperative interest associations, as well as by a group of cooperative audit associations.[3] Both groups have provided important advice during cooperative foundation processes and have connected young energy cooperatives with each other. Furthermore, they have provided an important connection between energy cooperatives and cooperative banks—one of the main founding partners for a number of cooperatives. As outlined in all three case studies, these intermediate cooperative alliances strongly supported the energy cooperative foundation process, especially in terms of the cooperative partners' expert knowledge of cooperative law and cooperative business management. As shown by case study one, other national operating intermediates existed, such as energiegenossenschaften-gruenden, which exclusively help energy cooperatives during their foundation process. From the very beginning of their business activities, the studied energy cooperatives were part of a well-established trans-local network of cooperative-focused intermediate associations. Thus, they were able to build upon already-existing trans-local interaction structures. Furthermore, energy cooperatives are actively connected across regions and are able to provide each other with support. The idea of founding an energy cooperative in

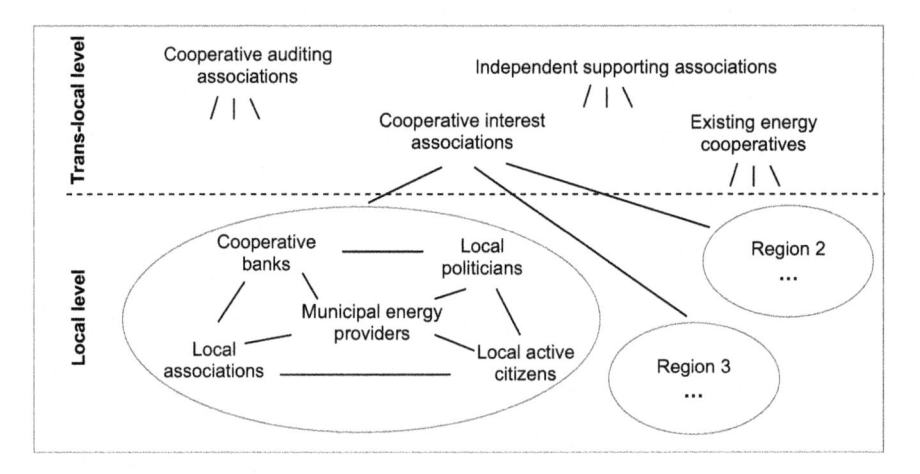

Fig. 8.3 Network set-up of energy cooperatives across time and space

---

[3] As outlined in Chap. 2, German cooperatives are, according to cooperative law (GenG), obligated to become members of a cooperative auditing association (genossenschaftlicher Prüfungsverband).

case study two was inspired by the fact that a similar energy cooperative already existed in a neighbouring region. In case study three, founders of NEW eG and Bürger eG travelled throughout the whole country in order to make their concept public.

Figure 8.3 outlines the network set-up which greatly contributed to the diffusion of energy cooperatives across time and space.

The networking activities observed during my research highlight how the scale of transitions is strongly influenced by the ways in which actors develop relationships across space, such as has been claimed by Coenen et al. (2012, p. 977).

8.2.2 Providing Decentralised Renewable Energy Production with a Technical and Societal Structure Across Time and Space

The community of energy cooperatives that has emerged in Germany has given the idea of decentralised renewable energy production and distribution a concrete technical as well as societal structure that is transferable across time and space. From a trans-local perspective, energy cooperatives can unfold their structuring forces by multiplying similar well-functioning collaborative interaction models across many different regions. The present strength of the enterprise network comes from having developed regional and renewable energy production into a new structural property of the national energy system by combining new rules with new resources throughout Germany (see annex 7 to 9). Energy cooperatives have not only implemented renewable energy production infrastructure in a number of different regions but have also concretely established a new societal understanding of regional and renewable energy, beyond the boundaries of local interaction. In this manner, they have demonstrated how decentralised renewable energy production and distribution can work across localities, despite being radically different compared to the conventional energy system. Energy cooperatives in Germany are functioning as a concrete proof of concept, after decades of abstract discussion regarding renewable energy.

The trans-local community surrounding energy cooperatives primarily focuses on the implementation of small renewable-power production units, followed by the implementation of small renewable heat-production units in combination with the operation of small district-level heating grids. As of December 2015, 88% of all energy cooperatives registered were active in the renewable energy sector and 73% produced renewable energy as their dominant business goal (all figures are explained in Chap. 6). Most of them—70%—concentrated on solar power, with 80% of all solar production cooperatives having installed not more than one megawatt of power capacity. Additionally, 23% of all energy cooperatives produced renewable heat or power with biomass and/or operated a small district-level heating system. Similarly to the figures on solar capacity, 76% of those cooperatives that produce biomass energy had installed not more than one megawatt of capacity. As outlined in Sect. 3.3, the production of renewable energy in small units and its distribution within particular regions is decentralising the energy production infrastructure in Germany. Instead of producing energy in a few large coal, gas or nuclear power plants that are connected to national transmission grids, energy is generated in a widely distributed net of small energy production units, connected to many distribution grids.

Special here is that communities now align the production of renewable energy with a very particular set of new norms and guidelines across many different regions, as shown in Sect. 8.1 (see also annex 9). These include:

(1) A strong sense of regional self-determination and a new regional self-confidence, derived by focussing on regional value creation (economic and societal values);
(2) a continual democratic feedback approach, achieved through diverse forms of participation and co-decision rights; and
(3) the importance of creating a collective will between local actors.

That these rights and guiding principles have not only been adopted by the three case-study energy cooperatives but by the overwhelming majority of renewable energy production cooperatives in Germany as indicated by a quantitative assessment of business goals: 71% of all renewable energy production cooperatives mention the participation of citizens[4] as a business goal and 74% claim to have a regional business focus.

These rules represent the normative basis for the whole community to implement renewable energy production projects within their focus region while simultaneously directing energy cooperatives and their partners towards acting similarly despite not being personally in contact with each other. They, thus, represent the collective identity of a new rising enterprise group that can help all involved actors to identify themselves as one community. The empirical results presented above underline the theoretical understanding outlined by the IBAST framework, according to which collective action is based upon a collective identity which can guide actors across time and space.

New values provide the basis for activating investments in small energy production units in multiple places, underlining once more the strong linkage between value creation and resource mobilisation. The community detaches renewable energy production from pure profit-maximisation—the most dominant guideline in the conventional energy sector. Energy cooperatives have evolved into one of the actor groups that has begun to palpably put the change towards renewable energy production into effect, across time and space. Established energy providers, including Vattenfall, e.on, EnBW and RWE, as well as municipally owned utilities, seem to have had difficulties or may not have been motivated to adapt to the new investment character of small production units (see Sect. 3.4), despite

---

[4]In Sect. 8.1, it was pointed out that energy cooperatives not only offer participation options for citizens but also for politicians, regional companies and associations.

implementation of a guaranteed feed-in tariff in 2000. As of 2012, only 12.5% of installed renewable energy capacity from on-shore wind, biomass and solar had actually been implemented by conventional energy providers, including the four largest companies operating in Germany (Vattenfall, e.on, EnBW and RWE) and municipal energy providers. By contrast, 87.5% of the capacity had been installed by strategic investors, private households and citizen-focused organisations, including energy cooperatives (trend:research GmbH & Leuphana Universität Lüneburg 2013, p. 42). By focussing on other values than on profit-maximisation, energy cooperatives have triggered a decentralisation of the investment structure dominated by the conventional energy sector, as many small to medium-sized member-focused companies are now investing in many small to medium-sized renewable energy production projects in multiple places. This is a completely different set-up compared to conventional energy production investments, where several large energy providers invest in a few large coal, gas or nuclear power plants.

Key here is that energy cooperatives are not only active in rural areas, such as has been claimed by earlier studies (for example, Maron and Maron 2012; Volz 2012), as my assessment has revealed that energy cooperatives are also well-represented in larger cities. About 68% of all German cities with a population of at least 100,000 already host energy cooperatives. Only one of the 14 largest German cities with a population of at least 500,000 does not host one. This diffusion across space seems to developed over time, as revealed by the fact that most energy cooperatives located at least partially in large cities (about 72%) have registered from 2011 onwards.

My empirical assessment seems to confirm that the multiplication of similar interaction models leads to diffusion of similar rule and resource patterns. Particular interaction models rise towards becoming dominant patterns within a community when they appeal to the majority of community members. Multiplied interaction patters are similar but not identical. Despite a strong degree of similarity throughout the whole cosmopolitan community, energy cooperatives have been triggering collaborations that are tailor-made to the individual context of the local level, as outlined in Sect. 8.1.

8.2.3 Distributing Proven Technologies and 'Secure' Project Types

Developing a future renewable energy system in Germany not only requires an enormous increase of renewable power and heat production. Whether and how renewable energy structures will be further elaborated also depends upon finding solutions for multiple challenges. Amongst other factors, I have indicated four action arenas that appear to be central for creating a successful energy-system transformation (see Sect. 3.2):

- Development and diffusion of new, not yet well-proven, technologies, in particular energy storage concepts and new heat production technologies;

- increase of energy efficiency in all areas, especially in the private dwelling sector;
- development of new mobility concepts; and
- greater coordination of energy demand and renewable supply.

Many best-practice examples exist through which energy cooperatives have demonstrated their capability of becoming involved in action areas other than renewable energy production. For example, Greenpeace Energy eG realised a power-to-gas project in order to make heat from access wind energy. Further, BEG Wolfhagen eG implemented an energy-efficiency advisory board and develops energy efficiency projects for its region (see case study one). Meanwhile, Personal Mobility Center NordWest eG was exclusively founded in order to support electro-mobility in the region around Oldenburg and Bremen. In fact, my empirical assessment has revealed that energy cooperatives have diversified business models along the energy value chain (see Table 6.1). However, from a trans-local perspective, the overwhelming majority of energy cooperative still concentrate on a limited number of technologies and project set-ups: mainly small to medium-sized photovoltaic and biomass plants, which are still being subsidised under the German Renewable Energy Act. These results complement the findings of Volz (2012), Maron and Maron (2012) and Holstenkamp and Müller (2013), though these earlier studies have failed to provide a detailed explanation of the reasons for this trend.

The empirical results of my research demonstrate that the evaluation and eventual selection of applicable projects within the renewable energy cooperative community has been strongly guided by a risk-averse business approach. Even though energy cooperatives may have the intention of getting involved in diverse and innovative renewable energy-related projects, for now they seem to generally prefer proven technologies and relatively secure project types. This conservative business approach can be understood as an unintended consequence of:

- new perceived responsibilities for citizen-members' capital;
- prevailing economic principles that remain unquestioned; and
- a lack of proper professionalisation.

This shows how the reproduction of similar interaction patterns across time and space detaches such unintended consequences of action from the particularity of the local level, as the diffusion of proven technologies and secure project types have become a fundamental characteristic of the entire renewable energy cooperative community. Because the majority of cooperatives have been interacting in similar ways, the consequences and conditions of their action have been manifested across time and space.

In all three case studies, energy cooperatives and their partners initially intended to realise many different kinds of business activities in order to support their focus region in developing renewable energy structures. Accordingly, each of the three energy cooperatives defined a diverse set of business purposes in their statues, including not only renewable energy production but also a variety of other investment fields, such as energy efficiency (all case studies) or the establishment of environmentally friendly mobility concepts (case study one). However, eventually

8.2 Influence of Energy Cooperatives from a Trans-local Perspective

all three energy cooperatives ended up focusing almost exclusively on the implementation of renewable energy production projects (photovoltaic, biomass and wind).

In all three cases, investment capital was primarily collected from regional citizens. In fact, the option to involve citizens and the opportunity to activate alternative capital sources were given as the main reasons for founding the energy cooperatives. In each case study, members provided a large amount of capital because they seemed to trust the energy cooperative to watch over it. Accordingly, all three energy cooperatives and their partners started to perceive a strong sense of responsibility for the collected capital from their citizen-members. Members of the steering boards felt particularly responsible, because they often knew regular members personally, as they live in the same regions. In addition, municipalities, cooperative banks and municipal energy providers—those mainly involved in founding and managing the energy cooperatives—feared loss of image for their own organisation if their energy cooperative were to run into serious problems. The reason here was a strong connection between cooperative members also being direct customers or voters. Members of BEG Wolfhagen eG, for example, are simultaneously customers of the municipal energy provider, while many members of Energie + Umwelt eG are also customers of the involved cooperative banks. Meanwhile, members of Bürger eG are, at the same time, the citizens who vote for the mayors of the involved municipalities. In consequence, the energy cooperatives and their partners have concentrated on preventing investment risks and have, therefore, selected projects with proven technology designs.

Even though profit maximisation is not the main intention of energy cooperatives, members still expect to receive at least a small dividend for their invested capital. Providing an opportunity to make a small amount of profit has, therefore, been central for mobilising new members and new capital. At the same time, cooperatives have to earn at least as much profit as required in order to cover their business costs and grow. Otherwise they would not be able to survive over the long term. Accordingly, a small level of profitability has become a dominant project selection criterion. As a result, 73% of surveyed renewable energy production cooperatives that were registered in 2010 had already generated revenues by 2012, 2 years after their registration. However, in none of the three cases presented here did involved actors seem to critically reflect on how the need to realise an economic return on investment may be limiting their choice of 'applicable' projects through which the overall goal of developing a 100% renewable energy supply for their focus region can be achieved. The prevailing economic principle of making profit thus appears to endure as a fundamental condition of action, remaining to a great extend unquestioned.

Case study three reveals that, for those energy cooperatives which do make an attempt to realise innovative business ideas, it is not always easy to motivate their members and other actors to become involved. NEW eG has been trying to create an immediate energy consumer–producer relationship by marketing the renewable power produced in its own plants directly to its members and other local actors. In order to achieve this goal, the energy cooperative initiated its own regional power tariff. However, by April 2016 only 8% of cooperative members had become customer of the power tariff—a very small minority. More research is necessary in order to understand the gap between the number of actors who may express demand for the power produced by their own energy cooperative (Sagebiel et al. 2014) and the number who actually go through with the required change of power tariff. Nevertheless, I assume that a general low-cost power orientation remains—even though in a limited sense—as a prevailing and to a great extent un-examined guiding principle for members and other energy customers when it comes to their own demand structures.

Another issue that I have identified in all three case studies is the unintentionally increased preference for secure project types. All three energy cooperatives are managed through voluntary engagement. This observation is in line with Volz (2012), who concludes, based upon a survey conducted in 2012, that about 94% of all energy cooperatives are managed on a voluntary basis (see also, Research Institute for Cooperation and Cooperatives 2012, p. 19). In all three cases, the voluntary set-up of the energy cooperatives became a challenging factor, because the steering boards had to cope with limited time resources. Thus, engagement in complex and insecure but innovative projects was avoided, because those responsible simply did not have the time to properly evaluate and analyse them. Only one of the three case study cooperatives has managed to professionalise its management by employing a managing director. It seems to be highly challenging to make such a step towards professionalisation, as it also means coping with additional labour costs, which further increases the need for continual profit-making. Lautermann (2016, pp. 97ff.) provides some initial, yet important, guidance for energy cooperatives with regard to how to organise their professionalisation process.

All in all, it seems that the studied energy cooperatives and their primary partners have become limited in their ability to support a variety of action arenas that are likely to be central for achieving a 100% renewable energy system. The development or distribution of innovative technologies, including energy storage, does not at present apply to the conservative business approach they are taking, because pilot projects often involve relatively higher investment risks than ones with established technologies do. Furthermore, becoming engaged in arenas such as energy-efficiency or electro-mobility is difficult for now, since such projects do not at present receive secured returns on investment through, for example, the German Renewable Energy Act. Coordination of their own energy supply with local demand seems to also be challenging for the cooperatives, because it has not proven to be easy to activate members and other actors to change their energy providers. In this sense, the potential of energy cooperatives to foster so-called prosumers may not be as high as expected (for example, Flieger 2014). Figure 8.4 provides an overview of the present level of trans-local engagement of German energy cooperatives in action fields that are considered to be central for Germany's energy system transition. Again, I do not mean to imply that no energy cooperatives have diversified their business activities, as various best-practice examples do exist, showing that energy cooperatives do have the potential to strive towards that direction but probably not reaching the same level of trans-local diffusion that has been achieved in the field of energy production.

8.2 Influence of Energy Cooperatives from a Trans-local Perspective

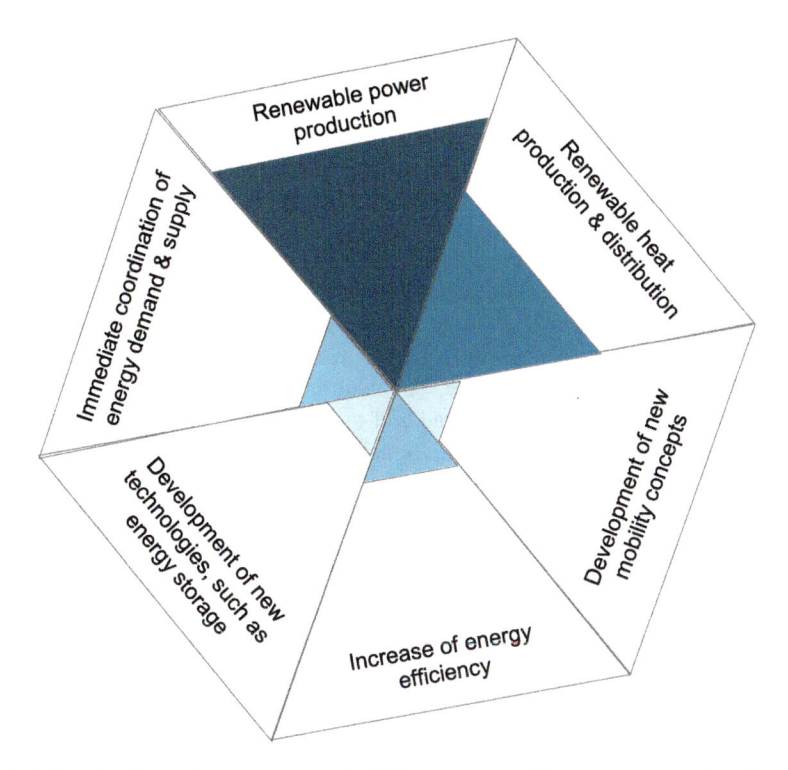

Fig. 8.4 Levels of trans-local engagement of German renewable energy cooperatives in action fields considered central for Germany's energy system transition

Thus far, energy cooperatives have focused on the decentralisation of renewable energy *supply*. From a trans-local perspective, however, energy cooperatives have not yet seriously pursued decentralised coordination of demand and supply or the decentralisation of flexibility options nor have they helped to connect different regions with each other by, for example, becoming engaged in extending grid infrastructure. Nevertheless, it seems to be too short-sighted to describe energy cooperatives as purely social innovations, such as has been done by, for example, the Research Institute for Cooperation and Cooperatives (2012, p. 21). Energy cooperatives only function because they combine societal and technical elements with each other and must, therefore, be seen as supporters of socio-technical innovations within energy transition processes. The probably unintended current preference of energy cooperatives for a conservative and risk-averse business approach and the limitations on action it entails illustrates quite well the co-evolution of prevailing regime principles, such as economic profit-making values, and new innovative niche structures.

8.3 Influence of Energy Cooperatives from a Systems Perspective

From the perspective of a transition process, institutionalisation of interaction at the wider system level means the *transfer* of similar (not identical) interaction patterns across multiple social actor groups and multiple social, spatial and physical settings over time, to the extent that aligned new rules and resource patterns become part of the practical routines of a whole socio-technical system. Exploring the influence of renewable energy cooperatives within transitions at the wider system level requires analysing their potential for maximising the distribution of the new structures that they stand for across time and space. The main interest here is to better understand this community's capability to expand and to achieve a stabilisation process of its aligned interaction patterns over time, a crucial aspect of which is to comprehend the community's ability to create reciprocity between community insiders and outsiders.

8.3.1 State of System Integration

As of 2015, all renewable energy production cooperatives, making up the overwhelming majority of the renewable energy community, mobilised 158,021 members and collected about 1.3 billion Euros in total capital, about 400 million Euros of which were provided by cooperative members. These financial resources empowered the community to install, amongst other technologies, about 500 megawatts peak solar power capacity in many different regions by May 2015, representing 1.4% of the total solar power capacity that had been installed in Germany by March 2015. It is assumed that the share of biomass energy or wind energy produced by energy cooperatives is even lower, because the number of energy cooperatives implementing these production technologies is lower compared to the number of energy cooperatives that install solar power.[5] Hence, one can say that the quantitative share of energy cooperatives on the distribution of renewable energy production capacity in Germany is still relatively small.

However, looking at the number of 158,021 mobilised citizens and organisations, one can certainly conclude that energy cooperatives were able to approach and activate a considerably high amount of actors for the development of renewable energy structures throughout Germany. In comparison, the German green political party Bündnis 90/Die Grünen only mobilised about 59,000 party members by the end of 2015 (see Sect. 6.2.1). Member development of energy cooperatives and political parties cannot be directly compared, however, since energy cooperatives are organisations with concrete business goals. Nevertheless, relating them to each other still gives an idea about the societal scope of energy cooperatives in time and space,

---

[5]The empirical approach is described in Sect. 5.2.

8.3 Influence of Energy Cooperatives from a Systems Perspective

because they also have an open membership approach and a socio-political position. Special here is that the group of community members not only includes citizens but also organisations, companies, politicians, associations and initiatives, as outlined in Sect. 6.3, Chap. 7 and Sect. 8.1.

Its new way of thinking and acting has been widely noticed beyond the boundaries of this enterprise community. Within their focus regions, energy cooperatives have been able to achieve much public attention. In the three case studies, each energy cooperative was mentioned in at least 50 to 194 articles published in leading local newspapers within a timeframe of 3–5 years. Between 26 and 120 articles per case study explicitly describe the energy cooperatives and point out their special values and operating guidelines. Since my empirical analysis only included assessment of one newspaper per case study within a certain timeframe, it is assumed that even more articles were actually published about the energy cooperatives and their activities. National politicians, such as the former Minister of the Federal Ministry for the Environment, Nature Conservation, Building and Nuclear Safety and the former Minister of the Federal Ministry for Economic Affairs and Energy, have clearly acknowledged the work of energy cooperatives (see Sect. 6.3.3). Leading societal associations, such as the Ethics Commission for a Safe Energy Supply or the German Advisory Council on Global Change, have started to explicitly praise energy cooperatives (Ethics Commission for a Safe Energy Supply 2011; German Advisory Council on Global Change 2011), describing them and their special mode of interaction as making an important contribution towards the development of a renewable energy system in Germany. These statements and descriptions are selected examples and do not present a systemised media analysis. However, they show that key societal and political opinion leaders have been granting the work of energy cooperatives a significance which goes far beyond the internal boundaries of the enterprise community and its focus regions. The importance of energy cooperatives for the whole energy system was further manifested in the EEG 2014 as, for the first time, the new diversity of actors in the energy sector was officially recognised by new national regulations, with §2, No. 5, sentence 3 of the EEG 2014 stating that the diversity of actors in producing renewable power deserves protection in the course of planned legal amendments, such as the introduction of a new tendering model. According to the official draft of the EEG 2014 from 8th April 2014, the government associates actor diversity with energy cooperatives and appreciates them as organisations that are needed in order to achieve the desired Energiewende.

> "When designing the concrete tender model, the actor diversity that has up to now been important for the success of the Energiewende is to be maintained, so that the interests of energy cooperatives and citizen projects are adequately taken into account throughout the process" (The Federal Government 2014, p. 161).[6]

---

[6] As described in Sects. 3.4 and 6.1.1, the tendering model for photovoltaic and wind power plants is one of the most central elements of EEG 2014 and EEG 2016, according to which all actors intending to install new wind and solar power capacity are required to participate in nation-wide tenders for power capacity.

("Bei der Ausgestaltung des konkreten Ausschreibungsdesigns soll auch die bisher für den Erfolg der Energiewende wichtige Akteursvielfalt aufrecht erhalten werden, so dass z. B. die Belange von Energiegenossenschaften oder Bürgerprojekten angemessen im weiteren Verfahren berücksichtigt werden"—The Federal Government 2014, p. 161.)

All in all it can be said that, from a systems perspective, the main current strength of energy cooperatives is their capability to motivate a great amount of actors to appreciate and even want to become involved in decentralised, regionally focused and democratically organised renewable energy production as well as to create a significance for their way of thinking beyond the boundaries of their own community, to the extent that key national political and societal opinion leaders have acknowledged their work. Being able to create such a level of reciprocity between community insiders and outsiders can facilitate the transfer of new rules and resource patterns among many societal groups and can help to further institutionalise aligned interaction patterns.

8.3.2 Expansion of Trans-local Practices

The community surrounding energy cooperatives has emerged over the past 10 years and is, thus, still very young. Energy cooperatives may further diffuse new energy structures across time and space if this community is able to expand, which is also fundamental for maximising attention for it beyond its own boundaries. In what ways is such expansion possible?

Individual energy cooperatives could grow in size. So far, the overwhelming majority have been small to medium-sized organisations, with 80% of all renewable energy production cooperatives having less than 200 members or less than two million Euros in total capital. Most energy cooperatives seem to stay within these ranges and do not grow larger (see Sect. 6.2.2). However, the growth rates of organisations with more members and more capital reveal that renewable energy production cooperatives do actually have the capacity to further increase their organisational scope. For example, the share of organisations that had more than five million Euros in total capital almost doubled: from 4% in 2010 to 7% in 2012.

Two internal growth factors were identified in analysing the case studies: (1) professionalization of internal management structures and (2) support from collaborative partners—communities, banks (in particular, cooperative banks) and energy-related companies (in particular, municipal energy providers)—even from before the beginning of official business operation. However, both internal growth factors also exhibit limitations. Energy cooperatives find themselves in a growth dilemma. They would have to grow in order to pay additional labour costs for professional management, but they also appear to need such professional management in order to grow. In consequence, most energy cooperatives do not take the step towards professionalisation. Strong collaborative partners can to some extent balance out the lack of a professional work force by providing know-how and working time from their own staff. However, heretofore dominant collaborative partners—particularly the group of cooperative banks and other energy-related companies—seem to be withdrawing from their general engagement in founding new energy cooperatives. This lack of professionalisation and decreased engagement of key collaborative partners in helping to found new energy cooperatives is likely to limit the number of new energy cooperatives with a high internal growth potential.

Energy cooperatives could also grow in numbers. By the end of 2015, 1055 of such organisations were registered. Energy cooperatives experienced a strong growth from 2006 until 2011. However, since 2011 annual registrations have continually declined, with 42 new organisations in 2015 reaching the lowest level since 2009.

It can, therefore, be concluded that energy cooperatives have been experiencing a decrease of their community growth dynamic, despite having received system-wide recognition for their innovative way of interaction.

8.3.3 Stability of Trans-local Practices Over Time

My empirical assessment has revealed that the growth of energy cooperatives has been directly correlated with the development of national regulations, in particular amendments to the GenG and the EEG as well as the introduction of the KAGB. National regulations represent regime principles and play a crucial role in organising interaction throughout a whole system.[7] During an ongoing socio-technical transition process, the interrelationships between innovative interaction patterns and national regulations can provide important indications with regard to the stability of new trans-local practices. Two perspectives must be taken into account here. On the one hand, new structures may indicate a positive stabilisation process at the wider system level if aligned interaction models are increasingly supported and legally guided through national regulations. On the other hand, new structures may indicate a strong process of stabilisation at the wider system level if aligned interaction models continue to exist even though support through national regulation decreases over time.

The guaranteed feed-in tariff for renewable energy under the EEG functioned as one of the main stabilisers for the interaction models of energy cooperatives and triggered the strong expansion of the community throughout Germany. It was particularly the positive development of the cost-value ratio with regard to photovoltaic plants, due to the combination of a guaranteed feed-in tariff and a strong technology cost reduction, which led to a system-wide stabilisation of their main business model—the installation of solar power plants—between 2006 and 2012.

From 2012 onwards, the seemingly intertwined development of the EEG and of energy cooperatives began to come apart, even though the EEG had begun to officially acknowledge the work of this rising trans-local community. The focus of

---

[7]Provided that system boundaries are equal to the boundaries of a nation state.

the EEG changed from mainly supporting increased power capacity to limiting the costs of energy system change, integrating renewable energy into the competitive energy market and exercising stronger capacity control. EEG amendments in 2012 and 2014 included strong reductions of feed-in tariffs for almost all renewable energy production technologies, in particular photovoltaic. Lower feed-in tariffs lowered project profitability and made it more difficult for cooperatives to find applicable projects. The unclear position of energy cooperatives under the KAGB between 2013 and 2015 further destabilised their main business model: investment in and installation of renewable energy production projects, especially of photovoltaic plants. Business activities covered by the EEG have become more complex, mainly due to the introduction of new market-oriented concepts, such as the tendering model or the new requirement for direct marketing. The introduction of new legal support for innovative concepts, such as decentralised energy self-supply, was formulated in such way that energy cooperatives could only apply respective rules to a limited extent. For example, energy self-supply receives the highest level of support if the power plant operator is, at the same time, owner of the building used, such as if a private household installs photovoltaic on their roof (see further down for more explanation on this topic).

The EEG 2014 and 2016 began to officially emphasise the work of small citizen-focused energy organisations. The EEG 2016 indicates that national politics want to support energy cooperatives and explicitly states that citizen-focused and locally oriented energy organisations or projects can contribute towards creating acceptance[8] for energy system change and are of great value for the Energiewende (Deutscher Bundestag 2016, p. 173). Even though the government announced the protection of energy cooperatives in the course of planned amendments (see previous section) and even though the EEG includes particular exceptions for citizen-focused energy organisations (see Sect. 6.1.1), the government and national regulations did not manage to capture the real needs of energy cooperatives.

The overall change of the EEG from a guaranteed feed-in tariff system to a market-oriented supporting model does not fit with the desire of energy cooperatives for legal security and limited complexity. Consequently, a reduced growth dynamic was a direct result. As shown in all three case studies, not only was the number of new registrations reduced but the project activities of already-existing energy cooperatives as well. Due to a lack of potentially acceptable projects, case studies one and two decreased the maximum member share for a single member to 1000 and 2500 Euros, respectively, leading to a strong limitation of further organisational resource empowerment. Without new investment capital, energy cooperatives are not able to realise new projects.

Another aspect of the problem shows that the government does not seem to understand the trans-local community and its special mode of interaction. The definition of citizen-focused energy organisations included in the EEG 2016 requires that the majority of votes lies in the hand of local citizens, that is, private persons.

---

[8]As outlined in Sect. 4.2.2.3, I differentiate between creating legitimation and creating acceptance.

8.3 Influence of Energy Cooperatives from a Systems Perspective

But such an understanding does not seem to appreciate the diversity of actor sets being fostered by energy cooperatives. A crucial finding revealed by my research is that energy cooperatives are much more than citizen-focused organisations and that they would lose part of their structural change potential if they reduce the actor diversity that they stand for just in order to apply for support under the new EEG. For example, NEW eG is exclusively formed by communities and, thus, cannot apply for the legal support being offered by the EEG 2016 for citizen-focused energy organisations, even though their main intention is to develop energy projects together with citizens that live in their region (case study three). It is thus most likely that the definition of citizens-focused organisations provided by the EEG 2016 will impact the set-up of newly registered energy cooperatives from 2017 onwards. Figure 8.5 outlines the interrelationships between legal developments and the growth of energy cooperatives in Germany.

Energy cooperatives have not yet managed to stabilise their special mode of interaction in the course of political changes over time. Such a stabilisation process would require a diversification of business activities in order to become more independent from the EEG. Energy cooperatives thus need to find ways to be able to better cope with greater complexity and project insecurity (see also, Lautermann 2016, pp. 97ff.). However, the lack of greater internal professionalisation—along with the extra managerial time and know-how entailed by it—as well as the preference for a conservative and risk-averse business approach has limited the flexibility of the whole community with regard to trying out new ideas, projects and business approaches.

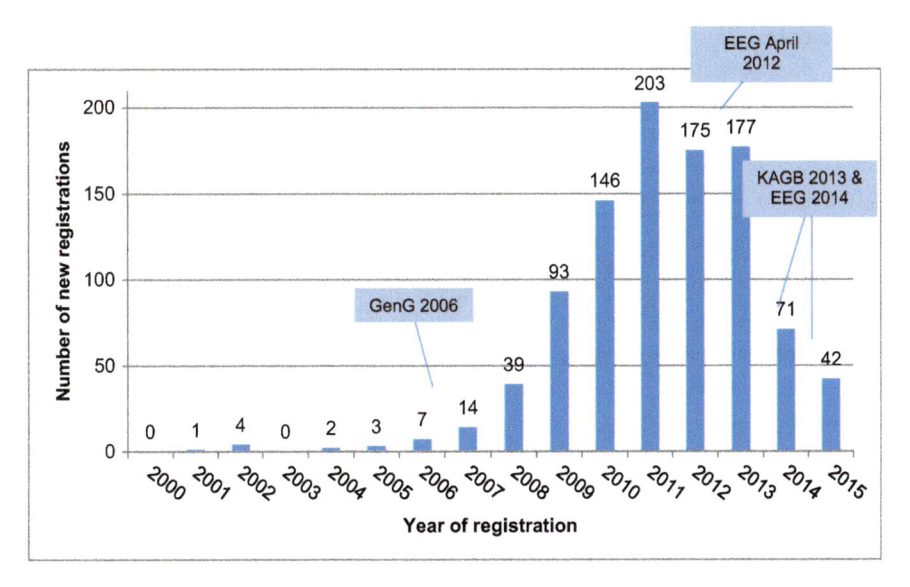

Fig. 8.5 Annual energy cooperative registrations between 2000 and 2014 in relation to amendments to the GenG, EEG and KAGB

Political support could help towards stabilising energy cooperatives in the future. The EEG, for example, could support more cooperative-friendly business models. Indications in this regard already exist, as the EEG 2016 includes a *Verordnungsermächtigung* (power to issue statutory instruments) according to which power self-supply organised by third parties (for example, renewable power production projects on rented apartments) has been put on par with self-supply organised by actors who at the same time own both the production plant and the building on which it operates (for example, renewable power production projects installed on privately owned homes). Accordingly, operators of renewable power plants may pay a reduced EEG contribution (*EEG-Umlage*) from 2017 onwards, if the production plant is installed on top or in their own building and if the power is supplied to third actors within the same building. The office for energy cooperatives of the German Cooperative and Raiffeisen Confederation has been continually emphasising that self-supply concepts organised through third parties could represent a promising business model for energy cooperatives, because it fits in well with their open membership approach and preference for small to medium-sized, local, decentralised energy production projects.

The connection between national regulations, such as the EEG and the KAGB, and the growth of energy cooperatives illustrate well the co-evolutionary process between new innovative niche structures and prevailing regime structures from the very beginning of niche creation. My empirical results, therefore, underline that describing niches as small spaces that are protected from a regime until they are strong enough to survive on their own is misleading.[9] They also reveal the fundamental influence of national policy on the development of innovative enterprise groups. The co-evolutionary development between niche and regime structures can be described as a wavelike process that is also quite contradictory, as political support of new interaction patterns can change into political limitations. From a systems perspective, national regulations are one of the most important conditions for the development of energy cooperatives. This result complements earlier studies such as (Hauser et al. 2015, p. 83; Kaphengst and Velten 2014; Research Institute for Cooperation and Cooperatives & Vienna University of Economics and Business 2012, p. 21).

All in all, following a period of strong stabilisation and growth, the interaction patterns of energy cooperatives in Germany have been experiencing a period of destabilisation and decrease of growth. The combination of internal and external challenges outlined above have destabilised the dynamic of the whole community. The well-functioning interaction model of energy cooperatives, which they have transferred into many different regions throughout Germany, directly depends upon

---

[9]It is often argued that niches are always protected by national regulations such as subsidies. In this thesis, I have argued that regulations are part of a regime, proposing that niches and regimes always co-evolve and can never be seen separately from each other (see Sect. 4.1.4.2 for critical reflection on this widely held idea about niches).

8.3 Influence of Energy Cooperatives from a Systems Perspective

the EEG, but changes to the EEG have led to a situation of conflict between their action principles and value principles.

The future will show whether the new EEG 2016, being put into force from 2017 onwards, and the clarification of the legal position of energy cooperatives under the KAGB, from 2015 onwards, will re-stabilise the interaction patterns of energy cooperatives so that their growth dynamic can recover. The future will also show whether energy cooperatives are able to emancipate themselves from the EEG over time. It also remains to be seen what kinds of effects diversification of business models could have on the stabilisation of this trans-local community as well as on the diffusion of socio-technical innovations within Germany's energy transition.

Chapter 9
Conclusion and Outlook

This work has explored the influence of renewable energy cooperatives within Germany's currently transitioning energy system. Section 9.1 presents an overall summary of my results and gives suggestions for the future, and Sect. 9.2 provides some reflection on the theoretical framework deployed. In closing, Sect. 9.3 discusses possibilities for future research.

9.1 Summary and Suggestions for the Future

Energy cooperatives function as an organisational roof under which local citizens as well as key local decision takers from civil society, politics and business can commit themselves to coming into a new kind of symbiotic relationship, with the concrete intention of developing renewable energy in a self-defined region. Energy cooperatives can empower local peers to change the very patterns of rules and resources making up the local energy structure that they draw upon. They trigger a far-reaching discursive reflection process and show how abstract energy goals can actually be transformed into concrete energy projects. In order to jointly realise projects, engaged citizens, local politicians, business and societal associations have to partly leave their established actor roles, forcing them to go beyond the boundaries of their own normal interaction routines. Here, consciously breaking with routines is the most important seed for structural change.

The main strength of energy cooperatives is that they combine the installation of renewable energy technologies with the implementation of new guiding principles, rights and values through immediate collaborative interaction. Special here is that their new and innovative mode of interaction can increase the **legitimacy, significance and dominance** of decentralised renewable energy production in regions. Energy cooperatives focus on the installation of small to medium-sized energy production plants, preferably solar or biomass power units with up to one megawatt capacity, in their specific focus regions. Energy cooperatives align the production of renewable energy with various new participation and co-decision rights, which have different intensity grades. They can be differentiated into financial, consultative, creative, and steering modes of participation. It is this new diversity of participation options which provides the basis for drawing together and unifying such heterogeneous groups of local peers in developing renewable energy, even though they all have different participation agendas and interests. In this way, energy cooperatives can help local actors from politics, business and society as well as an extended group of local citizens to provide each other with the **legal and normative legitimation** required to implement renewable energy projects in their focus region; energy cooperatives also contribute towards the creation of a common will to carry out such projects.

Energy cooperatives and their partners can increase the **significance** of renewable energy within a region because they have made taking a regional focus into one of their main guiding principles for all of their business activities. Consequently, they create and regulate new direct ownership arrangements between regional actors and regionally installed energy production projects, and their projects are to a great extent financed by local citizens. Energy cooperatives also combine economic value creation with the creation of new social values in their focus region. Taking matters into one's own hands, for example, has become a new dominant guiding principle for regionally embedded actors, triggering the development of regional renewable energy. As a result, energy cooperatives and their partners have built a new sense of regional self-confidence and strengthened the value of regional self-determination. Energy cooperatives are, thus, much more than vehicles for realising citizen-focused or community energy projects. Rather, energy cooperatives also stand for a new regional and renewable energy production structure: a totally new combination of rules and resources in Germany's energy sector, as displayed in Fig. 9.1.

The new structural properties outlined in Fig. 9.1 are completely based upon collaborative interaction. Energy cooperatives and their multiple local partners can turn into **dominant** actors in developing renewable energy at the local level by taking advantage of each other's individual and quite different actor backgrounds. Together they are able to activate necessary resources, including capital, supporters (members), technology projects, know-how, credibility or publicity. Their successful mobilisation of such key resources makes up their scope of local action.

Since 2006, energy cooperatives have been founded all over Germany, and the overwhelming majority have followed the same business focus and thus implemented the same pattern of rules and resources in many different regions. Energy cooperatives literally function as a proof of concept, because they have multiplied the same combination of rules and resources into so many different regions throughout Germany and demonstrate how decentralised, regionally embedded and democratically legitimated renewable energy production can actually work in society across time and space, despite being radically different compared to conventional energy system structures. The structuring potential of energy cooperatives lies, therefore, in the creation *and* multiplication of a very concrete interaction model that is not limited to particular times and spaces, leading to the diffusion of

9.1 Summary and Suggestions for the Future

Energy cooperatives stand for	
decentralised renewable energy production, particularly small to medium-sized production plants combined with	
• a new sense of regional self-confidenceand greater regional self-determination by focussing on regional value creation (economic, as well as societal values),	*Dimension of signification*
• a continual democratic feedback approach through diverse forms of participation and co-decision rights with different intensity grades, and • the importance of creating a common will between local actors,	*Dimension of legitimation*
based upon • financial resources from regional actors, especially from citizens as well as • regional know how, credibility and publicity.	*Dimension of domination*

Fig. 9.1 New structural properties within Germany's energy system created by renewable energy cooperatives

new structural properties within Germany's energy sector, as displayed in Fig. 9.1. Energy cooperatives have managed to create a strong collective identity, which has not only guided community members in interacting in similar ways but also provided the basis for new actors to join the community. The result is an innovative and rising trans-local enterprise community with a well-defined shared identity and a shared resource base across time and space. Energy cooperatives manage to provide a high level of abstraction at the trans-local level while, at the same time, maintaining a very individual character at the local level.

The development of a shared identity and a shared resource base across time and space was brought about through two central mechanisms: (1) **Focus on similar collaborative interaction models in many regions**: Three interaction models were found to function well in many different regions: (1a) collaboration with banks, in particular regionally embedded cooperative ones; (1b) collaboration with other energy companies, particularly with regionally embedded municipal energy providers; and (1c) collaboration with communities. Mostly local citizens are involved as founders and members in all three models. (2) **Access to established networks**: The trans-local enterprise group of energy cooperatives has not developed out of the blue but evolved out of a history of already-existing local and trans-local network activities. Regional founders of energy cooperatives are often already connected to each other before initiating the cooperative. Further, from the day of their initiation, energy cooperatives are embedded within a well-established and long-existing network of nation-wide cooperative associations. The role of established networks in the founding of energy cooperatives demonstrates quite well the co-evolution of new and already well-established interaction patterns.

Energy cooperatives still represent a minority of very small to medium-sized organisations in Germany. In fact, they have only been responsible for the installation of around 1.4% of renewable energy capacity (per technology) in Germany. However, despite being such a minority within Germany's energy sector, they have managed to become highly acknowledged well beyond the boundaries of their own community. They have, for example, been officially and explicitly named in the national EEG 2014 and 2016 as important actors in Germany's Energiewende. Accordingly, they have received special legal treatment with respect to the new tender model for renewable energy capacity, for example (see Sect. 3.4). It is their shared and well-defined identity, as well as their shared resource base, which has made German energy cooperatives into important structuring actors within Germany's energy system, despite their only presenting a small group. They have created attention and respect for their way of thinking and acting beyond their community, to the extent that the government aims to reference their modes of interaction within new national energy regulations. Being able to create a high level of reciprocity between community insiders and outsiders is the base for eventually transferring new rule and resource patterns into many different societal groups. It represents the start of institutionalisation and demonstrates quite well that structural change can take place through the efforts of small groups in a society.[1]

The establishment and multiplication of a very particular pattern of rules and resources throughout Germany has not only been the *greatest strength* of the community, however, but also its *greatest weakness*. The same mode of interaction pursued by many actors in many different regions has led to the circumstance that not only intended actions are being multiplied and spread more widely but also unintended ones, as unintended consequences and unknowledged conditions of action are being manifested and reinforced across time and space. As I have shown, the overwhelming majority of energy cooperatives have been pursuing a risk-averse business approach, despite their explicitly stated intention to become engaged in diverse and innovative renewable energy-related projects. Their overall risk adversity is mainly an unintended consequence of (a) new perceived responsibilities regarding citizens' capital; (b) prevailing economic principles that remain unquestioned; and (c) a lack of internal professionalisation. As a result, energy cooperatives have primarily focused on the installation of small to medium-sized renewable energy production units with a secured return on investment through a feed-in tariff under the EEG. However, many other activities are important in order to achieve a renewable energy system, with key areas being the diffusion of new mobility concepts, increase of energy efficiency, as well as direct connection and harmonisation of energy production and supply. However, only very few energy cooperatives have thus far become engaged in other business fields beyond renewable energy production. It can therefore be concluded that energy cooperatives have been having difficulties in transferring their structural change potential for creating new business activities by combining resource mobilisation with the creation of new

---

[1]For an overview on minority research see Maas and Clark (1984).

9.1 Summary and Suggestions for the Future

societal rules into other renewable energy-related activity areas. In addition, energy cooperatives tend to be critical of system developments which do not fit into their own perspective, in particular when it comes to centralised versus decentralised energy system structures. However, a mix of both system configurations is very likely required in order to achieve a 100% renewable energy system. With these findings, the present work demonstrates that other actors with the ability to take greater investment risks are probably also needed in order to achieve a fully renewable energy system.

Until a few years ago, the dominant and, despite some problems, seemingly well-functioning interaction model preferred by the majority of energy cooperatives had directly depended upon certain regulations in the EEG. In 2014, however, the EEG was substantially amended, with its overall approach being changed from a guaranteed feed-in tariff system to a market-oriented supporting model. As a result, after a period of strong increase between 2006 and 2011, the growth dynamic of German energy cooperatives has been continually decreasing from 2011 onwards, to the point where only 42 new energy cooperatives were registered in 2015. The new approach of the EEG literally does not fit the cooperatives' current risk-averse business approach. It seems to be almost worthless that the EEG has started to value the work of energy cooperatives by focussing on and integrating its exceptions, as the legal insecurity, lower project profitability, greater complexity and inadequate legal support for energy cooperatives under the new EEG rules has decreased the number of new registrations. Energy cooperatives seem to be having great problems in re-orienting themselves in the world of the new EEG and have lost their initial strength of being able to easily multiply. Moreover, they have not managed to create a reliable business base that would be robust enough to cope with external changes over time.

For the future I suggest that energy cooperatives may only be able to re-stabilise and continue to be structuring actors in Germany's energy system if (1) politicians begin to seriously examine and understand the needs of energy cooperatives and provide legal support for the community that can help it to recover and (2) if energy cooperatives emancipate themselves more decisively from being dependent on government regulations and support, such as with the EEG, by diversifying their business activities.

Politicians need to realise that the EEG, in its formulation prior to 2014, was not only the cornerstone for the development and diffusion of new renewable energy technologies but was also the basis for societal innovations. Thus, political responsibility here not only includes guidance of technology development and market integration but also a social dimension. In order to guide a socio-technical transition process, politicians have to learn to explicitly support new societal interaction models that can help new technologies to properly function in society. This can begin by understanding the needs of new rising actor groups so that they can unfold innovative interaction concepts. With regard to German energy cooperatives, existing support under the current EEG may not be enough.

At the same time, energy cooperatives need to find ways to be able to better deal with political change as well as conditions of greater complexity and project insecurity. Based upon the findings presented here, I believe it is important to coordinate a trans-local exchange of know-how about new business models, and it is crucial to find ways to implement successful processes of internal professionalisation. Existing networks of nationally operating intermediate actors from cooperative associations could play an important role in this regard. At the same time, energy cooperatives also need to consider how they can stabilise themselves and further grow without losing their value principles. Focussing on one dominant business model, to a great extent realised through voluntary engagement of local actors, has strongly characterised this community so far; thus, diversifying business activities and undergoing processes of professionalisation may also involve the risk of losing their special mode of acting.

Thus far, German energy cooperatives have shown that they can be 'Multiplying Mighty Davids'. The coming years and decades will show, however, in what ways energy cooperatives may develop within the long-term process of Germany's socio-technical energy transition. Time will also show how the relationship between energy cooperatives and national regulations will further evolve and what kinds of lasting influence this emerging business community may eventually have on Germany's changing energy system.

9.2 Reflection on Theoretical Framework Employed

The empirical assessment I have presented in this work was closely guided by the IBAST framework that I have developed. The discussion of my empirical results demonstrates how important it has been to apply such a holistic, wide-ranging and systematic analytical approach in order to explore and reflect upon the impact of innovative actors within an ongoing socio-technical transition. The IBAST framework has offered the necessary degree of analytical differentiation by providing the means for a comprehensive delineation of the interrelationships between structure and interaction within the current transition at three interrelated spatial levels: the local, the trans-local and the wider system level.

Through conducting the empirical analysis, I have come to the conclusion that the three dimensions of space highlighted by the framework have enabled the exploration of the interactions of a rising group of innovative actors within the present transition from different system angles. None of the three levels should be seen as autonomous areas, as they are strongly interlinked and can only be analytically separated. Thus, what happens at the wider system level is, in some way, a result of interaction at the local level. At the same time, local interaction is guided by dynamics at the trans-local and wider system levels. There is no ontological hierarchy between the levels, as interaction within and between them takes place simultaneously. This is because, according to Giddens, structure is a medium, as well as outcome, of interaction (Giddens 1984).

Empirical assessment carried out here has confirmed the most central aspects of the IBAST framework:

9.2 Reflection on Theoretical Framework Employed

- As proposed by the IBAST framework, interaction and structure must be conceptualised as two separate but recursively connected building blocks, in order to analyse the influence of actors in a socio-technical system. Interaction and structure are contextualised in time and space, and the dimension of space must be understood as combining the spatial, physical and social, as claimed by Giddens (1984).
- Innovative niche and regime structures co-evolve during the very process of interaction. Niches do not arise outside a regime but at its centre; existing structures to a certain extent form the ground for new structures; and routine interaction is an important basis for innovative activities. Therefore, radical change has a direct connection with prevailing established structures—they cannot be separated from each other.
- System change can be accelerated if similar interaction models and similar actor set-ups are multiplied in many regions. This can happen when a community is able to create a new pattern of rules and resources which is transferable and can provide a basis for collective action, which is itself based upon a shared identify and resource base. These guide interaction across time and across space, because they provide ontological security for the whole group with regard to doing 'the right thing'.
- Innovative communities have no control over their trans-local structural change potential. Unintended consequences and un-examined conditions of interaction being multiplied by many actors across time and space can create a dynamic which can neither be foreseen nor easily guided.

There is at least one dimension identified during my empirical assessment which I had not paid enough attention to in elaborating my theory. The co-evolutionary process of emergence, establishment, and replacement of new (niche) and old (regime) structures takes place in conflict-driven waves, entailing continual negotiation and balancing between prevailing and newly rising interests and needs. This was particularly revealed in the relationships that developed between changing national regulations and the rise, and later stagnation, of energy cooperatives. This is an imperfect process that changes over time, and more theory-based research seems to be necessary in this regard.

My empirical assessment has shown that energy cooperatives have become structuring actors, despite being a small minority in society. According to Mass and Clark (1984, p. 430), minority influence research holds that minorities can be influential when they exhibit consistent behaviour and when this consistency is understood by the majority as an expression of certainty and confidence. The findings of the present book appear to confirm this theory-driven view on minority influence, with the additional insight that consistent behaviour can be established with the help of a shared identity and resource base that exists across time and space. My findings also suggest that the majority only positively acknowledges the consistent behaviour of minorities—which can then be influential in re-structuring the established regime—as an expression of certainty and confidence if the minority community is able to create reciprocity between community insiders and outsiders. From a theoretical point of view, it could be very interesting and fruitful to discuss the results of the present book in more detail from a minority influence research perspective.

The IBAST framework is generalisable and therefore well be applicable for examining the impact of other innovative organisational groups within socio-technical transition processes, such as organic supermarkets, with a focus on changes in the food and agricultural system, or electro-mobility organisations, with a focus on changes in the societal mobility system. The framework could also present a fruitful approach for exploring the impact of prevailing organisational actor groups.

Gathering fundamental system knowledge requires a holistic and wide-ranging empirical approach as well as a clear analytical delineation of the phenomenon under study. I suggest that the present book reveals how important it is to combine quantitative and qualitative data, because only then it is possible to explore interrelationships between interaction and structure at different spatial levels with the necessary depth. However, such an analytical approach requires much time as well as know-how about a variety of empirical methods. Such an extensive, complex and time-consuming research design is a strength of the IBAST framework but also, at the same time, a limitation. Researchers applying the framework not only need to have appropriate resources and time but also access to large amounts of data.

Another analytical challenge with regard to exploring actor influence within transitions is to be able to define the point at which niche structures are not innovative or new anymore, because they reached a level at which they are continually reproduced across time and space to the extent that they have actually become a part of the regime structure that characterises a particular socio-technical system. Neither the multi-level perspective nor strategic niche management has been able to present a clear standpoint in this regard, and such challenges remain particularly evident when analysing ongoing transitions. Last but not least, it should be pointed out that, even though the IBAST framework strives to approach actor impact holistically, in relation to the greater dimensions of an entire transition process the area of focus can still only be seen as a 'small window' that has been explored.

9.3 Perspectives for Future Research

One of the central challenges brought to light by my research is how energy cooperatives can successfully make the step from a management structure that is almost exclusively based upon volunteering work towards a professional form of management. Scientific knowledge in this regard could strongly support further development of the whole community. As I have shown, energy cooperatives seem to have higher growth potential when they collaborate with cooperative banks, other energy companies or communities. However, the involvement of cooperative banks and other energy cooperatives has strongly decreased in the past few years. More research is necessary in order to understand the reasons for this decrease.

9.3 Perspectives for Future Research

Urban areas are fundamental for socio-technical transitions, since they are the global hot spots of multiple interactions. Consequently, the Energiewende may be eventually decided in Germany's cities. It is therefore crucial to pay more attention to actors and organisations seeking to develop innovative concepts in cities. In this sense, more knowledge is needed about energy cooperatives that are mainly active in cities. In what ways do urban-focused cooperatives and organisations differ from energy cooperatives that mostly work in rural areas? Is their foundation process different? What makes them special? How can energy cooperative in cities be supported?

The set-up of energy cooperatives and their subsequent management are highly influenced by individual actors. In each case study, the energy cooperative was initiated and developed by individuals who 'burn' for the project and are driven by a 'pioneering spirit'. It would appear, then, that an important pre-condition for the creation of new interaction models through such organisations is, therefore, the pro-active mind-set of such individuals. Thus it may be fruitful to give more attention to the activities, characters and biographies of such innovative and inspiring individuals who work in energy cooperatives or in other organisations. It could be interesting to examine in detail their internal motivations and personal drives and to analyse what they have required in order to unfold their pioneering spirit. It could also be interesting to analyse their biographies, in order to understand how and when such people turn into innovative leaders. Well-founded knowledge about such individuals would not only enrich the results of the present work but also the wider field of sustainability transition research. Research on entrepreneurship and didactics could provide important guidance in this regard (for example, Braukmann 2012; Braukmann et al. 2008, 2009).

During the case study analysis, I observed that not all of the values promoted and expressed by energy cooperatives and their partners seem to become distributed beyond their organisational boundaries. For example, BEG Wolfhagen eG in the first case study clearly pointed out that new participation options also entail a new sense of citizen responsibility with regard to the success of the Energiewende. However, whereas new participation options were widely discussed and highlighted as a new value in local newspapers, aligned citizen responsibility for the Energiewende were not mentioned. This example suggests that the selection and abstraction of new structural properties may not only take place within the community but also when rules are transferred from it into the wider public. It could be interesting to better understand how values change or are actually lost when they leave their home community and whether this process or dynamic can be further categorised. Such knowledge could help to better observe and even to a certain extent steer the transfer of rule and resource patterns into other societal groups.

This work has provided a detailed overview of the current status quo of the translocal community surrounding German energy cooperatives. However, it has only been able to provide a snapshot of a brief timeframe, since the community is only about 10 years old. Thus, research on the impact of energy cooperatives on Germany's Energiewende should continue over a time frame of 10 or 20 years, or even longer, so that their development and structural change potential can be evaluated in relation to Germany's ongoing energy transition.

Energy cooperatives not only exist in Germany but in many other countries as well. It could, therefore, be quite interesting to conduct a similar analysis in other countries and compare results. A key question to be tested should be whether the results of this work only apply to German energy cooperatives or whether they can be transferred to those in other countries.

Last but not least, energy cooperatives are only one of many kinds of organisational groups and initiatives that have been supporting the development of a renewable energy system in Germany. Consequently, comparing the activities and development of energy cooperatives with other pioneering actor groups, applying the IBAST framework, could help to more precisely embed the results of this work into the wider picture of a German socio-technical energy transition process that is being carried out by multiple groups of many different kinds of actors.

Annex

A.1 Overview of German Energy System Scenarios Until 2050 from the National Government and Leading German Institutions

Energy system goals for 2050	Government goals	(BMU 2012) Scenario A	(UBA 2010) Regionen-verbund	(WWF 2009) Scenario innovation without CCS	(Greenpeace 2009) Greenpeace scenario
Greenhouse gas reductions based upon 1990 values	80–95%	At least 80%	80–95%	95%	90%
Share of renewable energy in primary or end energy demand	60% of end energy demand	56% of end energy demand		More than 70% of primary energy demand	
Share of renewable energy in gross power consumption	80%	86%	100%	100%	100%
Share of renewable energy in end heat demand		52%		79%	
Share of renewable energy in the mobility sector		49%	Around 50%		Up to 100%
Share of new modes of driving in the personal driving sector		At least 50% electro-vehicles	50% electro-vehicles and plug-in hybrids	46% electro-vehicles and plug-in hybrids	Up to 100% electromobility

(continued)

Energy system goals for 2050	Government goals	(BMU 2012) Scenario A	(UBA 2010) Regionen-verbund	(WWF 2009) Scenario innovation without CCS	(Greenpeace 2009) Greenpeace scenario
		and plug-in hybrids			
Reduction of primary or end energy demand based upon 2005/2007/2008 values	−50% of primary energy demand	−50% of primary energy demand, −42% of end energy demand	−58% end energy demand	−58% end energy demand	−46% end energy demand
Reduction of power demand based upon 2005/2007/2008 values	−25%	−25%	−20%	−38%	−29%
Reduction of energy demand in the driving sector based upon 2005/2007/2008 values	−40%	−40%		−67% in the personal driving sector	−75% in the personal driving sector
Reduction of heat demand in the residential sector based upon 2005/2007/2008 values	−80%	−80%	−95%	−86%	At least −60%

Source: Based upon (Federal Ministry for the Environment, Nature Conservation, Building and Nuclear Safety 2012; Greenpeace & Friends of the Earth Germany 2013; Umweltbundesamt 2010; WWF Deutschland et al. 2009)

A.2 Envisioned Renewable Power Mix in 2050

	Based on 80% renewable power		Based on 100% renewable power					
	(BMU 2012) Scenario A		(UBA 2010) Regionenverbund		(WWF 2009) Innovation without CCS		(Greenpeace 2009) Greenpeace scenario	
Envisioned renewable power mix in 2050	TWh/a	Percent	TWh/a	Percent	TWh/a	Percent	TWh/a	Percent
Onshore wind	132	27%	170	32%	62	18%	90	19%
Offshore wind	128	26%	177	33%	147	43%	165	35%

(continued)

	Based on 80% renewable power		Based on 100% renewable power					
Envisioned renewable power mix in 2050	(BMU 2012) Scenario A		(UBA 2010) Regionenverbund		(WWF 2009) Innovation without CCS		(Greenpeace 2009) Greenpeace scenario	
	TWh/a	Percent	TWh/a	Percent	TWh/a	Percent	TWh/a	Percent
Photovoltaic	64	13%	104	19%	28	8%	50	11%
Biomass	59	12%	11	2%	41	12%	45	10%
Water	25	5%	22	4%	25	7%	25	5%
Geothermal energy	19	4%	50	9%	36	11%	93	20%
Other (imports, etc.)	61	13%	–	0%	–	0%	–	0%
Total	489	100%	534	100%	339	100%	468	100%

Source: Based upon (Federal Ministry for the Environment, Nature Conservation, Building and Nuclear Safety 2012, p. 311; Greenpeace & Friends of the Earth Germany 2013, p. 93; Umweltbundesamt 2010, p. 63; WWF Deutschland et al. 2009, p. 238)

A.3 Article Research on Regimes: Overview of the 10 Latest Published Articles (as of April 2016) That Apply Socio-technical Transition Theory in the Context of Regimes, Focussing on Energy

Year	Issue	Author	Title	Regime focus	Regime content	Regime boundary
2016	18	Fudge et al. (2016)	Local authorities as niche actors: the case of energy governance in the UK	Regulatory energy regime	Rules, technologies	National: United Kingdom
2016	18	Lockwood (2016)	Creating protective space for innovation in electricity distribution networks in Great Britain: The politics of institutional change	Regulatory energy regime	Not specified	National: Great Britain
2016	18	Lauber and Jacobsson (2016)	The politics and economics of constructing, contesting and restricting socio-political space for renewable: The	Electricity regime	Rules, artefacts and actors	National: United Kingdom, the Netherlands

(continued)

Year	Issue	Author	Title	Regime focus	Regime content	Regime boundary
			German Renewable Energy Act			
2016	18	Markard et al. (2016)	Socio-technical transitions and policy change: Advocacy coalitions in Swiss energy policy	Regulatory energy regime	Rules, artefacts and actors	National: Switzerland
2016	xx	Warren et al. (2016)	Australia's sustainable energy transition: The disjointed politics of decarbonisation	Electricity regime	Rules	National
2015	xx	Huijben et al. (2016)	Mainstreaming solar: Stretching the regulatory regime through business model innovation	Regulatory electricity regime	Not specified	National: The Netherlands, Belgium
2015	17	Hansen and Steen (2015)	Offshore oil and gas firms' involvement in offshore wind: Technological frames and undercurrents	Electricity regime; oil and gas regime	Actors, elements, technologies	National: Norway
2015	17	Wieczorek et al. (2015b)	Transnational linkages in sustainability experiments: A typology and the case of solar photovoltaic energy in India	Electricity regime	Rules	Transnational
2015	xx	Fontes et al. (2015)	The spatial dynamics of niche trajectory: The case of wave energy	Energy regime	Actors, rules, technology	National: Portugal
2015	xx	Faller (2016)	A practice approach to study the spatial dimensions of the energy transition	Heat regime	Rules and actors	National: Luxemburg

Comparison of Case Studies

A.4 Overview of Regional Details for Each Case Study

	Case study		
Regional details	Case Study 1 Cooperation between energy cooperative and another energy company	Case Study 2 Cooperation between energy cooperative and cooperative banks	Case Study 3 Cooperation between energy cooperative and communities
Focus region	Municipality of Wolfhagen	Districts of Neckar-Odenwald & Main-Tauber	Parts of the districts of Neustadt an der Waldnaab, Amberg-Sulzbach & Tirschenreuth
Reason for chossing this region as a focus region	The municipality of Wolfhagen became the focus region because of its political vision to achieve a 100% renewable power supply in the region and because the main purpose of BEG Wolfhagen eG is to be the owner of the municipal energy provider whose business focus is Wolfhagen.	Main-Tauber and Neckar-Odenwald became the focus region because of the districts' joint political goal to become a zero emission region and because the two districts are also the focus region of the cooperative banks that founded the energy cooperative Energie + Umwelt eG.	The focus region of NEW eG and the Bürger is formed by those communities that founded the energy cooperatives and that decided to jointly pursue a 100% renewable energy supply vision. Hence, the region includes parts of the districts of Neustadt an der Waldnaab, Amberg-Sulzbach and Tirschenreuth.
Size of foucs region	112 km^2 13,606 people 121 people/km^2	2431 km^2 271,689 people 113 people/km^2	919 km^2 97,379 people 106 people/km^2
Total regional power demand	66,000 MWh/a	1.9 million MWh/a	158,189 MWh/a
Share of regional power demand covered by energy cooperatives	61% of total regional power demand	1.6% of regional residential power demand	10% of regional household power demand
Share of cooperative members that live in the focus region	67%	100%	80%
Share of energy capacity installed in the focus region/ capital invested in other regions	200,000 Euros	93%	58%

A.5 Overview of New Actor Constellations Being Formed in the Three Case Studies

New actor compositions	Case study		
	Case Study 1 Cooperation between energy cooperative and other energy company	Case Study 2 Cooperation between energy cooperative and cooperative banks	Case Study 3 Cooperation between energy cooperative and communities
Citizens	*Energy customers* Cooperative members, cooperative steering board members, members of energy efficiency advisory committee	*Citizens of Neckar-Odenwald & Main-Tauber* Cooperative members	*Citizens* Cooperative members
Regional political decision takers	*Mayor of Wolfhagen* Member of the supervisory board of Stadtwerke Wolfhagen	*Administrative heads of the districts* Cooperative members, members of the cooperative's supervisory board	*17 mayors of the focus region* Cooperative members, members of cooperative's management and supervisory board
Regional companies	*Stadtwerke Wolfhagen* Members of its own supervisory board, members of energy efficiency advisory committee	*18 cooperative banks* Cooperative members, members of cooperative's management and supervisory board	*Stadtwerke Grafenwöhr & Stadtwerke Floß* Cooperative members, members of cooperative's supervisory board
Regional institutions/ associations	*Energie 2000 e.V.* Members of the supervisory board of Stadtwerke Wolfhagen, members of energy efficiency advisory committee	*Farmer's association of Neckar-Odenwald* Cooperative members, members of cooperative's supervisory board	

A.6 Overview of Business Goals of Energy Cooperatives in Each Case Study

	Case study		
Business goals	Case Study 1 Cooperation between energy cooperative and another energy company	Case Study 2 Cooperation between energy cooperative and cooperative banks	Case Study 3 Cooperation between energy cooperative and communities
Business goals (those that have actually been realised as of December 2014 are boldfaced)	**– BEG Wolfhagen eG holds a minority share in the municipal energy provider** **– Support of local energy efficiency projects** **– Development of own renewable energy projects** – Support the establishment of an environmentally friendly mobility concept	**– Initiation of renewable energy production projects** **– Shareholding in renewable energy production projects** – Initiation of projects that support renewable energy and climate protection in the region – Support energy efficiency in private households	**– Development and operation of own renewable energy projects** – Renewable energy advisory services, such as advise regarding energy efficiency – Bundling of purchasing activities for members as well as offering group contracts for insurance or cleaning services for photovoltaic plants

4.7 Overview of Mobilised Allocative Resources for Each Case Study

	Case study					
Allocated resources	Case Study 1 Cooperation between energy cooperative and another energy company		Case Study 2 Cooperation between energy cooperative and cooperative banks		Case Study 3 Cooperation between energy cooperative and communities	
	Registration year 2012	Year 2014	Registration year 2011	Year 2014	Registration year 2010	Year 2014
Total capital (millions of Euros)	2.4	3.1	4.5	11.9	9.2	22.6
Members (number)	265	675	882	1525	549	1303
Installed projects through December 2014	1 photovoltaic park (5 MW) 1 wind park (12 MW)		36 photovoltaic plants (6 MW) Wind park share (1.12 MW), Biomass plant (0.06 MW)		25 photovoltaic plants (16 MW)	

4.8 Overview of Authoritative Resources Mobilised in All Three Case Studies

Authoritative resources created through collaborative interaction		
Case Study 1 Cooperation between energy cooperative and another energy company	Case Study 2 Cooperation between energy cooperative and cooperative banks	Case Study 3 Cooperation between energy cooperative and communities
– Organisational identify – Image and Publicity – Trust in the business model of the energy cooperative – Trust between involved actors – Citizen acceptance of energy projects – Credibility of energy efficiency projects – Access to potential projects – Access to information and new ideas – Transparency – Expertise & know-how	– Organisational identity – Publicity – Image – Trust in the business model of the energy cooperative – Trust between involved actors – Transparency – Access to potential projects – Access to information and new ideas – Expertise & know-how	– Organisational identity – Publicity – Image – Courage – Trust in the business model of the energy cooperative – Trust between involved actors – Internal transparency – External project transparency – Acceptance of renewable energy projects – Access to potential projects – New project ideas – Expertise & know-how

A.9 Overview of the Mix of Newly Created Rules in All Three Case Studies

Rules created through collaborative interaction		
Case Study 1 Cooperation between energy cooperative and other energy company	Case Study 2 Cooperation between energy cooperative and cooperative banks	Case Study 3 Cooperation between energy cooperative and communities
Regulations – Direct ownership of citizens and energy customers over their own municipal energy provider – New co-decision and participation rights for citizens and energy customers – General limitation of individual-member investment volume and of profit distribution for the sake of the community.	*Regulations* – Official accreditation of cooperative banks for guiding citizen focused regional energy projects. – New co-decision and participation rights for citizens – New co-decision rights for politicians in the execution of renewable energy projects – General limitation of	*Regulations* – Municipal guidance and ownership of renewable energy projects – New co-decision and participation rights for citizens – New direct energy consumer–producer relationship – New forms of regional

(continued)

Rules created through collaborative interaction		
Case Study 1 Cooperation between energy cooperative and other energy company	Case Study 2 Cooperation between energy cooperative and cooperative banks	Case Study 3 Cooperation between energy cooperative and communities
– New forms of regional value creation *Interpretative schemes* – Taking matters into one's own hands' – New continual democratic feedback for municipal energy goals – Citizen identification with change process – New energy co-responsibility of citizens for regional energy issues – New co-responsibility of municipal energy provider for capital from citizens and customers – Symbiotic relationship between citizens and municipal energy provider	individual-member investment volume in renewable energy projects. – New forms of regional value creation *Interpretative schemes* – Taking matters into one's own hands' – Full citizen acceptance is a new pre-condition for getting involved in renewable energy projects – New co-responsibility of citizens for regional energy issues – New co-responsibility of cooperative banks for renewable energy project development – New symbiotic relationship between citizens and cooperative banks	value creation *Interpretative schemes* – Taking matters into one's own hands' – New regional self-confidence with respect to energy – The importance of a collective will between municipalities – Citizen identification with renewable energy projects – New co-responsibility of politicians for citizen-member capital – New symbiotic relationship between citizens and municipalities

A.10 Overview of Identified Participation Levels for Each Case Study

	Case study		
Activated participation levels	Case Study 1 Cooperation between energy cooperative and another energy company	Case Study 2 Cooperation between energy cooperative and cooperative banks	Case Study 3 Cooperation between energy cooperative and communities
Financial participation	**All** 675 members by acquiring 5997 member shares with a value of three million Euros by 2014	**All** 1525 members by acquiring 65,784 member shares with a value of 6.6 million Euros by 2014	**All** 1303 members by acquiring 28,460 member shares with a value of 14.4 million Euros by 2014
Consultive participation	**27–30%** of members attend general assemblies and make use of their voting rights, stay rather passive, and follow the line of the steering board	**10–13%** of members attend general assemblies and make use of their voting rights, stay rather passive, and follow the line of the steering board.	**15–30%** of members attend general assemblies and make use of their voting rights, stay rather passive, and follow the line of the steering board

(continued)

	Case study		
Activated participation levels	Case Study 1 Cooperation between energy cooperative and another energy company	Case Study 2 Cooperation between energy cooperative and cooperative banks	Case Study 3 Cooperation between energy cooperative and communities
Creative participation	**A group of citizens** is engaged in project development by being part of the energy efficiency supervisory committee and the aligned member working groups.	**Some citizens**, such as two farmers, engaged in project development because they use the energy cooperative in order to implement a project on their own property; additional working groups for members do not exist.	**No citizen** engagement in project development because additional working groups for members or projects on member property do not exist
Steering participation	**Seven** of seven steering board members are citizens.	**None** of the seven steering board members are citizens.	**Five** of the six steering board members of Bürger eG are citizens.

A.11 Overview of Project Acceptance in Each Case Study

	Case study		
Project acceptance	Case Study 1 Cooperation between energy cooperative and another energy company	Case Study 2 Cooperation between energy cooperative and cooperative banks	Case Study 3 Cooperation between energy cooperative and communities
Process of creating or strengthening acceptance	All three energy cooperatives align their business activities with new participation rights; they coordinate their groups of supporters over the long term and make them visible to the public. → Strong unity of supporters weakens opposing groups.		
Limits of creating acceptance	BEG Wolfhagen eG could not convince citizens who had positioned themselves against the Rödeser Berg wind park before the energy cooperative was founded.	Energie + Umwelt eG actively bypasses conflicts over project acceptance by using project acceptance as a pre-condition (rule) which must already exist before it becomes engaged in a project.	NEW eG could not convince the administrative head of Neckar-Odenwald, who already had positioned himself against wind power before NEW eG was founded.

A.12 Overview of Pre-existing Networking Activities in Each Case Study

Case Study 1 Cooperation between energy cooperative and another energy company	Case Study 2 Cooperation between energy cooperative and cooperative banks	Case Study 3 Cooperation between energy cooperative and communities
The citizen initiative Klimaoffensive Wolfhagen had sensitised political decision makers about taking more responsibility for climate protection and renewable energy in Wolfhagen. A result of the process was the official political decision to develop more renewable energy in the region and jointly found BEG Wolfhagen eG. The citizen initiative ProWind Wolfhagen supported wind park plans and became engaged in the foundation of the energy cooperative.	The cooperative banks Volksbank Mosbach, Volksbank, Main-Tauber and Volksbank Franken, which founded the energy cooperative Energie + Umwelt eG, were part a regional working group that met regularly in order to discuss how to support the development of the region. They describe themselves as a well-functioning alliance which realised the value-added of joint action.	The three main communities, which founded NEW eG, Grafenwöhr, Pressath and Eschenbach, have a long history of collaboration, which goes back to the early 1990s. Amongst other examples, they initiated VierStädtedreieck in order to support each other in realising the long-term potential of their region. The initiative, which eventually included all communities that also became part of NEW eG, formed a local action group under the EU LEADER programme.

A.13 Full Guidelines Used During Case Study Interviews (in German)

Block 1a: Gründungsprozess Energiegenossenschaft		
Wie kam es zu der Gründung der Energiegenossenschaft?		
Inhaltliche Aspekte	Nachfragen	Research question
☐ Änderungswünsche im regionalen Energiesystem ☐ Initiatoren ☐ Befürworter & Gegner	• Was war das Ziel der Gründung? • Wie war die Reaktion in der Region, z.B. der Bürger, Politik, Stadtwerke? • Welche Voraussetzungen waren wichtig?	• Wo sind Reibungspunkte im regionalen Energieregime? • Wie werden Nischenunternehmen gebildet?
Block 1b: Eigener Handlungsspielraum, Organisationsstrategie		
Beschreiben Sie doch einmal den Auftrag der [Organisation] in der Region.		
Inhaltliche Aspekte	Nachfragen	Research question
☐ Eigener Handlungsspielraum ☐ Eigene Grenzen ☐ Aufgabenbereich innerhalb	• Welche Aufgaben hat die Organisation? • Welche Vorteile bietet die [Organisation]?	• Wie ist die Organisation in Nischen bzw. in Regimestrukturen eingebettet?

(continued)

der Energiewertschöpfungskette (Produktion, Netz, Einkauf, Service) ☐ Organisationstrategie ☐ Zielgruppen ☐ Außenwahrnehmung	• Was kann die [Organisation] nicht? • Wie wird die [Organisation] wahrgenommen? • Welche Zielgruppen spricht sie vor allem [Organisation] an?	
Block 1c: Eigener Handlungsspielraum, Politik		
Beschreiben Sie doch einmal die energiepolitischen Ziele von [Ort].		
Inhaltliche Aspekte	Nachfragen	Research question
☐ Eigener Handlungsspielraum ☐ Eigene Grenzen ☐ Aufgabenbereich innerhalb der Energiewertschöpfungskette (Produktion, Netz, Einkauf, Service) ☐ Organisationstrategie ☐ Zielgruppen ☐ Außenwahrnehmung	• Welche energiepolitischen Themen werden derzeit behandelt? • Welche Rolle spielt das Thema Energie in der Politik von [Ort]? • Was erleichtert die Umsetzung? • Wo liegen Herausforderungen in der Umsetzung? • Wo liegen die Grenzen der Politik? • Wie werden die politischen Aktivitäten wahrgenommen?	• Wie ist die Organisation in Nischen bzw. in Regimestrukturen eingebettet?
Block 2a: Konstitution der Kooperation		
Welche Berührungspunkte gibt es zwischen [Organisation] und der Energiegenossenschaft?		
Inhaltliche Aspekte	Nachfragen	Research question
☐ Gründe für die Entscheidung, zu kooperieren ☐ Typische Merkmale der Beziehung ☐ Umgang mit Änderungsdruck/−wunsch ☐ Fehlende Ressourcen im eigenen Unternehmen ☐ Fehlende Regeln	• Wie kam es zu der Zusammenarbeit? • Was war das Ziel? • Welche Rolle spielen politische Gesichtspunkte? • Wie war die Reaktion in der Region, z.B. der Bürger, Politik, Stadtwerke	• Wie und warum haben sich Kooperations-akteure angenähert? • Welche Regeln und Ressourcen standen für die Kooperation zur Disposition?
Block 2b: Konstitution der Kooperation		
Welches ist denn der wichtigste Kooperationspartner der Energiegenossenschaft? Wie kam es zu der Zusammenarbeit zwischen [Organisation] und [Organisation]?		
Inhaltliche Aspekte	Nachfragen	Research question
☐ Gründe für die Entscheidung, zu kooperieren ☐ Typische Merkmale der Beziehung ☐ Umgang mit Änderungsdruck/−wunsch ☐ Fehlende Ressourcen im eigenen Unternehmen ☐ Fehlende Regeln	• Was war das Ziel der Zusammenarbeit? • Welche Voraussetzungen waren wichtig? • Welche Rolle spielen politische Gesichtspunkte? Extra Nachfragen für Politik: • Welche politischen Berührungspunkte gibt es?	• Wie und warum haben sich Kooperations-akteure angenähert? • Welche Regeln und Ressourcen standen für die Kooperation zur Disposition?

(continued)

Block 2c: Verlauf der Kooperation

Beschreiben Sie doch einmal ein typisches Beispiel der Zusammenarbeit.

Inhaltliche Aspekte	Nachfragen	Research question
☐ Gemeinsame Ressourcennutzung ☐ Schaffung neuer Ressourcen ☐ Verhandlung von bestehenden Regeln ☐ Schaffung von neuen Regeln ☐ Einfluss der Kooperation auf die Machtverteilung der Akteure ☐ Breaks, Veränderungen in der Zusammenarbeit	• Beschreiben sie doch einmal die Zusammenarbeit zwischen [Organisation] und [Organisation]? • Was waren besondere Situationen? • Welche Herausforderungen/ Konflikte gibt es? • Welche Rolle kommt der [Organisation] in der Zusammenarbeit zu? Wer hat das Sagen? • Welche Zugeständnisse wurden gemacht? • In wieweit sind [Organisation] und [Organisation] voneinander abhängig? Extra Nachfragen für die Politik: • Welche politische Funktion hat die Zusammenarbeit? • Existieren politischen Meinungsverschiedenheiten/ Konflikte? Wenn ja, welche?	• Welche Intensität hat die Interaktion? • Wie werden Ressourcen verteilt, entwickelt? • Wie werden Regeln neu verhandelt, entwickelt?

Block 4a: Bewertung der Organisation

Wie bewerten Sie die Arbeit der [Organisation]?

Inhaltliche Aspekte	Nachfragen	Research question
☐ Einfluss der Interaktion auf den eigenen Handlungsspielraum ☐ Vorteile ☐ Nachteile ☐ Rahmenbedingungen die eine Zusammenarbeit erleichtern	• Wo liegt der Mehrwert? • Wo sind die Grenzen der Arbeit? • Was die Arbeit erleichtert? • Was hat die Arbeit erschwert? Extra Nachfragen für die Politik: • Wo liegt der politische Mehrwert der Energiegenossenschaft?	• Welchen Einfluss hatte die Kooperation auf die Verbreitung von neuen Ideen? • Welchen Einfluss hatte die Kooperation auf die Vergrößerung des Unterstützerumfeldes?

Block 4b: Auswirkungen der Organisation

Was hat sich durch die Energiegenossenschaft verändert?

Inhaltliche Aspekte	Nachfragen	Research question
☐ Einfluss der Interaktion auf den eigenen Handlungsspielraum ☐ Zuwachs in der eG oder bei den Stadtwerken ☐ Bildung eines Wissenszentrums oder	• Was hat sie ermöglicht? • Was wurde durch sie entwickelt oder geschaffen? • Was hat man durch sie gelernt? • Wie wird sie von außen wahrgenommen?	• Welchen Einfluss hatte die Kooperation auf die Verbreitung von neuen Ideen? • Welchen Einfluss hatte die Kooperation auf die Vergrößerung des Unterstützerumfeldes?

(continued)

Netzwerken ☐ Verschiebung des Fokus auf Technologien und anderen Energieprojekten	• Wie ist die Reaktion der Bürger/der Politik? Extra Nachfragen für die Politik: • Was hat sich aus politischer Sicht geändert? • Was hat man aus politscher Sicht gelernt?	

Block 4c: Bewertung der Kooperation

Wie bewerten Sie die Zusammenarbeit (Verhältnis) zwischen [Organisation] und [Organisation]?

Inhaltliche Aspekte	Nachfragen	Research question
☐ Einfluss der Interaktion auf den eigenen Handlungsspielraum ☐ Vorteile ☐ Nachteile ☐ Rahmenbedingungen die eine Zusammenarbeit erleichtern	• Wo liegt der Mehrwert der Zusammenarbeit? • Wo sind die Grenzen der Zusammenarbeit? • Was die Zusammenarbeit erleichtert? • Was hat die Zusammenarbeit erschwert? • Was fehlte denn für eine (engere) Zusammenarbeit? • Wann wäre eine engere Zusammenarbeit interessant? Was wäre dazu nötig? Extra Nachfragen für die Politik: • Wo liegt der politische Mehrwert der Zusammenarbeit?	• Welchen Einfluss hatte die Kooperation auf die Verbreitung von neuen Ideen? • Welchen Einfluss hatte die Kooperation auf die Vergrößerung des Unterstützerumfeldes?

Block 4d: Auswirkungen der Kooperation →gemeinsamer Handlungsspielraum

Was hat sich durch die Zusammenarbeit zwischen [Organisation] und [Organisation] verändert?

Inhaltliche Aspekte	Nachfragen	Research question
☐ Einfluss der Interaktion auf den eigenen Handlungsspielraum ☐ Zuwachs in der eG oder bei den Stadtwerken ☐ Bildung eines Wissenszentrums oder Netzwerken ☐ Verschiebung des Fokus auf Technologien und anderen Energieprojekten	• Was hat die Zusammenarbeit ermöglicht? • Was wurde durch die Zusammenarbeit entwickelt oder geschaffen? • Was hat [Organisation] durch die Zusammenarbeit gelernt? • Wie wird die Zusammenarbeit von außen wahrgenommen? • Wie ist die Reaktion der Bürger/der Politik? Extra Nachfragen für die Politik: • Was hat sich aus politischer Sicht geändert? • Was hat man aus politscher Sicht gelernt?	• Welchen Einfluss hatte die Kooperation auf die Verbreitung von neuen Ideen? • Welchen Einfluss hatte die Kooperation auf die Vergrößerung des Unterstützerumfeldes?

(continued)

Block 5a: Perspektive der Organisation		
Wie bewerten Sie die Zukunft der [Organisation]?		
Inhaltliche Aspekte	Nachfragen	Research question
☐ Ressourcen eines regionalen Energiesystems ☐ Regeln eines regionalen Energiesystems	• Welche Pläne gibt es? • Was sind zukünftige Geschäftsziele? • Wo liegen die größten Herausforderungen? • Welche Rahmenbedingungen sind wichtig oder müssen sich ändern?	• Wie wird ein regionales Energiesystem definiert? • Wo liegen die Herausforderungen im Regime?
Block 5b: Perspektiven der Kooperation		
Welche Rolle spielt Ihrer Meinung nach die Zusammenarbeit zwischen [Organisation] und [Organisation] in der Zukunft?		
Inhaltliche Aspekte	Nachfragen	Research question
☐ Einschätzung zukünftiger Perspektiven für die Kooperation im regionalen Energiesystem	• Welches Potenzial sehen Sie in der Zusammenarbeit? • Wie definieren Sie Erfolg für diese Form der Kooperation? • Was wäre für Sie eine ideale Konstellation? • Was fehlt? Extra Nachfragen für die Politik: • Welche Rahmenbedingung müsste die Politik noch stellen?	• Welches Unterstützungspotenzial hat eine Kooperation in der Institutionalisierung neuer Ideen für die regionale Energiewende? • Wie wird die Institutionalisierung weiterverlaufen?
Block 5c: Verständnis Energiesystem und Wandel		
Was verstehen Sie unter einem regionalen Energiesystem?		
Inhaltliche Aspekte	Nachfragen	Research question
☐ Ressourcen eines regionalen Energiesystems ☐ Regeln eines regionalen Energiesystems	• Was beinhaltet ein regionales Energiesystem? • Welche Entwicklungen sind in der Zukunft wichtig? • Wo liegen die größten Herausforderungen? • Welche (politischen) Rahmenbedingungen sind wichtig?	• Wie wird ein regionales Energiesystem definiert? • Wo liegen die Herausforderungen im Regime?

A.14 Original German Quotes

Chapter 7.1, Original Quotes, Case Study I

1.1 "da haben wir dann halt eine sehr konstruktive Unterstützung von dem Vorstand der GLS-Bank [...] gehabt, der uns dann gesagt hat, haben Sie schon mal über eine

Genossenschaft nachgedacht? Und zwar nicht eine Genossenschaft am Windpark, sondern eine Genossenschaft an den Stadtwerken Wolfhagen. Und zwar, weil wir mit der Frage gekommen sind, wollen wir jetzt jedes Mal für jedes Projekt, was wir machen und wo wir Bürger dran beteiligen wollen, eine eigene Tochtergesellschaft gründen, wo dann Bürgerbeteiligung möglich ist und wir springen nur noch dann auf [...] Sitzungen rum, wo dann die Beteiligungsgenossenschaft Solarpark, die Beteiligungsgenossenschaft Windpark und die Beteiligungsgenossenschaft Biogasanlage sich trifft. Deswegen sind wir zu der Auffassung gekommen, die Erneuerbare-Energien-Projekte, die wir in Wolfhagen vorhaben, als direkte Tochtergesellschaft der Stadtwerke zu gründen [...] und eine Beteiligung an den Stadtwerken Wolfhagen zu ermöglichen" (R1N1, 48–50).

1.2 "Die politischen Mandatsträger haben gesagt, was mir persönlich noch wichtig war, wir wollen, dass alle Bürger die Chance haben sich zu beteiligen, nicht nur die, die Kohle haben" (R1N2, 32).

1.3 "Der Auftrag der BürgerEnergieGenossenschaft ist eben letztlich, die Beteiligung zu realisieren, die Gelder einzusammeln, was sie ja erfolgreich gemacht haben" (R1N2, 198).

1.4 "definieren wir halt das Thema Daseinsvorsorge noch etwas weiter im Sinne von, Daseinsvorsorge sollte uns auch interessieren, wo unser Strom herkommt. Und Daseinsvorsorge sollten wir auch drauf achten, wie wir diesen Strom verbrauchen—in Klammern wie viel wir davon verbrauchen [...]. Und das schließt für uns mit ein, dass wir uns halt auch um dezentrale Energieversorgung kümmern. Dass wir regionale und lokale Kraftwerke bauen aus Erneuerbaren Energien, unsere PV-Park, Windpark" (R1N1, 68, 72).

1.5 "Dass gesagt wurde, also diejenigen, die jetzt in der Energiegenossenschaft mitmachen wollen, ja auch davon profitieren wollen. Die sollen bitte aber auch dann Stromkunde im Moment der Stadtwerke sein und nicht jetzt bei FlexStrom, das gerade wieder in Diskussion ist, oder irgendeinem anderen Billiganbieter dann den Strom billig beziehen, aber vom Gewinn eher ein Stück abhaben wollen" (R1N4, 97).

1.6 "mittelfristig [ist] natürlich eine weitere Kundenbindung erfolgt wie es sonst kein anderer Versorger in Deutschland hat, ich wüsste nicht. Ich wüsste keinen anderen." (R1N1, 216).

1.7 "Und es hilft natürlich den Stadtwerken, den Kundenstamm zu behalten über eine längere Bindung" (R1N3a, 105).

1.8 "Mitglied kann nur werden, wer Kunde der Stadtwerke ist, und zwar Stromkunde [...] insofern kann man aus meiner Sicht—aus unserer Sicht—nicht mehr von einer BürgerEnergieGenossenschaft reden. Allenfalls von einer KundenEnergieGenossenschaft [...] Die haben auch Kunden aus ganz Deutschland. Und die können Mitglieder der BEG werden. Während andere Bürger der Stadt Wolfhagen ausgeschlossen sind, weil sie kein Stromkunde mehr sind" (R1N5, 55–59).

1.9 "Und wenn wir dann mal keine Lust mehr haben, dass es dann noch welche gibt, die nachkommen. Das ist ja das nächste Problem dann schon: Wer wird das dann übernehmen wollen? Zumindest nicht auf unentgeltlich-ehrenamtlich-Basis.

Ja, das ist immer so der Anspruch und Wirklichkeit. Also einfach diese Zeitfrage, das ist ja schon enorm" (R1N3b, 52).

1.10 *"Dass wir uns halt auch um dezentrale Energieversorgung kümmern. Dass wir regionale und lokale Kraftwerke bauen aus Erneuerbaren Energien, unseren PV-Park, Windpark. Und dass wir das hier gar nicht alleine können und dass das auch nicht Sache der Kommune alleine ist, [...]" (R1N1, 72).*

1.11 *"Ursache dieser Gründung ist—, dass der Rödeser Berg oder überhaupt diese vier Anlagen, egal wo sie stehen, finanziert werden müssen. Die Stadtwerke können es nicht aus eigenem Säckel machen, ja. Dafür ist die BürgerEnergieGenossenschaft gegründet, als Geldbeschaffungsmaschine" (R1N5, 143).*

1.12 *"Und dann war die Frage: Was machen wir jetzt mit den anderen Guthaben, die jetzt kommen? Ich meine, der Kuchen ist ja nur einmal da, letztes Jahresergebnis der Stadtwerke. Und es wird sich ja immer mehr verteilen dann. Und dann dadurch reduziert sich ja automatisch auch die Dividende der anderen. Also man muss jetzt schon was finden, dass wir unsere neuen Guthaben eben auch so anlegen, satzungsgemäß natürlich und so, dass es wirtschaftlich ist und noch eine schöne Dividende erwirtschaftet wird. Ist gar nicht so leicht, jetzt auch noch regional und so. Es steht kein Projekt konkret an, dass wir wirklich jetzt im Juli die nächsten [...] [Euros], die wir haben, schon irgendwo einsetzen können" (R1N3a, 273).*

1.13 *"Und müsste man eben schon befürchten, dass irgendwann, wenn dann wieder eine Hochzinsphase kommt und die Banken ganz andere Angebote an die Anleger machen können, dass dann auch einfach die Mitglieder abschwinden. Weil eben dann doch die Motivation auch die Dividende ist. Da will ich ja auch gar nicht dran denken" (R1N3a, 269).*

1.14 *"Dass die Mitglieder einfach merken: Hier tut sich mal inhaltlich was. Und wenn wir schon noch keine Dividende haben, aber es passiert mal was inhaltlich" (R1N3a, 73).*

1.15 *"Aber auf Dauer ist es nicht gut, weil ja auch immer nachgefragt wird von unseren Kritikern, den politischen: Was ist denn die Stadt als Gesellschafterin? Wird die nicht übern Tisch gezogen, wenn jetzt die andere Gesellschafterin Genossenschaft auf einmal noch Dienstleistungen kostenlos kriegt? Verdeckte Gewinnausschüttung durch Leistungen der Mitarbeiter und so" (R1N3a, 95).*

1.16 *"Oh, dass es langfristig aufgeht, unser Konzept. Dass es nicht so viele Schwierigkeiten gibt, jetzt durch das EEG, dass da zumindest noch Alternativen für uns sind, eigene Projekte zu machen. Und ich hoffe wirklich, dass wir nicht zu so einer reinen Beteiligungsgenossenschaft da verkommen. Also "verkommen" in Anführungszeichen! Aber dass man so beides haben kann. Dass man eben auch ein breiter aufgestelltes Portfolio hat, wo man einerseits halt Beteiligungen hat an größeren Projekten, die eben auch im Endergebnis wahrscheinlich viel effizienter sind natürlich, von der Energieerzeugung her und so betrachtet. Aber eben auch Eigenes hat. Dass man den Mitgliedern zeigt, wir haben auch selbst was getan, was wir vorweisen können" (R1N3b, 52).*

1.17 *"Ein sehr mutiger Schritt, denn damit werden die Bürger wirklich in das Unternehmen geholt. Zugleich ist dies ein substantieller Beitrag zur Finanzierung für kleinere Stadtwerke. Auch die Integration in die Gesamtstrategie ist gelungen"* (Stadtwerke Wolfhagen 2013).

1.18 *"Also das ist für ein kleines Stadtwerk ja wirklich schwierig, die eigene Identität darzustellen. Also man merkt das ja in Gesprächen mit Bürgern immer wieder, dass die eigentlich gar nicht mehr unterscheiden können zwischen Stadtwerken und großem Energieversorger. Das ist doch alles eins. Das passiert mir also auch hin und wieder mal, dass man das hört: ja, ach, E.ON und Stadtwerke, das ist doch sowieso alles eins. [...] Und das ist jetzt doch über die Energiegenossenschaft anders, denn da macht man sich Gedanken. Und 500 Euro haben doch viele zur Verfügung"* (R1N4, 79).

1.19 *"Man kannte ja manche wirklich gar nicht. [...] Und dann wusste ich gar nicht, was ist das für einer? Wer weiß was der für Interessen hat, [...]. Könnte ich mit denen überhaupt zusammenarbeiten und wie ticken die? Was haben die für wirkliche Motive oder so? [...]Und dass dann auch noch harmoniert, ist natürlich optimal, also bis jetzt noch, nach fast zwei Jahren [...]"* (R1N3b, 38).

1.20 *"Na ja, viele Risiken liegen halt bei den Stadtwerken. Jedenfalls, wenn es um größere Projekte geht. Und wir haben zwar unterm Strich dadurch weniger Ertrag—also weniger Dividende—als hätten wir es selbst als Investor betrieben, aber dafür auch eben eine breitere Streuung und mehr Sicherheit. Das ist ja eben dann auch ein Wert in sich"* (R1N3b, 44–46).

1.21 *"Und so gab es Befürworter des Windparks und Gegner des Windparks. Und auch dazu eine große schweigende Mehrheit auf beiden Seiten. Und von da aus, ja, haben wir einfach sozusagen, ich würde mal sagen, pro aktiv da noch mal auch in die durchaus schwierige Lage—von einer Pro-Bürgerinitiative und einer Gegen-Bürgerinitiative gegen den Windpark—haben wir einfach noch von einer ganz anderen Ebene—[abgeholt]"* (R1N1, 58).

1.22 *"Wir sind nicht gegen die Nutzung der Windkraft vor Ort, aber wir sind eindeutig nach wie vor gegen diesen geplanten Standort"* (R1N5, 19).

1.23 *"Man hat von Anfang an gesagt: Dorthin wollen wir, auf den Rödeser Berg. Und das war von Anfang an das Ziel. Man hat nach außen hin ... es hat Info-Veranstaltungen gegeben, erst nachdem wir das gefordert haben. Vorher ist der Bürger überhaupt nicht eingebunden worden in diese ganzen Planungen. Dann hat es Info-Veranstaltungen gegeben, da standen fünf Alternativstandorte zur Debatte. Innerhalb kürzester Zeit, ich glaube das Ganze hat etwa ein Vierteljahr gedauert. Mehrere Sitzungen und so weiter hat sich das dann auf dem Rödeser Berg nur noch herauskristallisiert. Mit allen möglichen Argumenten"* (R1N5, 45).

1.24 *"Gerade am Beispiel des Windparks wird deutlich, hätten wir die BEG früher gehabt ... Hätten wir die schon von Anfang an gehabt, dann hätten wir, glaube ich, das Projekt geschmeidiger hinbekommen"* (R1N1, 172).

1.25 *"Da sind wir im Laufe des Prozesses auch ganz neue Wege gegangen, wir haben da Workshops gemacht mit den Bürgern und so was. Das ist aber alles sehr spät gewesen. Das war sehr gut. Also heute würde ich das früher machen. [...] Also das war da, glaube ich, eine gute Möglichkeit das [Windpark Projekt] zu erläutern und zu erklären. Die Information zu transportieren"* (R1N2, 432–448).

1.26 *"Anfang letzten Jahres, ja, diese Zukunftskonferenz ... Ich war dabei, ja. Also grundsätzlich, ich habe an mehreren Diskussionsrunden da auch, war ich ... habe ich mit drin gesessen. Da war wirklich also, die Leute, die dort waren, die waren auch alle sehr interessiert am Thema. Haben auch teilweise selbst gute Ideen dabei gehabt. [...] wie gesagt, es kam eigentlich schon ein paar Jahre zu spät hier. So was hätte man ... so eine reine Informationsveranstaltung oder eine Beteiligungsveranstaltung in der Art hätte hier viel früher stattfinden müssen"* (R1N5, 471–479).

1.27 *"Und ich habe das auch gemerkt, die EnergieGenossenschaft hat also jetzt zum Frühjahr—oder im Winter war es eigentlich noch—eine Aktion gestartet zur Thermographie von Wohngebäuden. Also ich war sehr überrascht über die Resonanz, wie gut das angenommen wurde. Weil, ich kenne es von anderen Projekten, auch Banken haben das ja schon gemacht, dass es doch wesentlich verhaltener war"* (R1N4, 128).

1.28 *"Da werden wir auch immer wieder angesprochen oder sind wir angesprochen worden—wie kann denn ein Stadtwerk wirklich ernsthaft meinen, dass sie Energieeffizienz und Einsparungen nach vorne treiben wollen"* (R1N1, 170)?

1.29 *"Aber wenn es beispielsweise um das Thema Stromsparen geht, ist es schon sehr viel schwieriger, an die einzelnen Bürger ranzukommen und sie zu überzeugen, dass sie wirklich etwas tun können. Und auch weil es doch noch mehr in den gesamten persönlichen Bereich reingeht. So, und da, denke ich, hat die EnergieGenossenschaft schon den großen Vorteil, dass sie als Institution noch anders anerkannt ist. Die Bürger sind Mitglied. Das ist noch mal eine andere Vertrauensbasis da und ... ja, es ist vielleicht noch überzeugender, was jetzt von der BürgerEnergieGenossenschaft kommt, nach dem Motto: ja, die wollen uns ja nichts verkaufen, das sind wir ja eigentlich selbst und das glauben wir dann eher mal"* (R1N4, 122–124).

1.30 *"Das heißt der Auftrag hat sich ja auch geändert, der ist nicht so geblieben. 100 Prozent Erneuerbare sind dazu gekommen, dann eben die BürgerEnergieGenossenschaft ist dazu gekommen. Es ist schon auch ein Prozess, der da erkennbar wird"* (R1N2, 190).

1.31 *"Dass man sagt, oh, jetzt müssen wir 25 Prozent des Kuchens abgeben. [...] Aber wir brauchten auch das Geld, um die Anlagen zu bauen. Wir wären gar nicht in der Lage gewesen, das so aus kommunalen Geschichten zu sehen. Und von da aus muss man fragen letztlich, wollen wir den kleinen Kuchen ganz alleine essen, oder wollen wir die dicke Torte und geben davon ein Stück. Das ist so die Frage. Und ich meine, dann lieber die dicke Torte und davon Dreiviertel weiterhin behalten. Also unterm Strich gesehen, der Kuchen für die Stadt wird auch größer. Und von da aus ist es unterm Strich eine Win-Win-Situation. Die Stadt profitiert davon [...]"* (R1N1, 322–326).

1.32 *"Die Stadtwerke waren hundertprozentige Tochter der Stadt. Da war jeder Bürger beteiligt. Jetzt ist es ein exklusiver Kreis, diejenigen, die es sich leisten können, Mitglied in der BEG zu sein, die den Anteil zahlen können. Die sind jetzt zu 25 Prozent immer beteiligt"* (R1N5, 29).

1.33 "Ja, es ist sinnvoll, dass Ihr [die Bürger] mitbestimmen dürft und auch beteiligen dürft, dass Ihr also vorkommt in diesen Erneuerbaren Projekten" (R1N1, 134).

1.34 "Wo gesagt wurde, wir wollen auf jeden Fall eine breite Bürgerbeteiligung" (R1N2, 210).

1.35 "doch die Gegner, die im Windpark immer gesagt haben, Ihr macht nur Euer eigenes Ding und lasst keinen mitbestimmen. [. . .]Mit der Genossenschaft können wir dem aktiv begegnen im Sinne von Bürger, Ihr könnt tagtäglich mitbestimmen bei Energiethemen, indem Ihr euch eben beteiligt an den Stadtwerken und damit auch Einfluss gewinnt.[. . .] Also es ist eine deutlich umfangreichere Beteiligung halt als die Beteiligung halt nur an einer Erneuerbare-Energien-Anlage" (R1N1,50–54).

1.36 "Da gibt es natürlich dann schon bei Mitgliedern der Genossenschaft auch so die Stimmung: ja, eigentlich würden wir gerne noch ein bisschen mehr mitreden in Detail-Entscheidungen. Noch direkteren Einfluss beispielsweise auf die Geschäftspolitik nehmen. Aber das ist halt durch die Verträge recht klar geregelt, dass also Mitgliedschaft im Aufsichtsrat da ist und darüber zu laufen hat" (R1N4, 364).

1.37 "Eben eigentlich, um eine Geldanlage, denke ich, zu haben muss man wirklich sagen. Also ein Großteil—das ist leider so—, denke ich, wenn wir Gespräche haben da in den Sprechstunden, die fragen nicht: Und was machen dann die Stadtwerke genau mit dem Geld als Nächstes und so? Die fragen: Wie sieht denn die Dividende aus" (R1N3a, 131–133)?

1.38 "Das sind wirklich immer wenige die fragen. Es wird auch immer wenig Grundsätzliches gefragt oder diskutiert, jetzt über also als es um die Strompreise ging. [. . .] Ich glaube die meisten sind dann auch einfach erschlagen von so vielen Tagungsordnungspunkten" (R1N3c, 51).

1.39 "Also wir wollen halt nicht Leute, die sagen, ah, ich habe da in den Windpark mal investiert, weil, da gab es gute Rendite" (R1N2, 126).

1.40 "Wir hätten ja sonst ein paar Reiche gehabt, die hätten das Geld auf den Tisch gelegt und hätten dann sich gefreut. Und eine schöne Rendite gehabt, die sie bei keiner Bank bekommen" (R1N2, 34–36).

1.41 "Wir nennen es Gierbremse; das heißt mehr als sechs Prozent Kapitalausschüttung werden in den Stadtwerken oder werden der Genossenschaft nicht ausgeschüttet. Also monetär gibt es nicht mehr als sechs Prozent. Auf eine Maximalanlage von 10.000 Euro sind das also 600 Euro. Und alles was darüber hinausgehen sollte, sollte es wirklich ein gutes Jahr geben und die Projekte erfolgreich laufen, dann wird die Genossenschaft den Energiesparfonds füttern und damit natürlich das Thema Energieeffizienz noch mal ganz anders nach vorne bringen. Ich nenne es auch immer [. . .] Wolfhager Energiesparverein.—werden sie Mitglied, und es hat auch etwas Soziales" (R1N1, 160).

1.42 "Ja, es gibt ja so zwei Motive [für eine Mitgliedschaft]. [. . .] Das eine ist, dass man regional sein Geld anlegt" (R1N3a, 131).

1.43 "Zu wissen, das Geld, das wir ausgeben für Energie, das fließt jetzt nicht in irgendwelche große Konzerntöpfe rein und irgendwelche Aktionäre kriegen dann eine Auszahlung am Jahresende [...] sondern es bleibt in der Region, es bleibt vor Ort, wir selbst profitieren von dem" (R1N2, 133).

1.44 *"Ich meine, man hat ja immer das Gefühl, wir machen hier regionale Wertschöpfung, hervorragend. Natürlich kann es nicht Vorrang vor aller Wirtschaftlichkeit haben. Muss man auch wieder abwägen: Wie viel können wir uns leisten an Regionalität? [...] da muss man eben dann so eine Gewichtung finden, und da müssen wir auch viel Kompromisse machen, leider"* (R1N3a, 209).

1.45 *"Wir gehen jetzt den Weg, den wir gehen. Den halte ich auch absolut für den wichtigsten für die Zukunft. Wenn wir nur sagen, was wir alles nicht wollen, dann müssen wir halt auch eigene Wege gehen und deutlich machen, dass es funktioniert"* (R1N2, 86).

1.46 *"Zu sagen, wir finden einen Weg in Wolfhagen, wo man deutlich machen kann und aufzeigen kann, wir machen uns davon ein Stück weit unabhängig. Wir können es schaffen, unsere Energie selbst zu produzieren"* (R1N2, 123).

1.47 *"Es geht nicht drum, dass ein Akteur alles übernimmt in der Region, sondern dass diejenigen, die in der Region aktiv sind, im Interesse der Region zusammenarbeiten"* (R1N4, 392).

1.48 *"Dass die Erneuerbaren-Energie-Potenziale einer Region gemeinsam strategisch ganzheitlich geplant werden [...]dass wir sozusagen breite Mehrheiten in den Kommunen organisieren"* (R1N1, 274, 290).

1.49 *"Die Geschäftsführung sagt, wir haben da hier mal ein Projekt vor und wir wollen in die und die Richtung, dann kriegt das ein Aufsichtsrat, der ja dann auch mit Genossen—in Klammern, Vertretern von heute inzwischen 600 Bürgern—besetzt ist, ist das ja sozusagen für mich eine demokratische Rückkopplung in die Gesellschaft rein"* (R1N1, 60).

1.50 *"Und wenn wir dann im Aufsichtsrat darüber sprechen, dann wird es ja vom Aufsichtsrat auch in die BürgerEnergieGenossenschaft gespiegelt. Das heißt wiederum in die Bürgerschaft. Also ich [Stadt Wolfhagen] kriege dadurch ein ganz anderes Feedback"* (R1N2, 420).

1.51 *"Und da ist sozusagen eine BEG auch immer hilfreich im Sinne von, dass sie halt auch noch mal mit danach guckt. [...]. Und mit darauf achtet, dass wir dann sozusagen an solchen Stellen—also auch bei kritischen Sachen, [...] auch sensibel mit den Bürgerinteressen dann umgeht"* (R1N1, 294).

1.52 *"Muss ich halt sagen, das, was wir für unnütz in Netzinvestitionen halten und wo wir uns auf Boden des Gesetzes befinden, also sprich wo sozusagen die Netzausbaukosten nicht in einem sinnvollen Verhältnis zu dem Mehrwert an Stromertrag stehen, dann wehren wir uns auch dagegen"* (R1N2, 296).

1.53 *"Dass die BürgerEnergieGenossenschaft schon sehr wach engagiert mitgestaltet, mitwirkt. Und auch frühzeitig Dinge wahrnimmt und Dinge mit einbringt. [...] Also man ist doch durch die BürgerEnergieGenossenschaft sehr viel näher am Bürger dran"* (R1N2, 296–300).

1.54 *"der Stil, wie die Entscheidungen getroffen werden dann. Jetzt gerade im Aufsichtsrat, wenn die Diskussionskultur geweckt wird, dann ist das eben vielleicht auch dadurch gekommen, dass eben andere sind, die anders fragen, anders gucken"* (R1N3a, 231).

1.55 "Es ist jetzt einfach so, dass im Aufsichtsrat die beiden Vertreter mehr mit am Tisch sitzen. Und natürlich ihre Gedanken mit einfließen lassen. Ihre Bedenken, ihre Anregungen, ihre Ziele, was auch immer. Und das macht es natürlich noch mal, ja—schwieriger ist das falsche Wort—aber das macht es noch mal anspruchsvoller" (R1N2, 270–272).

1.56 "Und ich habe das Gefühl, dass wir viel mehr mitdenken als wir müssten. Also, und das ist vielleicht wieder was [...] nervt" (R1N3a, 187–191).

1.57 "Die dürfen eigentlich gar nicht erzählen, was da besprochen wird und abgestimmt wird im Aufsichtsrat normalerweise. Aber wie soll denn sonst das mit der Demokratisierung funktionieren, [...]" (R1N3a, 151).

1.58 "die Identifikation sozusagen mit diesen Veränderungen den Bürgern zu ermöglichen, und von da aus haben wir die Zielrichtung, den Menschen [...]—wie soll es kommen, warum und wieso; warum bauen wir gerade Windräder und nicht irgendwas anderes; und warum ausgerechnet auf dem Berg?—dass wir die Menschen auch daran beteiligen" (R1N1, 72).

1.59 "Ohnmächtiger oder Mitverantwortlicher, das heißt Ohnmächtiger guckt nur zu und lamentiert rum. Der Mitverantwortliche trägt auch Verantwortung" (R1N1, 344).

1.60 "Das ist jetzt auch unser Stadtwerk und wir sind da auch ein bisschen verantwortlich, dass da nichts schief läuft" (R1N3a, 191).

1.61 "Wir hängen da ja so mittendrin. Wir müssen ja auch unser Gesicht und unseren Kopf hinhalten den Leuten [gegenüber], die uns dann auch kennen, hier und so. Und die sagen: Habt Ihr das nicht gewusst" (R1N3a, 191)?

1.62 "Da achten wir eben auch natürlich drauf, dass vielleicht nicht irgendwelche Risiken eingegangen werden. [...] weil ja auch ihr eigenes Geld da drin ist [...] und dass da auch möglichst nichts Riskantes mit gemacht wird. Das würde man sonst auch nicht machen, aber das merkt man schon, dass da im Aufsichtsrat drauf geachtet wird. Was ja im Grunde auch ganz gut ist" (R1N2, 326–328).

1.63 "Ich sehe uns in der Symbiose. Ich freue mich, wenn die BEG sozusagen wächst und gedeiht. Weil jeder weitere Genosse einerseits Geld [...] bei uns in den Stadtwerken [...] reinbringt als frisches Eigenkapital" (R1N1, 216).

1.64 "Wenn es den Stadtwerken gut geht, geht es uns gut. Und umgekehrt, [...]" (R1N3a, 73).

1.65 "Zu Anfang hatten wir das Gefühl, wir sind ja alle wir. Aber wir sind ja nicht die Stadtwerke, und die Stadtwerke sind nicht die Genossenschaft" (R1N3a, 77).

1.66 "Eigentlich sind wir ja doch auch so auf diese Symbiose da zugeschnitten, da hat man auch wenig Freiheiten. Muss man gucken, wie sich das in der Realität weiterentwickelt" (R1N3b, 42).

Chapter 7.2, Original Quotes, Case Study II

2.1 "So, und da ging es einfach da drum zu spinnen: Wie könnte man in dem Bereich, dem Geschäftsfeld Erneuerbare Energien, ein Konzept aufstellen? Um etwas zu schaffen, sage ich mal, so eine kleine Volksbewegung, viele kleine Beiträge, um etwas Großes zu erreichen" (R2N3, 18).

2.2 "Wir waren aber dann der Meinung—und das können wir am besten, wir Genossen—, dass wir das in einer Genossenschaft auf die Beine stellen" (R2N1, 7).

2.3 "Wir wollten einfach zeigen: Okay, Genossenschaften sind stark, ja. Und wenn man sich bündelt, sind wir noch viel stärker" (R2N3, 108).

2.4 "Ich meine das aber wirklich, ich bin davon überzeugt: Wie wir das aufgebaut haben, mit mehreren Instituten, einem größeren Raum, nicht nur eine Bank für sich alleine—haben wir auch diskutiert! [...] Und ich glaube, das tut auch der E + U und auch den Mitgliedern in der E + U und auch den Banken gut, dass man da nicht klein-klein, sondern dass wir da gesagt haben: Komm, packen wir es größer an, dann wird es auch eine größere Sache" (R2N1, 90)!

2.5 "Wir sind alles Volksbanken, wir gehören zur gleichen Familie" (R2N1, 15).

2.6 "Die Banken arbeiteten schon [...], bei den einen oder anderen die Region betreffenden Überlegungen zusammen. Und die Zusammenarbeit für die E + U, die ist schon Kraft dieser Organschaft gegeben" (R2N1, 34).

2.7 "Als aktiver und zuverlässiger Partner bekennen wir uns zu unserer regionalen Verantwortung" (Volksbank Franken eG 2014).

2.8 "Unser Antrieb ist unsere Vision, der erste Ansprechpartner für die Menschen der Region zu sein—[...]" (Volksbank Mosbach eG 2014).

2.9 "Als genossenschaftliche Bank vor Ort haben wir eine enge Bindung zur Region" (Volksbank Main-Tauber eG 2014).

2.10 "Ohne dieses Vertriebsnetz wären sie nie so gigantisch gewachsen" (R2N3, 59).

2.11 "Das ist gut, das wird ankommen, da kann ich meinen Menschen, meinen Mitgliedern in der Bank was bieten, wenn sie auch Mitglied dort werden" (R2N1, 19).

2.12 "Wir haben auch Kunden für die Bank durch die Energiegenossenschaft gewonnen. Es sind auch Kunden gekommen, die auf der Homepage auf uns aufmerksam gemacht worden sind, die jetzt nicht in dem Vertrieb der Volks- und Raiffeisenbanken stecken. So, und die dann angerufen haben über unser Info-Telefon. Und dieses Info-Telefon geht automatisch auf mich. [...] Und dann kamen die drauf: Also dann ist es ja hochinteressant auch vielleicht Kunde bei der Genossenschaftsbank zu sein. Und dann habe ich zu denen gesagt: So, und dann probieren wir es doch mal gemeinsam" (R2N3, 310–318).

2.13 "Professionalität zu haben, die verwaltungsmäßig nichts kostet" (R2N1, 17).

2.14 "Ich denke, wir haben da sehr gute Vorstände, die da wirklich das auch ehrenamtlich machen und sehr viel Zeit und Energie da investieren. Und ich denke—das sieht man ja auch im Ergebnis jetzt von den letzten zwei Jahren—, dass das sehr gut läuft" (R2N2, 128).

2.15 "Wir müssen da umstrukturieren. [...] Sie können das nicht mehr bewerkstelligen. [...] Also es nimmt Unmengen Zeit weg—Unmengen. Und irgendwann—das sage ich Ihnen ganz offen—ist auch das erreicht, man will auch dann nicht mehr. Weil, irgendwo muss man auch noch leben. Und ich sage es ganz unverfroren: Keiner von uns dreien geht unter 60 Stunden nach Hause, die Woche—keiner! Keiner. Keiner. Also ich sage eher 70. Und dann ist jetzt die Schmerzgrenze erreicht" (R2N3, 278–282).

2.16 *"Dann hat man überlegt: Welche Persönlichkeiten könnten uns denn helfen in der Vermarktung? [...] Und daraufhin ist man natürlich gekommen, dass man sagt: Okay, wir nehmen zwei Landräte, [...], wir nehmen jeweils einen Vorstand aus den Gründungsbanken und wir nehmen [...] [eine Person] von der Grünen-Liste"* (R2N3, 26).

2.17 *"Da macht es Sinn, diese Energiegenossenschaften erstens mal mit Leben zu füllen, aber zweitens auch hier unseren Banken—unseren Volks- und Raiffeisenbanken—auch zu helfen, Erträge zu generieren. [...] die Abhängigkeit ist einfach da"* (R2N3, 402).

2.18 *"Für uns war der Hinterhalt da, [...] wenn Sie als kleinere Genossenschaft 100.000 brauchen, was Sie erst mal vorlegen müssten. Und das war alles weg! In der Hinsicht einfach"* (R2N3, 59).

2.19 *"So, also alle Kosten wurden von den jeweiligen Häusern übernommen. Ob das jetzt, wo wir ein Prospekt gemacht haben und und und, hat alles die Bank bezahlt. Alle Marketingauftritte, alle Banken bezahlt. Also die haben uns das Leben da schon leichter gemacht und alle Experten auch zur Verfügung gestellt"* (R2N3, 65).

2.20 *"Und wir sind sehr wirtschaftlich denkend, wir legen immer ein Wirtschaftsprogramm über jede Maßnahme. Und wenn die nicht die entsprechende Eigenkapitalrendite ausweist, dann ist das für uns schon im Endeffekt—außer die Windmühle—gestorben"* (R2N3, 380).

2.21 *"Theoretisch könnten wir einen Windpark nehmen, wir könnten einen Windpark bauen—wir würden nicht kaputtgehen, auch wenn es schiefläuft"* (R2N3, 361).

2.22 *"Und dass jeder auch einen kleinen wirtschaftlichen Vorteil ... das ist halt jetzt, weil die normalen Zinsen sehr niedrig sind, ist das sehr gut, die Ausschüttung, was die Energiegenossenschaft da machen kann—im Vergleich zum laufenden oder zum Sparbuch oder so was"* (R2N2, 130).

2.23 *"Und wenn die Bundesregierung nicht jetzt aus nicht bezahlbaren Gründen die Einspeisevergütung zurückgenommen hätte, wären wir noch mehr gewachsen als das, was bis jetzt da ist"* (R2N1, 62).

2.24 *"Ja, die größten Herausforderungen liegen in der Umsetzung des EEGs—brauchen wir gar nicht rum machen. Wo wir selber noch gar nicht alles wissen, wie wir es richtig eigentlich gestalten werden. So, und das ist der Bremsschuh Nummer Eins, das neue EEG"* (R2N3, 378).

2.25 *"Heute haben Sie mal etwas anders beleuchtet. Ja, eigentlich eine Bankgenossenschaft. [...]"* (R2N3, 418).

2.26 *"Man muss natürlich eins wissen: Die Genossenschaft Energie + Umwelt wird immer mit Banken in Verbindung gebracht. Und so wird sie auch in der Öffentlichkeit wahrgenommen. Aber so vertreten wir sie auch mit nach außen"* (R2N3, 57–59).

2.27 *"Der genossenschaftliche Grundgedanke ist: Selbsthilfe, Selbstverwaltung und Selbstverantwortung. Erst mal so drei Begriffe. Und im Paragraph eins steht: Förderung der Mitglieder. So, das heißt diesen Anspruch muss man auch gewährleisten. Und wenn die Stimme der Genossen ruft und sagt: Leute, wir müssen uns ein bisschen verändern, wir müssen sehen, dass wir gemeinsam etwas auf den*

Weg schaffen, was jetzt vom klassischen Grundgedanken nur rein bankspezifisch da war, dann konnte man das nur, wenn man den genossenschaftlichen Ursprungssinn vertritt: so [eine Energiegenossenschaft] gründen und sich nichts [an externer Energieexpertise] einkaufen" (R2N3, 98–100).

2.28 *"Also Potenziale ist ein gegenseitiges auf-sich-aufmerksam-machen. Es schadet der Bank nicht, wenn sie in den E + U– Veranstaltungen dadurch, dass die Menschen, die dort auftreten, Bänker aus der Nicht-E + U sind. Das dadurch mit bewerben oder dass wir auch als eine Bank, wenn wir unterwegs sind oder wenn wir Mitgliederversammlungen haben oder Vertreterversammlungen haben, auch über die E + U "richten"—ist natürlich in Anführungsstrichen—ein gegenseitiges Befruchten, was wir auch gerne haben" (R2N1, 86).*

2.29 *"Wir mussten als regionale Genossenschaftsbanken auf dem Geschäftsfeld Energie, Thema Umwelt, uns sowieso platzieren. Weil man sagt, aus rein finanzierungstechnischen Gründen: Wir mussten Know-how eigentlich mit bieten, dass man sagen kann, wir können mehr als Geld und Zinsen [...]das bringt einfach viel. Die Kombination Energiegenossenschaft und damit Kompetenz in dem Feld Erneuerbare Energie" (R2N3, 78, 90).*

2.30 *"Wir haben den Landrat mit dabei—das war uns auch ganz wichtig, ihn in der Aufbauphase hier mit einzubinden. Einfach gibt es eine hervorragende Reputation [...]" (R2N1, 46).*

2.31 *"Wenn da etwas mal—in Anführungsstrichen—aus dem Ruder laufen würde, würden wir ja das Image und die Reputation der Bankhäuser belasten" (R2N1, 36).*

2.32 *"Und das sagen wir auch immer: Leute, wir sind nicht Bank, wir sind Genossenschaft. Energiegenossenschaft Aufsichtsratssitzung. Wir können nicht immer alles bis zur Perfektion haben, wir können nicht in jedes Risiko komplett abchecken und das schon durchgepaukt haben. Wir haben nicht jeden Tag einen Staranwalt neben uns sitzen, ja, der uns alle rechtlichen Bedenken ... aber das ist einfach so geprägt" (R2N3, 218).*

2.33 *"Das packt man mal an und so, wie wir uns das am Grünen Tisch überlegt haben in der Gründungsphase, wussten wir einfach voneinander, dass wir uns aufeinander verlassen können, uns da nicht irgendwelche Dinge ausmalen, die wir nicht umsetzen können" (R2N1, 62).*

2.34 *"Und was sich einfach gezeigt hat in der kurzen Zeit: dass wir so viel Mitglieder und so viel Projekte und so viel Mittel zur Verfügung gestellt bekommen haben, wo es sich schnell gezeigt hat: In diese Genossenschaft stecken wir unser gesamtes Vertrauen, weil wir wissen, da sind Profis am Werke" (R2N3, 358).*

2.35 *"Was auch wichtig war, dass die einfach gesehen haben: Okay, die haben schon Projekte, die kommen jetzt zügig voran, hier können wir Geld investieren und das Vertrauen ist einfach da" (R2N3, 76).*

2.36 *"Also akquiriert, wie gesagt, haben wir jeweils über unser Beziehungsmanagement laufen lassen. Also wir sind in den großen Banken—man muss immer die drei großen sehen; das sind schon ein paar Brocken—gesagt: Wir haben alle Berater—Firmenkundenberater—zusammen und haben gesagt: So, schreibt bitte eine Liste, wo vielleicht interessante Projekte für uns sein könnten. Wo können wir ansiedeln? Welche Dächer müssen saniert werden? Wo könnten wir denn entsprechend mit Einmal-Dachmieten entsprechend hantieren? Und und und. So, und aufgrund dessen hatten wir Ruckzuck—ich möchte jetzt nicht lügen, aber—ungefähr 70, 80 Projekte, die uns ran getragen wurden. [...] Also wir haben alle vertrieblichen Wege genommen, die gingen" (R2N3, 136–142).*

2.37 *"Und mittlerweile ist das, wie gesagt, die Energie + Umwelt kennt man. So, und dadurch kriegen sie auch viele Angebote" (R2N1, 152).*

2.38 *"Also wie gesagt, wir sehen auch, dass aufgrund unserer Größenordnung natürlich auch viel an uns herangetragen wird aus Insolvenzmasse. Ja, also Insolvenzverwalter kümmern sich da drum, die rufen uns auch an: Seid Ihr in der Lage, das zu stemmen" (R2N3, 361)?*

2.39 *"Und deshalb auch aufgrund der Größe, die wir [als Energiegenossenschaft] haben, laufen Projektanfragen auf diese drei Häuser wie geschnitten Brot" (R2N3, 90).*

2.40 *"Dass man frühzeitig an Informationen kommt, das ist Bank, ist Verwaltungsseite, also Landratsämter hier. Wo man dann auch wieder einwirken kann auf jetzt zum Beispiel Regionalplan Wind oder so, wo man dann Informationen einfach besser bekommt" (R2N2, 150).*

2.41 *"Die Bänker stehen für die finanzielle abgewogene Ausrichtung und Belastbarkeit und können die Konzepte jeweils prüfen" (R2N1, 46).*

2.42 *"Beide Landräte sind drin, die auch die rechtliche Situation und das da gut beurteilen können" (R1N2, 144).*

2.43 *"Ich brauche Menschen vor Ort, die [...] gemeinsam was entwickeln wollen. Und da ist keine prädestinierter als die Menschen, die das im Blut haben bei den Volks- und Raiffeisenbanken. [...] Dann kann man auch sagen: Das ist geradezu angesagt, wenn in einer Region keine Genossenschaft da ist, keine Initiative für eine Energiegemeinschaft, dann sollte es von der Volks- und Raiffeisenbank kommen" (R2N1, 94).*

2.44 *"Die Generalversammlungen sind proppenvoll" (R2N3, 292).*

2.45 *"Wichtig ist immer—ich sage es noch einmal ... was kommt unterm Strich raus" (R2N3, 304–306).*

2.46 *"Also ja, und dass bei der Generalversammlung von der E&U gibt es denn mal die eine oder andere Frage, aber das ist auch nicht mehr, und das wird dann mehr oder weniger ausführlich beantwortet. Was vielleicht auch damit zusammenhängt, ich meine, nicht jeder steht auf und sagt was, wenn 400 Leute ihn angucken, das spielt natürlich auch eine Rolle" (R2N4, 97).*

2.47 *"Wir stellen auch vor, was wir konkret umsetzen wollen in der Zukunft. Es kommt aber nie—ich habe jetzt drei mitgemacht—zu klassischen Diskussionen. [...] Und wenn die dann sehen, wie kostenarm wir arbeiten, also wenn wir uns die Personalkosten angucken oder andere Kosten, dann sagen die schon: Oh, oh! Dann sind die schon mehr als zufrieden. Und auch Projekte. Aber es kommt nicht, dass sie sagen: Okay, beschäftigt euch mal mit dem oder dem Thema. Da fehlt—aber auch gewollt; auch gewollt! Deswegen ... da fehlt so der Saft aus der Wurzel. Der ist nicht gegeben, weil wir auf dieser Ebene begonnen haben. Wir haben eigentlich schon einen Standard. Ich sage mal, einen Wald ausgebaut. Aber wir hatten schon wirklich einen festen Baum stehen, wo jeder sich an diesem Baum beteiligen konnte, weil wir genau wussten: Der wirft Früchte ab. So, und da haben sie keine Ideen von unten" (R2N3, 292, 306).*

2.48 *"Hier ist es so, die Leute wollen mitmachen: Jawohl, die sind dabei. Aber die sagen immer noch: Ja, also im Hintergrund sind ja noch große Apparate, die sorgen dafür, dass wir vorne eine gute Dividende bekommen. Es ist ja auch so, es ist ja nicht falsch gedacht" (R2N3, 254).*

2.49 *"Die hätte das auch alleine machen können, aber deren ihrer Betrieb ist so groß, sage ich jetzt mal, flächenmäßig und auch viehmäßig, dass sie sich dort nur . . . das Grund-Ding zur Verfügung stellen wollten und dass sie sich nicht selber damit belasten müssen. Und auch die Kreditlinie von den Betrieben und so, das sind ja Riesenbeträge, die da investiert werden müssen, dass die nicht zusätzlich belastet werden" (R2N2, 124).*

2.50 *"Das hängt einmal zusammen gewissen politischen Restriktionen. Wohl wirtschaftlich gut, gibt es Bedenken. Weil—ich sag's noch mal—zwei Landräte drin sind, die wissen viel im Leben, was da drin ist, packen natürlich nicht alles aus—ja—und geben da einen Stein vor. Wir wollten uns zum Beispiel auch an dem Projekt, wo Sie waren, beteiligen. [. . .] Und das ist durch den Aufsichtsrat geflogen, weil entsprechende politische Gremien ihr Veto eingelegt haben. Also das passiert, wenn man Hochkaräter drin hat. Hätten wir nur sechs [Aufsichtsratsmitglieder] drin gehabt, die nicht politisch organisiert sind, wäre das Ding durchgegangen" (R2N3, 196–204).*

2.51 *"Eine [Situation], die mich besonders negativ bewegt hat, das war eben ein "Erpressungsversuch"—in Anführungsstrichen von jemanden, der über die zulässige Summe Geld einzahlen wollte und mich dann bedroht hat: Wenn wir das nicht machen, dann wird er seine Gelder [. . .] wegziehen. Und das ist eine Erfahrung gewesen, die hat nicht so einen großen Spaß gemacht, aber wir haben das in den zwei Gesprächen dann—na, ich sage mal—befrieden können" (R2N1, 56).*

2.52 *"Grenzen sehe ich jetzt im normalen Bereich, dass einfach, wir haben uns ein bisschen regional begrenzt hier, dass wir nur in unserer Region da investieren wollen. Und, ja, (seufzt) dass einfach die Wirtschaftlichkeit immer schlechter wird von Photovoltaik, und Wind ist auch nicht überall möglich. Und deswegen, ja, kommt man so langsam an Grenzen" (R2N1, 132).*

2.53 *"Oder wir selber, wir reden nicht nur, wir machen es auch vor [. . .]. Und bei dem Thema E + U war der Gedanke eben mitzuhelfen, auch dem Bürger die Möglichkeit zu geben für kleines Geld mit dabei zu sein und zu sagen: Ich habe hier auch mitgeholfen, dass wir hier autarker werden in der Region" (R2N1, 9).*

2.54 *"Die Genossenschaftsbanken gibt es seit über 150 Jahren schon. Raiffeisen haben die gegründet, damit die Menschen in den Regionen für sich etwas tun, was der Einzelne nicht für sich voranbringen kann. So leben wir Volks- und Raiffeisenbanken. [. . .] Und deswegen ist für mich kein anderer Teilnehmer aus den anderen beiden Bankensäulen geeigneter als Volks- und Raiffeisenbanken sich des Themas anzunehmen, E + U oder Genossenschaften für Energie" (R2N1, 92).*

2.55 *"Also—um es eigentlich sehr kurz zu machen—wir machen nur ein Projekt, wo die Bürgerinnen und Bürger dahinterstehen. [. . .] also sind wir keine Bewegung in dem klassischen Sinne, die trotz enormer Widerstände das durchsetzt, sondern wir hören auf Volkes Stimme" (R2N3, 330–332).*

2.56 "Aber wir beteiligen uns wirklich nur da, wo Volkes Stimme sagt: Ja. Mehr machen wir nicht" (R2N3, 356).

2.57 "Kein einfaches Thema, aber das darf man auch nicht verschweigen, dass da eventuell mal gewisse Interessenkollisionen auftauchen könnten. Und da würden wir sicherlich die Banksicht, sprich mitberücksichtigen wollen in einer Region, in der wir vertreten sind, mit der wir einen hohen Marktanteil—[...]—beispielsweise von fünfzig Prozent haben in einem Gebiet. Und da gegen die Bevölkerung dann über die E + U zu gehen, könnte uns im Bankgeschäft richtig wehtun" (R2N1, 40).

2.58 "Wenn wir merken, Volkes Stimme ist sehr massiv, dann klinken wir uns da erst mal ganz schön raus" (R2N3, 340, 352).

2.59 "Leute, wenn's soweit ist und wir haben gemeinsam das Bedürfnis, wir machen's jetzt, wir setzen es jetzt um und machen da was draus—dann stehen wir zur Verfügung. Aber wir würden nie etwas durchdrücken" (R2N3, 340).

2.60 "Hier gibt es ja Raubritter—Großprojektierer, Tschuldigung" (R2N3, 330)!

2.61 "Wir wollen die Energiewende für den kleinen Mann. Weil, Volksbanken steht nicht für Größenwahn, sondern für den kleinen Mann [...]" (R2N3, 116).

2.62 "Für uns spielt Rendite steht nicht im Vordergrund. Sondern wir haben immer gesagt: Viel Eigenkapital, wenig Fremdkapital—wir wollen nicht hebeln" (R2N3, 24).

2.63 "Wir gehen ja da nicht ran also mit dem Bank-Auge und womöglich eine höhere Rendite zu haben, sondern wir haben eine Verantwortung gegenüber den 1.400 Mitgliedern, die sich auf uns Menschen, die Bänker sind [...]. Die verlassen sich auf uns als Personen, dass wir in der E + U seriöses, überschaubares, berechenbares Geschäft machen" (R2N1, 27).

2.64 "Erweiterung des Aufgabenspektrums. Heller und wacher zu sein, wenn irgendwelche Themen aufschlagen, die man sieht und die man hört, nachzudenken: Könnte das auch was für uns als E + U sein? Einfach auf der Verantwortung des ehrenamtlichen Auftrages bei der E + U da, im Aufsichtsrat zu sitzen, da mitzudenken, Informationen oder andere Dinge, was man erzählt, da miteinzubringen" (R2N1, 76).

Chapter 7.3, Original Quotes, Case Study III

3.1 "Ich habe halt gesagt, dass letztendlich über diese interkommunale Kooperation wir die Möglichkeit haben die Zukunft energiepolitisch zu gestalten. Ich habe halt gesagt, was wir alles als Kommune an Zahlungen leisten an die Energiekonzerne, was wir für Probleme haben mit Beleuchtung. Wir werden wirklich diktatorisch von oben herab behandelt von den großen Energiekonzernen. Und das ist einfach der Sprung gewesen, um davon loszukommen, der erste Schritt zu mindestens" (R3N3, 27).

3.2 "Wir haben gesagt also, wir wollen weder besondere Leute oder Kapitalgesellschafter oder sonst was drin haben, sondern wir wollen unsere Bürger dafür gewinnen" (R3N5, 31).

3.3 "Die Mitglieder kennen sich gegenseitig, sind alles Nachbarn, das ist einfach ein ganz anderes Verhältnis, weil, wenn der Umfang zu groß ist, werden die Beziehungen nicht mehr so eng sein. Also das ist eigentlich der Hauptgrund, warum das nicht so groß werden soll" (R3N1, 215).

3.4 "Dass sich halt eigentlich jeder—ja—eigentlich fast jeder beteiligen kann. Das war dann eigentlich der Hintergrund" (R3N1, 9).

3.5 "Und siehe da, dann ist halt die sogenannte Zustimmung auch erfolgt von unseren Bürgern in dieser Region und dann ist das so ein Selbstläufer geworden. Heute kann man sagen, das ist ein Selbstläufer. Damals haben wir mit wenig begonnen und wir haben gesagt: Ja, wenn wir eine oder zwei so kleine Anlagen bauen, dann sind wir glücklich, dann sind wir zufrieden. Aber wir sind wirklich dann überrollt worden. Wenn Sie sich vorstellen, wir haben zwischenzeitlich in gut fünf Jahren 14 Millionen einkassiert" (R3N5, 75).

3.6 "Das Paradoxe ist immer gewesen, und das haben wir festgestellt: Überall, wo wir erschienen sind und haben unser Modell vorgestellt, haben etliche Leute gesagt, bei euch treten wir bei, da werden wir Mitglied. Das war paradox. Wir haben uns in Dresden, [...] wir haben uns in Oberbayern, Neuötting, wir waren in Regensburg, überall wo wir waren, haben wir Mitglieder gewonnen." (R3N5, 79).

3.7 "Und ich meine, ein Bürgermeister hat halt ganz andere Möglichkeiten als ein Privatbürger. Also es geht auch vor allem um die Initiierung von solchen Projekten wie große PV-Parks und so weiter, da ist natürlich der Bürgermeister der ideale Ansprechpartner" (R3N3, 21).

3.8 "Also für mich war ein Konflikt, dass der Herr [...] jetzt zweiter als Vorstand ist, weil er nicht mehr Bürgermeister ist. Das habe ich nicht so gut empfunden, weil das dann so den Ruf hat, so wie [...] bei den Stadtwerken. Da gab es mal einen Spiegel-Artikel zu dem Thema, dass die Stadtwerke dann die—wie sagt man das?—das Abstellgleis ist für Politiker für bestimmte Pöstchen. Den Touch möchte ich nicht" (R3N3, 127).

3.9 "Oder dann haben wir einfach gesehen, dass der Arbeitsaufwand so umfangreich wird, dass der Bürgermeister das nicht mehr leisten kann, das geht nicht" (R3N2, 89).

3.10 "Und inzwischen haben wir, glaube ich, 20 Millionen Euro investiert und das kann man halt so nebenbei nicht mehr bewerkstelligen" (R3N1, 27).

3.11 "Grenzen gibt es nur momentan, dass man sagt: Man kann es von der Personalstärke nicht mehr überarbeiten, sonst haben wir keine Grenzen" (R3N4, 115).

3.12 "Aber wenn wir dann das Modell vorgestellt hatten, meistens war der Vorstandsvorsitzender von der Stadt, vertreten in den Sitzungen und der hat das dann erläutert und siehe da, dann kam die Resonanz, ja, das ist eine tolle Sache, da machen wir mit, da sind wir dabei" (R3N5, 55).

3.13 "Ganz einfach zum Geldsammeln ist die Genossenschaft [Bürger eG] gegründet worden" (R3N1, 9).

3.14 "Für die Bürgergenossenschaft sprach eigentlich, sage ich jetzt mal, die Möglichkeit, dass wir eben mehr Kapital bekommen. Also langfristig. Weil, die Kommunen müssen dann letztendlich jedes Mal, wenn wir ein neues Projekt starten, müssten wir jedes Mal wieder Kapitalerhöhungen durchführen. Und das ist natürlich auch in der Kommune nicht ganz so einfach, weil wir auch viele kleine Kommunen haben, die eben nicht so viel Eigenkapital haben, um jetzt—sage ich einmal—Ruckzuck mal 50.000 Euro Kapitalerhöhung in die Genossenschaft einzubringen. Und deswegen war so die Idee, dass wir praktisch unser Kapital im Wesentlichen über die Bürgergenossenschaft—ja—organisieren" (R3N2, 81).

3.15 "Und dass sich das dann so schnell entwickelt hat, das haben wir auch nicht gedacht. Wenn man in einer Dachanlage anfängt und dann hat sich das so schnell entwickelt, dass auf einmal so viel Geld da war, dass wir uns doch an einen Park getraut haben" (R3N1, 9).

3.16 "Die werden von der Kommune aus schon Beziehungen da sein oder Kredite bei den Banken und alles. Das ist halt einfach, wenn die Kommunen dahinterstehen, ist vieles einfacher" (R3N1, 83).

3.17 "Wir haben also Versprechen gemacht schon, bevor wir das eigentlich zum Beschluss erklärt haben, dass Mindestausschüttung drei Prozent sind pro Jahr" (R3N5, 73).

3.18 "Das ist ja nicht bloß der gute Wille, sondern da steckt schon ein bisschen Profit auch dahinter. Und das ist ja gut, wenn erst mal eine gewisse Rendite vorhanden ist, sonst funktioniert das Ganze nicht. Dass man mal schaut, dass man das soweit wie möglich nach unten bringt, aber das muss sich irgendwie rechnen, sonst funktioniert das Geschäft nicht" (R3N1, 143).

3.19 "Ja, wir haben schon Grenzen erfahren in der Form, wenn wir von den Bürgern oder von der Bürgergenossenschaft sehr viel Geld zur Verfügung hatten, aber keine Projekte, die wir realisieren konnten. [...] dann wird es natürlich auch ein Problem die drei Prozent auszuschütten. Und wenn dann diese Finanzscheibe dann rückwärts gedreht wird, wird das dazu führen, dass die Kapitalanleger wieder das Weite suchen" (R3N2, 85).

3.20 "Natürlich ist momentan jeder davon überzeugt aufgrund der aktuellen Zinsstruktur. Was passiert aber in fünf, sechs, sieben Jahren? Oder auch in kürzerer Zeit schon? Wenn die Zinsen wieder über unsere aktuelle Rendite steigen. Dann ist natürlich die Frage: [...] lässt man es trotzdem liegen als grünes Gewissen" (R3N4, 151)?

3.21 "Ja. Gerade im Jahr 2014 eben durch das, dass das EEG gerade für die großen Anlagen zu schnell abgeschmolzen ist, die Einspeisevergütung, da hat es ja immer sehr, also wurden immer sogenannte Timesteps gegeben: Wann muss ich eben spätestens den Vertrag unterschreiben? Wann muss alles endverhandelt und geprüft sein? Damit halt gerade noch gebaut werden kann, damit man gerade noch am letzten Tag halt dieses Monats das EEG fertig wird? Und das war also schon die eine oder andere schlaflose Nacht" (R3N4, 67).

3.22 "Also die Parks, die wir gekauft haben in den jeweiligen Gemeinden, das war halt nur dort möglich, weil die räumlichen Voraussetzungen dort da waren" (R3N2, 123).

3.23 "Ja, die Zukunft, das hängt von der Windkraft ab. Wenn wir die verwirklichen können, das ist ja eigentlich im Moment das A und O. Weil, das ist das nächste größere Projekt. Dass wir die Windkraft forcieren können [...]" (R3N1, 231).

3.24 *"Weil ja die Hürden, die planrechtlichen Hürden auch ziemlich hoch bei uns sind. Und weil unser Landratsamt da auch sehr restriktiv vorgeht, meiner Meinung nach. Also wir haben kein windkraftfreundliches Landratsamt, ist meine persönliche Meinung. Das werden bestimmt Ihnen die anderen Bürgermeister auch bestätigen, [...] Dann im Regionalverband war ich auch schon bei den Sitzungen mit drin gesessen, habe ich ganz einfach gemerkt, dass der Vorsitzende—damalige Vorsitzende—Landrat [...] einfach nur Verzögerungstaktik. Der wollte das einfach rausziehen [...]. Und dann wird sich das eben dann selbst erübrigen"* (R3N3, 53).

3.25 *"Also die aktuelle Gesetzeslage. Gerade das EEG, gerade eben das KAGB, das sind für uns Riesenthemen, wo man sagt: Wie können wir das umsetzen? Wie können wir das hier konform eben machen"* (R3N4, 103)?

3.26 *"Also natürlich da noch das eigene Netz zu übernehmen, wäre schon ein Traum von mir. Das Netz zu übernehmen und dann wirklich einen Genossenschaftstarif für alle zu bezahlen und versuchen wirklich die Energieautarkheit hier zu mindestens in Nordbayern voranzubringen oder in den Landkreisen, in diesen drei Landkreisen sprich: Tirschenreuth, Neustadt/Kulm, Bayreuth. Das wäre schon toll"* (R3N3, 121).

3.27 *"Die Elektromobilität fördern zum Beispiel. [...] Das ist wieder eine andere interkommunale Kooperation, die aber mit der NEW zusammenarbeiten will in der Form: Diese möchte die Elektromobilität stärken und braucht die NEW als Stromlieferant eventuell"* (R3N3, 11–15).

3.28 *"Thema Wärme, Wärmenetze - momentan in Bayern ein großer Renner. Wir haben es bei uns angeschaut, wir haben es probiert aufgrund der räumlichen Entfernungen oder Distanzen. Also ich finde bisher noch kein Projekt, wo wir es umsetzen können"* (R3N4, 111).

3.29 *"Das ist ein Wahnsinn, wie die unterwegs sind, wie die noch mit Leidenschaft da für unsere Genossenschaft hier Werbung machen"* (R3N4, 125)!

3.30 *"Das war für mich einer der Auslöser, durch diese Informationswelle, die wir ausgelöst haben. Wir sind von Kommune zu Kommune getingelt, haben dort mehrfach im Jahr Vorträge gehalten zu unserem Konzept und dann kamen die Anmeldungen, also die Zeichnungen der Anteile, das war ein Strom ohne Ende. Also ich war total überrascht, wie ich die Zahlen gehört hab [...]"* (R3N3, 37).

3.31 *"Wir haben uns ja auch die absolute Seriosität auf die Fahnen geschrieben"* (R3N3, 45).

3.32 *"Wir trauen uns einfach, das ist eigentlich unser größter Vorteil"* (R3N1, 119).

3.33 *"Ja, die Bürgerinnen und Bürger haben uns ganz einfach das zugetraut [...]"* (R3N1, 139).

3.34 *"Ich kann bloß sagen, tagtäglich kommen Leute zu uns und wollen Geld einzahlen – das ist wirklich so"* (R3N5, 133).

3.35 *"Wenn das mein Bürgermeister macht, dann muss das was Gutes sein. Also diese Vorbildfunktion der NEW gegenüber der Bürgergenossenschaft, ja, das ist so viel gerade eben Leute. Wir haben, ich sage mal, die Generation 50+ bei uns überwiegend, die hier bei uns die große Menge des Geldes hat und die vertrauen wirklich auf das Wort ihres Bürgermeisters. Und durch das muss man wirklich sagen, ist hier ein wahnsinniges Vertrauen entstanden, das sind nicht irgendwelche Leute, sondern die Bürgermeister. Die Kommunen stehen dahinter" (R3N4, 73).*

3.36 *"Das ist eine ganz bunte Truppe, politisch betrachtet. Die haben aber zu gewissen Punkte wirklich die gleiche Wellenlänge, das ist eigentlich das Geheimnis, warum das eigentlich funktioniert, denke ich. Und halt das nötige Vertrauen auch füreinander" (R3N3, 99).*

3.37 *"Die Leute kennen sich untereinander. Das ist fast ein blinder Zuruf, man verlässt sich aufeinander. [...] Also wirklich diese Absprachen, diese Verlässlichkeit, das ist eigentlich der riesen Vorteil" (R3N4, 91).*

3.38 *"Bei einem neuen Bürgermeister, der muss ja das erst wieder kennenlernen, der muss erst wieder rein wachsen. Und da sieht man halt schon, dass, der andere, der fünf Jahre dabei war jetzt seit Gründung, der weiß wie es abläuft, der weiß wie es funktioniert. Und da ist halt oftmals ein neuer, der erst mal etwas mit Recht kritischer ist, der Fragen stellt, der sagt: Wie funktioniert das? Warum ist das so" (R3N4, 95)?*

3.39 *"Oftmals wird den drei Vorständen freie Hand gegeben bei Entscheidungen, die letztendlich über siebenstellige Beträge zu fällen sind. Da herrscht schon ein gewisses Vertrauen mittlerweile, muss man echt sagen." (R3N3, 57).*

3.40 *"Ja, am Anfang war das komisch, weil man halt gemeint hat, die Bürgermeister sind ein bisschen weit weg. Aber im Endeffekt waren die genauso wie jeder andere auch. [...] Also wir werden ernstgenommen, wir können auch mitreden und alles. Wir machen auch oft Vorschläge." (R3N1, 95).*

3.41 *"Also die spielen im Endeffekt da doppeltes Spiel [...]. Ja, was habe ich gelernt? Wir hätten es wahrscheinlich damals nicht aufnehmen sollen, alle beide. Dann hätten wir das Problem nicht" (R3N2, 105, 111).*

3.42 *"Wir spielen immer mit offenen Karten, da denke ich, also: Ist nicht so, dass man sagt, eine und die andere Genossenschaft, sondern es ist immer miteinander. Und das ist die große Stärke—ja" (R3N4, 101).*

3.43 *"Die Bürgermeister haben alle versucht Projekte zu verwirklichen in der Kommune und jeder in seiner Kommune geschaut: Da können wir ein paar Dachanlagen machen oder können das machen, können dies machen" (R3N1, 91).*

3.44 *"Weil er [Bürgermeister] erst mal seine Verwaltung auch mit ins Boot holen kann, also er kennt alle Informationen, er hat Zugriff zu Immobilien, zu den Leuten, die bereit sind überhaupt Flächen zur Verfügung zu stellen. Er kennt Ansprechpartner auch in der Industrie oder Firmen, die letztendlich regenerative Anlagen bauen. Und er hat natürlich einen guten Kontakt zum Landratsamt, um sich eventuell die Genehmigungen einzuholen" (R3N3, 23).*

3.45 *"Wir sind auch besonders von der Bürgergenossenschaft aktiver. Also das wird auch ernstgenommen, aber es ist nicht, dass das einfach irgendwie untergebuttert wird. Na ja, von der Bürgergenossenschaft, wir sind auf jeden Fall, wir versuchen schon Projekte zu verwirklichen oder neue Projekte zu bringen oder neue Ideen einzubringen. Also das wird schon wahrgenommen und dann, es funktioniert auch oft" (R3N1, 221–223).*

3.46 *"Das ist auch wieder so ein Vorteil: Wenn eine Kommune in einem gewissen Sektor voranschreitet, weil sie ganz einfach die Möglichkeiten hat das erst mal allein* zu tun und wir schauen uns dann das Positive ab. Und die lassen sich auch mit Absicht das Positive abschauen. Das ist auch wieder so ein Schritt weg vom Kirchturmdenken, das ist auch so ein Vorteil dieser interkommunalen Genossenschaft" (R3N3, 123).

3.47 "Ja, das Ziel ist eigentlich, dass wir unsere Energieversorgung in die Hand der Bürger oder Kommunen bringen" (R3N1, 65).

3.48 "Ja, in dem Umfang hätte es mit unserem Stadtwerk nicht funktioniert. [...] Alle anderen haben ja keine Stadtwerke, sind meistens kleinere Kommunen, auf dem Gebiet ist das für uns nicht ganz so einfach. Das ginge mit einer größeren Stadt vielleicht, aber für uns, da ist ja die Struktur gar nicht vorhanden" (R3N1, 151).

3.49 "Wir planen, wir suchen uns die Standorte aus und wir verwirklichen dann die Vorhaben" (R3N5, 107).

3.50 "Das ist vorausgehend der Kommunen und dann hier diese Möglichkeit eben, dass sich jeder private Einwohner der Landkreise hier beteiligen kann" (R3N4, 77).

3.51 "Also ganz einfach, dass wir die Bürger ins Boot holen [...] ich kann die NEW nicht ohne die Bürger sehen [...]" (R3N3, 109, 111).

3.52 "Wir [NEW eG] können frei entscheiden und das ist der Vorteil" (R3N5, 211).

3.53 "Und wir wollten das also bewusst trennen, dass man sagt, also "die Bürger sind hier als Beteiligungsgesellschaft", sagen wir es mal so, "und auf der anderen Seite sind die Kommunen beziehungsweise die NEW und wir können dann frei hantieren" (R3N5, 211).

3.54 "Das entscheidet nicht der Bürger, der Bürger hat überhaupt keinen Einfluss. Wir als Vertreter von der NEW und die Vertreter von der Bürgergenossenschaft, wir bestimmen eigentlich. Bürger brauchen wir nicht fragen. Das würde zu weit gehen, wenn wir jeden Einzelnen fragen müssten, was wir tun müssen mit dem Geld. Das ist extra so gewollt und so gemacht worden mit dieser Genossenschaft. Da sind wir also frei. Der nimmt sein Geld, legt es an und wir lassen es arbeiten" (R3N5,193).

3.55 "Das Sagen, würde ich sagen, hat die NEW eigentlich." (R3N3, 105).

3.56 "Im Endeffekt in der Arbeit ist alles eins. Weil, wir haben Vorstandssitzungen, wenn die Bürger- und die kommunalen Vorstände [...] das wird im Endeffekt alles gemeinsam beschlossen. Also im Moment funktioniert die Zusammenarbeit tadellos. Es gibt schon mal Streitpunkte hier und da, aber es funktioniert" (R3N1, 35).

3.57 "Ja, abhängig sind wir ja eigentlich komplett, weil wir—ja, wir haben zwar die Anteile, aber direkt an den Anlagen sind wir nicht beteiligt gewesen, sage ich mal so" (R3N1, 105).

3.58 "Auf der Jahreshauptversammlung kommen dann schon mal wieder Fragen und dann werden Themen angesprochen, aber die Mehrheit ist eigentlich inaktiv" (R3N1, 121).

3.59 "Gut, ich denke natürlich auch, dass gerade im Bereich der Bürgergenossenschaft der Anreiz natürlich auch die Verzinsung war" (R3N2, 79).

3.60 "Die meisten, die große Mehrheit ist sehr zufrieden, ist sehr umgänglich" (R3N4, 129).

3.61 "Ja, ja, das war das Ziel, dass wir praktisch einen eigenen Strom produzieren und dann auch an die Bürger, an die Mitglieder der Genossenschaft verkaufen. Und, ich sage mal, diese Begeisterung, denke ich, ist dann übergeschwappt auf die anderen Kommunen, das sind halt immer mehr der Reihe nach beigetreten—ja" (R3N2, 29).

3.62 "Der Gemeinschafts- und Nachhaltigkeitsgedanke sowie der regionale Gedanke stehen dabei im Vordergrund. Dies bedeute, dass die Mitglieder von der Arbeit der Genossenschaft profitieren. [...] Außerdem soll das Geld, das in der Region eingesammelt wird, den hier lebenden Bürgern und angesiedelten Betrieben zugutekommen" (Neue Energien West eG 2009b).

3.63 "Eben, das Geld bleibt entweder in der Kommune oder bei den BürgerInnen, das bleibt bei uns, wird nicht Gott weiß wo ... ja, die Steuern, alles, das bleibt alles bei uns" (R3N1, 149).

3.64 "Und es können da wieder Projekte verwirklicht werden und wäre halt dem nicht möglich ist, weil, wir sind ja meistens kleinere Kommunen, und da wir fast keine Gewerbesteueraufkommen haben, da ist dann so ein Projekt schon gut, wenn es in der Kommune ist" (R3N1, 167).

3.65 "Ich habe halt gesagt, dass letztendlich über diese interkommunale Kooperation wir die Möglichkeit haben die Zukunft energiepolitisch zu gestalten" (R3N3, 27).

3.66 "Das ist der Klimaschutz und das sind die Kosten, die da sind. Und die zwingen uns einfach, dass wir da neue Wege gehen. Das sind für mich die zwei ausschlaggebenden Faktoren" (R3N2, 17).

3.67 "Ja, die Energiegenossenschaft ist eigentlich unser Medium zur Umsetzung unserer energiepolitischen Ziele, unserer Klimaschutzziele. Das haben wir uns eigentlich auf die Fahne geschrieben, dass wir aktiv zum Klimaschutz beitragen. Und wenn die Kommune nichts macht, dann wird der Bürger in der Kommune nichts machen. Wir sind die Beispielgeber" (R3N2, 39).

3.68 "Wir sind ein Wagnis eingegangen, das Wagnis hat sich gelohnt und ich würde es sofort wieder tun. Das kann ich also wirklich sagen" (R3N5, 229).

3.69 "Ja, wenn wir unsere Ziele verwirklichen können. Ganz einfach, dass wir unsere Anlagen dort bauen, wo wir unsere Energie brauchen. Dass wir praktisch unabhängig werden, das wird für die Region, in welcher Forma auch immer ... Das wird noch etwas dauern, aber die Voraussetzungen sind vorerst hier" (R3N1, 245).

3.70 "Wir denken regional [...]" (R3N5, 225).

3.71 "Und selbst, wenn jetzt die Energiekonzerne saubere Energie machen, aber die Abhängigkeit hat man trotzdem noch, und das ist auch natürlich, die können ja eigentlich machen was sie wollen. Und deshalb wollen wir ja die Unabhängigkeit, dass wir selbst bestimmen können dann und das Geld bleibt bei uns. Und da ist eben nicht die Dividende im Vordergrund, sondern, [...] unsere eigene Energie zu einem bezahlbaren Preis. Das ist dann eigentlich der Profit" (R3N1, 65).

3.72 "Wir haben jetzt gerade zum Glück—in Anführungsstrichen; eigentlich ist es ja kein Glück—diese Monstertrassen-Diskussion bei uns. Jetzt kommen die Monstertrassen, warum kommen die? Weil wir dann den Windstrom von Norddeutschland bekommen sollen—ja—, den wir eigentlich selber hier erzeugen könnten. Und dann frage ich immer ... Wenn wir jetzt eine Umfrage machen müssten hier in Nordbayern, was möchten die Leute lieber haben? Monstertrasse oder Windkräder? Da würde eindeutig, denke ich mal, das Züglein Richtung Windkraft ausschlagen" (R3N3, 49).

3.73 "Aber da geht auch jeder mit, weil die das auch wissen. Die sagen, unsere Arbeit ist nur dann erfolgreich, wenn wir wirklich Strategien gemeinsam entwickeln [...]" (R3N5, 203).

3.74 "Dass wir alle an einem Strang ziehen und unsere Arbeit, die wir begonnen haben, auch weiterhin forcieren und diese Ziele nicht aus dem Auge verlieren" (R3N5, 283).

3.75 "Ich bin schon der Meinung, es ist ein anderes Denken erkennbar. Man schaut also über den sogenannten Tellerrand hinaus und die einseitige Kirchturmpolitik ist nicht mehr gefragt" (R3N5, 227).

3.76 "Vorteile? Wir haben mehr Geld" (R3N1, 81).

3.77 "Der eine Bürgermeister will dies, der andere dieses, je nachdem was für seine Gemeinde passt. [...] die Mehrheit geht trotzdem wieder in die Richtung von der Genossenschaft, ja. Am Ende funktioniert es dann schon wieder. Bis jetzt hat es sich gezeigt, das funktioniert" (R3N1, 71–73).

3.78 "Dass dann schon manchmal so kleine so Streitgespräche entstehen: warum, warum nicht? Und dann halt: hier ein kleinerer Bürgermeister—also kleiner im Sinne von der Kommune her, sich locker zurücklehnen kann, während die großen Kommunen halt hier immer wieder auf ihr Recht und auf ihre Stärke und auf ihre Größe hinweisen, wo die dann sagen: Was wollt Ihr denn" (R3N4, 62)?

3.79 "Dass sich die Bürger damit identifizieren können und das Ganze einfach überschaubar ist" (R3N1, 259).

3.80 "Wir haben gestern Morgen zufällig einen getroffen, den kenne ich auch gut, der hat gesagt, jetzt hast du mich überzeugt, ich habe also 10.000 eingezahlt für mich und 10.000 für meine Frau. Der hat bisher noch nichts gemacht, aber er hat gesagt, ist eine gute Sache, jetzt begreife ich es, ich bin dabei" (R3N5, 135).

3.81 "Also für 17 Bürgermeister eine ganz schöne Stange Geld. Und es ist ja auch die Konstellation hier das Besondere, dass wir ja Privatvermögen verwalten" (R3N3, 61).

3.82 "Das hat sich am Anfang schon ein bisschen komisch gefühlt, weil es da um Millionen gegangen ist und vor allem Geld für die Bürgerinnen und Bürger, und da ist es nicht ganz so einfach, weil, wir können da ja nichts in den Sand setzen, weil, wir wohnen ja alle da in der Gegend und die haben uns ja auch das Geld gegeben" (R3N1, 9).

3.83 "Ich kann nicht mit dem Geld der Bürger pokern, sondern das müssen ganz klare Projekte sein, die finanziell auf soliden Beinen stehen und die letztendlich dann auch den Ertrag abwerfen, den wir brauchen, um dann diese Anteile der Bürger zu bedienen mit den Zinsen" (R3N2, 93).

3.84 "Ja gut, wie ich schon eigentlich gesagt habe: Die interkommunale braucht den Geldgeber der BürgerEnergieGenossenschaft und die BürgerEnergieGenossenschaft braucht die Bürgermeister, die anschieben, Projekte verwirklichen in möglichst schneller Zeit. Also das ist genau die Abhängigkeit zueinander" (R3N3, 103).

References

50Hertz Transmission GmbH, Amprion GmbH, TenneT TSO GmbH, & TransnetBW GmbH. (Eds.). (2014). *Netzentwicklungsplan 2014, zweiter Entwurf.* Retrieved 09.10.2016, from http://www.netzentwicklungsplan.de

50Hertz Transmission GmbH, Amprion GmbH, TenneT TSO GmbH, & TransnetBW GmbH. (Eds.). (2016). *Netzentwicklungsplan Strom 2025, Version 2015.* Retrieved 09.10.2016, from www.netzentwicklungsplan.de

Amschler, H. (2014). "NEW" – Neue Energien West eG und Bürger-Energiegenossenschaft West eG. Zwei Genossenschaften – ein Ziel. In H. Bauer, C. Büchner, & F. Markmann (Eds.), *KWI Schriften: Vol. 8. Kommunen, Bürger und Wirtschaft im solidarischen Miteinander von Genossenschaften* (pp. 57–62). Potsdam: Universitätsverlag Potsdam.

Arapostathis, S., Pearson, P. J., & Foxon, T. J. (2014). UK natural gas system integration in the making, 1960–2010: Complexity, transitional uncertainties and uncertain transitions. *Environmental Innovation and Societal Transitions, 11*, 87–102.

Audretsch, D. B., Bönte, W., & Keilbach, M. (2008). Entrepreneurship capital and its impact on knowledge diffusion and economic performance. *Journal of Business Venturing, 23*(6), 687–698.

Audretsch, D. B., Bönte, W., & Keilbach, M. (2011). Determinants and impact of entrepreneurship capital: The spatial dimension and a comparison of different econometric approaches. In S. Desai, P. Nijkamp, & R. R. Stough (Eds.), *New directions in regional economic development. The role of entrepreneurship theory and methods, practice and policy* (pp. 41–59). Cheltenham, Glos: Edward Elgar.

Augenstein, K. (2014). *e-Mobility as a sustainable system innovation. Insights from a captured niche.* PhD thesis, Bergische Universität Wuppertal, Wuppertal.

BaFin Bundesanstalt für Finanzdienstleistungsaufsicht (Ed.). (2015). *Auslegungsschreiben zum Anwendungsbereich des KAGB und zum Begriff des "Investmentvermögens".* Retrieved 01.10.2016, from https://www.bafin.de

Bakker, S., van Lente, H., & Meeus, M. T. H. (2012). Credible expectations – The US Department of Energy's Hydrogen Program as enactor and selector of hydrogen technologies. *Contains Special Section: Actors, Strategies and Resources in Sustainability Transitions, 79*(6), 1059–1071.

Barney, J. (1991). Firm resources and sustained competitive advantage. *Journal of Management, 17*(1), 99–120.

Barney, J. B. (2001). Is the resource-based "view" a useful perspective for strategic management research? Yes. *The Academy of Management Review, 26*(1), 41–56.

Bartosch, U., Hennicke, P., & Weiger, H. (Eds.). (2014). *Gemeinschaftsprojekt Energiewende. Der Fahrplan zum Erfolg*. München: oekom verlag.

Battilana, J., Leca, B., & Boxenbaum, E. (2009). How actors change institutions: Towards a theory of institutional entrepreneurship. *The Academy of Management Annals, 3*(1), 65–107.

Bauknecht, D., & Funcke, S. (2013). Dezentralisierung oder Zentralisierung der Stromversorgung: Was ist darunter zu verstehen? *Energiewirtschaftliche Tagesfragen, 63*(8), 14–17.

Bauknecht, D., Vogel, M., & Funcke, S. (2015). *Energiewende – Zentral oder dezentral? Diskussionspapier im Rahmen der Wissenschaftlichen Koordination des BMBF Förderprogramms: "Umwelt – und Gesellschaftsverträgliche Transformation des Energiesystems"*. In Öko-Institut e.V. (Ed.). Retrieved 01.10.2016, from https://www.oeko.de/publikationen/p-details/energiewende-zentral-oder-dezentral/

Bauman, Z. (1989). Hermeneutics and modern social theory. In D. Held & J. B. Thompson (Eds.), *Social theory of modern societies. Anthony Giddens and his critics* (pp. 34–55). Cambridge [England], New York: Cambridge University Press.

Bauwens, T., Gotchev, B., & Holstenkamp, L. (2016). What drives the development of community energy in Europe? The case of wind power cooperatives. *Energy Transitions in Europe: Emerging Challenges, Innovative Approaches, and Possible Solutions, 13*, 136–147.

Baxter, P., & Jack, S. (2008). Qualitative case study methodology: Study design and implementation for novice researchers. *The Qualitative Report, 13*(4), 544–559.

Bayerisches Landesamt für Statistik und Datenerhebung (Ed.). (2014). *Statistik kommunal 2013. Eine Auswahl wichtiger statistischer Daten für den Regierungsbezirk Oberpfalz*. Retrieved 01.10.2016, from https://www.statistik.bayern.de/statistikkommunal/

Becker, A. (1996). *Rationalität strategischer Entscheidungsprozesse: Ein strukturationstheoretisches Konzept. DUV: Wirtschaftswissenschaft*. Wiesbaden: DUV, Dt. Univ.-Verl.

Becker, S., & Kunze, C. (2014). Transcending community energy: Collective and politically motivated projects in renewable energy (CPE) across Europe. *People, Place and Policy Online, 8*(3), 180–191.

Becker, S., Gailing, L., & Naumann, M. (2012). *Neue Energie-Landschaften-neue Akteurs-Landschaften. Eine Bestandsaufnahme im Land Brandenburg*. Rosa-Luxemburg-Stiftung (Ed.). Retrieved 01.10.2016, from www.rosalux.de

BEG Wolfhagen eG (Ed.). (2012a). *Quartalsbericht (28.3.2012 bis 30.6.2012)*. Retrieved 01.10.2016, from www.beg-wolfhagen.de

BEG Wolfhagen eG. (2012b). *Gelungener Start der BürgerEnergieGenossenschaft Wolfhagen eG i.G. (BEG): Pressemitteilung zur Gründungsveranstaltung am 28.März 2012*. Retrieved 01.10.2016, from http://www.beg-wolfhagen.de/index.php/presse

BEG Wolfhagen eG (Ed.). (2013a). *Mitgliederinformation Juli 2013*. Retrieved 01.10.2016, from http://www.beg-wolfhagen.de/

BEG Wolfhagen eG. (2013b). *Satzung der BürgerEnergieGenossenschaft Wolfhagen*. Wolfhagen. Retrieved 01.10.2016, from www.beg-wolfhagen.de

BEG Wolfhagen eG. (2014a). *Anlage zu TOP 10 der Tagesordnung vom 22.11.2014 (4. Generalversammlung) (Seite 1)*.

BEG Wolfhagen eG. (2014b). *BaFin Registrierung am 01.12.2014 erfolgt*. Wolfhagen. Retrieved 01.10.2016, from http://www.beg-wolfhagen.de/

BEG Wolfhagen eG (Ed.). (2015). *Photovoltaik Park Wolfhagen*. Retrieved 01.10.2016, from http://www.beg-wolfhagen.de/

Bergek, A. (2002). *Shaping and exploiting technological opportunities: The case of renewable energy technology in Sweden*. PhD thesis, Chalmers University of Technology, Goeteburg.

Bergek, A., Jacobsson, S., Carlsson, B., Lindmark, S., & Rickne, A. (2008). Analyzing the functional dynamics of technological innovation systems: A scheme of analysis. *Research Policy, 37*, 407–429.

Berger, P. L., & Luckmann, T. (1966). *The social construction of reality: A treatise in the sociology of knowledge. Anchor books* (Vol. 589). Garden City, NY: Doubleday.

References

Berlo, K., & Wagner, O. (2013). *Stadtwerke-Neugründungen und Rekommunalisierungen. Energieversorgung in kommunaler Verantwortung.* Retrieved 01.10.2016, from http://wupperinst.org/uploads/tx_wupperinst/Stadtwerke_Sondierungsstudie.pdf

Berlo, K., Wagner, O., Merten, F., Richter, N., & Thomas, S. (2008). *Perspektiven dezentraler Infrastrukturen im Spannungsfeld von Wettbewerb, Klimaschutz und Qualität. Ergebnisse für die Energiewirtschaft.* Wuppertal: Wuppertal Institut für Klima, Umwelt, Energie GmbH.

Bertelsmann Stiftung, Deutscher Städtetag, & Deutscher Städte- u. Gemeindebund (Eds.). (2008). *Beruf Bürgermeister/in. Eine Bestandsaufnahme für Deutschland.* Retrieved 01.10.2016, from http://www.bertelsmann-stiftung.de/de/presse-startpunkt/presse/pressemitteilungen/pressemitteilung/pid/buergermeister-geniessen-hohes-ansehen-in-der-bevoelkerung-und-haben-freude-an-ihrem-amt/

Best, B., Prantner, M., & Augenstein, K. (2012). The concept of regime and 'flat ontologies': Empirical potential and methodological implications. Proceedings of the 3rd international conference on sustainability transitions. August 29th–31st 2012, Lyngby, Denmark; Track E. – Lyngby: Dänemarks Technical University.

Bijker, W. E., & Law, J. (Eds.). (1992). *Inside technology. Shaping technology/building society: Studies in sociotechnical change.* Cambridge, MA: MIT Press.

Bijker, W. E., Hughes, T. P., & Pinch, T. J. (Eds.). (1987). *The Social construction of technological systems: New directions in the sociology and history of technology.* Cambridge, MA: MIT Press.

Binz, C., & Truffer, B. (2011). Technological innovation systems in multiscalar space. Analyzing an emerging water recycling technology with social network analysis. *Geographica Helvetica, 66*(4), 254–260.

Binz, C., Truffer, B., & Coenen, L. (2014). Why space matters in technological innovation systems – Mapping global knowledge dynamics of membrane bioreactor technology. *Research Policy, 43*(1), 138–155.

Bioenergie-Region Hohenlohe-Odenwald-Tauber GmbH. (Ed.). (2014). *Über uns.* Retrieved 01.10.2016, from http://www.bioenergie-hot.de/

Bioenergie-Region Hohenlohe-Odenwald-Tauber GmbH. (Ed.). (2015). *Drei Kreise – ein Energiemanagement.* Retrieved 01.10.2016, from http://www.bioenergie-region-hot.de/

bjp. (2009, October 11). Raus aus der Abhängigkeit. Bürger-Energie-Genossenschaft informiert über Alternativen zu Großkonzernen – Neue Jobs, Nr. 233. *der Neue Tag,* p. 23.

Bolton, R., & Foxon, T. J. (2015). A socio-technical perspective on low carbon investment challenges – Insights for UK energy policy. *Environmental Innovation and Societal Transitions, 14*, 165–181.

Bönte, W., Procher, V. D., & Urbig, D. (2016). Biology and selection into entrepreneurship – The relevance of prenatal testosterone exposure. *Entrepreneurship Theory and Practice, 40*(5), 1121–1148.

Borup, M., Brown, N., Konrad, K., & van Lente, H. (2006). The sociology of expectations in science and technology. *Technology Analysis & Strategic Management, 18*(3/4), 285–298.

Braukmann, U. (2012). Zur "Didaktik der Entwicklung unternehmerischer Persönlichkeit" – Genese, kursorischer Überblick und referenztheoretische Bezüge zur problemorientierten Didaktik. In S. Seufert & C. Metzger (Eds.), *Kompetenzentwicklung in unterschiedlichen Lernkulturen. Festschrift für Dieter Euler zum 60. Geburtstag* (pp. 465–486). Paderborn: Eusl.

Braukmann, U., & Bartsch, D. (2014). Entrepreneurship Education im Spannungsfeld interessenspolitischer Instrumentalisierung und bildungstheoretischer Legitimität. In U. Braukmann, B. Dilger, & H.-H. Kremer (Eds.), *Wirtschaftspädagogische Handlungsfelder. Festschrift für Peter F. E. Sloane zum 60. Geburtstag* (pp. 41–72). Detmold: Eusl.

Braukmann, U., & Schneider, D. (2010). Zum Bild des mittelständischen Unternehmens – Analyse des Status quo anhand einer empirischen Vollerhebung von Schulbüchern des Landes Nordrhein-Westfalen und Plädoyer für ein "aufgeklärtes" Unternehmensbild. In W. Baumann, U. Braukmann, & W. Matthes (Eds.), *Innovation und Internationalisierung* (pp. 201–230). Wiesbaden: Springer.

Braukmann, U., Bijedic, T., & Schneider, D. (2008). *"Unternehmerische Persönlichkeit"* – *eine theoretische Rekonstruktion und nominaldefinitorische Konturierung*. Wuppertal: Schumpeter School of Business and Economics.

Braukmann, U., Bijedic, T., & Schneider, D. (2009). Von der Mikro- zur Makrodidaktik in der Entrepreneurship Education – Zum Paradigmawechsel der Förderung unternehmerischen Denkens und Handelns in der Aus- und Weiterbildung. In Bertelsmann Stiftung (Ed.), *Generation unternehmer? Youth entrepreneurship education in Deutschland* (pp. 231–268). Gütersloh: Verlag Bertelsmann-Stiftung.

Braukmann, U., Fischedick, M., & Lindfeld, C. (2012). Zur programmatischen Neuausrichtung der Gründungs- und Innovationsförderung aus Universitäten und Forschungseinrichtungen mittels CEODD und SCTGIZ. In S. Armutat & A. Seisreiner (Eds.), *Differentielles management* (pp. 254–284). Wiesbaden: Springer.

Breschi, S., & Malerba, F. (1997). Sectoral innovation systems: Technological regimes, schumpeterian dynamics, and spatial boundaries. In B. Carlsson (Ed.), *Technological systems and industrial dynamics* (pp. 130–152). Boston [u.a.]: Kluwer Academic.

Breyer, C., Müller, B., Möller, C., Gaudchau, E., Schneider, L., Resch, M., ... (2013). In Reiner Lemoine Institut (Ed.), *Vergleich und Optimierung von zentral und dezentral orientierten Ausbaupfaden zu einer Stromversorgung aus Erneuerbaren Energien in Deutschland*. Reiner Lemoine Institut.

Bridge, G., Bouzarovski, S., Bradshaw, M., & Eyre, N. (2013). Geographies of energy transition: Space, place and the low-carbon economy. *Energy Policy, 53*, 331–340.

Brinckerhoff, P. C. (2000). *Social entrepreneurship: The art of mission-based venture development. Wiley nonprofit law, finance, and management series*. New York: Wiley.

Brown, H. S., & Vergragt, P. J. (2008). Bounded socio-technical experiments as agents of systemic change: The case of a zero-energy residential building. *Technological Forecasting and Social Change, 75*(1), 107–130.

Brummer, V., Herbes, C., Rognli, J., Gericke, N., & Blazejewski, S. (2016). *Conflict handling in Renewable Energy Cooperatives (RECs): Organizational effects and member well-being*.

Brundtland, G. H., & Khalid, M. (Eds.). (1987). *Report of the world commission on environment and development: Our common future*. Retrieved 01.10.2016, from http://www.un-documents.net/wced-ocf.htm

Bryant, C. G. A., & Jary, D. (1991). Coming to terms with Anthony Giddens. In C. G. A. Bryant & D. Jary (Eds.), *International library of sociology. Giddens' theory of structuration. A critical appreciation* (pp. 1–31). London, New York: Routledge.

Budde, B., Alkemade, F., & Weber, K. M. (2012). Expectations as a key to understanding actor strategies in the field of fuel cell and hydrogen vehicles. *Contains Special Section: Actors, Strategies and Resources in Sustainability Transitions, 79*(6), 1072–1083.

Bührle, B. (2010). *Bürgerenergiegenossenschaften – Formen zukunftsträchtiger Energiewirtschaft?* Bachelor Thesis, Hochschule für Wirtschaft und Umwelt Nürtingen-Geislingen.

Bundesnetzagentur für Elektrizität, G. T. P. U. E. (Ed.). (2013). *Monitoringbericht 2013*. Retrieved 01.10.2016, from www.bundesnetzagentur.de

Bundesverband Erneuerbare Energie e.V. (2014). *Neustart für Energiewende geht nach hinten los*. Berlin. Retrieved 01.10.2016, from http://www.bee-ev.de/home/presse/mitteilungen/detailansicht/neustart-fuer-energiewende-geht-nach-hinten-los/

Bundesverband Erneuerbare Energien e.V. (2016). Bilanz zum EEG 2017: Deutliche Drosselung der Energiewende, leichte Verbesserungen im Detail. Berlin. Retrieved 01.10.2016, from http://www.bee-ev.de/home/presse/mitteilungen/detailansicht/bilanz-zum-eeg-2017-deutliche-drosselung-der-energiewende-leichte-verbesserungen-im-detail/

BÜNDNIS 90/DIE GRÜNEN. (2015). *Genossenschaften nicht vom Energiemarkt ausschließen! Anlässlich des gemeinsamen Besuchs der Energiegenossenschaft Rehfelde e.V. erklären Simone Peter, Bundesvorsitzende von BÜNDNIS 90/DIE GRÜNEN und Jan Hinrich Glahr, Vizepräsident des Bundesverbandes Windenergie*. Retrieved 01.10.2016, from http://www.gruene.de/themen/klima-schuetzen/genossenschaften-nicht-vom-energiemarkt-ausschliessen.html

References

Bündnis Bürger Energie e.V. (2015). *Eckpunktepapier zu Ausschreibungen: Bundesregierung nimmt Akteursvielfalt und Bürgerenergie nicht ernst.* Retrieved 01.10.2016, from https://www.buendnis-buergerenergie.de/presse/pressemitteilungen/?newsid=60&cHash=994c1c0627a2a1c7b61b65cc2eff316c

Bürgerenergiegenossenschaft West eG. (Ed.). (2013). *Satzung Bürger-Energiegenossenschaft West eG.* Retrieved 01.10.2016, from http://www.neue-energien-west.de/buerger-energiegenossenschaft-west

Bürgerenergiegenossenschaft West eG. (Ed.). (2015). *Bürger eG.* Retrieved 01.10.2016, from http://www.neue-energien-west.de/buerger-energiegenossenschaft-west

Bürgerinitiative (BI) Wolfhager Land – Keine Windkraft in unseren Wäldern. (2011). *BI legt Bericht zur Bedeutung des Rödeser Berges für den Naturschutz vor.* Wolfhagen. Retrieved 01.10.2016, from http://www.kein-windrad-im-wald.de/pressemitteilungen_der_bi0.html

Bürgerinitiative (BI) Wolfhager Land – Keine Windkraft in unseren Wäldern. (2012). *Der Bürgerentscheid kommt!* Wolfhagen. Retrieved 01.10.2016, from http://www.kein-windrad-im-wald.de/791.html

Bürgerinitiative (BI) Wolfhager Land – Keine Windkraft in unseren Wäldern. (2013). *Bewusste Lärmverschmutzung am Rödeser Berg.* Wolfhagen. Retrieved 01.10.2016, from http://www.kein-windrad-im-wald.de/pressemitteilungen_der_bi0.html

Busco, C. (2009). Giddens' structuration theory and its implications for management accounting research. *Journal of Management & Governance, 13*(3), 249–260.

Byrne, R. P. (2009). *Learning drivers. Rural electrification regime building in Kenya and Tanzania.* PhD, University of Sussex, East Sussex.

Callinicos, A. (1985). Anthony Giddens: A contemporary critique. *Theory and Society, 14*(2), 133–166.

Caniëls, M. C. J., & Romijn, H. A. (2008). Actor networks in strategic niche management: Insights from social network theory. *Futures, 40*(7), 613–629.

Capallo, S. (2008). Die Strukturationstheorie der Strategischen Managementforschung. In T. Wrona (Ed.), *Gabler-Edition Wissenschaft. Strategische Managementforschung. Aktuelle Entwicklungen und internationale Perspektiven* (1st ed., pp. 105–126). Wiesbaden: Gabler.

Carlsson, B., & Jacobsson, S. (1997). Diversity creation and technological system: A technology policy perspective. In C. Edquist (Ed.), *Systems of innovation. Technologies, institutions, and organizations* (pp. 266–290). London, Washington: Pinter.

Clean Energy Sourcing, Elektrizitätswerke Schönau, Greenpeace Energy, MVV Energie, & Naturstrom (Eds.). (2015). *Das Grünstrom-Markt-Modell. EEG-Strom als Grünstrom in den Markt integrieren.* Retrieved 01.10.2016, from http://www.gruenstrom-markt-modell.de/

Club of Rome. (2515). *The story of the Club of Rome.* Retrieved 01.10.2016, from http://www.clubofrome.org/

Coenen, L., Benneworth, P., & Truffer, B. (2012). Toward a spatial perspective on sustainability transitions. *Special Section on Sustainability Transitions, 41*(6), 968–979.

Collins, R. (1981). Mirco-translation as a theory-building strategy. In K. Knorr-Cetina & A. V. Cicourel (Eds.), *Advances in social theory and methodology. Toward an integration of micro- and macro-sociologies* (pp. 81–108). Boston: Routledge & Kegan Paul.

Creswell, J. W., & Plano Clark, V. L. (2011). *Designing and conducting mixed methods research* (2nd ed.). Los Angeles: Sage.

Dagasan, P., Trockel, S., & Schulz, S. (2014). In EnergieAgentur.NRW (Ed.), *Das neue EEG 2014 Aktuelle Änderungen für Wind-, Solar- und Biomasseanlagen.* EnergieAgentur.NRW.

Das, T. K., & Teng, B.-S. (1998). Resource and risk management in the strategic alliance making process. *Journal of Management, 24*(1), 21–42.

de Carolis, D. M., & Saparito, P. (2006). Social capital, cognition, and entrepreneurial opportunities: A theoretical framework. *Entrepreneurship Theory and Practice, 30*(1), 41–56.

Dean, T. J., & McMullen, J. S. (2007). Toward a theory of sustainable entrepreneurship: Reducing environmental degradation through entrepreneurial action. *Journal of Business Venturing, 22*(1), 50–76.

Debor, S. (2014). *The socio-economic power of renewable energy production cooperatives in Germany. Results of an empirical assessment*. Wuppertal Papers No. 178. Wuppertal.

Debor, S. (2017). Gesellschaftspolitische Gestaltungsmöglichkeiten durch Kooperation von Energiegenossenschaften und Stadtwerken: Erfahrungen aus der Praxis. In J. Rückert-John & M. Schäfer (Eds.), *Governance für Gesellschaftstransformation. Herausforderungen des Wandels in Richtung nachhaltige Entwicklung* (pp. 109–132). Springer VS: Wiesbaden.

Degenhardt, H., & Holstenkamp, L. (2011). Genossenschaftlich organisierte Bürgerbeteiligung als Finanzierungs- und Nachhaltigkeitsmodell. In W. George & T. Berg (Eds.), *Regionales Zukunftsmanagement: Vol. 5. Regionales Zukunftsmanagement. Band 5: Energiegenossenschaften gründen und erfolgreich betreiben* (pp. 47–55). Lengerich [u.a.]: Pabst Science.

Deuten, J. J. (2003). *Cosmopolitanising technology: A study of four emerging technological regimes*. PhD, Twente University, Enschede.

Deutsche Bundesbank. (Ed.). (2013). *Ertragslage und Finanzierungsverhältnisse deutscher Unternehmen im Jahr 2012: Monatsbericht Dezember 2013*. Retrieved 01.10.2016, from www.bundesbank.de

Deutsche Gesellschaft für Sonnenenergie e.V. (Ed.). (2014a). *EnergyMap: Main-Tauber Kreis*. Retrieved 01.10.2016, from http://www.energymap.info/energieregionen/DE/105/110/162/597.html

Deutsche Gesellschaft für Sonnenenergie e.V. (Ed.). (2014b). *EnergyMap: Neckar-Odenwald-Kreis*. Retrieved 01.10.2016, from http://www.energymap.info/energieregionen/DE/105/110/160/357.html

Deutsche Gesellschaft für Sonnenenergie e.V. (Ed.). (2015). *Bundesrepublick Deutschland 25% EEG-Strom*. Retrieved 01.10.2016, from http://www.energymap.info/energieregionen/DE/105.html

Deutscher Bundestag (Ed.). (2015). *Jenseits der Tagespolitik – die Enquete-Kommissionen: Teil 2*. Retrieved 01.10.2016, from https://www.bundestag.de/dokumente/textarchiv/22015969_enquete2/199444

Deutscher Bundestag. (Ed.). (2016). *Entwurf eines Gesetzes zur Einführung von Ausschreibungen für Strom aus erneuerbaren Energien und zu weiteren Änderungen des Rechts der erneuerbaren Energien*. Retrieved 01.10.2016, from https://www.bmwi.de/DE/Themen/Energie/Erneuerbare-Energien/eeg-2017-wettbewerbliche-verguetung.html

Dewald, U., & Truffer, B. (2012). The local sources of market formation: Explaining regional growth differentials in German photovoltaic markets. *European Planning Studies, 20*(3), 397–420.

DiMaggio, P. J., & Powell, W. W. (1983). The Iron cage revisited: Institutional isomorphism and collective rationality in organizational fields. *American Sociology Review, 48*(2), 147–160.

Docherty, I., & Shaw, J. (2012). The governance of transition policy. In F. W. Geels, R. Kemp, G. Dudley, & G. Lyons (Eds.), *Automobility in transition? A socio-technical analysis of sustainable transport* (pp. 104–122). New York: Routledge.

Dolata, U. (2009). Technological innovations and sectoral change. *Research Policy, 38*(6), 1066–1076.

Dosi, G. (1982). Technological paradigms and technological trajectories: A suggested interpretation of the determinants and directions of technical change. *Research Policy, 6*(3), 147–162.

Dosi, G., Freemann, C., Nelson, R., & Soete, L. (Eds.). (1988). *Technical change and economic theory*. London, New York: Pinter.

dpa. (2015, February 20). Gläubiger müssen sich entscheiden. *Handelsblatt*. Retrieved from http://www.handelsblatt.com/finanzen/steuern-recht/recht/prokon-pleite-glaeubiger-muessen-sich-entscheiden/11399150.html

Dubielzig, F. S. S. (2004). In Centre for Sustainability Management (Ed.), *Methoden transdisziplinärer Forschung und Lehre. Ein zusammenfassender Überblick*. Centre for Sustainability Management.

Dudenverlag. (2015a). *Rechtschreibung: der Mut (die Bedeutung)*. Berlin: Dudenverlag.

Dudenverlag. (2015b). *Seriös*. Berlin: Dudenverlag.

Duschek, S. (1996). *Miszelle zur Dualität von Struktur*. Wuppertal: unveröff. Man.

Ebert, T., & Henke, K. (2012, October). *Energiewende Nordhessen. Szenarien für den Umbau der Stromversorgung auf eine dezentrale und erneuerbare Erzeugungsstruktur*. Kassel.

Edquist, C. (2004). Reflections on the systems of innovation approach. *Science and Public Policy, 31*(6), 485–489.
Ehrnberg, E., & Jacobsson, S. (1997). Technological discontinuities and incumbents' performance: An analytical framework. In C. Edquist (Ed.), *Systems of innovation. Technologies, institutions, and organizations* (pp. 318–341). London, Washington: Pinter.
Energie + Umwelt eG. (Ed.). (2013). *Satzung Energie + Umwelt eG*. Retrieved 01.10.2016, from http://www.epueg.de
Energie + Umwelt eG. (Ed.). (2014a). *Bürgerprojekte*. Retrieved 01.10.2016, from http://www.epueg.de/buerger-projekte/index.html
Energie + Umwelt eG. (Ed.). (2014b). *Geschäftsbericht 2013*. Retrieved 01.10.2016, from www.epueg.de
Energie + Umwelt eG. (Ed.). (2014c). *Unsere Vision*. Retrieved 01.10.2016, from http://www.epueg.de/wir-ueber-uns/unsere-vision/index.html
Energie + Umwelt eG. (Ed.). (2014d). *Unsere Ziele*. Retrieved 01.10.2016, from http://www.epueg.de/wir-ueber-uns/unsere-ziele/index.html
Energie + Umwelt eG. (Ed.). (2015). *4. Generalversammlung der Energie + Umwelt eG*. Retrieved 01.10.2016, from http://www.epueg.de/wir-ueber-uns/aktuelles/aktuelles/2015-05-21-generalversammlung-2015.html
Energie 2000 e.V. (Ed.). (2015). *Wolfhagen 100% EE – Entwicklung einer nachhaltigen Energieversorgung für die Stadt Wolfhagen*. Retrieved 01.10.2016, from http://www.energieoffensive-wolfhagen.de/
Energieagentur Main-Tauber-Kreis GmbH. (Ed.). (2011). *Potenzialanalyse Erneuerbare Energien Gesamt Main-Tauber-Kreis*. Retrieved 01.10.2016, from http://www.ea-main-tauber-kreis.de/erneuerbarkomm/diagramm.php?gemeinde=Gesamt+Main-Tauber-Kreis#
Energieagentur Main-Tauber-Kreis GmbH. (Ed.). *Über uns*. Retrieved 01.10.2016, from http://www.ea-main-tauber-kreis.de/ueber-uns.html
EnergieAgentur Neckar-Odenwald-Kreis. (Ed.). (2014). *Über uns*. Retrieved 01.10.2016, from http://www.eanok.de/ueber-uns/
Englund, H., & Gerdin, J. (2014). Structuration theory in accounting research: Applications and applicability. *Critical Perspectives on Accounting, 25*(2), 162–180.
Erneuerbare-Energien-Gesetz – EEG. *Bundesgesetzblatt* 1066, Federal Ministry of Justice and Consumer Protection 21.07.2014.
Erstes Gesetz zur Änderung des Erneuerbare-Energien-Gesetz. *Bundesgesetzblatt* 1170, Federal Ministry of Justice and Consumer Protection 2010.
Essletzbichler, J. (2012). Renewable energy technology and path creation: A multi-scalar approach to energy transition in the UK. *European Planning Studies, 20*(5), 791–816.
Ethics Commission for a Safe Energy Supply. (Ed.). (2011). *Germany's energy transition – A collective project for the future*. Retrieved 01.10.2016, from https://www.bundesregierung.de/ContentArchiv/DE/Archiv17/_Anlagen/2011/05/2011-05-30-abschlussbericht-ethikkommission_en.pdf?__blob=publicationFile&v=2
European Central Bank. (Ed.). *Key ECB interest rates*. Retrieved 01.10.2016, from https://www.ecb.europa.eu/stats/monetary/rates/html/index.en.html
European Commission. (Ed.). (2011a). *Energy roadmap 2050*. Retrieved 01.10.2016, from https://ec.europa.eu/energy/en/topics/energy-strategy/2050-energy-strategy
European Commission. (Ed.). (2011b). *Energy roadmap 2050. Impact assessment. Part 2 including Part II of Annex 1 'Scenarios – assumptions and results' and Annex 2 'Report on Stakeholders scenarios'*. Retrieved 01.10.2016, from https://ec.europa.eu/energy/en/topics/energy-strategy/2050-energy-strategy
European Commission. (Ed.). (2015). *Leader. Introduction*. Retrieved 01.10.2016, from http://ec.europa.eu/agriculture/rur/leaderplus/index_en.htm
Faller, F. (2016). A practice approach to study the spatial dimensions of the energy transition. *Environmental Innovation and Societal Transitions, 16*, 85–95.

Farla, J., Markard, J., Raven, R. P. J. M., & Coenen, L. (2012). Sustainability transitions in the making: A closer look at actors, strategies and resources: Contains special section: Actors, strategies and resources in sustainability transitions. *Technological Forecasting and Social Change, 79*(6), 991–998.

Federal Ministry for Economic Affairs and Energy. (Ed.). (2014a). *Das Erneuerbare-Energien-Gesetz 2014. Die wichtigsten Fakten zur Reform des EEG.* Retrieved 01.10.2016, from http://www.bmwi.de/DE/Themen/Energie/Erneuerbare-Energien/eeg-2014.html

Federal Ministry for Economic Affairs and Energy. (Ed.). (2014b). *Erneuerbare Energien in Zahlen. Nationale und internationale Entwicklung im Jahr 2013.* Retrieved 01.10.2016, from http://www.erneuerbare-energien.de/EE/Navigation/DE/Service/Erneuerbare_Energien_in_Zahlen/erneuerbare_energien_in_zahlen.html

Federal Ministry for Economic Affairs and Energy. (Ed.). (2014c). *Smart Energy made in Germany. Erkenntnisse zum Aufbau und zur Nutzung intelligenter Energiesysteme im Rahmen der Energiewende.* Retrieved 01.10.2016, from http://www.e-energie.info/

Federal Ministry for Economic Affairs and Energy. (Ed.). (2015a). *Die Energiewende – ein gutes Stück Arbeit. Erneuerbare Energien 2014. Daten der Arbeitsgruppe Erneuerbare Energien Statistik (AGEE-Stat).* Retrieved 01.10.2016, from http://www.erneuerbare-energien.de/EE/Navigation/DE/Service/Erneuerbare_Energien_in_Zahlen/Arbeitsgruppe/arbeitsgruppe_ee.html

Federal Ministry for Economic Affairs and Energy. (Ed.). (2015b). *Die Energiewende gemeinsam zum Erfolg führen. Auf dem Weg zu einer sicheren, sauberen und bezahlbaren Energieversorgung.* Retrieved 01.10.2016, from http://www.bmwi.de/DE/Themen/Energie/Energiewende/gesamtstrategie.html

Federal Ministry for Economic Affairs and Energy. (Ed.). (2015c). *Ein Strommarkt für die Energiewende. Ergebnispapier des Bundesministeriums für Wirtschaft und Energie (Weißbuch).* Retrieved 01.10.2016, from http://www.bmwi.de/DE/Themen/Energie/Strommarkt-der-Zukunft/strommarkt-2-0.html

Federal Ministry for Economic Affairs and Energy. (Ed.). (2015d). *Eine Zielarchitektur für die Energiewende: Von politischen Zielen bis zu Einzelmaßnahmen.* Retrieved from http://www.bmwi.de/DE/Themen/Energie/Energiewende/zielarchitektur.html

Federal Ministry for Economic Affairs and Energy. (Ed.). (2016a). *Erneuerbare Energien in Deutschland. Das Wichtigste im Jahr 2014 auf einen Blick. BMWi nach Arbeitsgruppe Erneuerbare Energien-Statistik (AGEE-Stat).* Retrieved 01.10.2016, from http://www.erneuerbare-energien.de

Federal Ministry for Economic Affairs and Energy. (Ed.). (2016b). *Referentenentwurf des BMWi (IIIB2). Entwurf eines Gesetzes zur Einführung von Ausschreibungen für Strom aus erneuerbaren Energien und zu weiteren Änderungen des Rechts der erneuerbaren Energien.* Retrieved 01.10.2016, from https://www.bmwi.de

Federal Ministry for Economic Affairs and Energy. (Ed.). (2016c). *Zeitreihen zur Entwicklung der erneuerbaren Energien in Deutschland.* Retrieved 01.10.2016, from http://www.erneuerbare-energien.de/

Federal Ministry for Economic Affairs and Technology. (Ed.). (2009). *E-Energy. IKT-basiertes Energiesystem der Zukunft.* Retrieved 01.10.2016, from https://www.bmwi.de/BMWi/Redaktion/PDF/Publikationen/Technologie-und-Innovation/e-energy-ikt-basiertes-energiesystem-der-zukunft,property=pdf,bereich=bmwi,sprache=de,rwb=true.pdf

Federal Ministry for Economic Affairs and Technology, & Federal Ministry for the Environment, Nature Conservation, Building and Nuclear Safety. (Eds.). (2010). *Energiekonzept für eine umweltschonende, zuverlässige und bezahlbare Energieversorgung.* Retrieved 01.10.2016, from www.bmu.de

Federal Ministry for the Environment, Nature Conservation, Building and Nuclear Safety. (Ed.). (2012). *Langfristszenarien und Strategien für den Ausbau der erneuerbaren Energien in Deutschland bei Berücksichtigung der Entwicklung in Europa und global.* Retrieved 01.10.2016, from http://www.erneuerbare-energien.de/

Flick, U. (2011). *Triangulation: Eine Einführung*. Wiesbaden: VS Verlag für Sozialwissenschaften/ Springer Fachmedien Wiesbaden, Wiesbaden.

Flieger, B. (2008). Energiegenossenschaften. *Eine andere Energiewirtschaft ist möglich Einstieg in das Tagungsthema. Beitrag zur Tagung "Energiegenossenschaften"*. Retrieved 01.10.2016, from http://www.innova-eg.de/produkte/publikationen/

Flieger, B. (2011a). Economic participation in urban climate protection – Energy cooperatives: Citizen participation in the municipally-organised energy turnaround. In Heinrich-Böll-Foundation Brandenburg e.V. (Ed.), *Participation in urban climate protection. Answers of European Municipalities* (pp. 58–67). Potsdam.

Flieger, B. (2011b). Energiegenossenschaften. Eine klimaverantwortliche, bürgernahe Energiewirtschaft ist möglich. In S. Elsen (Ed.), *Münchener Hochschulschriften für angewandte Sozialwissenschaften: Vol. 244. Ökosoziale Transformation. Solidarische Ökonomie und die Gestaltung des Gemeinwesens; [Perspektiven und Ansätze von unten]* (1st ed., pp. 305–328). Neu-Ulm: AG-SPAK-Bücher.

Flieger, B. (2011c). Lokale Wertschöpfung durch Bürgerbeteiligung. *Verbands-Management, 37* (1), 50–57.

Flieger, B. (2014). Was können (Energie-)Genossenschaften zur nachhaltigen Entwicklung beitragen. Genossenschaft als Rechtsform, als Organisationsform und als Narratives Konzept. In R. Pfriem (Ed.), *5. Spiekerooger KlimaGespräche. Was können gemeinschaftsorientierte Formen des Wirtschaften s zu nachhaltiger Entwicklung beitragen? Dokumentation 5. Spiekerooger KlimaGespräche vom 7.–9. November 2013*. Oldenburg: dbv, Dt. Buchverlag.

Flieger, B., Klemisch, H., & Radtke, J. (2015). Bürgerbeteiligung in und durch Energiegenossenschaften. *eNewsletter Netzwerk Bürgerbeteiligung, 03*.

Foester, H. v. (1992). Entdecken und Erfinden – Wie läßt sich Verstehen verstehen? In H. Gumin & H. Meier (Eds.), *Einführung in den Konstruktivismus* (pp. 41–88). München, Zürich: Pieper.

Fontes, M., Sousa, C., & Ferreira, J. (2016). The spatial dynamics of niche trajectory: The case of wave energy. *Environmental Innovation and Societal Transitions, 19*, 66–84.

Forrest, N., & Wiek, A. (2015). Success factors and strategies for sustainability transitions of small-scale communities – Evidence from a cross-case analysis. *Environmental Innovation and Societal Transitions, 17*, 22–40.

Foxon, T. J. (2013). Transition pathways for a UK low carbon electricity future. *Special Section: Transition Pathways to a Low Carbon Economy, 52*, 10–24.

Fränkische Nachrichten (Main-Tauber). (2010, October 23). Volks-und Raiffeisenbanken: Gemeinsame Energiegenossenschaft steht vor der Gründung. Beteiligung der Bürger möglich. *Fränkische Nachrichten, Region Main-Tauber*.

Fränkische Nachrichten (Neckar-Odenwald). (2010, October 21). In der Region: Volksbanken und Raiffeisenbanken wollen Anfang 2011 eine Energiegenossenschaft gründen/Bürger können sich beteiligen. Start in ein neues Energiezeitalter steht bevor. *Fränkische Nachrichten, Region Neckar Odenwald*.

Fraunhofer Institute for Building Physics IBP. (2010a). *Wolfhagen 100% EE – Entwicklung einer nachhaltigen Energieversorgung für die Stadt Wolfhagen*. Kassel.

Fraunhofer Institute for Building Physics IBP. (2010b). *Wolfhagen 100% EE. Anhang*. Kassel.

Fraunhofer Institute for Solar Energy Systems. (Ed.). (2012). *100% erneuerbare Energien für Strom und Wärme in Deutschland*. Retrieved 01.10.2016, from https://www.ise.fraunhofer.de/de/veroeffentlichungen/studien-und-positionspapiere/studie-100-erneuerbare-energien-fuer-strom-und-waerme-in-deutschland

Fraunhofer Institute for Wind Energy and Energy System Technology & Stadtwerke Union Nordhessen (SUN). (2012). *Energiewende Nordhessen Abschlussbericht. Szenarien für den Umbau der Stromversorgung auf eine dezentrale und erneuerbare Erzeugungsstruktur*. Kassel.

Freemann, C., & Perez, C. (1988). Structural crises of adjustment, business cycles and investment behaviour. In G. Dosi, C. Freemann, R. Nelson, & L. Soete (Eds.), *Technical change and economic theory* (pp. 38–66). London, New York: Pinter.

Frey Martin. Ein Stück Stadtwerke bitte, Nr. 09/2012. *Sonne, Wind & Wärme*, pp. 40–43.

Fudge, S., Peters, M., & Woodman, B. (2016). Local authorities as niche actors: The case of energy governance in the UK. *Environmental Innovation and Societal Transitions, 18*, 1–17.

Fuenfschilling, L., & Truffer, B. (2014). The structuration of socio-technical regimes – Conceptual foundations from institutional theory. *Research Policy, 43*(4), 772–791.

Gabriel, S. (2014). *Rede des Bundesministers für Wirtschaft und Energie, Sigmar Gabriel, zum Entwurf eines Gesetzes zur grundlegenden Reform des Erneuerbare-Energien-Gesetzes und zur Änderung weiterer Bestimmungen des Energiewirtschaftsrechts vor dem Deutschen Bundestag am 8. Mai 2014 in Berlin*. Retrieved 01.10.2016, from https://www.bundesregierung.de/Content/DE/Bulletin/2010-2015/2014/05/50-1-bmwi-bt.html

Geels, F. W. (2002). Technological transitions as evolutionary reconfiguration processes: A multi-level perspective and a case-study, Nr. 8/9. *Research Policy*, pp. 1257–1274.

Geels, F. W. (2004). From sectoral systems of innovation to socio-technical systems: Insights about dynamics and change from sociology and institutional theory. *Research Policy, 33*(6–7), 897–920.

Geels, F. W. (2005). *Technological transitions and system innovations*. Cheltenham, Northampton, MA: Edward Elgar.

Geels, F. W. (2006). The hygienic transition from cesspools to sewer systems (1840–1930): The dynamics of regime transformation. *Research Policy, 35*(7), 1069–1082.

Geels, F. W. (2011). The multi-level perspective on sustainability transitions: Responses to seven criticisms. *Environmental Innovation and Societal Transitions, 1*, 24–40.

Geels, F. W. (2014). Reconceptualising the co-evolution of firms-in-industries and their environments: Developing an inter-disciplinary Triple Embeddedness Framework. *Research Policy, 43*(2), 261–277.

Geels, F., & Deuten, J. J. (2006). Local and global dynamics in technological development: A socio-cognitive perspective on knowledge flows and lessons from reinforced concrete. *Science and Public Policy, 33*(4), 265–275.

Geels, F. W., & Kemp, R. (2012). The multi-level perspective as a new perspective for studying socio-technological transitions. In F. W. Geels, R. Kemp, G. Dudley, & G. Lyons (Eds.), *Automobility in transition? A socio-technical analysis of sustainable transport* (pp. 49–82). New York: Routledge.

Geels, F., & Raven, R. P. J. M. (2006). Non-linearity and expectations in niche-development trajectories: Ups and downs in Dutch biogas development (1973–2003). *Technology Analysis & Strategic Management, 18*(3–4), 375–392.

Geels, F. W., & Schot, J. (2007). Typology of sociotechnical transition pathways. *Research Policy, 36*(3), 399–417.

Geels, F. W., & Schot, J. (2010). The dynamics of transitions: A socio-technical perspective. In J. Grin, J. Rotmans, & J. Schot (Eds.), *Transitions to sustainable development. New directions in the study of long term transformative change* (pp. 11–93). New York: Routledge.

Geels, F. W., Kemp, R., Dudley, G., & Lyons, G. (Eds.). (2012). *Automobility in transition?: A socio-technical analysis of sustainable transport*. New York: Routledge.

Gemeinde Speinshart. (Ed.). (2015). *Herzlich Willkommen Gemeinde Speinshart*. Retrieved 01.10.2016, from http://speinshart.de/frameset.htm

Gemeinde Wolfhagen. (Ed.). (2014). *Einwohnerzahlen (Haupt- und Nebenwohnungen – 31.12.2014)*. Retrieved 01.10.2016, from http://www.wolfhagen.de/de/rathaus/zahlen_fakten/einwohnerzahlen.php?navanchor=1110036.

German Advisory Council on Global Change. (2003). *World in transition – Towards sustainable energy systems (Flagship report)*. Berlin: German Advisory Council on Global Change.

German Advisory Council on Global Change. (2011). *World in transition – A social contract for sustainability (Flagship report)*. Berlin: German Advisory Council on Global Change.

German Cooperative and Raiffeisen Confederation – reg. assoc. (Ed.). (2013a). *Energiegenossenschaften. Ergebnisse der Umfrage des DGRV und seiner Mitgliederverbände im Frühsommer 2013*. Retrieved 01.10.2016, from http://www.genossenschaften.de/dgrv-stellt-aktuelle-umfrage-zu-energiegenossenschaften-vor

German Cooperative and Raiffeisen Confederation – reg. assoc. (Ed.). (2013b). *Was ist eine Genossenschaft?*. Retrieved 01.10.2016, from http://www.genossenschaften.de/was-ist-eine-genossenschaft

German Cooperative and Raiffeisen Confederation – reg. assoc. (Ed.). (2014). *Energiegenossenschaften. Ergebnisse der Umfrage des DGRV und seiner Mitgliedsverbände*. Retrieved from https://www.dgrv.de/de/dienstleistungen/energiegenossenschaften/jahresumfrage.html

German Cooperative and Raiffeisen Confederation – reg. assoc. (Ed.). (2015). *Zahlen und Fakten 2015*. Retrieved 15.06.2015, from https://www.dgrv.de/de/publikationen/nneuzahlenundfakten.html

German Institute for Economic Research. (2013). *Energiewende und Versorgungssicherheit: Deutschland braucht keinen Kapazitätsmarkt*. Berlin. Retrieved 01.10.2016, from http://www.diw.de/de/diw_01.c.432384.de/themen_nachrichten/energiewende_und_versorgungssicherheit_deutschland_braucht_keinen_kapazitaetsmarkt.html

German Solar Association. (2014). *EEG-Reform: Energiewende-Bremse statt Strompreis-Bremse*. Berlin. Retrieved 01.10.2016, from https://www.solarwirtschaft.de/de/presse/pressemeldungen/pressemeldungen-im-detail/news/eeg-reform-energiewende-bremse-statt-strompreis-bremse.html

Gerring, J. (2007). *Case study research: Principles and practices*. New York: Cambridge University Press.

Gesetz betreffend die Erwerbs-und Wirtschaftsgenossenschaften (Genossenschaftsgesetz – GenG). *Bundesgesetzblatt* 2230, Federal Ministry of Justice and Consumer Protection 2006.

Gesetz für den Vorrang Erneuerbarer Energien (Erneuerbarer-Energien-Gesetz-EEG) sowie zur Änderung des Energiewirtschaftsgesetzes und des Mineralölsteuergesetzes. *Bundesgesetzblatt* 305, Federal Ministry of Justice and Consumer Protection 2000.

Gesetz zur Änderung des Rechtsrahmens für Strom aus solarer Strahlungsenergie und zu weiteren Änderungen im Recht der erneuerbaren Energien. *Bundesgesetzblatt* 1754, Federal Ministry of Justice and Consumer Protection 2012.

Gesetz zur Umsetzung der Richtlinie 2011/61/EU über die Verwalter alternativer Investmentfonds. *Bundesgesetzblatt* 1981, Federal Ministry of Justice and Consumer Protection 2013.

Giddens, A. (1976). *New rules of social method. A positive critique of interpretative sociologies*. New York: Polity Press.

Giddens, A. (1979). *Central problems in social theory: Action, structure and contradiction in social analysis*. Berkeley: University of California Press.

Giddens, A. (1981). *A contemporary critique of historical materialism. Volume 1: Power, property and the state*. London: Macmillan, Polity Press.

Giddens, A. (1982). Power, the dialectic of control and class structuration. In A. Giddens & G. MacKenzie (Eds.), *Social class and the division of labour: Essays in honour of Ilya Neustadt* (pp. 29–45). Cambridge: Cambridge University Press.

Giddens, A. (1984). *The constitution of society: Outline of the theory of structuration*. Berkeley, CA: University of California Press.

Giddens, A. (1989). A reply to my critics. In D. Held & J. B. Thompson (Eds.), *Social theory of modern societies. Anthony Giddens and his critics* (pp. 249–301). Cambridge [England], New York: Cambridge University Press.

Giddens, A. (1995). *Konsequenzen der Moderne. Suhrkamp-Taschenbuch Wissenschaft* (Vol. 1295). Frankfurt am Main: Suhrkamp.

Gottschick, M. (2015). Reflexive capacity in local networks for sustainable development: Integrating conflict and understanding into a multi-level perspective transition framework. *Journal of Environmental Policy & Planning*, 1–22.

Governing coalition of Germany. (Ed.). (2015). *Eckpunkte für eine erfolgreiche Umsetzung der Energiewende. Politische Vereinbarungen der Parteivorsitzenden von CDU, CSU und SPD vom 1. Juli 2015*. Retrieved 01.10.2016, from http://www.bmwi.de/DE/Themen/Energie/Energiewende/gesamtstrategie.html

Gray, B., & Wood, D. J. (1991). Collaborative alliances: Moving from practice to theory. *The Journal of Applied Behavioral Science, 27*(1), 3–22.

Greenpeace (Ed.). (2009). *Klimaschutz: Plan B 2050. Energiekonzept für Deutschland*. Retrieved 01.10.2016, from https://www.greenpeace.de/publications

Greenpeace, & Friends of the Earth Germany. (Eds.). (2013). *Flexibilität erhöhen, Versorgung sichern, Energiewende vorantreiben. Zur Debatte um ein neues Strommarktdesign in Deutschland*. Retrieved 01.10.2016, from www.bund.de

Greenpeace Energy eG. (2014). *Bundesregierung hält an Einschnitten für die Bürgerenergie fest*. Hamburg. Retrieved 01.10.2016, from http://www.greenpeace-energy.de/presse/artikel/article/bundesregierung-haelt-an-einschnitten-fuer-die-buergerenergie-fest.html

Gregory, D. (1989). Presences and absences: Time-space relations and structuration theory. In D. Held & J. B. Thompson (Eds.), *Social theory of modern societies. Anthony Giddens and his critics* (pp. 185–214). Cambridge [England], New York: Cambridge University Press.

Gregson, N. (1989). On the (ir)relevance of structuration theory to empirical research. In D. Held & J. B. Thompson (Eds.), *Social theory of modern societies. Anthony Giddens and his critics* (pp. 235–248). Cambridge [England], New York: Cambridge University Press.

Grin, J. (2008). The multi-level perspective and design of system innovations. In J. C. van den Bergh & F. R. Bruinsma (Eds.), *Managing the transition to renewable energy. Theory and practice from local, regional and macro perspective* (pp. 47–80). Cheltenham, Northampton, MA: Edward Elgar.

Grin, J. (2010). Understanding transitions from a governance perspective. In J. Grin, J. Rotmans, & J. Schot (Eds.), *Transitions to sustainable development. new directions in the study of long term transformative change* (pp. 223–338). New York: Routledge.

Grin, J., Rotmans, J., & Schot, J. (2010a). From persistent problems to system innovations and transitions. In J. Grin, J. Rotmans, & J. Schot (Eds.), *Transitions to sustainable development. New directions in the study of long term transformative change* (pp. 1–4). New York: Routledge.

Grin, J., Rotmans, J., & Schot, J. (2010b). Preface. In J. Grin, J. Rotmans, & J. Schot (Eds.), *Transitions to sustainable development. New directions in the study of long term transformative change* (pp. xvii–xxix). New York: Routledge.

Grin, J., Rotmans, J., & Schot, J. (Eds.). (2010c). *Transitions to sustainable development. New directions in the study of long term transformative change*. New York: Routledge.

Grosskopf, W., Münkner, H.-H., & Ringle, G. (2012). *Unsere Genossenschaft: Idee – Auftrag – Leistungen*. Wiesbaden: DG-Verlag.

Hall, R. (1992). The strategic analysis of intangible resources. *Strategic Management Journal, 13*(2), 135–144.

Hansen, T., & Coenen, L. (2015). The geography of sustainability transitions: Review, synthesis and reflections on an emergent research field. *Environmental Innovation and Societal Transitions, 17*, 92–109.

Hansen, G. H., & Steen, M. (2015). Offshore oil and gas firms' involvement in offshore wind: Technological frames and undercurrents. *Environmental Innovation and Societal Transitions, 17*, 1–14.

Hasanov, I. (2010). *Konsumentenverhalten bei Ökostromangeboten*. PhD thesis, Universität Duisburg-Essen.

Hauptmann, S. (2015). Structure and action in organisational social media settings: Genre and speech act analysis combined – Exemplified on a case of strategic action. *Die Betriebswirtschaft: DBW = Business Administration Review, 75*(6), 431–444.

Hauschildt, J., & Salomo, S. (2011). *Innovationsmanagement*. München: Vahlen.

Hauser, E. H. J., Dröschel, B., Klann, U., Heib, S., & Grashof, K. (2015). In Institut für ZukunftsEnergieSysteme (Ed.), *Nutzeneffekte von Bürgerenergie. Eine wissenschaftliche Qualifizierung und Quantifizierung der Nutzeneffekte der Bürgerenergie und ihrer möglichen Bedeutung für die Energiewende*. Institut für ZukunftsEnergieSysteme.

Heesen, B., & Gruber, W. (2011). *Bilanzanalyse und Kennzahlen: Fallorientierte Bilanzoptimierung*. Wiesbaden: Gabler Verlag.

Hegger, D. L., van Vliet, J., & van Vliet, B. J. (2007). Niche management and its contribution to regime change: The case of innovation in sanitation. *Technology Analysis & Strategic Management, 19*(6), 729–746.

Heinrich Böll Stiftung. (2011). *Stellungnahme von Ralf Fücks zur Einladung, in der Atomkommission mitzuarbeiten*. Berlin. Retrieved from http://www.boell.de/presse/pressestellungnahme-ralf-fuecks-atomkommission-11612.html

Hekkert, M. P., Suurs, R. A., Negro, S. O., Kuhmann, S., & Smits, R. E. (2007). Functions of innovation systems: A new approach for analysing technological change. *Technological Forecasting and Social Change, 74*, 413–432.

Hendriks, C. (2008). On inclusion and network governance: The democratic disconnect of Dutch energy transitions. *Public Administration, 86*(4), 1009–1031.

Hennicke, P., & Welfens, P. J. J. (2012). *Energiewende nach Fukushima: Deutscher Sonderweg oder weltweites Vorbild?* München: oekom verlag.

Hermans, F., van Apeldoorn, D., Stuiver, M., & Kok, K. (2013). Niches and networks: Explaining network evolution through niche formation processes. *Research Policy, 42*(3), 613–623.

Herriott, R. E., & Firestone, W. A. (1983). Multisite qualitative policy research: Optimizing description and generalizability. *Educational Researcher, 12*(2), 14–19.

Hielscher, H. (2014, November 29). Prokon-Anleger verlieren Kapital. *Wirtschaftswoche*. Retrieved from http://www.wiwo.de/finanzen/boerse/windkraft-prokon-anleger-verlieren-kapital/11048548.html

Hieschler, S. (2011). In University of Sussex (Ed.), *Community energy in the UK. A review of the research literature*. University of Sussex.

Hildebrand, J., Rau, I., & Schweizer-Ries, P. (2012). Die Bedeutung dezentraler Beteiligungsprozesse für die Akzeptanz des Ausbaus erneuerbarer Energien. Eine umweltpsychologische Betrachtung. *Informationen zur Raumentwicklung,* (9/10), 491–501.

Höfer, H.-H., & Rommel, J. (2015). Internal governance and member investment behavior in energy cooperatives: An experimental approach. *Utilities Policy, 36*, 52–56.

Hofman, P. S., & Elzen, B. (2010). Exploring system innovation in the electricity system through sociotechnical scenarios. *Technology Analysis & Strategic Management, 22*(6), 653–670.

Holstenkamp, L. (2015). *The rise and fall of electricity distribution cooperatives in Germany*. Retrieved 01.10.2016, from www.leuphana.de/businessandlaw

Holstenkamp, L., & Müller, J. R. (2013). *Zum Stand von Energiegenossenschaften in Deutschland – Eine Typologie, Paper Präsentation auf der AGI-Nachwuchswissenschaftler-Tagung 2013*. Münster.

Holstenkamp, L., & Müller, J. R. (2015). In Leuphana Universität Lüneburg (Ed.), *Zum Stand von Energiegenossenschaften in Deutschland. Aktualisierter Überblick über Zahlen und Entwicklungen zum 31.12.2014*. Leuphana Universität Lüneburg.

Holstenkamp, L., & Ulbrich, S. (2010). *Bürgerbeteiligungsmodelle mittels Fotovoltaikgenossenschaften: Marktüberblick und Analyse der Finanzierungsstruktur*. Arbeitspapierreihe Wirtschaft & Recht, No. 3. Lüneburg, pp. 31–33.

Holtz, G., Brugnach, M., & Pahl-Wostl, C. (2008). Specifying "regime" – A framework for defining and describing regimes in transition research. *Technological Forecasting and Social Change, 75* (5), 623–643.

Hoogma, R., Kemp, R., Schot, J., Truffer, B., Hoogma, R., & Schot, J. (2002). *Experimenting for sustainable transport. The approach of strategic niche management*. London: Spon.

Hoppenbrock, C., & Fischer, B. (2012). In Institut dezentrale Energietechnologien (IdE) (Ed.), *Was ist eine 100ee-Region und wer darf sich so nennen? Information zur Aufnahme und Bewertung*. Institut dezentrale Energietechnologien (IdE). Retrieved 01.10.2016, from http://www.100-ee.de

Hughes, T. P. (1983). *Networks of power: Electrification in Western society, 1800–1930*. Baltimore, MD, London: Johns Hopkins University Press.

Hughes, T. P. (1986). The seamless web: Technology, science, etcetera, etcetera. *Social Studies of Science, 16*(2), 281–292.

Hughes, T. P. (1987). The evolution of large technology systems. In W. E. Bijker, T. P. Hughes, & T. J. Pinch (Eds.), *The Social construction of technological systems. New directions in the sociology and history of technology* (pp. 51–82). Cambridge, MA: MIT Press.

Huijben, J., Verbong, G., & Podoynitsyna, K. (2016). Mainstreaming solar: Stretching the regulatory regime through business model innovation. *Environmental Innovation and Societal Transitions, 20*, 1–15.

Intergovernmental Panel on Climate Change. (Ed.). (2014). *Climate change 2014: Mitigation of climate change.* Retrieved 01.10.2016, from http://www.ipcc.ch

Intergovernmental Panel on Climate Change. (Ed.). (2015a). *Climate change 2014. Synthesis report.* Retrieved 01.10.2016, from http://www.ipcc.ch/report/ar5/syr/

Intergovernmental Panel on Climate Change. (Ed.). (2015b). *Climate change 2014. Synthesis report. Summary for policymakers.* Retrieved 01.10.2016, from http://www.ipcc.ch

International Energy Agency. (Ed.). (2015a). *Energy and climate change.* Retrieved 01.10.2016, from https://www.iea.org

International Energy Agency. (Ed.). (2015b). *Energy and climate change. World energy outlook special report.* Retrieved 01.10.2016, from http://www.iea.org/

International Energy Agency. (Ed.). (2015c). *World energy outlook 2015.* Retrieved 01.10.2016, from http://www.iea.org/

International Energy Agency. (Ed.). (2016). *Energy technology perspectives 2016.* Retrieved 01.10.2016, from http://www.iea.org/

Jackson, M. O. (2011). *Social and economic networks.* Princeton, NJ, Woodstock: Princeton University Press.

Jacobsson, S., & Bergek, A. (2004). Transforming the energy sector: The evolution of technological systems in renewable energy technology. *Industrial and Corporate Change, 13*(5), 815–849.

Jacobsson, S., & Bergek, A. (2011). Innovation system analyses and sustainability transitions: Contributions and suggestions for research. *Environmental Innovation and Societal Transitions, 1*, 41–57.

Jakobsen, I. (2008). In University of Oslo, & University of Aalborg (Eds.), *The road to renewables: A case study of wind energy, local ownership and social acceptance at Samsø.* University of Oslo, & University of Aalborg.

Jepperson, R. L. (1991). Institutions, institutional effects and institutionalism. In W. W. Powell & P. J. DiMaggio (Eds.), *The new institutionalism in organizational analysis* (pp. 143–163). Chicago: University of Chicago Press.

Jones, M. (2011). Structuration theory. In R. Galliers & W. L. Currie (Eds.), *The Oxford handbook of management information systems* (pp. 113–137). Oxford: Oxford University Press.

Kaehlert, G. (2011). Regionale Energieversorgung: Von der Vision zur funktionierenden Praxis. In W. George & T. Berg (Eds.), *Regionales Zukunftsmanagement: Vol. 5. Regionales Zukunftsmanagement. Band 5: Energiegenossenschaften gründen und erfolgreich betreiben* (pp. 32–39). Lengerich [u.a.]: Pabst Science.

Kalkbrenner, B. J., & Roosen, J. (2016). Citizens' willingness to participate in local renewable energy projects: The role of community and trust in Germany. *Energy Research & Social Science, 13*, 60–70.

Kaphengst, T., & Velten, K. E. (2014). In WWWforEurope project (Ed.), *Energy transition and behavioural change in rural areas. The role of energy cooperatives.* WWWforEurope project.

Kayser, L. (2014). In Klaus Novy Institut e.V. (KNi) (Ed.), *Energiegenossenschaften – ein Bürgerbeteiligungsmodell zur Umsetzung der Energiewende? Möglichkeiten und Grenzen einer politischen Beteiligung.* Klaus Novy Institut e.V. (KNi).

Kemfert, C., & Traber, T. (2013). In Deutsches Institut für Wirtschaftsforschung e.V. (Ed.), *Verteilungseffekte von Kapazitätsmechanismen: Auf den Typ kommt es an.* Deutsches Institut für Wirtschaftsforschung e.V.

Kemp, R. (1994). Technology and the transition to environmental sustainability. The problem of technological regime shifts. *Futures, 26*(10), 1023–1046.

Kemp, R., Schot, J., & Hoogma, R. (1998). Regime shifts to sustainability through processes of niche formation: The approach of strategic niche management. *Technology Analysis & Strategic Management, 10*(2), 175–198.

Kemp, R., Geels, F. W., & Dudley, G. (2012). Sustainability transitions in the automobility regime and the need for a new perspective. In F. W. Geels, R. Kemp, G. Dudley, & G. Lyons (Eds.), *Automobility in transition? A socio-technical analysis of sustainable transport* (pp. 3–28). New York: Routledge.

Kern, F., & Smith, A. (2008). Restructuring energy systems for sustainability? Energy transition policy in the Netherlands. *Transition Towards Sustainable Energy Systems, 36*(11), 4093–4103.

Khan, J. (2013). What role for network governance in urban low carbon transitions? *Special Issue: Advancing Sustainable Urban Transformation, 50*, 133–139.

Kieser, A., & Ebers, M. (Eds.). (2006). *Organisationstheorien*. Stuttgart: Kohlhammer.

Kießling, B. (1988). Die "Theorie der Strukturierung" interview mit Anthony Giddens. *Zeitschfrift für Soziologie, 17*(4), 286–295.

Klemisch, H. (2014a). Die Transformation der Energiewirtschaft. Die Rolle von Genossenschaften in der Energiewende. *Ökologisches Wirtschaften, 29*, 22–23.

Klemisch, H. (2014b). Energiegenossenschaften als regionale Antwort auf den Klimawandel. In C. Schröder & H. Walk (Eds.), *Genossenschaften und Klimaschutz. Akteure für zukunftsfähige, solidarische Städte* (pp. 149–166). Springer VS: Wiesbaden.

Klemisch, H. (2014c). Energiegenossenschaften. Chance für den Strukturwandel. *Fachzeitschrift für Alternative Kommunalpolitik* (1), 33–35.

Klemisch, H. (2016). Genossenschaftliche Energiewende in den Kommunen. Nach der Euphorie ist vor der Energiegenossenschaft 2.0. *Fachzeitschrift für Alternative Kommunalpolitik* (3), 46–48.

Klemisch, H., & Maron, H. (2010). Genossenschaftliche Lösungsansätze zur Sicherung der kommunalen Daseinsvorsorge. *Zeitschrift für das gesamte Genossenschaftswesen; Organ für Forschung und Praxis genossenschaftlicher Kooperation, 6*(1), 3–13.

Knoke, D., & Yang, S. (2008). *Social network analysis. Quantitative applications in the social sciences* (Vol. 154). Los Angeles: Sage.

Kompetenznetzwerk Dezentrale Energietechnologien e.V. (Ed.). (2009). *Leitfaden Sieben Schritte auf dem Weg zur klimaneutralen Kommune. Erfahrungen aus dem Projekt "Strategien von Kommunen zur Erreichung von Klimaneutralität"*. Retrieved 01.10.2016, from www.deenet.org

Kondratieff, N. D., & Stolper, W. F. (1935). The long waves in economic life. *The Review of Economics and Statistics, 17*(6), 105–115.

Konrad, K., Truffer, B., & Voß, J.-P. (2008). Multi-regime dynamics in the analysis of sectoral transformation potentials: Evidence from German utility sectors. *Journal of Cleaner Production, 16*, 1190–1202.

Krause, F., Bossel, H., & Müller-Reißmann, K.-F. (1981). *Energie-Wende. Wachstum und Wohlstand ohne Erdöl und Uran*. Freiburg.

Kristof, K. (2010). *Models of change: Einführung und Verbreitung sozialer Innovationen und gesellschaftlicher Veränderungen in transdisziplinärer Perspektive*. Zürich: vdf Hochschulverlag.

Kruse, J. (2008). *Reader "Einführung in die Qualitative Interviewforschung"*. Freiburg.

Kühl, S., Strodtholz, P., & Taffertshofer, A. (2009). Qualitative und quantitative Methoden der Organisationsforschung – ein Überblick. In S. Kühl, P. Strodtholz, & A. Taffertshofer (Eds.), *Handbuch Methoden der Organisationsforschung. Quantitative und qualitative Methoden* (pp. 13–31). Wiesbaden: Verlag für Sozialwissenschaften.

Landratsamt Neckar-Odenwald-Kreis. (Ed.). (2014). *Klimaschutzorientiertes Investitionsprogramm für den Neckar-Odenwald-Kreis*. Retrieved 01.10.2016, from http://www.neckar-odenwald-kreis.de/Aktuelle+Themen/Projekte/Klimaschutzkonzept.html

Landratsamt Neustadt a.d.Waldnaab. (Ed.). (2015). *Der Kooperationsraum VierStädtedreieck*. Retrieved 01.10.2016, from http://www.regionalmanagement.neustadt.de/default.aspx?ID=ccdc05ac-fdc1-42ef-896e-d60aebbc1bbc

Lauber, V., & Jacobsson, S. (2016). The politics and economics of constructing, contesting and restricting socio-political space for renewables – The German Renewable Energy Act. *Environmental Innovation and Societal Transitions, 18,* 147–163.

Lautermann, C. (2016). *Handlungsorientierungen für Energiegenossenschaften.* Retrieved 01.10.2016, from http://engeno.net/index.php/ergebnisse/handlungsorientierungen/

Law, J. (1987). Technology and heterogeneous engineering: The case of Portuguese expansion. In W. E. Bijker, T. P. Hughes, & T. J. Pinch (Eds.), *The social construction of technological systems. New directions in the sociology and history of technology* (pp. 111–134). Cambridge, MA: MIT Press.

Law, J., & Callon, M. (1992). The life and death of an aircraft: A network analysis of technical change. In W. E. Bijker & J. Law (Eds.), *Inside technology. Shaping technology/Building society. Studies in sociotechnical change* (pp. 21–52). Cambridge, MA: MIT Press.

lgc. (2010, February 10). Informationen über neue Energien, Nr. 133, p. 20.

Lockwood, M. (2016). Creating protective space for innovation in electricity distribution networks in Great Britain: The politics of institutional change. *Environmental Innovation and Societal Transitions, 18,* 111–127.

Loorbach, D. (2007). *Transition management: New mode of governance for sustainable development.* Utrecht: International Books.

Loorbach, D., & Rotmans, J. (2010). The practice of transition management: Examples and lessons from four distinct cases. *Futures, 42*(3), 237–246.

Loorbach, D., & Verbong, G. (2012). Conclusion. In G. Verbong & D. Loorbach (Eds.), *Governing the energy transition. Reality, illusion or necessity?* (pp. 317–335). New York: Routledge.

Lopolito, A., Morone, P., & Sisto, R. (2011). Innovation niches and socio-technical transition: A case study of bio-refinery production. *Futures, 43*(1), 27–38.

Lovell, H., Bulkeley, H., & Owens, S. (2009). Converging agendas? Energy and climate change policies in the UK. *Environment and Planning C: Government and Policy, 27*(1), 90–109.

Maas, A., & Clark, R. d. (1984). Hidden impact of minorities: Fifteen years of minority influence research. *Psycholgical Bulletin, 95*(3).

Malerba, F. (2002). Sectoral systems of innovation and production. *Research Policy, 31*(2), 247–264.

Markard, J., & Truffer, B. (2008). Technological innovation systems and the multi-level perspective: Towards an integrated framework. *Research Policy, 37*(4), 596–615.

Markard, J., Raven, R., & Truffer, B. (2012). Sustainability transitions: An emerging field of research and its prospects. *Research Policy, 41*(6), 955–967.

Markard, J., Suter, M., & Ingold, K. (2016). Socio-technical transitions and policy change – Advocacy coalitions in Swiss energy policy. *Environmental Innovation and Societal Transitions, 18,* 215–237.

Markt Floß (Ed.). (2015). *Solarpark Floß.* Retrieved 01.10.2016, from http://www.floss.de/nav/default.asp?SID=N57LN57N11

Maron, B. (2009). *Energiegenossenschaften und ihr Beitrag zu einer Nachhaltigen Energieversorgung.* KNi-Paper No. 01.

Maron, B., & Maron, H. (2012). *Genossenschaftliche Unterstützungsstrukturen für eine sozialräumlich orientierte Energiewirtschaft. Machbarkeitsstudie.* Köln. Retrieved 01.10.2016, from http://www.kni.de/media/pdf/Machbarkeitsstudie_Unterstuetzungsstrukturen_Geno.pdf

Mayer, N. J., & Burger, B. (2014). In Fraunhofer-Institut für Solare Energiesysteme ISE (Ed.), *Kurzstudie zur historischen Entwicklung der EEG-Umlage.* Fraunhofer-Institut für Solare Energiesysteme ISE. Retrieved 01.10.2016, from https://www.ise.fraunhofer.de/de/downloads/pdf-files/data-nivc-/kurzstudie-zur-historischen-entwicklung-der-eeg-umlage.pdf

Mayntz, R. (2009). *Sozialwissenschaftliches Erklären: Probleme der Theoriebildung und Methodologie. Schriften aus dem Max-Planck-Institut für Gesellschaftsforschung* (Vol. 63). Frankfurt/Main: Campus.

Mayring, P. (2010). *Qualitative Inhaltsanalyse: Grundlagen und Techniken, Beltz Pädagogik* (11th ed.). Weinheim: Beltz.

References

McKenna, R., Herbes, C., & Fichtner, W. (2015). In Karlsruher Institut für Technologie (Ed.), *Energieautarkie: Definitionen, Für-bzw. Gegenargumente, und entstehende Forschungsbedarfe*. Karlsruher Institut für Technologie.

Meadows, D., Meadows, D., Zahn, E., & Milling, P. (1972). *Die Grenzen des Wachstums. Bericht des Club of Rome zur Lage der Menschheit*. New York: Universe Books.

Monstadt, J., & Wolff, A. (2015). Energy transition or incremental change? Green policy agendas and the adaptability of the urban energy regime in Los Angeles. *Energy Policy, 78*, 213–224.

Morris, C., & Pehnt, M. (2015). In Heinrich Böll Foundation (Ed.), *Energy transition. The German Energiewende*. Heinrich Böll Foundation. Retrieved 01.10.2016, from www.energytransition.de

Moss, T. (2009). Intermediaries and the governance of sociotechnical networks in transition. *Environment and Planning A, 41*(6), 1480–1495.

Müller, N. (2013, December 18). Rotoren auf dem Rödeser Berg sind genehmigt, Nr. Resort Lokales. *Wolfhagener Allgemeine*, p. 1.

Müller, N. (2014, January 21). BI will gegen Windräder klagen. *HNA Hessische/Niedersächsische Allgemeine Zeitung*.

Müller, M. O., Stämpfli, A., Dold, U., & Hammer, T. (2011). Energy autarky: A conceptual framework for sustainable regional development. *Energy Policy, 39*(10), 5800–5810.

Münkner, H.-H. (2011). Regionale und kooperative Ökonomie. In W. George & T. Berg (Eds.), *Regionales Zukunftsmanagement: Vol. 5. Regionales Zukunftsmanagement. Band 5: Energiegenossenschaften gründen und erfolgreich betreiben* (pp. 56–72). Lengerich [u.a.]: Pabst Science.

Musiolik, J., Markard, J., & Hekkert, M. (2012). Networks and network resources in technological innovation systems: Towards a conceptual framework for system building. *Technological Forecasting and Social Change, 79*(6), 1032–1048.

Nelson, R. R., & Winter, S. G. (1982). *An evolutionary theory of economic change*. Cambridge, MA: Belknap Press of Harvard University Press.

Neue Energien West eG. (Ed.). (2009a). *Satzung und Geschäftsordnung*. Retrieved 01.10.2016, from http://www.neue-energien-west.de/neue-energien-west

Neue Energien West eG. (2009b). *Stadtwerke und Stadt stellen Energiegenossenschaft "NEWeG" vor*. Grafenwöhr. Retrieved 01.10.2016, from http://www.neue-energien-west.de/2009/10/07/stadtwerke-und-stadt-stellen-energiegenossenschaft-new-eg-vor/8

Neue Energien West eG. (2009c). *Wir haben die Sonne auf unserer Seite*. Grafenwöhr. Retrieved 01.10.2016, from http://www.neue-energien-west.de/2009/10/07/neue-energien-wir-haben-die-sonne-auf-unserer-seite/4

Neue Energien West eG. (2013). *NEW eG kauft Solarpark in Hütten nicht!* Grafenwöhr. Retrieved 01.10.2016, from http://www.neue-energien-west.de/2013/01/17/new-eg-kauft-solarpark-in-hutten-nicht/964

Neue Energien West eG. (Ed.). (2014). *NEW-Anlagen*. Retrieved 01.10.2016, from http://neumann-solar.solarlog-web.de/

Nilsson, M. (2012). Energy governance in the European Union: Enabling conditions for a low carbon transition? In G. Verbong & D. Loorbach (Eds.), *Governing the energy transition. Reality, illusion or necessity?* (pp. 296–316). New York: Routledge.

Nitsch, J. (2014). In Bundesverband Erneuerbare Energien e.V. (Ed.), *Szenarien der deutschen Energieversorgung vor dem Hintergrund der Vereinbarungen der Großen Koalition*. Bundesverband Erneuerbare Energien e.V. Retrieved 01.10.2016, from http://www.bee-ev.de/fileadmin/Publikationen/Studien/20140205_BEE-Szenarien_GROKO_Nitsch.pdf

Nord, U. (2012, October). *Die Redebeiträge von Uwe Nord, stv. Fraktionsvorsitzender des Bündnis Wolfhager Bürger bei der Stadtverordnetenversammlung am 11.10.2012*. Wolfhagen. Retrieved from http://www.bwb-wolfhagen.de/aktuelles-infos.html

Ortmann, G. (1995). *Formen der Produktion: Organisation und Rekursivität*. Opladen: Westdt. Verl.

Ortmann, G., & Sydow, J. (Eds.). (2001). *Strategie und Strukturation: Strategisches Management von Unternehmen, Netzwerken und Konzernen* (1st ed.). Wiesbaden: Gabler.

Ortmann, G., Sydow, J., & Windeler, A. (2000). Die Organisation als reflexive Strukturation. In G. Ortmann, J. Sydow, & K. Türk (Eds.), *Organisation und Gesellschaft. Theorien der Organisation. Die Rückkehr der Gesellschaft* (pp. 315–354). Wiesbaden: VS Verlag für Sozialwissenschaften.

Orum, A. M., Feagin, J. R., & Sjoberg, G. (1991). The nature of the case study. In J. R. Feagin, A. M. Orum, & G. Sjoberg (Eds.), *A case for the case study* (pp. 1–26). Chapel Hill: University of North Carolina Press.

Parker, J. (2000). *Structuration.* Buckingham: Open University Press.

Parson, T. (1951). *The social system.* Glencoe, IL: Free Press.

Penna, C. C. R., & Geels, F. W. (2012). Multi-dimensional struggles in the greening of industry: A dialectic issue lifecycle model and case study. *Contains Special Section: Actors, Strategies and Resources in Sustainability Transitions, 79*(6), 999–1020.

Perez, C. (1983). Structural change and assimilation of new technologies in the economic and social systems. *Futures, 15*(5), 357–375.

Peter, S. (2013). In Umweltbundesamt (Ed.), *Modellierung einer vollständig auf erneuerbaren Energien basierenden Stromerzeugung im Jahr 2050 in autarken, dezentralen Strukturen.* Umweltbundesamt.

Peteraf, M. A. (1993). The cornerstones of competitive advantage: A resource-based view. *Strategic Management Journal, 14*(3), 179–191.

Polanyi, K. (1944). *The great transformation: The political and economic origins of our time.* Boston, MA: Beacon Press.

Porter, M. E. (1990). *Competitive strategy: Techniques for analyzing industries and competitors.* New York: The Free Press.

Porter, M. E. (1991). Towards a dynamic theory of strategy. *Strategic Management Journal, 12,* 95–117.

Powell, W. W. (1991). Expanding the scope of institutional analysis. In W. W. Powell & P. J. DiMaggio (Eds.), *The new institutionalism in organizational analysis* (pp. 183–203). Chicago: University of Chicago Press.

Powell, W. W., & DiMaggio, P. J. (Eds.). (1991). *The new institutionalism in organizational analysis.* Chicago: University of Chicago Press.

Pozzebon, M. (2004). The influence of a structurationist view on strategic management research. *Journal of Management Studies, 41*(2), 247–272.

Pro Wind Wolfhagen – Energiewende jetzt. (2010). *Film "Wind des Wandels".* Wolfhagen. Retrieved 01.10.2016, from http://www.prowindwolfhagen.de/pages/posts/film-wind-des-wandels5.php

PROKON regenerative Energien eG. (2016). In Bundesministerium der Justiz und für Verbraucherschutz (Ed.), *Jahresabschluss zum Geschäftsjahr vom 01.01.2015 bis zum 31.07.2015.* Bundesministerium der Justiz und für Verbraucherschutz. Retrieved 01.10.2016, from https://www.bundesanzeiger.de/ebanzwww/wexsservlet

Raab, U. (2015). In nordbayern.de (Ed.), *Hinter den Kulissen des Energiedialogs brodelt es.* nordbayern.de. Retrieved 26.01.2015, from http://www.nordbayern.de/region/pegnitz/hinter-den-kulissen-des-energiedialogs-brodelt-es-1.4108921

Radtke, J. (2014). In Bundesnetzwerk Bürgerschaftliches Engagement e.V. (Ed.), *Die Energiewende in Deutschland und die Partizipation der Bürger.* Bundesnetzwerk Bürgerschaftliches Engagement e.V.

Radtke, J. (2016). *Bürgerenergie in Deutschland. Partizipation zwischen Rendite und Gemeinwohl.* Thesis, Universität Siegen.

Raiffeisenbank eG Elztal. (Ed.). (2013). *Bericht des Vorstandes zum Geschäftsjahr 2013.* Retrieved 01.10.2016, from https://www.raiffeisenbank-elztal.de/wir-fuer-sie/ueber-uns/zahlen-fakten.html

Raiffeisenbank Kraichgau eG. (Ed.). (2013). *Geschäftsbericht 2013.* Retrieved 01.10.2016, from https://www.raiba-kraichgau.de/wir-fuer-sie/ueber-uns/zahlen-fakten.html

Raiffeisenbank Neudenau-Stein-Herbolzheim eG. (Ed.). (2013). *Jahresbericht 2013*. Retrieved 01.10.2016, from https://www.raiffeisenbank-neudenau.de/Ihre_Raiffeisenbank/ueber-uns/zahlen-fakten.html

Rammert, W. (1997). New rules of sociological method: Rethinking technology studies. *British Journal of Sociology, 48*(2), 171–191.

Ranson, S., Hinings, B., & Greenwood, R. (1980). The structuring of organizational structures. *Administrative Science Quarterly, 25*(1), 1.

Rau, I., Schweizer-Ries, P., & Hildebrand, J. (2012). Participation: The silver bullet for the acceptance of renewable energy? In S. Kabisch, A. Kunath, P. Schweizer-Ries, & A. Steinführer (Eds.), *Vulnerability, risks, and complexity. Impacts of global change on human habitats* (pp. 177–192). Cambridge, MA: Hogrefe.

Rauschmayer, F. C. S., & Masson, T. (2015). In Helmholtz-Zentrum für Umweltforschung GmbH – UFZ (Ed.), *Ergebnisse der EnGeno Mitgliederbefragung von Energiegenossenschaften*. Helmholtz-Zentrum für Umweltforschung GmbH – UFZ. Retrieved 01.10.2016, from http://engeno.net/index.php/ergebnisse-der-engeno-mitgliederbefragung-von-energiegenossenschaften/

Raven, R. (2004). Implementation of manure digestion and co-combustion in the Dutch electricity regime: A multi-level analysis of market implementation in the Netherlands. *Energy Policy, 32*(1), 29–39.

Raven, R. P. J. M. (2005). *Strategic Niche Management for Biomass. A comparative study on the experimental introduction of bioenergy technologies in the Netherlands and Denmark*. PhD thesis, Eindhoven University, Eindhoven.

Raven, R. P. (2006). Towards alternative trajectories? Reconfigurations in the Dutch electricity regime. *Research Policy, 35*(4), 581–595.

Raven, R. (2007). Niche accumulation and hybridisation strategies in transition processes towards a sustainable energy system: An assessment of differences and pitfalls. *Energy Policy, 35*(4), 2390–2400.

Raven, R. P. J. M. (2012). Analyzing emerging sustainable energy niches in Europe. In G. Verbong & D. Loorbach (Eds.), *Governing the energy transition. Reality, illusion or necessity?* (pp. 123–152). New York: Routledge.

Raven, R. P. J. M., & Geels, F. W. (2010). Socio-cognitive evolution in niche development: Comparative analysis of biogas development in Denmark and the Netherlands (1973–2004). *Technovation, 30*(2), 87–99.

Raven, R., & Verbong, G. (2007). Multi-regime interactions in the Dutch energy sector: The case of combined heat and power technologies in the Netherlands 1970–2000. *Technology Analysis & Strategic Management, 19*(4), 491–507.

Raven, R., & Verbong, G. (2009). Boundary crossing innovations: Case studies from the energy domain. *Technology in Society, 31*(1), 85–93.

Raven, R. P. J. M., Heiskanen, E., Lovio, R., Hodson, M., & Brohmann, B. (2008). The contribution of local experiments and negotiation processes to field-level learning in emerging (niche) technologies: Meta-analysis of 27 new energy projects in Europe. *Bulletin of Science, Technology & Society, 28*(6), 464–477.

Raven, R., Schot, J., & Berkhout, F. (2012). Space and scale in socio-technical transitions. *Environmental Innovation and Societal Transitions, 4*, 63–78.

Reckwitz, A. (2007). Anthony Giddens. In D. Kaesler (Ed.), *Klassiker der Soziologie. Von Talcott Parsons bis Anthony Giddens* (5th ed., pp. 311–337). München: Verlag C.H. Beck.

Regierung der Oberpfalz. (Ed.). (2015a). *Aktuelle Einwohnerzahlen und Fläche*. Retrieved 01.10.2016, from http://www.regierung.oberpfalz.bayern.de/dbGden/ew_fl.php

Regierung der Oberpfalz. (Ed.). (2015b). *Oberpfalzkarten – Grundkarten*. Retrieved 01.10.2016, from http://www.regierung.oberpfalz.bayern.de/leistungen/landesplanung/karten/rokali/rokali_opfkarten.php

Renewable Energy Policy Network for the 21st Century. (Ed.). (2016). *Renewable 2016. Global Status Report*. Retrieved 01.10.2016, from www.ren21.net/

Research Institute for Cooperation and Cooperatives, & Vienna University of Economics and Business. (Eds.). (2012). *Energy cooperatives and local ownership in the field of renewable energy technologies as social innovation processes in the energy system.* Retrieved 01.10.2016, from https://www.wu.ac.at/en/

Rhein-Neckar-Zeitung. (2010, December 21). Energiegenossenschaft in Gründung Beteiligen kann sich jeder Bürger – Volks und Raiffeisenbanken stehen für Energie und Umwelt. *Rhein-Neckar-Zeitung.*

Rip, A. (1992). A quasi-evolutionary model of technological development and a cognitive approach to technology policy. *RISESST. Rivista di studi epistemologici e sociale sulla scienza e la technologia* (2), 69–102.

Rip, A., & Kemp, R. (1998). Technological change. In S. Rayner & E. L. Malone (Eds.), *Human choice & climate change. Resources and technology* (Vol. 2, 1st ed., pp. 327–392). Ohio: Batelle Press.

Rogers, E. M. (1962). *Diffusion of innovation.* New York: Free Press.

Rommel, K., & Meyerhoff, J. (2009). Empirische Analyse des Wechselverhaltens von Stromkunden. Was hält Stromkunden davon ab, zu Ökostromanbietern zu wechseln? *Zeitschrift für Energiewirtschaft, 33*(1), 74–82.

Rosenberg, N., & Frischtalk, C. R. (1984). Technological innovation and long waves. *Cambridge Journal of Economics, 8*(1), 7–27.

Rosenfeld, S. A. (1996). Does cooperation enhance competitiveness? Assessing the impacts of inter-firm collaboration. *Evaluation of Industrial Modernization, 25*(2), 247–263.

Rothaermel, F. T., & Boeker, W. (2008). Old technology meets new technology: Complementarities, similarities, and alliance formation. *Strategic Management Journal, 29*(1), 47–77.

Rothaermel, F. T., & Hill, C. W. L. (2005). Technological discontinuities and complementary assets: A longitudinal study of industry and firm performance. *Organization Science, 16*(1), 52–70.

Rotmans, J., & Loorbach, D. (2010). Towards a better understanding of transitions and their governance. A systemic and reflexive approach. In J. Grin, J. Rotmans, & J. Schot (Eds.), *Transitions to sustainable development. New directions in the study of long term transformative change* (pp. 105–222). New York: Routledge.

Rotmans, J., Kemp, R., & van Asselt, M. (2001). More evolution than revolution: Transition management in public policy. *Foresight, 3*(1), 15–31.

Röttgen, N. (2011). *Inhalt Rede des Bundesministers für Umwelt, Naturschutz und Reaktorsicherheit, Dr. Norbert Röttgen, zum Haushaltsgesetz 2012 vor dem Deutschen Bundestag am 6. September 2011 in Berlin.* Retrieved 01.10.2016, from https://www.bundesregierung.de/Content/DE/Bulletin/2010-2015/2011/09/85-3-bmu-bt-haushalt.html

Rühl, M. (2013, May). *Beteiligung einer Bürgergenossenschaft an einem kommunalen Stadtwerk – ein Vorteil für alle.* Präsentation der Stadtwerke Wolfhagen auf der 17. Euroforum-Jahrestagung, Berlin.

Rühl, M. (2015, February). *100%ige Versorgung aus Erneuerbaren Energien erreichen – ist das möglich? Ein Erfahrungsbericht der Stadtwerke Wolfhagen.* Wolfhagen.

Rutschmann, I. (2009). Genossenschaften auf dem Vormarsch. Bürgerliche Energieerzeuger entdecken die Vorteile einer bisher wenig genutzten Rechtsform. *Photon, Februar* 78–88.

Sagebiel, J., Müller, J. R., & Rommel, J. (2014). Are consumers willing to pay more for electricity from cooperatives? Results from an online Choice Experiment in Germany. *Energy Research & Social Science, 2*, 90–101.

Saunders, P. (1989). Space, urbanism and the created environment. In D. Held & J. B. Thompson (Eds.), *Social theory of modern societies. Anthony Giddens and his critics* (pp. 215–234). Cambridge [England], New York: Cambridge University Press.

Schaffland, H.-J., & Korte, O. (2006). Novellierung des Genossenschaftsgesetzes. *PerspectivePraxis* (1).

Schaltegger, S. (2002). A framework for ecopreneurship. Leading bioneers and environmental managers to ecopreneurship. *Greener Management International, 38*, 45–58.

Schaltegger, S., & Wagner, M. (2011). Sustainable entrepreneurship and sustainability innovation: Categories and interactions. *Business Strategy and the Environment, 20*(4), 222–237.
Schaper, M. (2002). The essence of ecopreneurship. *Greener Management International, 38*, 26–30.
Schneidewind, U. (1998). *Die Unternehmung als strukturpolitischer Akteur.* Marburg: Metropolis-Verlag.
Schneidewind, U. (2011). "Embedded technologies": The case for a deeper understanding of innovation. *Exzellenz – The Cluster Magazine for North Rhine-Westphalia,* (4), 14–15.
Schneidewind, U. (2015). Warum die Energiewende auch eine Wissenschaftswende braucht. In Landesregierung Nordrhein-Westfalen (Ed.), *Transformationsforschung NRW. Wege in ein nachhaltiges Energieversorgungssystem* (pp. 8–9).
Schneidewind, U., & Petersen, H. (1998). Changing the rules: Business-NGO partnerships and structuration theory. *Greener Management International, 24*, 105–115.
Schneidewind, U., & Scheck, H. (2012). Zur Transformation des Energiesektors – ein Blick aus der Perspektive der Transition-Forschung. In H.-G. Servatius, U. Schneidewind, & D. Rohlfing (Eds.), *Smart Energy. Wandel zu einem nachhaltigen Energiesystem* (pp. 45–61). Berlin, Heidelberg: Springer.
Schneidewind, U., Augenstein, K., & Scheck, H. (2013). The transition to renewable energy systems: On the way to a comprehensive transition concept. In D. Stolten & V. Scherer (Eds.), *Transition to renewable energy systems* (pp. 119–136). Weinheim: Wiley-VCH.
Schot, J. (1998). The usefulness of evolutionary models for explaining innovation. The case of the Netherlands in the nineteenth century. *History and Technology, 14*(3), 173–200.
Schot, J., & Geels, F. W. (2008). Strategic niche management and sustainable innovation journeys: Theory, findings, research agenda, and policy. *Technology Analysis & Strategic Management, 20*(5), 537–554.
Schot, J., Hoogma, R., & Elzen, B. (1994). Strategies for shifting technological systems. The case of the automobile system. *Futures, 26*(10), 1060–1076.
Schröder, C., & Walk, H. (Eds.). (2014). *Genossenschaften und Klimaschutz: Akteure für zukunftsfähige, solidarische Städte.* Wiesbaden: Springer VS.
Schultz, S. (2014, July 17). Gutachten für Regierung: Experten warnen Gabriel vor Kapazitätsmarkt. *Spiegel online.* Retrieved from http://www.spiegel.de/wirtschaft/unternehmen/energiewende-gutachten-warnen-gabriel-vor-kraftwerk-kapazitaetsmaerkten-a-981653.html
Schulze, C. (2015). *Motive und Beitrittsgründe zu einer Bürgerenergiegenossenschaft.* KNi-Paper No. 01.
Schumpeter, J. A. (1934). *The theory of economic development. Harvard economic studies.* Cambridge, MA: Harvard University Press.
Schumpeter, J. A. (1942). *Capitalism, socialism and democracy* (1st ed.). New York: Harper & Brothers.
Schumpeter, J. A. (1954). *History of economic analysis.* London: Allen & Unwin.
Schütz, A. (1974). *Der sinnhafte Aufbau der sozialen Welt. Eine Einleitung in die verstehende Soziologie.* Frankfurt am Main: Suhrkamp.
Schwendter, R. (Ed.). (1986). *AG-SPAK-Bücher: Reihe Selbstverwaltung: Vol. 72. Die Mühen der Berge. Grundlegungen zur alternativen Ökonomie. Teil 1* (1st ed.). München: AG-SPAK.
Scott, J. (1994). Conceptualizing organizational fields: Linking organizations and societal systems. In H.-U. Derlien, U. Gerhardt, & F. W. Scharpf (Eds.), *Systemrationalität und Partialinteresse. Festschrift für Renate Mayntz* (pp. 203–211). Baden-Baden: Nomos.
Scott, W. R. (1995). *Institutions and organizations* (1st ed.). Thousand Oaks, CA: Sage.
Scott, J. (2013). *Social network analysis* (3rd ed.). Los Angeles, London, New Delhi, Singapore, Washington, DC: Sage.
Seawright, J., & Gerring, J. (2008). Case selection techniques in case study research: A menu of qualitative and quantitative options. *Political Research Quarterly, 61*(2), 294–308.
Sensfuß, F., & Ragwitz, M. (2007). *Analyse des Preiseffektes der Stromerzeugung aus erneuerbaren Energien auf die Börsenpreise im deutschen Stromhandel – Analyse für das Jahr 2006.* Karlsruhe. Retrieved 01.10.2016, from http://publica.fraunhofer.de/documents/N-61386.html

Servatius, H.-G., Schneidewind, U., & Rohlfing, D. (Eds.). (2012). *Smart Energy: Wandel zu einem nachhaltigen Energiesystem*. Berlin: Springer.

Sewell, W. H. (1992). A theory of structure: Duality, agency, and transformation. *American Journal of Sociology, 89*(1), 1–29.

Seyfang, G., & Haxeltine, A. (2012). Growing grassroots innovations: Exploring the role of community-based initiatives in governing sustainable energy transitions. *Environment and Planning C: Government and Policy, 30*(3), 381–400.

Seyfang, G., & Smith, A. (2007). Grassroots innovations for sustainable development: Towards a new research and policy agenda. *Environmental Policy, 16*(4), 584–603.

Seyfang, G., Hielscher, S., Hargreaves, T., Martiskainen, M., & Smith, A. (2014). A grassroots sustainable energy niche? Reflections on community in the UK. *Environmental Innovation and Societal Transitions, 13*, 21–44.

Shackley, S., & Green, K. (2007). A conceptual framework for exploring transitions to decarbonised energy systems in the United Kingdom. *Energy, 32*(3), 221–236.

Shove, E. (2012). Energy transition in practice. The case of global indoor climate change. In G. Verbong & D. Loorbach (Eds.), *Governing the energy transition. Reality, illusion or necessity?* (pp. 51–75). New York: Routledge.

Smith, A. (2007). Translating sustainabilities between green niches and socio-technical regimes. *Technology Analysis & Strategic Management, 19*(4), 427–450.

Smith, A. (2012). Civil society in sustainable energy transition. In G. Verbong & D. Loorbach (Eds.), *Governing the energy transition. Reality, illusion or necessity?* (pp. 180–203). New York: Routledge.

Smith, A., & Raven, R. P. J. M. (2012). What is protective space? Reconsidering niches in transitions to sustainability. *Research Policy, 41*(6), 1025–1036.

Smith, A., Stirling, A., & Berkhout, F. (2005). The governance of sustainable socio-technical transitions. *Research Policy, 34*(10), 1491–1510.

Smith, A., Voß, J.-P., & Grin, J. (2010). Innovation studies and sustainability transitions: The allure of the multi-level perspective and its challenges. *Special Section on Innovation and Sustainability Transitions, 39*(4), 435–448.

Späth, P., & Rohracher, H. (2010). 'Energy regions': The transformative power of regional discourses on socio-technical futures. *Research Policy, 39*(4), 449–458.

Späth, P., & Rohracher, H. (2012). Local demonstrations for global transitions—Dynamics across governance levels fostering socio-technical regime change towards sustainability. *European Planning Studies, 20*(3), 461–479.

Stadt Wolfhagen. (2011a). *Bürgermeister unterschreibt Charta der 100ee-Regionen*. Retrieved 01.10.2016, from http://www.wolfhagen.de/de/wirtschaft/energiewende/link_aktuelle_texte/B__rgermeister_unterschreibt_Charta_der_100ee.pdf

Stadt Wolfhagen. (Ed.). (2011b). *Wahlen und Abstimmungen in Wolfhagen-Wahlergebnis der Wahl 2011*. Retrieved 01.10.2016, from http://www.wolfhagen.de/de/rathaus/wahlen.php?navanchor=1110038

Stadtanzeiger Mosbach. (2013, May 17). Zweite Generalversammlung der "Energie plus Umwelt eG". Entlastung von Vorstand und Aufsichtsrat, Nr. 20. *Stadtanzeiger Mosbach*, p. 14.

Stadtverordnetenversammlung Wolfhagen. (2008). *Protokoll zur Stadtverordnetenversammlung vom 17.04.2008: Windkraft in Wolfhagen (Grundsatzbeschluss zur erneuerbaren Stromversorgung bis 2015)*. Wolfhagen.

Stadtwerke Grafenwöhr. (Ed.). (2011). *Integriertes Klimaschutzkonzept für den westlichen Landkreis Neustadt a.d. Waldnaab*. Retrieved 01.10.2016, from http://www.neue-energien-west.de/integriertess-klimaschutzkonzept

Stadtwerke Grafenwöhr. (Ed.). (2015). *Unternehmensdaten*. Retrieved 01.10.2016, from http://stadtwerke-grafenwoehr.de/web/daten.php

Stadtwerke Wolfhagen. (2013). *Stadtwerke-Award 2013: Stadtwerk Wolfhagen erhält den 2. Preis für Bürgerbeteiligungsprojekt*. Wolfhagen. Retrieved 01.10.2016, from https://www.stadtwerke-wolfhagen.de/index.php?option=com_content&view=article&id=230:stadtwerke-erhalten-2-platz-fuer-buergerbeteiligungsmodell&catid=23:aktuelles&Itemid=29

Stadtwerke Wolfhagen. (2014). *Jahresabschluss zum Geschäftsjahr vom 01.01.2012 bis zum 31.12.2012 und Tätigkeitsabschluss.* Bundesministerium der Justiz und für Verbraucherschutz (Ed.). Retrieved from https://www.bundesanzeiger.de/ebanzwww/wexsservlet

Stadtwerke Wolfhagen. *Stadtwerke Wolfhagen verleihen VW e-Up für kostenlose Testfahrten.* Wolfhagen. Retrieved 01.10.2016, from http://www.stadtwerke-wolfhagen.de/index.php?option=com_content&view=article&id=300&Itemid=119

Stake, R. E. (1995). *The art of case study research.* Thousand Oaks, CA: Sage.

Stake, R. E. (2005). *Multicase research methods: Step by step cross-case analysis.* New York: Guilford Press.

Statistisches Bundesamt. (Ed.). (2015). *Städte in Deutschland nach Fläche, Bevölkerung und Bevölkerungsdichte.* Retrieved 01.10.2016, from https://www.destatis.de/DE/ZahlenFakten/LaenderRegionen/Regionales/Gemeindeverzeichnis/Administrativ/AdministrativeUebersicht.html

Statistisches Landesamt Baden-Württemberg. (Ed.). (2014). *Regionaldaten.* Retrieved 01.10.2016, from http://www.statistik.baden-wuerttemberg.de

Steven, N., Beckmann, M., Gräbnitz, D., & Mirkovic, R. (2014). Social entrepreneurs and social change. Tracing impacts of social entrepreneurship through ideas, structures, and practices. *International Journal of Entrepreneurial Venturing, 6*(1), 51–68.

Sustainability Transitions Research Network. (Ed.). (2015). *17th newsletter.* Retrieved 01.10.2016, from http://www.transitionsnetwork.org

Suurs, R. A., & Hekkert, M. P. (2012). Motors of sustainable innovation. understanding transtions from a technological innovation system's perspective. In G. Verbong & D. Loorbach (Eds.), *Governing the energy transition. Reality, illusion or necessity?* (pp. 153–179). New York: Routledge.

Swedberg, R. (2007). *Rebuilding Schumpeter's theory of entrepreneurship.* Conference on Marshall, Schumpeter and Social Science, Hitotsubashi University, 17–18th March 2007

Sydow, J. (2005). *Strategische Netzwerke. Evolution und Organisation, Neue Betriebswirtschaftliche Forschung* (Vol. 100, 6th ed.). Wiesbaden: Gabler.

Sydow, J. (Ed.). (2010a). *Management von Netzwerkorganisationen: Beiträge aus der "Managementforschung"* (5th ed.). Wiesbaden: Gabler.

Sydow, J. (2010b). Management von Netzwerkorganisationen – zum Stand der Forschung. In J. Sydow (Ed.), *Management von Netzwerkorganisationen. Beiträge aus der "Managementforschung"* (5th ed., pp. 373–470). Wiesbaden: Gabler.

Sydow, J., & Duschek, S. (2010). *Management interorganisationaler Beziehungen. Netzwerke – Cluster – Allianzen.* Stuttgart: Kohlhammer.

Sydow, J., & Windeler, A. (1998). Organizing and evaluating interfirm networks: A structurationist perspective on network processes and effectiveness. *Organization Science, 9*(3), 265–284.

The Federal Government. (Ed.). (2011). *Der Weg zur Energie der Zukunft – sicher, bezahlbar und umweltfreundlich.* Retrieved 01.10.2016, from http://www.bmwi.de/DE/Themen/Energie/Energiewende/zielarchitektur.html

The Federal Government. (Ed.). (2014). *Entwurf eines Gesetzes zur grundlegenden Reform des Erneuerbare-Energien-Gesetzes und zur Änderung weiterer Bestimmungen des Energiewirtschaftsrechts.* Retrieved 01.10.2016, from http://www.bmwi.de/DE/Themen/energie,did=634382.html

The Federal Government. (Ed.). (2016). *Gesetzesentwurf der Bundesregierung. Entwurf eines Gesetzes zur Einführung von Ausschreibungen für Strom aus erneuerbaren Energien und zu weiteren Änderungen des Rechts der erneuerbaren Energien (Erneuerbare-Energien-Gesetz – EEG 2016).* Retrieved 01.10.2016, from www.bmwi.de

The German Advisory Council on the Environment. (Ed.). (2011). *Wege zur 100% erneuerbaren Stromversorgung. Sondergutachten.* Retrieved 01.10.2016, from http://www.umweltrat.de/SharedDocs/Downloads/DE/02_Sondergutachten/2011_07_SG_Wege_zur_100_Prozent_erneuerbaren_Stromversorgung.pdf?__blob=publicationFile

The International Co-operative Alliance. (Ed.). (2014). *What is a cooperative.* Retrieved 01.10.2016, from http://ica.coop/en/whats-co-op/co-operative-identity-values-principles

Theurl, T. (2008). Klimawandel. Herausforderungen und Tätigkeitsfelder für Genossenschaften. *IFG intern Newsletter, (1)*, 19–22.
Thrift, N. (1983). On the determination of social action in space and time. *Environment and Planning D: Society and Space, (1)*, 23–57.
Thue, L. (1995). Electricity rules. The formation and development of the Nordic electricity regime. In A. Kaijser & M. Hedin (Eds.), *Nordic energy systems. Historical perspectives and current issues* (pp. 11–30). Canton, MA: Science History Publications.
Tigabu, A. D., Berkhout, F., & van Beukering, P. (2015). The diffusion of a renewable energy technology and innovation system functioning: Comparing bio-digestion in Kenya and Rwanda. *Technological Forecasting and Social Change, 90*, 331–345.
Toften, K., & Hammervoll, T. (2010). Strategic orientation of niche firms. *Journal of Research in Marketing and Entrepreneurship, 12*(2), 108–121.
Toften, K., & Hammervoll, T. (2013). Niche marketing research: Status and challenges. *Marketing Intelligence & Planning, 31*(3), 272–285.
trend:research GmbH, & Leuphana Universität Lüneburg. (Eds.). (2013). *Definition und Marktanalyse von Bürgerenergie in Deutschland*. Retrieved 01.10.2016, from http://www.buendnis-buergerenergie.de
trend:research GmbH, & Leuphana Universität Lüneburg. (Eds.). (2014). *Marktrealität von Bürgerenergie und mögliche Auswirkungen von regulatorischen Eingriffen*. Retrieved 01.10.2016, from http://www.buendnis-buergerenergie.de
Truffer, B., & Coenen, L. (2012). Environmental innovation and sustainability transitions in regional studies. *Regional Studies, 46*(1), 1–21.
Trutnevyte, E., Stauffacher, M., Schlegel, M., & Scholz, R. W. (2012). Context-specific energy strategies: Coupling energy system visions with feasible implementation scenarios. *Environmental Science & Technology, 46*(17), 9240–9248.
Turnheim, B., & Geels, F. W. (2012). Regime destabilisation as the flipside of energy transitions: Lessons from the history of the British coal industry (1913–1997). *Energy Policy, 50*, 35–49.
uli. (2014, March 22). Bau des Windparks: Petition eingestellt. *HNA Hessische/Niedersächsische Allgemeine Zeitung*.
Umweltbundesamt. (Ed.). (2010). *Energieziel 2050: 100% Strom aus erneuerbaren Quellen*. Retrieved 01.10.2016, from http://www.umweltbundesamt.de/publikationen/energieziel-2050
Umweltbundesamt. (Ed.). (2014). *Treibhausgasneutrales Deutschland im Jahr 2050*. Retrieved 01.10.2016, from http://www.umweltbundesamt.de/publikationen/treibhausgasneutrales-deutschland-im-jahr-2050-0
United Nations. (Ed.). (2013). *Cooperatives in social development and the observance of the International Year of Cooperatives. Report of the Secretary-General*. Retrieved 01.10.2016, from https://daccess-ods.un.org/home.html
United Nations. (Ed.). (2015). *Paris Agreement*. Retrieved 01.10.2016, from http://unfccc.int/paris_agreement/items/9485.php
Urry, J. (1981). Localities, regions and social class. *International Journal of Urban and Regional Research, 5*(4), 455–473.
Urry, J. (1991). Time and space in Gidden's social theory. In C. G. A. Bryant & D. Jary (Eds.), *International library of sociology. Giddens' theory of structuration. A critical appreciation* (pp. 160–175). London, New York: Routledge.
van Bree, B., Verbong, G., & Kramer, G. (2010). A multi-level perspective on the introduction of hydrogen and battery-electric vehicles. *Technological Forecasting and Social Change, 77*, 529–540.
van de Poel, I. (2003). The transformation of technological regimes. *Research Policy, 32*(1), 49–68.
van den Bergh, J. C., & Bruinsma, F. R. (Eds.). (2008). *Managing the transition to renewable energy. Theory and practice from local, regional and macro perspective*. Cheltenham; Northampton, MA: Edward Elgar.
van den Bergh, J. C. J. M., Truffer, B., & Kallis, G. (2011). Environmental innovation and societal transitions: Introduction and overview. *Environmental Innovation and Societal Transitions, 1*, 1–23.

References

van der Brugge, R. (2009). *Transition dynamics in social-ecological systems. The case of Dutch water management*. PhD thesis, Erasmus University Rotterdam, Rotterdam.

van der Loo, F., & Loorbach, D. (2012). The Dutch Energy Transition Project (2000–2009). In G. Verbong & D. Loorbach (Eds.), *Governing the energy transition. Reality, illusion or necessity?* (pp. 220–250). New York: Routledge.

van der Schoor, T., & Scholtens, B. (2015). Power to the people: Local community initiatives and the transition to sustainable energy. *Renewable and Sustainable Energy Reviews, 43*, 666–675.

van der Schoor, T., van Lente, H., Scholtens, B., & Peine, A. (2016). Challenging obduracy: How local communities transform the energy system. *Energy Transitions in Europe: Emerging Challenges, Innovative Approaches, and Possible Solutions, 13*, 94–105.

van der Vleuten, E., & Högselius, P. (2012). Resisting change? The transnational dynamics of European energy regimes. In G. Verbong & D. Loorbach (Eds.), *Governing the energy transition. Reality, illusion or necessity?* (pp. 75–100). New York: Routledge.

van Eijck, J., & Romijn, H. (2008). Prospects for Jatropha biofuels in Tanzania: An analysis with Strategic Niche Management. *Energy Policy, 36*(1), 311–325.

Verbong, G., & Geels, F. (2007). The ongoing energy transition: Lessons from a socio-technical, multi-level analysis of the Dutch electricity system (1960–2004). *Energy Policy, 35*(2), 1025–1037.

Verbong, G., & Geels, F. W. (2012). Future electricity systems. Visions, scenarios and transition pathways. In G. Verbong & D. Loorbach (Eds.), *Governing the energy transition. Reality, illusion or necessity?* (pp. 203–219). New York: Routledge.

Verbong, G., & Loorbach, D. (Eds.). (2012a). *Governing the energy transition: Reality, illusion or necessity?* New York: Routledge.

Verbong, G., & Loorbach, D. (2012b). Introduction. In G. Verbong & D. Loorbach (Eds.), *Governing the energy transition. Reality, illusion or necessity?* (pp. 1–23). New York: Routledge.

Verbong, G., Geels, F. W., & Raven, R. (2008). Multi-niche analysis of dynamics and policies in Dutch renewable energy innovation journeys (1970–2006): Hype-cycles, closed networks and technology-focused learning. *Technology Analysis & Strategic Management, 20*(5), 555–573.

Vergragt, P. J. (2012). Carbon capture and storage. Sustainable solution or reinforced carbon lock-in? In G. Verbong & D. Loorbach (Eds.), *Governing the energy transition. Reality, illusion or necessity?* (pp. 101–124). New York: Routledge.

Verheul, H., & Vergragt, P. J. (1995). Social experiments in the development of environmental technology: A bottom-up perspective. *Technology Analysis & Strategic Management, 7*(3), 315–326.

Volkmann, C. K., & Tokarski, K. O. (2010). Soziale innovationen und social entrepreneurship. In W. Baumann, U. Braukmann, & W. Matthes (Eds.), *Innovation und Internationalisierung* (pp. 151–170). Wiesbaden: Springer.

Volksbank Franken eG. (Ed.). (2013). *Jahresabschluss 2013*. Retrieved 01.10.2016, from https://www.volksbank-franken.de/wir-fuer-sie/ueber-uns/zahlen-fakten.html

Volksbank Franken eG. (Ed.). (2014). *Unser Unternehmensleitbild*. Retrieved 01.10.2016, from https://www.volksbank-franken.de/wir-fuer-sie/ueber-uns/leitbild.html

Volksbank Kirnau eG. (Ed.). (2013). *Kurzbericht des Jahresabschlusses 2013*. Retrieved 01.10.2016, from https://www.volksbank-kirnau.de/wir-fuer-sie/ueber-uns/zahlen-fakten.html

Volksbank Krautheim eG. (Ed.). (2013). *Offenlegungsbericht per 31.12.2013 nach § 26a KWG (i.V.m. §§ 319 ff. SolvV) und nach § 7 Instituts-Vergütungsverordnung*. Retrieved 01.10.2016, from https://www.vobak.de/wir-fuer-sie/ueber-uns/zahlen-fakten.html

Volksbank Limbach eG. (Ed.). (2013). *Geschäftsbericht 2013*. Retrieved 01.10.2016, from https://www.vb-limbach.de/wir-fuer-sie/ueber-uns/zahlen-fakten.html

Volksbank Main-Tauber eG. (Ed.). (2013). *Jahresbericht 2013*. Retrieved 01.10.2016, from https://www.vobamt.de/

Volksbank Main-Tauber eG. (Ed.). (2014). *Unser Leitbild*. Retrieved 01.10.2016, from https://www.vobamt.de/

Volksbank Mosbach eG. (Ed.). (2013). *Ausgezeichnet-Menschlich-Fair. Geschäftsbericht 2013*. Retrieved 01.10.2016, from http://pdfblaetter.dgverlag.de/mosbach/geschaeftsbericht_2013/#/4

Volksbank Mosbach eG. (Ed.). (2014). *Unternehmensleitbild*. Retrieved 01.10.2016, from https://www.vb-mosbach.de/wir-fuer-sie/ueber-uns/leitbild.html

Volksbank Neckartal eG. (Ed.). (2013). *Geschäftsbericht 2013*. Retrieved 01.10.2016, from https://www.volksbank-neckartal.de/meine-bank/ueber-uns/zahlen-fakten.html

Volksbank Vorbach-Tauber eG. (Ed.). (2013). *Geschäftsbericht 2013*. Retrieved 01.10.2016, from https://www.voba-vorbach-tauber.de/service/rechtliche-hinweise/impressum.html

Volz, R. (2012). *Genossenschaften im Bereich erneuerbarer Energien: Status quo und Entwicklungsmöglichkeiten eines neuen Betätigungsfeldes*. Thesis, Universität Hohenheim, Hohenheim.

Wagner, O., & Berlo, K. (2011). Zukunftsperspektiven kummunaler Energiewirtschaft. *RaumPlanung,* (158/159), 236–241.

Walgenbach, P. (2006). Neo-institutionalistische Ansätze in der Organisationstheorie. In A. Kieser & M. Ebers (Eds.), *Organisationstheorien* (pp. 353–402). Stuttgart: Kohlhammer.

Warren, B., Christoff, P., & Green, D. (2016). Australia's sustainable energy transition: The disjointed politics of decarbonisation. *Environmental Innovation and Societal Transitions, 21*, 1–12.

Wasserman, S., & Faust, K. (1994). *Social network analysis: Methods and applications. Structural analysis in the social sciences* (Vol. 8). Cambridge, New York: Cambridge University Press.

Wernerfelt, B. (1984). A resource-based view of the firm. *Strategic Management Journal, 5*(2), 171–180.

Whittington, R. (1992). Putting Giddens into action: Social systems and managerial agency. *Journal of Management Studies, 29*(6), 693–712.

Whittington, R. (2010). Giddens, structuration theory and strategy as practice. In D. Golsorkhi, L. Rouleau, D. Seidl, & E. Vaara (Eds.), *Cambridge handbook of strategy as practice* (pp. 109–127). Cambridge, New York: Cambridge University Press.

Wieczorek, A. J., Hekkert, M. P., Coenen, L., & Harmsen, R. (2015a). Broadening the national focus in technological innovation system analysis: The case of offshore wind. *Environmental Innovation and Societal Transitions, 14*, 128–148.

Wieczorek, A. J., Raven, R., & Berkhout, F. (2015b). Transnational linkages in sustainability experiments: A typology and the case of solar photovoltaic energy in India. *Environmental Innovation and Societal Transitions, 17*, 149–165.

Wirth, H. (2016). *Aktuelle Fakten zur Photovoltaik in Deutschland*.

Witkamp, M. J., Royakkers, L. M., & Raven, R. P. J. M. (2009). *From Cowboys to Diplomats: Why the growth of social entrepreneurship requires a different attitude than its creation: 2nd EMES International Conference on Social Enterprise, 1–4 July 2009, Trento, Italy*. Retrieved 01.10.2016, from http://www.euricse.eu/sites/default/files/db_uploads/documents/1254752389_n166.pdf

Witkamp, M. J., Raven, R. P., & Royakkers, L. M. (2011). Strategic niche management of social innovations: The case of social entrepreneurship. *Technology Analysis & Strategic Management, 23*(6), 667–681.

Witt, P. (2016). Gründerteams. In: Faltin, G. (Ed.), *Handbuch entrepreneurship*. Wiesbaden: Gabler.

Wüstenhagen, R. (1998). *Greening Goliaths versus multiplying Davids: Pfade einer Coevolution ökologischer Massenmärkte und nachhaltiger Nischen: IWÖ-Diskussionsbeitrag Nr. 61*.

Wüstenhagen, R., Wolsink, M., & Bürer, M. J. (2007). Social acceptance of renewable energy innovation: An introduction to the concept. *Energy Policy, 35*(5), 2683–2691.

WWF Deutschland, Öko-Institut e.V., & Prognos AG. (Eds.). (2009). *Modell Deutschland – Klimaschutz bis 2050. Vom Ziel her denken*. Retrieved 01.10.2016, from http://www.wwf.de/themen-projekte/klima-energie/modell-deutschland/klimaschutz-2050/

References

Yildiz, Ö., Rommel, J., Debor, S., Holstenkamp, L., Mey, F., Müller, J. R., et al. (2015). Renewable energy cooperatives as gatekeepers or facilitators? Recent developments in Germany and a multidisciplinary research agenda. *Energy Research & Social Science, 6*, 59–73.

Yin, R. K. (2009). *Case study research: Design and methods, Applied social research methods series* (Vol. 5, 4th ed.). Los Angeles, CA: Sage.

Zentes, J., Swoboda, B., & Morschett, D. (Eds.). (2005a). *Kooperationen, Allianzen und Netzwerke: Grundlagen – Ansätze – Perspektiven* (2nd ed.). Wiesbaden: Gabler.

Zentes, J., Swoboda, B., & Morschett, D. (2005b). Kooperationen, Allianzen und Netzwerke – Grundlagen, "Metaanalyse" und Kurzabriss. In J. Zentes, B. Swoboda, & D. Morschett (Eds.), *Kooperationen, Allianzen und Netzwerke. Grundlagen – Ansätze – Perspektiven* (2nd ed., pp. 3–34). Wiesbaden: Gabler.

Zerche, J., Schmale, I., & Blome-Drees, J. (1998). *Einführung in die Genossenschaftslehre: Genossenschaftstheorie und Genossenschaftsmanagement*. München, Wien: Oldenbourg Verlag.

Zimmer, M., & Ortmann, G. (1996). Strategisches management, strukturationstheoretisch betrachtet. In H. H. Hinterhuber, A.-A. Ayad, & G. Handlbauer (Eds.), *Das neue strategische Management. Perspektiven und Elemente einer zeitgemässen Unternehmensführung* (pp. 87–114). Gabler: Wiesbaden.

Zimmermann, J.-R. (2015). Das. *Thema Photovoltaik ist nicht tot! neue energie, 2015*(5), 58–59.

Zweite Verordnung zur Änderung der Energieeinsparverordnung. *Bundesgesetzblatt* 3951, Federal Ministry of Justice and Consumer Protection 21.11.2013.

Printed by Printforce, the Netherlands